Handbook on Stiffness & Damping in Mechanical Design

Handbook on Stiffness & Damping in Mechanical Design

Eugene I. Rivin

ASME
PRESS

Library of Congress Cataloging-in-Publication Data

Rivin, Eugene I.
Handbook on stiffness & damping in mechanical design / by Eugene I. Rivin.
    p. cm.
  Includes bibliographical references and index.
  ISBN 978-0-7918-0293-9
  1. Engineering design.
  2. Dynamics, Rigid.
  3. Damping (Mechanics)
  I. Title.
TA174.R545 2010
  624.1'76—dc22   2009022512

CONTENTS

Foreword   xvii

Chapter 1   Introductions and Definitions   1

1.1   Basic Notions   3

    1.1.1   Stiffness   3

    1.1.2   Damping   8

1.2   Influence of Stiffness and Damping on Strength   10
    and Length of Service

    1.2.1   Influence of Stiffness on Uniformity of   10
         Stress Distribution

    1.2.2   Influence of Stiffness and Damping on   13
         Vibration/Dynamics

1.3   Negative Stiffness and Damping   14

    1.3.1   Elastic Instability   16

    1.3.2   Stick-Slip   19

    1.3.3   Mechanical Linkages   20

    1.3.4   Mechanisms with Nonlinear   23
         Position Functions

    1.3.5   Electromechanical Systems   31

    1.3.6   Stiffness and Damping of Cutting Process   35

         1.3.6a   Influence of Machining System   37
                Stiffness and Damping on Accuracy
                and Productivity

1.4   Stiffness and Damping of Some Widely   42
    Used Materials

1.5 General Comments on Stiffness in
Mechanical Design   50

References   51

Chapter 2 Modes of Loading and Stiffness of Structural
Components   53

2.1 Influence of Mode of Loading on Stiffness   55

2.2 Influence of Beam Design on its Bending Stiffness
and Damping   64

    2.2.1 Round versus Rectangular Cross Section   65

    2.2.2 Stiffness and Damping of Helically Patterned
Tubular Beams   67

    2.2.3 Composite/Honeycomb Beams and Plates   70

2.3 Torsional Stiffness   72

2.4 Influence of Stress Concentrations   76

2.5 Stiffness of Frame/Bed Components   76

    2.5.1 Local Deformations of Frame Parts   81

2.6 General Comments on Stiffness Enhancement of
Structural Components   83

2.7 Stiffness-Critical Metal Elastic Elements (Springs)   85

    2.7.1 Coil Springs   86

    2.7.2 Slotted Springs   92

    2.7.3 Friction Springs   92

    2.7.4 Miniature Tabular Elastic Elements
Utilizing Giant Superelasticity Effect   94

2.8 Static Deformation Characteristics of Quasi-Linear
Rubber Elements   97

    2.8.1 Stiffness of Bonded Rubber Blocks   102

References   111

**Chapter 3 Nonlinear and Variable Stiffness Systems; 113
         Preloading**

3.1 Definitions   115

  **3.1.1 "Constant Natural Frequency"   116
         Nonlinear Characteristic**

**3.2 Embodiments of Mechanical Elements   119
     with Nonlinear Stiffness**

  3.2.1 Material-Related Nonlinearity   120

  3.2.2 Geometry-Related Nonlinearity   122

       3.2.2a Coil and Leaf Springs   122

       3.2.2b Belleville (Disc) Springs   124

  **3.2.3 Nonlinear Spring Elements with Softening 126
         Nonlinear Characteristics**

3.3 Statically Nonlinear Rubber Elements   130

  **3.3.1 Compressed Elements with   131
         Controlled Bulging**

  **3.3.2 Streamlined Nonlinear Rubber   135
         Flexible Elements**

  3.3.3 Thin-Layered Rubber-Metal Laminates   142

**3.4 Stiffness Management by Preloading   146
     (Strength-to-Stiffness Transformation)**

  3.4.1 Embodiments of the Preloading Concept   147

  3.4.2 Antagonist Actuators   154

  **3.4.3 Preloaded Flexible Elements with   155
         Variable Stiffness**

  **3.4.4 Some Dynamic Effects Caused by   159
         Variable Stiffness**

3.5 Assembled Frame-Like and Beam-Like Structures 161

  3.5.1 Integrity of Assembled Structures   161

       3.5.1a Integrity of an Assembled Beam   161

       **3.5.1b Integrity of an Assembled Preloaded 163
               Structure**

3.5.2   Stiffness of Assembled Structures   166

3.5.3   Dynamics of Assembled Structures   172

References   175

Chapter 4   Contact (Joint) Stiffness and Damping   177

4.1   Introduction   179

4.2   Contact Deformations Between   180
      Non-Conforming Surfaces

4.3   Contact Deformations Between   183
      Conforming and Quasi-Comforming Surfaces

4.4   Contact Stiffness in Structural Analysis   186

4.5   Quasi-Conforming Contact Deformations in   193
      Cylindrical/Conical Connections

      4.5.1   Cylindrical Connections   194

              4.5.1a   Connections with Clearance Fits   194

              4.5.1b   Interference-Fit Cylindrical   202
                       Connections

      4.5.2   Elastic Displacements in Conical (Tapered)   203
              Connections

              4.5.2a   Test Data   204

              4.5.2b   Computational Evaluation of   210
                       Contact Deformations in
                       Tapered Connections

              4.5.2c   Influence of Manufacturing Errors   213

              4.5.2d   Finite Element Modeling of   217
                       7/24 Taper Connection

              4.5.2e   Short Taper Connections   219

              4.5.2f   Some General Comments on Tapered   221
                       Connections

4.6   Tangential Contact Compliance   223

      4.6.1   Experimental Study of Tangential   224
              Compliance of Flat Joints

4.6.2   Dynamic Model of Tangential Compliance and Damping   234

4.7   Practical Case: Study of a Modular Tooling System   242

4.8   Damping of Mechanical Contacts   246

     4.8.1   Damping in Flat Joints   248

     4.8.2   Damping in Cylindrical and Tapered Connections   249

     4.8.3   Energy Dissipation in Power Transmission Components   250

References   251

Chapter 5   Supporting Systems/Foundations   253

5.1   Influence of Support Characteristics   255

5.2   Rational Location of Supporting/ Mounting Elements   263

5.3   Overconstrained (Statically Indeterminate) Systems   273

5.4   Influence of Foundation on Structural Deformations   280

     5.4.1   General Considerations   280

     5.4.2   Machines Installed on Individual Foundations or on Floor Plate   284

5.5   Deformations of Long Machine Bases   288

References   292

Chapter 6   Stiffness and Damping of Power Transmission Systems and Drives   293

6.1   Basic Notions   294

6.2   Compliance of Mechanical Power Transmission and Drive Components   296

     6.2.1   Basic Power Transmission Components   296

         6.2.1a   Stiffness of Ball Screws   307

6.2.2   Compliance of Pneumatic System Components   **311**

6.2.3   Compliance of Hydraulic System Components   **313**

6.2.4   Dynamic Parameters of Electric Motors (Actuators)   **316**

6.3   **Parameter Reduction in Mathematical Models**   **318**

6.4   **Practical Examples of Structural Compliance Breakdown**   **329**

6.4.1   A Hydraulically Driven Robot   **330**

6.4.2   Electromechanically Driven Robot of Jointed Structure   **335**

6.4.3   Electromechanically Driven Parallelogram Robot with Harmonic Drive   **339**

6.4.4   Electromechanically Driven Spherical Frame Robot   **342**

6.4.5   Summary   **347**

6.5   **More on Stiffness and Damping of Antifriction Bearings and Spindles**   **348**

6.5.1   Stiffness of Spindles   **348**

6.5.2   Stiffness and Damping of Antifriction Bearings   **353**

6.6   **Damping in Power Transmission Systems**   **364**

References   **369**

Chapter 7   **Design Techniques for Reducing Structural Deformations (Stiffness Enhancement Techniques)**   **371**

7.1   **Structural Optimization Techniques**   **373**

7.2   **Compensation of Structural Deformations**   **379**

7.2.1   Passive Compensation Techniques   **380**

7.2.2   Active (Servo-Controlled) Systems for Stiffness Enhancement   **389**

7.3 **Stiffness Enhancement by Reduction of Stress Concentrations** — 400

7.4 **Strength-to-Stiffness Transformation** — 401

    7.4.1 **Buckling and Stiffness** — 402

    7.4.2 **"Reverse Buckling" Concept** — 405

    7.4.3 **Stiffening of Slender Parts by Axial Tension during Machining** — 412

    7.4.4 **Self-Contained Stiffness Enhancement Systems** — 416

7.5 **Temporary Stiffness Enhancement Techniques** — 425

7.6 **Performance Enhancement of Cantilever Components** — 427

    7.6.1 **General Comments** — 427

    7.6.2 **Stationary and Rotating Around Longitudinal Axis Cantilever Components** — 430

    7.6.3 **Cantilever Components Rotating Around Transverse Axis** — 433

        7.6.3a **Solid Component** — 433

        7.6.3b **Combination Link** — 435

7.7 **Damping Enhancement Techniques** — 440

    7.7.1 **Introduction** — 440

    7.7.2 **Dampers** — 441

    7.7.3 **Dynamic Vibration Absorbers** — 448

References — 450

Chapter 8 Use of "Managed Stiffness" in Design — 455

8.1 **Cutting Edge/Machine Tool Structure Interface** — 457

    8.1.1 **Introduction** — 457

    8.1.2 **Techniques for Reduction of Cutting Forces** — 459

    8.1.3 **Influence of Stiffness and Damping in the Cutting Zone on Cutting Forces and Tool Life** — 461

8.1.4   Machining Systems with Intentionally   463
        Reduced Tool Stiffness

        8.1.4a   Cutting Tools with Reduced   463
                 Normal Stiffness

        8.1.4b   Cutting Tools with Reduced   465
                 Tangential Stiffness

        8.1.4c   Trading-off the Stiffness for   471
                 Damping to Improve the Overall
                 Machining Performance

8.2   Stiffness of Clamping Devices   481

      8.2.1   Introduction   481

      8.2.2   General Purpose Clamping Devices   482

      8.2.3   "Solid State" Tool Clamping Devices   487

8.3   Modular Tooling   492

8.4   Tool/Machine Interfaces. Tapered Connections   496

      8.4.1   Managed Stiffness Connections to   497
              Reduce Friction-Induced Position
              Uncertainties

      8.4.2   7/24 Steep Taper Connections   498

              8.4.2.1   Definition of the Problem   498

              8.4.2.2   Tapered Toolholder/Spindle   500
                        Interfaces for Machine Tools.
                        Practical Sample Cases

      8.4.3   Other Tapered and Geared Toolholder/   512
              Spindle Interfaces

              8.4.3.1   Curvic Coupling Connection   513

              8.4.3.2   KM System   514

              8.4.3.3   HSK System   516

8.5   Benefits of Intentional Stiffness Reduction   519
      in Design Components

      8.5.1   Hollow Roller Bearings   520

      8.5.2   Stiffness Reduction in Power   523
              Transmission Gears

8.5.3   **Stiffness Reduction of Chain Transmissions**   **530**

8.5.4   **Compliant Bearings for High-Speed Rotors**   **531**

8.6   **Constant Force (Zero Stiffness) Vibration Isolation Systems**   **534**

8.7   **Anisotropic Elastic Elements as Limited Travel Bearings (Flexures)**   **541**

8.7.1   **Elastic Kinematic Connections (Flexures)**   **543**

8.7.1a   **Elastic Connections for Rotational Motion**   **543**

8.7.1b   **Elastic Connections for Translational Motion**   **545**

8.7.1c   **Elastic Motion Transformers**   **546**

8.7.2   **Elastic Kinematic Connections Using Thin-Layered Rubber-Metal Laminates**   **547**

8.7.2a   **Rubber-Metal Laminates as Anisotropic Elastic Elements**   **548**

8.7.2b   **Use of Rubber-Metal Laminates as Limited Travel Bearings**   **550**

8.7.2c   **Wedge Mechanisms**   **555**

8.7.2d   **Use of Rubber-Metal Laminates as Compensators**   **560**

8.8   **Modification of Parameters in Dynamic Models**   **560**

8.8.1   **Evaluation of Stiffness and Inertia Components in Multi-Degrees-of-Freedom Systems**   **561**

8.8.2   **Modification of Structure to Control Vibration Responses**   **564**

References   **567**

Appendix 1   **Single-Degree-of-Freedom Dynamic Systems with Damping**   **573**

References   **585**

Appendix 2	**Stiffness/Damping/Natural Frequency Criteria**	**587**
A2.1	**Introduction**	**589**
A2.2	**Self-Excited Vibrations/Dynamic Stability Criterion**	**590**
A2.3	**Vibration Isolation of Mechanical Objects**	**596**
	A2.3.1 Influence of Isolation on Chatter Resistance	**596**
	A2.3.2 Forced Vibrations	**599**
	A2.3.2a Vibration Level Criteria	**599**
	A2.3.2b Vibration Isolation Criteria for Vibration- Sensitive Objects	**601**
A2.4	**Use of Stiffness-Damping Criteria**	**603**
A2.5	**Discussion**	**605**
References		**606**
Appendix 3	**Influence of Axial Force on Beam Vibrations**	**607**
Reference		**611**
Appendix 4	**Characteristics of Elastomeric (Rubberlike) Materials**	**613**
A4.1	**Basic Notions**	**615**
A4.2	**Static Deformation Characteristics of Rubberlike Materials**	**618**
A4.3	**Elastic Stability of Rubber Parts**	**623**
A4.4	**Dynamic Characteristics of Rubberlike Materials**	**627**
A4.5	**Fatigue Resistance of Elastomeric Elements**	**633**
A4.6	**Creep of Rubberlike Materials**	**638**
References		**640**

Appendix 5   Power Transmission Couplings   643

A5.1   Introduction   645

A5.2   General Classification of Couplings   645

A5.3   Rigid Couplings   647

A5.4   Misalignment-Compensating Couplings   649

A5.5   Torsionally Flexible Couplings and Combination   663
Purpose Couplings

    A5.5.1   Roles of Torsionally Flexible Coupling   664
    in Transmission

    A5.5.2   Compensation Ability of Combinations   670
    Purpose Couplings

    A5.5.3   Comparison of Combination   673
    Coupling Designs

References   677

Appendix 6   Systems with Multiple Load-Carrying   679
Components

A6.1   Introduction   681

A6.2   Load Distribution Between Rolling Bodies   681
and Stiffness of Antifriction Bearings

A6.3   Loading of Spoked Wheels   685

A6.4   Analytical Solution for Bicycle Wheel   688

A6.5   Torsional Systems with Multiple   689
Load-Carrying Connections

References   694

Appendix 7   Compliance Breakdown for a Cylindrical   695
(OD) Grinder

Reference   706

About the Author   707
Index   709

FOREWORD

The most important attributes of many a mechanical system/structure are strength, stiffness, and stability. Scores of professional books had been published on various aspects of strength which are important for designers, many books had been published on stability of mechanical and civil engineering structures. Also, the strength and stability issues are heavily represented in all textbooks on machine elements. On the other hand, the stiffness-related issues until recently were practically neglected, with a few exceptions. The author could not find any comprehensive professional publication addressing various important aspects of stiffness in mechanical design, besides his sold-out book on "*Stiffness and Damping in Mechanical Design*" (Marcel Dekker, 1999, 512 pp.) The present book is a result of its significant rework and expansion. The author has an extensive experience and expertise in conceptual mechanical design, design components, vibration control, use of elastomers in design, design of production equipment (machine tools and tooling systems, robotics), etc. This expertise touched many practical aspects of stiffness consideration in the design process. The author wants to think that this expertise allows him to write this "*Handbook on Stiffness and Damping in Mechanical Design*."

Also, these experiences led to two observations. The first is an observation that there is a need for a comprehensive monograph/handbook addressing various aspects of stiffness in mechanical design. The other observation is that in many important real life cases, wherein the dynamic behavior of the system is important, stiffness and damping attributes cannot and should not be separated. While dynamics as well as forced and self-excited vibrations of mechanical systems are becoming increasingly important, damping and stiffness are usually considered separately. However, frequently, damping and stiffness are closely interrelated, and efforts to improve one parameter while neglecting another are usually ineffective or even counterproductive.

Computers are becoming more and more powerful tools assisting the design process. Finite Element Analysis (FEA) and other software packages constituting Computer-Aided Design (CAD) allow quick and realistic visualization and optimization of stresses and deformations inside the component of a structure as well as in the whole structure. However, the results of such analyses are useful only if the adapted analytical models are correct. But they are often not correct, especially for complex systems with critical role of contact stiffness, parallel kinematics machines, etc., if the designer does not possess a broader view of the system. The computer technology, which frees the designer from the tedious drafting and computational chores, not only allows but also forces him to concentrate on general, conceptual issues of design. Some of these issues are so-called "conceptual design," reliability, energy efficiency, accuracy, use of advanced materials in the appropriate parts of the system/structure. One of the most important conceptual issues is stiffness of mechanical structures and their components.

The book cited above was intended to start correcting the absence of a comprehensive source on stiffness-related issues by addressing various aspects of structural stiffness and structural damping and their roles in design. Several typical cases in which stiffness is closely associated with damping had been addressed. Since stiffness, especially in interaction with damping, is a very large subject, only the basic conceptual issues related to stiffness had been presented, rather than detailed analytical techniques. The same approach is adopted in the present Handbook. A more detailed analytical treatment is given only in a few cases where the results were never published before or had been published in hard-to-obtain sources (e.g., in languages other than English). Many of these concepts are illustrated by practical results either in the text or in Appendices. The first book was based on materials prepared for *"Stiffness in Design"* tutorial successfully presented at several Annual Meetings of the American Society for Precision Engineering (ASPE). These materials are, in turn, based to a substantial degree on personal professional experiences and research results of the author.

This book is advancing the concepts addressed in the first book on stiffness cited above. The author believes that its contents are much more comprehensive and justify adding the word "Handbook" in its title. Several conceptual issues are added, such as negative stiffness and damping; close-form analytical expressions for calculating stiffness of typical stiffness critical design elements, both metal springs and elastomeric elements; extensive introduction is given to static and dynamic properties of elastomeric materials; description of behavior of superelastic materials under structural (compressive) loading is added; etc. Interrelation of damping with stiffness is addressed much more extensively than before. The important issues of stiffness and damping specific to Micro Electromechanical Systems (MEMS) and nanosystems are not specially addressed since there is not much published material and since they are very specific. The analytical techniques related to consideration of contact stiffness and to constructing 3-D models of effective stiffness of mechanisms, while useful for specific applications, will, undoubtedly, also direct the reader to development of analytical models for other mechanical systems.

Many important stiffness- and damping-related issues were studied in-depth in the former Soviet Union. The results are still very relevant, but they were published in Russian and, practically, are not available to the engineering community in the non-Russian speaking countries. Some of these results are reflected in the Handbook.

A general introduction to the subject is given in *Chapter 1*. General design areas are described for which the stiffness criterion is critical. Practical embodiments of mechanical systems containing sources of negative stiffness and negative damping are described. It also addresses selection of structural materials for stiffness- and damping-critical applications.

Information on influence of the mode of loading and of the component design on stiffness is provided in *Chapter 2*.

Chapter 3 is dedicated to an important subject of nonlinear and variable stiffness systems. Specially addressed is an important issue of preloading. An interesting class of nonlinear elastomeric elements is described, including elements whose preloading does not significantly increase resistance to limited travel motions.

Design and performance information on various aspects of normal and tangential contact stiffness, as well as of damping associated with mechanical contacts, is given in *Chapter 4*. Information on these subjects is very scarce in the English language technical literature.

Some important issues related to influence on stiffness of mechanical components by their supporting conditions and devices, as well as by foundations for machines, are addressed in *Chapter 5*.

Chapter 6 concentrates on very specific issues of stiffness (and damping) in power transmission and drive systems which play a significant role in various mechanical systems.

Numerous useful design techniques, both passive and active, aimed on enhancing structural stiffness (i.e., reducing structural deformations) are described In *Chapter 7*.

Special cases in which performance of stiffness-critical system can be improved by reduction or a proper tuning of the components' stiffness are described in *Chapter 8*. A special attention is given here to tooling sub-systems/structures in machining systems.

The Handbook also contains seven Appendices.

Appendix 1 gives introductory information on dynamic systems with non-viscous (hysteretic) damping. While the viscous damping model is very convenient for analytical studies of linear dynamic systems, it does not represent real life systems. The hysteretic damping model more properly represents real dynamic systems but, unfortunately, only seldom described in vibration textbooks.

Appendix 2 presents the first attempt to consider stiffness and damping in their interrelation, as useful practical criteria.

Appendix 3 illustrates a technique complementary to similar techniques addressed in Chapter 7 for increasing effective stiffness of mechanical components (shafts) without changing their material or geometry.

Appendix 4 describes basic mechanical characteristics of elastomeric (rubber-like) materials, whose wider use in mechanical design is hampered by difficulties in finding such information in mechanical design literature.

Appendix 5 provides principles and criteria for designing and selection of power transmission couplings, important elements of power transmission systems. The proposed analytical approach is expected to help designers of power transmission systems in selecting optimal couplings from a huge variety of coupling designs on the market.

Some important issues related to systems with multiple load-carrying components are addressed in *Appendix 6*.

Appendix 7 describes in detail a practical case of constructing a compliance breakdown for a complex precision mechanical system - an OD grinder.

The issues related to the *"Stiffness in Design"* topic are numerous and very diverse. This Handbook does not pretend to cover all the issues related to stiffness and damping in mechanical design, but it expands significantly compared with the book cited above.

It is expected that the book will be useful not only to mechanical designers, especially ones working with precision devices, but also to vibration and dynamic specialists. I hope that instructors in the mechanical engineering field will use numerous examples in the book in the teaching process.

CHAPTER

1

Introduction and Definitions

1.1 BASIC NOTIONS

1.1.1 Stiffness

Stiffness is the capacity of a mechanical system to sustain external loads without excessive changes of its geometry (deformations). It is one of the most important design criteria for mechanical components and systems. While strength is considered the most important design criterion, there are many cases when stresses in components and their connections are significantly below the allowable levels, and dimensions as well as performance characteristics of mechanical systems and their components are determined by stiffness requirements. Typical examples of such mechanical systems are aircraft wings and frames and beds of production *machinery* (machine tools, presses, etc.), in which stresses frequently do not exceed 3 MPa to 7 MPa (500 psi to 1,000 psi), a small fraction of the ultimate strength of the material. Another stiffness-critical group of mechanical components is power transmission components, especially shafts, whose excessive deformations may lead to failures of gears and belts, while stresses in the shafts caused by the payload are relatively low.

Recently, there were achieved great advances in improving strength of mechanical systems and components. These advances are based on development of high-strength structural metals, composites, other materials, better understanding of fracture/failure phenomena, and development of better techniques for stress analysis and computation, which resulted in reducing of stress concentrations and safety factors. But these advances often result in reduction of cross-sections of structural components. Since the payload-induced loads in the structures usually do not change, structural deformations in the systems using high-strength materials and/or designed with reduced safety factors are becoming more pronounced. It is important to note that while strength of structural metals can be greatly improved by selection of alloying materials and/or of heat treatment procedures (as much as five to seven times improvements for steel and aluminum), modulus of elasticity (Young's modulus) is not very sensitive to alloying and to heat treatment. For example, the Young's modulus of stainless steels is even 5% to 15% lower than that of carbon steels, e.g., see Table 1.1 below. As a result, stiffness can be modified (enhanced) only by proper selection of the component geometry (shape and size) and by optimizing interactions between components of the system.

Stiffness effects on performance of mechanical systems are due to influence of deformations on static and fatigue strength, wear resistance, efficiency (friction losses), accuracy, dynamic/vibration stability, manufacturability. To summarize, importance of the stiffness criterion is currently increasing due to:

1. Increasing *accuracy requirements* which call for reduction of deformations;
2. Increasing *use of high-strength materials* resulting in reduced cross-sections of the components and, accordingly, in increasing structural deformations;
3. Better analytical techniques resulting *in smaller safety factors*, which also result in the reduced cross sections and increasing deformations.
4. Increasing *importance of dynamic characteristics* of machines since their increased speed and power, combined with lighter structures, may result in either intense resonances or in self-excited vibrations, or both (chatter, stick-slip, etc.).

Factors 2, 3, and 4 are especially pronounced for surface and flying vehicles (cars, airplanes, rockets, etc.) in which the strength resources of the materials are utilized to the maximum to reduce weight.

Generically, stiffness k is defined as a ratio of force P to deformation Δ,

$$k = P / \Delta, \qquad (1.1.1)$$

where P can be force proper in Newtons (N) or pounds (lbs) for translational deformation, Δ is in meters (m) or inches (in.), or can be a moment or a torque in Nm or in.-lbs for angular deformations Δ in degrees (°) or radians (rad). Actually, stiffness is a complex parameter of a system. At each point, there are generally different values of translational stiffness k_{xx}, k_{yy}, k_{zz} in three orthogonal directions of a selected coordinate frame X, Y, Z; three values of interaxial (cross) stiffness k_{xy}, k_{xz}, k_{yz} related to deformations caused by forces acting along other orthogonal axes since usually $k_{xy} = k_{yx}$, $k_{xz} = k_{zx}$, $k_{yz} = k_{zy}$; three values of angular stiffness $k_{\alpha\alpha}$, $k_{\beta\beta}$, $k_{\gamma\gamma}$ about X, Y, and Z axes, respectively, and three values of cross-angular stiffness $k_{\alpha\beta}$, $k_{\alpha\gamma}$, $k_{\beta\gamma}$, since usually $k_{\alpha\beta} = k_{\beta\alpha}$, $k_{\alpha\gamma} = k_{\gamma\alpha}$, $k_{\beta\gamma} = k_{\gamma\beta}$. If the cross (interaxial) stiffness values vanish, $k_{xy} = k_{xz} = k_{yz} = 0$ and $k_{\alpha\beta} = k_{\alpha\gamma} = k_{\beta\gamma} = 0$, then X, Y, Z axes of this coordinate frame are the *principal stiffness axes*. These definitions are important; in special cases, ratios of the stiffness values in the orthogonal directions may determine dynamic stability of the system. Such is the case of chatter instability of some machining operations, e.g., [1]. Chatter stability in these operations increases if either the cutting or the friction force vector, or both are oriented in a certain way relative to principal stiffness axes X and Y. Improper stiffness ratios in vibration isolators and machinery mounts may cause undesirable intermodal coupling in vibration isolation systems [2].

Main effects of an inadequate stiffness are excessive absolute deformations of some components of the system and/or excessive relative displacements between two or several components. Such excessive deformations/displacements can cause:

1. Geometric distortions (inaccuracies);
2. Change of actual loads and friction conditions which may lead to reduced efficiency, accelerated wear or fretting corrosion, etc.;
3. Dynamic instability (self-excited vibrations);
4. Increased amplitudes of forced vibrations.

Inadequate stiffness of transmission shafts may cause some specific effects. Linear and angular bending deformations of the shafts caused by the payloads determine behavior of other components. Angular bending deformations cause stress concentrations and increased vibrations in antifriction bearings and may distort lubrication and friction conditions in sliding bearings. Angular and linear deformations lead to distortions of the meshing process in gears and worm transmissions resulting in stress concentrations and variations of the instantaneous transmission ratios, thus causing increasing dynamic loads. Angular deformations cause stress concentrations and changing friction conditions in traction drives, etc.

It is worthwhile to introduce some definitions related to stiffness:

1. *Structural proper* (*"structural"*) *stiffness* associated with deformations of a part or a component considered as a beam, a plate, a shell, etc.;
2. *Structural contact* (*"contact"*) *stiffness* associated with deformations in a connection between two components (contact deformations may exceed structural deformations in precision systems);

3. *Compliance e = 1/k*, defined as a reciprocal parameter to stiffness k (ratio of deformation to force causing this deformation);
4. *Linear stiffness* versus *nonlinear stiffness;*
5. *Hardening* versus *softening* nonlinear stiffness;
6. *Static stiffness* k_{st} (stiffness measured during a very slow loading process, or under a periodic loading with a frequency lower than 0.05 Hz to 0.5 Hz) versus *dynamic stiffness*, k_{dyn}, which is measured under faster changing loads. Dynamic stiffness can be characterized by a *dynamic stiffness coefficient*, $K_{dyn} = k_{dyn}/k_{st}$. Usually $K_{dyn} > 1$ and depends on frequency and/or amplitude of load and/or amplitude of the vibratory displacement (e.g., see [2]). In many cases, especially for fibrous and elastomeric materials, K_{dyn} as a function of vibration amplitude is usually inversely correlated with damping.

The above "dynamic stiffness" and the "dynamic stiffness coefficient K_{dyn}" usually are properties of the material. However, the "dynamic stiffness" term is also used in application to dynamic systems comprising, besides deformable (elastic) components, also inertia and damping components. The simple one-degree-of-freedom dynamic system is shown in Fig. 1.1.1. Here mass m is connected to the supporting surface via deformable element having stiffness, k, and damping element having damping coefficient, c. A time-varying force $F(t)$ is applied to mass m. Equation of motion of this system for a sinusoidal excitation is

$$m\ddot{x}(t) + c\dot{x}(t) + kx(t) = F \sin \omega t \qquad (1.1.2)$$

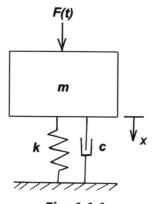

Fig. 1.1.1.

A Laplace transform of Eq. (1.1.2) with the damping term absent is

$$F(s) = Ms^2X(s) + KX(s). \tag{1.1.3}$$

The dynamic stiffness $K(s)$ of this system is the relationship between the amplitude of the applied sinusoidal force and the resulting displacement amplitude of the force application point. It is dependent on frequency, ω, of the force and on the physical parameters of the system,

$$K_d(s) = \frac{F(s)}{X(s)}\Big|_{s=j\omega}. \tag{1.1.4}$$

Fig. 1.1.2 shows the frequency dependence of dynamic stiffness of the system in Fig. 1.1.1 on the non-dimensional frequency. The smallest value of the dynamic stiffness is at the resonant (natural) frequency of the system

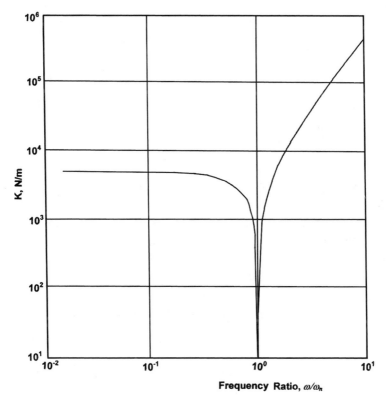

Fig. 1.1.2.

$\omega_n = \sqrt{\dfrac{k}{m}}$, when a small force can excite large displacement amplitudes. The dynamic systems are, generally, not considered in this book.

1.1.2 Damping

Damping is a capacity of a vibrating system to transform a fraction of energy of the vibratory process during each cycle of vibrations into another form of energy. This energy transformation leads to reducing intensity (amplitude) of a forced vibratory process or to a gradual decay and fading of a free vibration process. If the vibratory process is of a self-exciting type (e.g., chatter vibrations in metal cutting operations), then the damping capacity of the system may partially suppress or completely prevent development of the self-excited vibrations. Some damping effects and definitions are addressed in Appendix 1.

Since vibratory processes can be observed in various systems (mechanical, electrical, biological, astronomical, etc.), the damping mechanisms vary accordingly. Damping capacity in mechanical systems is, usually, a capacity to transform (dissipate) the vibratory energy into thermal energy (heat). This transformation often occurs due to external friction in the mechanical system or due to internal hysteretic losses ("internal friction") in materials of constitutive components of the vibratory system.

Vibrations are undesirable in majority of mechanical systems, since they cause increased stresses, frequent repetition of high stresses thus reducing fatigue life, deterioration of performance characteristics (e.g., deteriorated productivity, surface finish, and accuracy of metal cutting due to chatter), deterioration of environmental conditions (e.g., reduced comfort), etc. In such cases, damping is a very desirable property, and its increase can be achieved by using high damping materials, judiciously designed sliding (or micro-sliding, see Section 4.8) contact connections, and specially constructed dampers.

A common misconception about using high-damping elements in mechanical systems is undesirable heat generation associated with high damping. However, if a damping element is not subjected to forced movements of a predetermined magnitude, such as in misalignment-compensating couplings, e.g., as shown in Fig. A5.6, but just for reducing the resonance amplitude, then the heat generation is decreasing with increasing damping. This statement can be easily proven for the system in

Fig. 1.1.1. The maximum potential energy stored in the flexible element during one cycle of vibrations is

$$V = k\frac{A^2}{2},$$ (1.1.5)

where k = generalized translational or torsional stiffness, and A = generalized amplitude. A fraction ψV of V is transformed into heat. The relative energy dissipation ψ can be expressed as $\psi = 2\delta$, where δ is the log decrement of the system. The most intense heat generation is at resonance, when the vibration amplitude is the greatest,

$$A_{res} = A_{ex}\frac{\pi}{\delta},$$ (1.1.6)

where A_{ex} is generalized excitation amplitude. Accordingly, energy dissipation per cycle at resonance, responsible for heat generation, is

$$\Delta V_{res} = \psi k\frac{A_{res}^2}{2} = \psi k A_{ex}^2 \frac{\pi^2}{2\left(\psi/2\right)^2} = \frac{2\pi^2 k A_{ex}^2}{\psi},$$ (1.1.7)

So that the *heat generation is reduced* inversely proportionally to the *damping increase*. The reason for this paradox is the fact that the change of damping, first of all, changes the dynamic system and its characteristics, such as vibratory amplitudes of its components, and the heat generation is a secondary effect. If the system is prone to self-excited vibrations (chatter), then again the increased damping would change the system making it more dynamically stable, obviously with no increase in amplitudes/heat generation.

In some cases, the vibratory process is used as a production process (vibratory technology including vibratory conveyance, vibratory screening, vibration- and ultrasonic-assisted cutting, etc.). In such cases, the damping capacity of the system usually should be reduced to enhance the effectiveness of the respective equipment. Accordingly, materials with low hysteretic losses are used in such systems, connections in which sliding and friction can develop are avoided, and solid-state elements (flexures), are used in some cases, e.g., see Chapter 8.

Hysteretic losses are also often undesirable in precision systems as well as in micro-electro-mechanical systems (MEMS) and in nano-systems.

In these applications, solid-state systems, often using low-hysteresis materials, are widely employed.

Although stiffness and damping are very different characteristics of mechanical systems, their effects are often interrelated. Just increasing or decreasing stiffness or damping characteristics is not as effective as coordinated changes of both characteristics so that an appropriate stiffness/damping criterion (see Appendix 2) is increasing or decreasing.

1.2 INFLUENCE OF STIFFNESS AND DAMPING ON STRENGTH AND LENGTH OF SERVICE

This influence can materialize in several ways:

1. *Inadequate or excessive stiffness* of parts may lead to *overloading* of associated parts or to a non-uniform stress distribution, especially in statically indeterminate systems; see Chapter 5;
2. *Impact/vibratory loads* are significantly dependent on stiffness and/or damping;
3. *Inadequate stiffness* may significantly influence strength if loss of structural stability *(buckling)* of some component occurs.

1.2.1 Influence of Stiffness on Uniformity of Stress Distribution

It is known that fatigue life of a component depends on a high power (five to nine) of maximum (peak) stresses. Thus, uniformity of the stress distribution is very important.

Fig. 1.2.1 [3] shows influence of stiffness of rims 1, 2 of meshing gears on load distribution in their teeth. In Fig. 1.2.1a, left sides of both gear

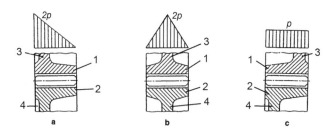

Fig. 1.2.1. Contact pressure distribution in meshing gears as influenced by design of gears.

rims have higher stiffness than their right sides due to positioning of the stiffening disc/spokes 3, 4 on their hubs. This leads to concentration of the loading in the stiff area so that the peak contact stresses in this area are about two times higher than the average stress, p, between the meshing profiles. In Fig. 1.2.1b, the gear hubs are symmetrical but again, the stiff areas of both rims work against each other. Although the stress distribution diagram is different, the peak stress is still about twice as high as the average stress. The design shown in Fig. 1.2.1c results in a more uniform stiffness along the tooth width and, accordingly, in much lower peak stresses, about equal to the average stress magnitude. The diagrams in Fig. 1.2.1 are constructed with an assumption of absolutely rigid shafts. If shaft deformations are significant, they can substantially modify the stress distributions and even reverse the characteristic effects shown in Fig. 1.2.1.

Another example of influence of stiffness on load distribution is shown in Fig. 1.2.2. It is a schematic model of tightened threaded connection between bolt 1 and nut 2. Deformations of the nut body can be neglected. Fig. 1.2.2a shows a threaded connection not tightened, thus little beams simulating thread coils are not deformed. Fig. 1.2.2b shows a connection wherein compliance of the thread coils is much greater than compliance of the bolt body. In this case all coils have the same deformations and, accordingly, are uniformly loaded. In the connection in Fig. 1.2.2c, compliance of the bolt body is commensurate with compliances of the thread coils. Thus, deformation of the upper coil Δ_1 is less than deformation of the lower coil $\Delta_1 + \Delta H$, where ΔH is deformation of the bolt body having length H before tightening.

This results in a very non-uniform load distribution between the coils. Theoretically, for an accurately fabricated ten-coil thread, the first coil takes $0.3P$ to $0.35P$, where P is the total axial load on the bolt, while the eighth coil takes only $0.04P$. In real-life threaded connections, the load distribution may be less uniform due to inaccuracies, but more uniform

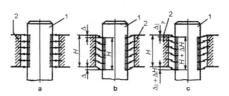

Fig. 1.2.2. Contact pressure distribution (b) in threaded connection (a).

due to possible yielding of the highest loaded coils, due to contact deformations in the thread, and due to higher compliances of the contacting coils because of their inaccuracies and less than perfect contact. Thus, the first coil may take only $0.25P$ to $0.3P$ instead of $0.3P$ to $0.35P$. However, it is still a very dramatic non-uniformity which can cause excessive plastic deformations of the most loaded coils or their fatigue failure, or both. Such a failure may cause a chain reaction of failures in the threaded connection.

Redistribution and concentration of loading influencing the overall deformations and the effective stiffness of the system can be observed in various mechanical systems. Fig. 1.2.3 [3] shows a pin connection of a rod 1 with a tube 2. Since the tube is much stiffer than the rod, a large fraction of the axial load, P, is acting on the upper pin which can be overloaded (Fig. 1.2.3a). The simplest way to equalize loading of the pins is by loosening the hole for the upper pin (Fig. 1.2.3b). It leads to the load being applied initially only to the lower pin. The upper pin takes the load only after some stretching of the rod had occurred. The clearance between the rod and the upper pin can be optimized for a more-or-less equal load distribution considering actual stiffness values of the tube and of the rod. Another way to achieve the same effect is by *prestressing* (*preloading*) the system by creating an initial (compressive) loading (during assembly) to counteract the loading by force P (Fig. 1.2.3c). This effect also can be achieved by simultaneously drilling holes in the rod and in the tube, while the rod is heated to the specified temperature (Fig. 1.2.3d), and then inserting the pins. If only the lower pin is inserted, then after the rod cools down, it would shrink (Fig. 1.2.3e). However, if both pins are inserted while the rod is still hot, the shrinkage cannot occur,

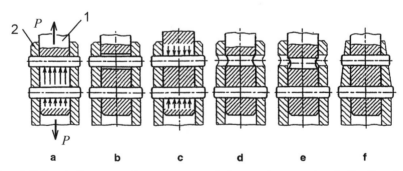

Fig. 1.2.3. Influence of component deformations on load distribution.

and the system becomes prestressed. The load equalization effect can also be achieved by local reduction of the tube stiffness (Fig. 1.2.3f).

1.2.2 Influence of Stiffness and Damping on Vibration/Dynamics

This effect of stiffness can be due to several mechanisms:

At an *impact*, kinetic energy of the impacting mass is transformed into potential energy of elastic deformation; accordingly, dynamic overloads are stiffness-dependent. For a simple model in Fig. 1.2.4, kinetic energy of mass, m, moving with velocity v, and impacting a structure having stiffness k, is

$$E = \frac{1}{2}mv^2 \qquad (1.2.1)$$

After the impact, this kinetic energy transforms into potential energy of the structural impact-induced deformation x,

$$V = \frac{1}{2}kx^2 = E = \frac{1}{2}mv^2 \qquad (1.2.2)$$

Since the impact force $F = kx$, from (1.2.2)

$$x = v\sqrt{\frac{m}{k}}, \text{ and } F = v\sqrt{km} \qquad (1.2.3)$$

Thus, in the first approximation, the impact force is proportional to the square root of stiffness.

For *forced vibrations*, a *resonance* can cause significant overloads. The resonance frequency can be shifted by a proper choice of either stiffness and mass values, or of their distribution, across the system or both. Shifting of the resonance frequencies may help to avoid the excessive

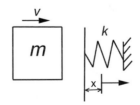

Fig. 1.2.4. Impact interaction between moving mass and stationary spring.

resonance displacement amplitudes and overloads, but only if the forcing frequencies are specified and cannot shift. In many cases, this is not a realistic assumption. For example, the forcing (excitation) frequencies acting on a machine tool during a milling operation vary within a broad range with changing number of cutting inserts in the milling cutter and changing spindle speed (rpm). Thus, a shift of the natural (resonance) frequency may result not in reduction but even in increase of probability of a resonance at the new natural frequency. A much more effective way to reduce resonance amplitudes is by *enhancing damping* in the vibratory system. The best results can be achieved if the stiffness and damping changes are considered simultaneously (see Appendix 2 and discussion on loudspeaker cones in Section 1.4).

Variable stiffness of shafts, bearings, and mechanisms (in which stiffness may be orientation-dependent) may cause quasi-harmonic (*parametric*) vibrations and overloads, but also can reduce vibrations, e.g., see Section 3.4. While variability of the stiffness can be reduced or increased by design modifications, these modifications can also be combined with *damping enhancement*.

Chatter resistance (stability in relation to self-excited vibrations) of machine tools and other processing machines is often determined by the criterion $K\delta$ (K-effective stiffness, δ-damping, e.g., logarithmic decrement) [4] and Appendix 2. Since in many cases, such as in mechanical joints (Chapter 4) and in materials, dynamic stiffness and damping are interrelated, the stiffness increase can be counterproductive if it is accompanied by reduction of damping. In some cases, even stiffness reduction can be beneficial if it is accompanied by a greater increase in damping (Appendix 2).

Deviation of the vector of cutting (or friction) forces from a principal stiffness axis may cause self-exciting vibrations (*coordinate coupling*) [1].

Low stiffness of the drive system may cause stick-slip vibration of the driven unit on its guideways.

1.3 NEGATIVE STIFFNESS AND DAMPING

When the terms "stiffness" and "damping" are used, it is generally assumed that the respective coefficients have positive signs. For "stiffness," it means that the resulting displacement will be in the same direction as the force. For "damping," it means that the sign of the damping term in the equation of motion, e.g., in Eq. (A1.1a) in Appendix 1, is positive, thus causing a gradual decaying of free vibrations or reduction of the resonance amplitudes with increasing magnitude of the damping coefficient.

However, *negative stiffness* and *negative damping* concepts can be useful for analyzing potential instabilities in mechanical and, especially, in electro- and hydro-mechanical systems. Such systems include systems with elastic instabilities (e.g., buckling), frictional interactions ("stick-slip" phenomenon); cutting process (chatter phenomenon); mechanical linkage systems; electro-mechanical systems (e.g., sensors, induction motors); aerodynamic systems (e.g., self-excited oscillations of high-rise structures, smoke stacks, and transmission lines due to interaction with wind); hydrodynamic systems (e.g., vibrations of pipes in a fluid flow); etc. These systems are briefly described below, with exception of the latter two which have extensive literature coverage, e.g., [5, 6].

The negative stiffness coefficient describes a system wherein the external force and the resulting displacement are in the opposite directions. Obviously, the systems with the negative stiffness are unstable. Usually, such a system exists as a subsystem in a larger mechanical system where the negative stiffness component can be stabilized (compensated) within an otherwise positive stiffness matrix.

The negative stiffness can be realized in purely passive mechanical systems loaded beyond their stability limit. The transition from positive to negative stiffness can be controlled by varying forces, potentially causing instability. For practical applications, not only cases in which the negative stiffness is realized are important, but also marginal cases when the instability-causing force is approaching the loss-of-stability magnitude thus resulting in reduced or zero values of stiffness ("quasi-negative stiffness"). The "negative stiffness" terms can also appear in equations of motion of "active", usually electromechanical, systems comprising external sources of energy.

The positive damping is associated with dissipation of vibratory energy in the system. On the contrary, the negative damping represents a mechanism for adding vibratory energy to the system, thus increasing amplitudes of vibratory processes already present in the system, or excited by external impacts or vibrations, or bringing the system to a condition of dynamic instability thus starting self-excited vibrations. Obviously, the negative damping is always associated with injection into the system of some sort of external energy, either mechanical motion energy or its other forms, such as electric energy, hydraulic energy, etc.

1.3.1 Elastic Instability

Typical cases of stiffness control by applying a destabilizing force are mechanical systems approaching buckling conditions.

"Buckling" of an elongated structural member loaded by a compressive axial force P is loss of stability (collapse) of the structural member when the compressive force reaches a certain critical magnitude P_{cr}, which is also called the *Euler force*. Often, the buckling process is presented as a discrete situation: *stable/unstable*. However, the process of instability development is a gradual continuous process during which bending stiffness of the structural member is monotonously decreasing with increasing axial compressive force P. The member collapses at $P = P_{cr}$, when its bending stiffness becomes zero.

This process can be illustrated on the example of a cantilever column in Fig. 1.3.1a. Bending moment, M, causes deflection of the column. If the axial compressive force $P = 0$, bending stiffness is

$$k_o = M/x \qquad (1.3.1)$$

where x is deflection at the end of the column due to moment M. If $P \neq 0$, it creates additional bending moment, Px, which further increases the bending deformation and, thus, reduces the effective bending stiffness.

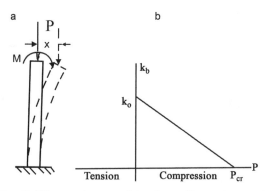

Fig. 1.3.1. Stiffness change (b) of cantilever beam loaded by compression force, P (a).

The overall bending moment becomes [7]

$$M_{ef} = M/(1 - P/P_{cr}),\qquad(1.3.2)$$

and the resulting bending stiffness is approximately

$$k_b = (1 - P_{cr})k_o.\qquad(1.3.3)$$

Fig. 1.3.1b illustrates dynamics of the stiffness change for the column in Fig. 1.3.1a with increasing compressive force, P. Equation (1.3.3) was experimentally validated in [8]. The stiffness-reducing effect presented by this equation must be considered in many practical applications. For example, supporting a machined part (e.g., for turning or grinding) between the centers involves application of a substantial axial force which may result in a significant reduction of bending stiffness for slender parts having relatively low P_{cr}. As it is shown in Appendix 3, natural frequencies of the beam are also decreasing with increasing axial compressive force. If $\omega_{n,o}$ is the initial (without axial force applied) n-th natural frequency of a beam, then the natural frequency ω_n modified by the axial compression force P, is

$$\omega_n = \omega_{n,o}\sqrt{1 - \frac{1}{n^2}\frac{P}{P_{cr}}}\qquad(1.3.4)$$

Thus, preloading by a compressive force of a structural member loaded in bending results in reduction of its bending stiffness. If the collapsing event and the accompanying shape change of the beam at $P > P_{cr}$ is prevented, e.g., by constraining, then its stiffness becomes negative per Eq. (1.3.3).

The effect of bending stiffness reduction for a slender structural components loaded in bending can be very useful in cases when stiffness of a component has to be adjustable or controllable. An application of this effect for vibration isolators is proposed in [9], also see Chapter 8, and in [10]. Figure 1.3.2 [10] shows a device which protects object 14 from horizontal vibrations transmitted from supporting structure 24. The isolation between 14 and 24 is provided by several isolating elements 18 and 16. Each isolator 18, 16, is a thin stiff (metal) post, 60, 32, respectively. The horizontal stiffness of the isolation system is determined by bending stiffness of posts 60 and 32. This stiffness can be adjusted by changing compression force applied to posts 60 and 32 (in series) by loading bolt

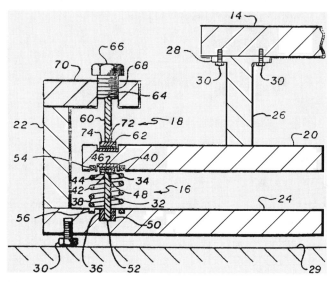

Fig. 1.3.2. Vibration isolator utilizing negative stiffness.

66. The horizontal stiffness can be made extremely low (even negative if the compressive force applied by bolt 66 exceeds the critical force). In the latter case, the stability can be maintained by springs 38 providing positive stiffness.

Similar effects accompany buckling of rubber-metal laminated columns, Appendix 4.

Fig. 1.3.3 shows a more complex case of instability of a thin-walled cylinder (ring) under radial loading between flat loading surfaces (Fig. 1.3.3a). The initial loading is characterized by flat contact areas between the loading platens and the deformed cylinder (Fig. 1.3.3b) and by a slightly non-linear load-deflection plot (1 in Fig. 1.3.3d), where $\beta = \dfrac{12PR^2}{E\delta^3}; \alpha = \dfrac{\Delta}{R}$ [11]. At the relative deformation $\alpha \approx 0.285$, instability develops, and the contact area splits into two parallel strips (Fig. 1.3.3c). The short instability event is associated with the negative stiffness of segment 2 in the load-deflection plot in Fig. 1.3.3d. Compression of the buckled ring/tube in Fig. 1.3.3c might be recoverable (the initial shape of the ring in Fig. 1.3.3a is restored after removal of the loading force, P). It was shown in [12] that the recoverable radial compression deformation of a thin-walled superelastic tubing can be as great as 60% to 65%.

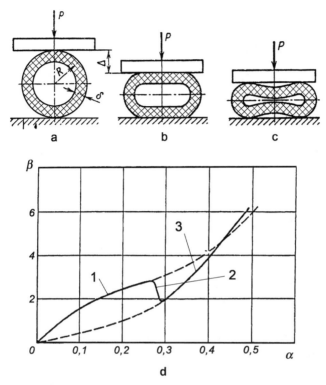

Fig. 1.3.3. Deformation phases of a radially loaded tube (a, b, c) and load-deflection plot (d).

1.3.2 Stick-Slip

The simplest example of mechanical energy transformation into negative damping is a frictional self-excited system. In Fig. 1.3.4a, rigid block (mass, m) is held on belt moving with constant speed v_o. Belt is supported by two pulleys 1 and 2 and driven by pulley 1. Block m is attached by spring (stiffness, k) to stationary base. The friction force $F(\dot{x})$ between block m and belt depends on relative block-belt velocity (sliding velocity), $v_{rel} = \dot{x} - v_o$ or on $\dot{x}$, since $v_o = $ const (Fig. 1.3.4b). At $v_{rel} = 0$, F becomes a "stiction" (or "static friction") force F_o and it is decreasing with increasing $|v_{rel}| > 0$ up to $v_{rel} = v_1$, after which F is increasing (or becoming steady) with increasing v_{rel}. The equation of motion for such system is

$$m\ddot{x} + F(\dot{x}) + kx = 0 \cdot \qquad (1.3.5)$$

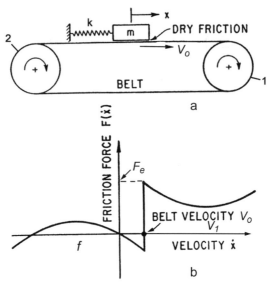

Fig. 1.3.4. Belt-block stick-slip system (a) and friction force vs. velocity characteristic (b).

or, in the first approximation,

$$m\ddot{x} + \frac{dF(\dot{x})}{d\dot{x}}\dot{x} + kx = 0,\tag{1.3.6}$$

where $dF(\dot{x})/d\dot{x}$ is the damping coefficient. It is clear from Fig. 1.3.4b that the damping coefficient is positive at higher relative velocities, $|v_{rel}| > v_1$, but negative at lower velocities, $|v_{rel}| < v_1$. The *negative friction-induced damping* at slow relative motions may cause self-excited vibrations in the sliding connection, so-called "stick-slip" vibrations if the (positive) structural damping in spring k, is not adequate to compensate for the friction-induced negative damping. These vibrations can also be abated by increasing the sliding velocity and by changing the negative slope of the friction plot in Fig. 1.3.4b (e.g., by reducing the friction coefficient difference between $v_{rel} = 0$ and $v_{rel} = v_1$, which can be achieved by using specially formulated lubricants).

1.3.3 Mechanical Linkages

It was demonstrated in [13] that inevitable presence of structural compliances in joints of moving linkages results in superposition of vibratory motions on the programmed "gross motions". Some dynamic equations of the vibratory motions contain in their stiffness and damping coefficients terms determined not only by structural parameters, such as stiffness values and inertias of the system components, but also by kinematic parameters, such as programmed velocities and accelerations of the gross motions of the links. These kinematically induced terms can have both *positive* and *negative* signs and thus influence effective structural stiffness and damping parameters.

Figure. 1.3.5 [13] shows a dynamic model of a two-degree-of-freedom (planar *x-y*) *Cartesian robot with compliant joints*. Two translational and one rotational (angular) stiffness elements are shown for each joint. Of course, the springs shown in Fig. 1.3.5 are active only for incremental (vibratory) components of links' motion and do not obstruct their gross motions. The structural damping is not reflected in this model to make it less cumbersome.

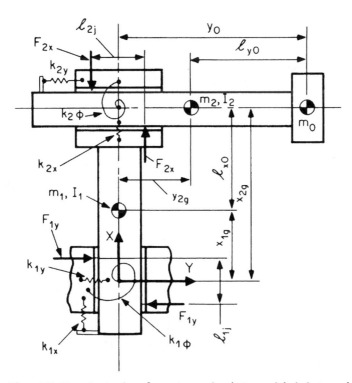

Fig. 1.3.5. Cartesian frame manipulator with joint and actuator compliances.

Coordinates x and y of the centers of gravity of links 1 and 2 can be presented as sums of gross coordinates x_{1g}, x_{2g}, y_{2g}, and of small incremental displacements caused by deformations of six springs. Deformations of translational springs k_{1x}, k_{1y}, k_{2x}, k_{2y} are designated as ψ_{11}, ψ_{12}, ψ_{21}, ψ_{22}, and deformations of angular springs $k_{1\phi}$, $k_{2\phi}$ are designated as ϕ_{11}, ϕ_{22}. Since both translational and rotational incremental motions are considered, both masses m_1, m_2 and moments of inertia I_1, I_2 of the links are used in the equations of motion. Usually, radial stiffness of the joints is quite high, thus it can be assumed that $\psi_{11} = \psi_{22} = \psi_{12} = \psi_{21} \approx 0$. All six equations of motion are dynamically coupled. Two equations of motion containing kinematically induced stiffness and damping terms are given as Eqs. (1.3.7) and (1.3.8) which are, essentially, equations of motion for angular coordinates ϕ_{11}, ϕ_{22}, coupled with other coordinates.

$$[m_2 x_{1g}^2 + I_1 + m_2 y_{2g}^2 + (l_{x0} + x_{1g})^2 + I_2 + m_0(y_{2g} + l_{y0})^2$$
$$+ m_0(x_{1g} + l_{x0})^2]\ddot{\phi}_{11} + [m_2 y_{2g}^2 + I_2 + m_0(l_{y0} + y_{2g})^2]\ddot{\phi}_{22}$$
$$+ 2[m_1 x_{1g}\dot{x}_{1g} + m_2 y_{2g}\dot{y}_{2g} + m_2 \dot{x}_{1g}(l_{x0} + x_{1g}) + m_0\dot{y}_{2g}(l_{y0} + y_{2g})$$
$$+ m_0\dot{x}_{1g}(l_{x0} + x_{1g})]\dot{\phi}_{11} \qquad\qquad (1.3.7)$$
$$+ 2[m_2 y_{2g}\dot{y}_{2g} + m_0\dot{y}_{2g}(l_{y0} + y_{2g})]\dot{\phi}_{22}$$
$$+ [k_{1\phi} + m_1 x_{1g}\ddot{x}_{1g} + m_2 \ddot{x}_{1g}(l_{x0} + x_{1g}) + m_2 y_{2g}\ddot{y}_{2g} + m_0(l_{y0} + y_{2g})\ddot{y}_{2g}]\phi_{11}$$
$$= m_2 \ddot{x}_{1g} y_{2g} + \ddot{y}_{2g}(l_{x0} + x_{1g}) + m_0[(y_{2g} + l_{y0})\ddot{x}_{2g} + (x_{1g} + l_{x0})\ddot{y}_{2g}]$$

$$-[m_2 y_{2g}^2 + I_2 + m_0(l_{y0} + y_{2g})^2]\ddot{\phi}_{11} + [m_2 y_{2g}^2 + I_2 + m_0(l_{y0} + y_{2g})^2]\ddot{\phi}_{22}$$
$$+ 3[m_2 y_{2g}\dot{y}_{2g} + m_0(y_{2g} + l_{y0})\dot{y}_{2g}]\dot{\phi}_{11} + 3[m_2 y_{2g}\dot{y}_{2g} + m_0(y_{2g} + l_{y0})\dot{y}_{2g}]\dot{\phi}_{22}$$
$$+ [(m_2 + m_0)y_{2g}\ddot{y}_{2g} + m_0 l_{y0}\ddot{y}_{2g}]\phi_{11} + [k_{2\phi} + (m_2 + m_0)y_{2g}\ddot{y}_{2g} + m_0 l_{y0}\ddot{y}_{2g}]\phi_{22}$$
$$= [(m_2 + m_0)y_{2g} + m_0 l_{y0}]\ddot{x}_{2g} \qquad\qquad (1.3.8)$$

In Eqs. (1.3.7) and (1.3.8), the *stiffness* terms (for vibratory coordinates ϕ_{11} and ϕ_{22}) and the *damping* terms (for vibratory velocities ϕ_{11} and ϕ_{22}) contain, besides terms related to inertias, stiffness values, and geometric parameters of the system, also time-varying gross coordinates x_g, y_g, and their derivatives (velocities and accelerations). Thus, these terms contain *kinematically induced* stiffness and damping components. Depending on magnitudes and signs of these components, they can effectively increase or decrease the overall stiffness and damping values, and even make them negative. The influence of the kinematically induced components can be commensurate with or even exceed the influence of the structural stiffness and damping. These elements reduce and/or increase natural frequencies and may dangerously reduce (or even make negative) the effective damping. The influence of the kinematically induced stiffness and damping components is decreasing with increasing structural stiffness and damping, respectively, and increasing with increasing travel distances and gross velocities of the links.

This was demonstrated by computer simulation of two different patterns of programming of a motion cycle for a model in Fig. 1.3.5 used as a "pick-and-place" manipulator. The structural parameters are: $x_{1g} = 0.3$ m; $y_{2g} = 0.5$ m; $l_{x0} = 1.2$ m; $l_{y0} = 1.2$ m; $m_1 = 10$ kg; $m_2 = 5$ kg; $m_0 = 2$ kg; $I_1 = 1$ kg·m^2; $I_2 = 0.5$ kg·m^2; $k_{1\phi} = 10^6$ N·m/rad; $k_{2\phi} = 5 \times 10^5$ N·m/rad. The initial conditions are: $x_{1g} = y_{2g} = 0$; $\dot{x}_{1g} = \dot{y}_{2g} = 0$; $\phi_{11} = 5 \times 10^{-3}$ rad; $\phi_{22} = 0$. The programmed motion cycles are shown in Figs. 1.3.6a and 1.3.7a, the time histories of the angular vibratory coordinates ϕ_{11} *and* ϕ_{22} are shown in Figs. 1.3.6b,c and 1.3.7b,c. Since vertical scales are different for all plots, the maximum amplitudes are marked. For the programmed cycle in Fig. 1.3.6a, vibrations of ϕ_{11} *and* ϕ_{22} have a constant frequency and are decaying. However, for the motion cycle in Fig. 1.3.7a, the frequencies of the vibratory processes are changing (due to a greater influence of the kinematically induced negative stiffness), and very large excursions of the angular vibrations sustain during the whole duration of the motion cycle due to the kinematically induced negative damping. The maximum amplitude of ϕ_{22} is ~40% larger than for the first programmed cycle and 40% larger than the initial deflection of ϕ_{11}. These dynamic effects can become even more pronounced for faster and/or less structurally stiff systems.

The similar effects develop also for other types of manipulator structures, such as polar and jointed manipulators.

1.3.4 Mechanisms with Nonlinear Position Functions

Negative damping terms can also develop in equations of motion describing *mechanisms with nonlinear position functions*. In such mechanisms, a uniform motion (usually, rotation) of the input (driving) shaft is transformed into a non-uniform motion of the output member. Such mechanisms can be designed as linkages (e.g., in metal forming presses and other production and automation machines) or as cam mechanisms.

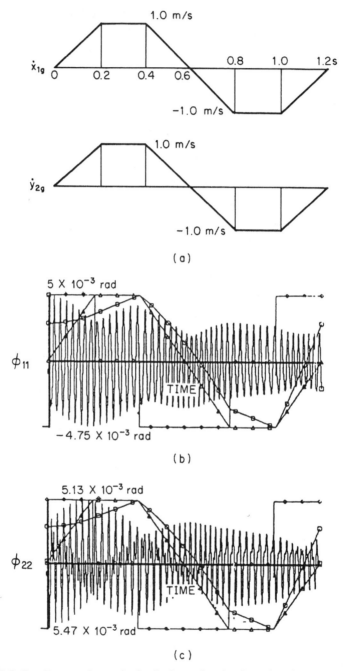

Fig. 1.3.6. Dynamic path deviations (b, c) of a compliant Cartesian manipulator for programmed velocity cycle a; lines: □ — x_{1g}; △ — $\dot{x}_{1g}$; ◊ — $\ddot{x}_{1g}$.

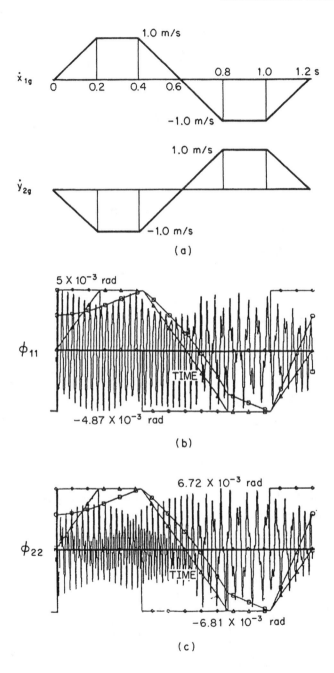

Fig. 1.3.7. Dynamic path deviations (b, c) of a compliant Cartesian manipulator for programmed velocity cycle *a*; lines: □ — x_{1g}; △ — $\dot{x}_{1g}$; ◊ — $\ddot{x}_{1g}$.

The position function describes position coordinate x of the output link as a function of coordinate ϕ of the driving shaft,

$$x = \Pi(\phi) \qquad (1.3.9)$$

"Transfer functions" of the mechanism are derivatives of the function Π [14],

$$\Pi' = \frac{d\Pi}{d\varphi}; \Pi'' = \frac{d^2\Pi}{d\varphi^2}; \Pi''' = \frac{d^3\Pi}{d\varphi^3}. \qquad (1.3.10)$$

The transfer functions are purely geometric characteristics, which can be correlated with kinematic characteristics of the mechanism. Thus, velocity of the output link is

$$\dot{x} = \frac{dx}{dt} = \frac{dx}{d\phi}\frac{d\phi}{dt} = \Pi'(\phi)\,\dot{\phi}\,; \qquad (1.3.11)$$

acceleration and jerk, respectively, are

$$\ddot{x} = \Pi''(\phi)\dot{\phi}^2 + \Pi'(\phi)\ddot{\phi}\,; \; \dddot{x} = \Pi'''(\phi)\dot{\phi}^3 + 3\Pi''(\phi)\dot{\phi}\ddot{\phi} + \Pi'(\phi)\dddot{\phi} \qquad (1.3.12)$$

With an assumption of a constant speed of the drive shaft (presses are usually equipped with large flywheels), $\dot{\phi} = \omega = $ const, and transfer functions can be expressed as

$$\Pi'(\phi) = \frac{\dot{x}}{\omega}; \quad \Pi''(\phi) = \frac{\ddot{x}}{\omega^2}; \quad \Pi'''(\phi) = \frac{\dddot{x}}{\omega^3}. \qquad (1.3.13)$$

A dynamic analysis of heavily loaded and/or high-speed mechanisms should definitely consider elastic deformations (compliances) in the linkage. A dynamic model of a nonlinear position function mechanism with a compliant linkage is shown in Fig. 1.3.8. Since the main driving shaft

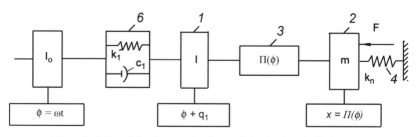

Fig. 1.3.8. Mathematical model of elastic linkage.

usually carries a large flywheel, its moment of inertia can be considered as infinite, $I_o = \infty$, thus its rotational speed ω is constant. The driven linkage 1 has effective moment of inertia, I, and is connected to the main shaft via elastic connection (stiffness k_1, damping coefficient c_1). The output (working) member 2 (ram or slide) is sliding in guideways, and its mass can be referred to the shaft 1 as an effective moment of inertia I_s by multiplying it with the square of the first transfer function (see also Chapter 6). Force, F, is the resistance of the workpiece which can have stiffness k_n. Vibrations can develop in this one-degree-of-freedom, variable parameter system. If the vibratory coordinate is q_1, the dynamic equation of motion can be written as [14]

$$(I + m\Pi'^2)\ddot{q}_1 + (c_1 + 2\Pi'\Pi''m\omega)\dot{q}_1 + k_1 q_1 + \Pi'\Pi''m\dot{q}_1^2 = \\ - \Pi'\Pi''m\omega^2 - k_n\Pi\Pi' - F\Pi'.$$

(1.3.14)

It is clear that the effective damping coefficient of the moving system is quite different from the structural damping coefficient of the same system if it is vibrating but is not driven (stationary, $\omega = 0$). The additional *kinematically induced damping term* $c_k = 2\Pi'\Pi''m\omega$ varies during each cycle of the mechanism motion. Since transfer functions Π' and Π'' change signs during the cycle, this term may become negative. Its importance is increasing with the increased speed ω of the mechanism, with increasing (by design) mass, m, of the output member, and of course, it is dependent on the kinematic parameters, Π' and Π''. In high speed mechanisms, the probability of the total damping term becoming negative is very real. If it happens, very significant overloads may develop in the mechanism, as demonstrated by the example below.

Example. Fig. 1.3.9 [15] shows a drive unit of an extrusion press. The extrusion operation requires a long stroke of the slide while loaded by the rated force. Also, the slide velocity should not change significantly along the stroke. Because of these requirements, conventional drives using simple crank mechanisms are not optimal for performing extrusion operations. The required uniformity of the slide velocity and the load are achieved in the extrusion press by adding additional links. In Fig. 1.3.9, the driving motor 16 drives, via belt transmission 18, the main driving shaft with the flywheel 19. This press has two parallel independent slides, both driven from the main driving shaft. Each slide 13 is associated with eccentric shaft 27 (axis E) driven by reduction gear pair (gear 25 on

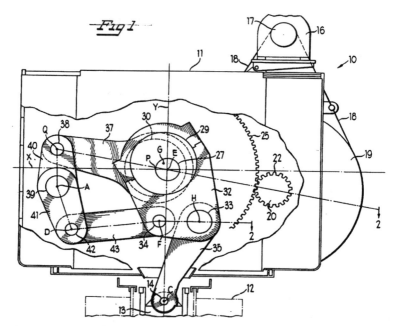

Fig. 1.3.9. Linkage schematic of the drag-link extrusion press.

eccentric shaft is engaged with pinion 22 mounted on the main driving shaft). Eccentric shaft 27 has two eccentric journals 29 and 30, with their axes at G and P, respectively. Eccentric journal 29 is engaged with link 32 which is, in its turn, jointed with pitman (connecting rod) 35 which drives slide 13 through joint 14. The linkage modifying the position function consists of strap 37 supported by eccentric journal 30; rocking link 40-41 connected to 37 by joint 38; and cross link 43 joined by joint 42 with rocking link 40-41 and by joint 34 with link 32.

The described linkage delivers the position function characterized by a uniform rated load and a uniform slow speed on a long segment of the stroke. Fig. 1.3.10a shows the slide velocity (first transfer function Π') as

Fig. 1.3.10. First (a) and second (b) transfer functions of the extrusion press linkage; indicator of kinematically induced damping (c); statically calculated (1) and recorded (2—at the first run after repair at 37.5 spm, 3—after 15,000 extrusion cycles) load in cross-link (d); recorded load in pitman at the first run after repair at 37.5 spm, 4, and after 15,000 extrusions, 5 (e).

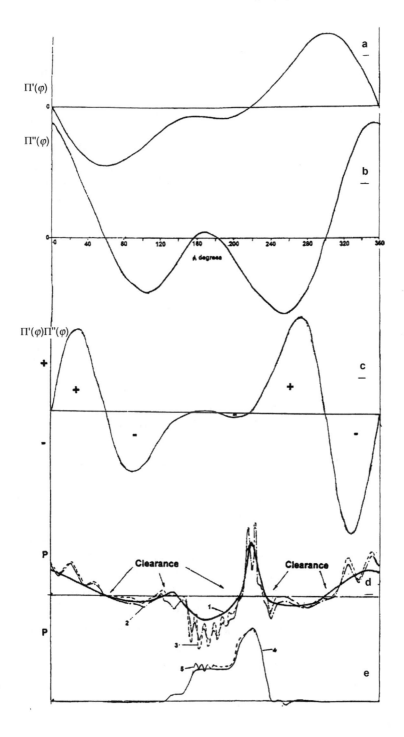

a function of the eccentric shaft rotation angle ϕ for an extrusion press built with the linkage shown in Fig. 1.3.9. Each slide has 600-ton capacity, and 22-in. (550 mm) working stroke. The press was initially built to run at 50 spm (strokes per min).

Fig. 1.3.10a,b,c shows schematic (without numerical values) plots of Π', Π'', $\Pi'\Pi''$, respectively, for the linkage mechanism of this press. Fig. 1.3.10c indicates that the product $\Pi'\Pi''$ becomes negative with a large magnitude just before the slide reaches the Top Dead Center (TDC). The induced negative damping value for the kinematically induced damping

$$\eta_k = \frac{\delta_k}{2\pi} = \frac{2\Pi'\Pi''m\omega}{2m\Pi'^2\omega_n} = \frac{\Pi''\omega}{\Pi'\omega_n} = \frac{\ddot{x}}{\dot{x}}\frac{1}{\omega_n}, \qquad (1.3.15)$$

where ω_n-natural frequency of the system and δ_k-log decrement (see Appendix 1). The instanteneous value of the kinematically induced damping depends on the inertia and stiffness of the drive implicitly, via natural frequency ω_n; increase in stiffness and/or reduction of mass result in higher ω_n and, consequently, in smaller η_k. For a given linkage (given Π' and Π'') the induced relative damping value depends on the input rpm, ω.

The peak of negative damping magnitude in Fig. 1.3.10c is at $\phi = \sim340°$; for this mechanism at 50 spm $dx/dt = 1.175$ m/sec at $\phi = 335°$; 0.952 m/sec at $\phi = 340°$; 0.715 m/sec at $\phi = 345°$, thus $(\Delta^2 x/\Delta t^2)_{340°} \approx 13.3$ m/sec^2. Substituting these quantities and $\omega_n = 2\pi(8.8) = 55.3$ rad/sec into Eq. (1.3.15), the maximum value of the kinematic damping at $\phi = 340°$ is $(\eta_k)_{max} = -0.25$. The average value of η_k between $\phi = 300°$ and $\phi = 360°$ is $(\eta_k)_{av} = \sim -0.17$, much higher in magnitude then the structural (positive) damping, $\eta_x = 0.025$. As a result, the amplitude of the dynamic vibratory load excited by impact in the joints of the linkage by reversal of acceleration of the slide at $\phi = 300°$ is not decaying (or even slightly increasing) during the first two cycles (Fig. 1.3.10d), until the system enters the positive kinematically induced damping zone. Additional dynamic load excitation events occur at $\phi = 141°$ and $\phi = 204°$ when reversals of static load are occurring simultaneously in several links (links 32, 37, and 43 in Fig. 1.3.9). These second and third excitation zones overlap with other zones of relatively light kinematically induced negative damping. The slow decay of the intense dynamic loads caused several catastrophic failures of the press [15].

1.3.5 Electromechanical Systems

An example of an electro-mechanical system with a negative stiffness component is shown in Fig. 1.3.11 (*capacitance displacement sensor*) [16]. Such sensors are widely used in MEMS. Displacement $y(t)$ of upper plate 1 relative to stationary lower plate 2 is determined by charge $Q = \varepsilon AU/[d-y(t)]$ of upper plate 1 (voltage U is applied between plates 1 and 2). A mechanical spring 3 is stabilizing positioning of measuring plate 2. While mechanical spring 3 is pushing plates 1 and 2 apart with the force $P_{mech} = ky$, the oppositely charged plates 1 and 2 experience electrical attraction to each other with the force

$$P_{el} = -\frac{\varepsilon AU^2}{2[d - y(t)]^2}. \tag{1.3.16}$$

Thus, the equation of motion of plate 1 is

$$m\ddot{y} + ky - P_{el} = P_{ex}. \tag{1.3.17}$$

Here m is mass of the upper plate; d-initial distance between the plates at $U = 0$; A-effective surface area of each plate; $\varepsilon = 8.85 \times 10^{-9}$ F/m-permittivity of vacuum (about equal to permittivity of air); $P_{ex} = 0$ – external force (e.g., caused by pressure).

While $ky = P_{mech}$ in Eq. (1.3.17) is the mechanical (positive) linear restoring force dependent only on displacement $y(t)$, P_{el} is the electrical (negative) nonlinear destabilizing force, also dependent only on displacement $y(t)$. Expression (1.3.17) can be expanded into a Taylor series as an explicit function of y.

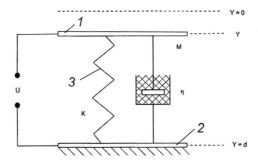

Fig. 1.3.11. Electrostatic distance sensor with a moving upper plate.

The stabilizing effect of the mechanical spring and the destabilizing effect of the electrical negative spring-like attraction can be harmonized and tuned by varying voltage, U, and by varying stiffness, k, of the mechanical spring.

Induction motors can generate very intense negative damping [17]. An electromagnetic torque T_{em} versus slippage, $s = (\omega_o - \omega)/\omega_o$, characteristic of an induction motor is shown in Fig. 1.3.12. Here ω_o is the synchronous rotational frequency (frequency of electric current ω_{el} divided by the number p of pairs of poles in the motor stator), and ω-the actual rotational frequency of the rotor. The solid line 1 represents the actually measured $T_{em}(s)$ characteristic. The broken line 2 in the lower part of the plot indicates the "nominal" characteristic in this part, as it can be found in a catalog. Fig. 1.3.12 shows the dimensionless ratio T_{em}/T_r, where T_r-the rated (nominal) torque of the motor. The $T_{em}(s)$ characteristic has two distinct parts: stable part A wherein the derivative $dT_{em}/ds >$ 0, and unstable (starting) part B, where $dT_{em}/ds < 0$, as shown by line 3. If the motor operates in part B of the torque-slippage characteristic, then increasing speed of the rotor leads to increasing torque, providing a desirable start-up regime for the motor. This is also equivalent to negative

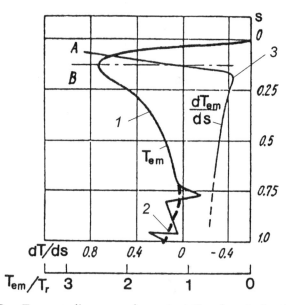

Fig. 1.3.12. Torque-slippage characteristic of an induction motor (rated power 14 KW, 1500 rpm).

damping, since increase in a vibratory component $\dot{\varphi}_r$ of the rotor velocity leads to a further increase in $\dot{\varphi}_r$. In the other half of the vibration period, $\dot{\varphi}_r$ is diminishing, thus reducing torque T_{em} which again contributes to increasing amplitude of the rotor vibration. The opposite is true for the part A of the characteristic.

The "electromagnetic damping" coefficient associated with the induction motor is equal to slope of the $T_{em}(s)$ plot taken with the opposite sign,

$$c_{em} = -\frac{dT(\dot{\varphi}_r)}{d\dot{\varphi}_r} = \frac{1}{\omega_o} \frac{dT(s)}{ds}. \tag{1.3.18}$$

A dynamic system "motor–driven mechanical system" can be, in the first approximation, modeled as a two-mass system in Fig. 1.3.13, where I_m is the total effective moment of inertia of the mechanical system referred to the motor rotor shaft, $e_m = 1/k_m$ and c_m are the effective torsional compliance and damping coefficients of the driven mechanical system referred to the rotor shaft (see Chapter 6), I_r is moment of inertia of the rotor, e_{em}, c_{em} are torsional compliance and damping coefficients of the electromagnetic field. The equivalent inertia of the electric supply network is assumed to be infinitely large, $I_s \approx \infty$. Within the unstable part B of the torque-slippage characteristic, e_{em} is large, thus k_{em} is low and can be neglected in the first approximation. Equations of motion of this model can be written as

$$\begin{aligned} I_r\ddot{\varphi}_r + c_{em}\dot{\varphi} + c_m(\dot{\varphi}_r - \dot{\varphi}_m) + k_m(\varphi_r - \varphi_m) &= T \\ I_m\ddot{\varphi}_m + c_m(\dot{\varphi}_m - \dot{\varphi}_r) + k_m(\varphi_m - \varphi_r) &= 0, \end{aligned} \tag{1.3.19}$$

where T - the accelerating torque applied to the rotor. As shown in [17], if in the unstable area, B, when $c_{em} < 0$,

$$\left| \left(\frac{I_m}{I_m + I_r} \right)^2 c_{em} \right| > c_m, \tag{1.3.20}$$

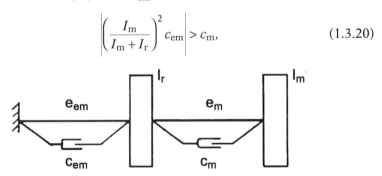

Fig. 1.3.13. Two-mass dynamic model of an electromechanical drive system.

then the electromechanical system becomes unstable, and any excitation would excite oscillations with the natural frequency of the mechanical system with the increasing amplitude. The lower part of solid line 2 (actually measured characteristic) in Fig. 1.3.12 shows that some possible excitation sources can be "kinks" on the $T(s)$ plot of a real-life induction motor. With high-inertia mechanical systems ($I_m \gg I_r$), in Eq. (1.3.20) magnitude of effective electromagnetic negative dumping straight bracket becomes stronger, acceleration time is long, and the drive stays longer within the area B (thus negative damping is acting for a longer time). Consequently, amplitudes of the self-excited vibratory process can build up to a dangerous level. Fig. 1.3.14 shows oscillograms of torque in the mechanical system driven by an induction motor (14 KW, $\omega_o = 1500$ rpm) when $I_m = 11 I_r$ (Fig. 1.3.14a) and $I_m = 4.5 I_r$ (Fig. 1.3.14b). Time is displayed on the horizontal axis, torque on the vertical axis; T_{st} - response of the system to the starting torque of the motor. Amplitudes of the torque oscillation whose frequency is the natural frequency of the mechanical drive system, $f_n \approx 20$ Hz, are increasing with the *logarithmic increment* of vibration $\delta \approx -0.12$. The negative logarithmic increment of the electromagnetic system of the motor is $\delta \approx -0.27$, while the (positive) logarithmic decrement of the mechanical drive is $\delta \approx 0.15$ (see Chapter 6). The peak torque T_{max} in the case of Fig. 1.3.14a exceeded $4.5 T_r$ and thus led to multiple catastrophic failures in a milling machine gearbox. The overloads in Fig. 1.3.14b, representing a mechanical drive with a smaller inertia, are

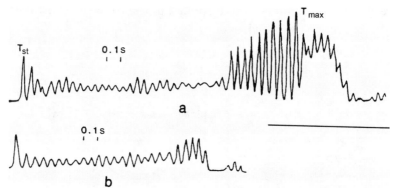

Fig. 1.3.14. Transient starting torques in electromechanical drive system driven by induction motor of Fig. 1.3.12; $a - I_m = 11 I_r$; $b - I_m = 4.5 I_r$.

significantly lower, as predicted by Eq. (1.3.20). The process of gradually increasing amplitudes breaks when the vibratory velocity "trespasses" into high positive damping area, A, during the vibratory process.

1.3.6 Stiffness and Damping of Cutting Process

Deformations in the machining system are developing not only due to the finite stiffness of the structural components, but also due to finite stiffness of the cutting process itself. While stiffness of the cutting process is always positive, its damping can become negative thus creating dynamic (chatter) instability. The cutting process can be modeled as a spring representing effective cutting stiffness and a damper representing effective cutting damping. The stiffness and damping parameters can be derived from the expression describing the dynamic cutting force. Various approximate expressions for dynamic cutting forces had been suggested. The most convenient basic expression for deriving the stiffness and damping parameters of the cutting process is one given in [1]. The dynamic increment of the cutting force dP_z in the z-direction for turning operation can be written as

$$dP_z = K_1[z(t) - \mu z(t - T)] + K_2\, \dot{z}(t). \qquad (1.3.21)$$

Here z is vibratory displacement between the tool and the workpiece, whose direction is perpendicular to the axis of the workpiece and also to the cutting speed direction in the horizontal plane; $\mu = 0$ to 1 is an overlap factor between two subsequent tool passes in the z-direction; K_1 is *cutting stiffness coefficient* in the z-direction; K_2 is *penetration rate coefficient* due to the tool penetrating the workpiece in the z-direction; and $T = 2\pi/\Omega$, where Ω rad/sec is the rotating speed of the workpiece in turning.

By assuming displacement z as

$$z(t) = A \cos \omega t \qquad (1.3.22)$$

where A = an indefinite amplitude constant, and ω = chatter frequency. Equation (1.3.21) can be rearranged as

$$dP_z = K_{cz}dz + C_{cz}\frac{dz}{dt}dt, \qquad (1.3.23)$$

where

$$K_{cz} = K_1[1 - \mu \cos 2\pi(\omega/\Omega)] \tag{1.3.24}$$

$$C_{cz} = K_1(\mu/\omega) \sin 2\pi(\omega/\Omega) + K_2\dot{z}(t). \tag{1.3.25}$$

K_{cz} and C_{cz} can be defined as effective cutting stiffness and effective cutting damping coefficients, respectively (since only the z-direction is considered, the subscript z is further omitted). The effective cutting stiffness and damping coefficients are functions not only of the cutting conditions but also of the structural parameters of the machining system (its effective stiffness and mass), which enter Eqs. (1.3.24) and (1.3.25) via frequency ω. The dynamic cutting force P_z depends not only on displacement $z(t)$ but also on velocity $\dot{z}(t) = dz/dt$. The velocity-dependent term may bring the system instability when effective cutting damping $C_{cz} < 0$, and the magnitude of C_{cz} is so large that it cannot be compensated by positive structural damping.

Experimental determination of the cutting process stiffness can be illustrated on the example of a cantilever workpiece [18]. A cantilever workpiece with a larger diameter segment at the end (Fig. 1.3.15) can be modeled as a single degree of freedom system with stiffness K_w without cutting and with stiffness $K = \dfrac{K_w K_c}{K_w + K_c}$ during cutting, where K_w = the stiffness of the workpiece at the end, and K_t = the effective cutting stiffness. Since stiffness of the cantilever workpiece is relatively low compared with structural stiffness of the machine tool or of the clamping

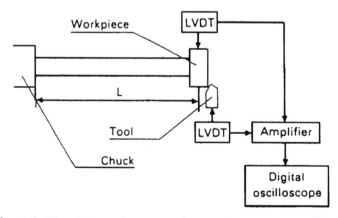

Fig. 1.3.15. Measuring setup for cutting process stiffness.

chuck, chatter conditions are often determined by the workpiece and the cutting process only. Thus, if the natural frequency f_w of the workpiece (without cutting) and the frequency of the tool or workpiece vibration at the chatter threshold were measured, then the effective cutting stiffness can be using the following equation, considering that the effective mass of the system is the same in both cases:

$$K_c = \left(\frac{f_c^2}{f_w^2} - 1 \right) K_w. \tag{1.3.26}$$

The frequency f_w can be measured using an accelerometer, while the chatter frequency can be measured by markings on the workpiece or on the tool using a linear variable differential transformer (LVDT) during cutting as shown in Fig. 1.3.15.

A cantilever bar with overhang L = 127 mm (5 in.) having stiffness (as measured) K_w = 1,850 N/mm (10,400 1b/in.) was used for the tests. The natural frequency is f_w = 200 Hz, and the equivalent mass is about 1.2 kg (0.0065 1b·sec²/in.). The values of the effective cutting stiffness and vibration amplitude under different cutting conditions are given in Fig. 1.3.16a-c. It can be seen that smaller vibration amplitudes are correlated with higher effective cutting stiffness values. This validates representation of the effective cutting stiffness as a spring.

Damping of the cutting process was studied in [19]. It was established that damping of the cutting process is generally decreasing with the increasing cutting speed until it becomes negative as shown in Fig. 1.3.17, although there are some speed intervals when the process damping is not decreasing. The chatter starts when the magnitude of the cutting process-induced negative damping exceeds the magnitude of positive damping of the machine structure.

1.3.6a Influence of Machining System Stiffness on Accuracy and Productivity

Elastic deformations of the production (machining) system: *machine tool-fixture-tool-machined part* under cutting forces are responsible for a significant fraction of the part inaccuracy. These deformations also influence productivity of the machining system, either directly by slowing the process of achieving the desired geometry or indirectly by causing self-excited chatter vibrations.

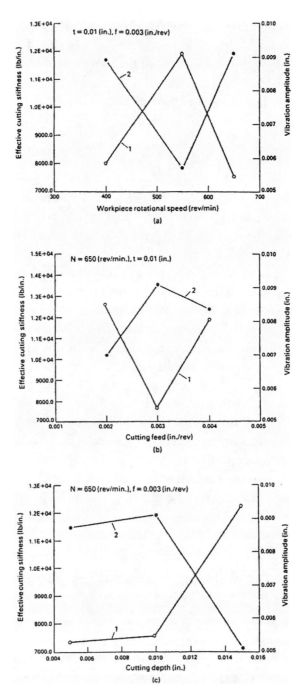

Fig. 1.3.16. Effective cutting stiffness (line 1) and workpiece vibration amplitude (line 2) vs. a-cutting speed; b-feed; c-depth of cut.

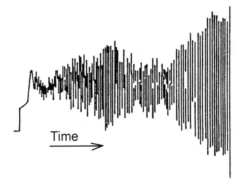

Time

Fig. 1.3.17. Increasing vibration amplitudes when negative damping is generated in the cutting zone.

In a process of machining a precision part from a roughly shaped blank workpiece, there is the task to reduce deviation Δb of the blank surface from the desired geometry to a smaller allowable deviation Δp of the part surface (Fig. 1.3.18). This process can be modeled by introduction of an *accuracy enhancement factor,* ζ

$$\zeta = \Delta b/\Delta p = (t_1 - t_2)/z_1 - z_2) \qquad (1.3.27)$$

where t_1, t_2 = max, min depth of cut; z_1, z_2 = displacements of the cutter which are normal to the machined surface due to structural deformations caused by the cutting forces. If the cutting force is

$$P_z = C_m ts^q, \qquad (1.3.28)$$

then

$$\zeta = (k/C_m)/s^q, \qquad (1.3.29)$$

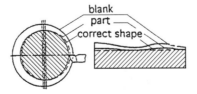

blank
part
correct shape

Fig. 1.3.18. Evolution of geometry of machined parts when machining system has finite stiffness.

where C_m = material coefficient; k = stiffness of the machining system; t = depth of cut; s = feed; q = 0.6 to 0.75. For the process of turning medium hardness steel with s = 0.1 to 0.75 mm/rev on a lathe with k = 20 N/μm,

$$\zeta = 150 \text{ to } 30. \tag{1.3.30}$$

Knowing the shape deviations of the blanks and the required accuracy, the above formula for ζ allows to estimate the required k and the allowable s or decide on the number of passes required to achieve the desired accuracy.

Inadequate stiffness of the machining system may result in various distortions of the machining process. Some examples of such distortions are shown in Fig. 1.3.19. The total cross sectional area of the cut is smaller during the transient phases of cutting (when the tool enters into and exits from the machined part) than during the steady cutting. As a result, deflection of the workpiece is smaller during the transient phases thus resulting in deeper cuts (Fig. 1.3.19 a, b).

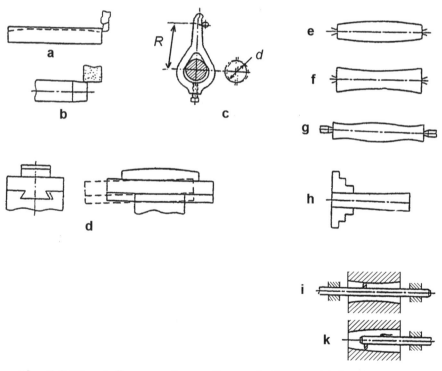

Fig. 1.3.19. Influence of compliances in the machining system on geometry of machined parts.

Turning of a workpiece supported between two centers requires driving of the workpiece by a driving yoke clamped to it (Fig. 1.3.19c). Asymmetry of the driving system results in an eccentricity (runout) of the part with the magnitude

$$\delta = P_z d / k_c R, \tag{1.3.31}$$

where k_c = stiffness of the supporting center closest to the driving yoke.

Heavy travelling tables supporting workpieces on milling machines, surface grinders, etc., may change their angular orientation due to changing contact deformations in the guideways caused by shifting of the center of gravity during the travel (Fig. 1.3.19d). This also results in geometrical distortions of the part surface.

A surface deviation Δ caused by a variable stiffness of the machining system can be expressed as

$$\Delta = P_z (1/k_{min} - 1/k_{max}) \tag{1.3.32}$$

where k_{min}, k_{max} = low and high stiffness of the machining system, and P_y = cutting force.

Fig. 1.3.19e shows a "barrel" shape generated in the process of turning a slender elongated part between the rigid supporting centers while Fig. 1.3.19f shows a "corset" shape when a rigid part is supported by the compliant centers. The part in Fig 1.3.19g is itself slender and was supported by compliant centers. Fig. 1.3.19h shows the shape of a cantilever part clamped during machining in a non-rigid chuck. Fig. 1.3.19i shows the shape of the hole bored by a slender boring bar moving in two stationary rigid supports, while Fig. 1.3.19k illustrates the shape of the hole machined by a cantilever slender boring bar guided by one stationary support.

The role of stiffness enhancement is to reduce these distortions. When they are repeatable, corrections which would compensate for these errors can be commanded to a CNC machine by its controller. However, the highest accuracy is still obtained when the error is small, and it is always preferable to have an appropriate stiffness to avoid complications of this compensation procedure.

An extensive experimental study on influence of stiffness on accuracy and productivity of grinding very hard silicon nitride ceramics is described in [20].

Manufacturing requirements for stiffness of parts often determine the possibility of their fabrication with high productivity (especially for mass-production). Sometimes, shaft diameters for mass-produced machines are determined not by the required strength but by a need to increase itheir stiffness to create a possibility of productive multi-cutter machining of either the shaft or of the associated components (e.g., gears), or both. Machining a low-stiffness shaft may lead to chatter, to a need to reduce regimes, and to copying of inaccuracies of the original blank.

Stiffness of the production equipment influences not only its accuracy and productivity. For example, stiffness characteristics of a stamping press also influence its energy efficiency (since the deformation of a low-stiffness frame absorbs a significant fraction of energy contained in one stroke of the moving ram); dynamic loads and noise generation (due to the same reasons); product quality (since large deformations of the frame cause misalignments between the punch and the die and thus, distortions of the stampings); die life (due to the same reasons). In a crank press, wherein the maximum force is developing at the end of the stroke, the amount of energy spent on the elastic structural deformations can be greater than the amount of useful energy (e.g., spent on the punching operation). Abrupt unloading of the frame after the breakthrough event causes dangerous dynamic loads/noise increasing with increasing structural deformations.

In mechanical measuring instruments/fixtures, a higher stiffness is sometimes needed to reduce their deformations by the measuring (contact) force.

Deformations at the tool end caused by the cutting forces result in geometric inaccuracies and in a reduced dynamic stability of the machining process. It is important to understand that there are many factors causing deflections at the tool end. For example, in a typical boring mill, deformation of the tool itself represents only ~11% of the total deflection, while deformation of the spindle and its bearings is responsible for ~37%, and the tapered interface between the toolholder and the spindle hole is responsible for ~52% of the total deflection.

1.4 STIFFNESS AND DAMPING OF SOME WIDELY USED MATERIALS

Stiffness of a structural material is characterized by its elastic (Young's) modulus E for tension/compression. However, there are many cases when knowledge of just Young's modulus is not enough for a judicious selection of the structural material. Another important material parameter is shear modulus G, which can be calculated as

$$G = \frac{E}{2(1+v)},$$ (1.4.1)

where v = the Poisson's ratio. *For most metals $G \approx 0.4E$.*

Frequently, stiffer materials (materials with higher E) are heavier. Thus, use of such materials may result in structures having smaller cross-sections but not lower weight (in some cases even greater weight, which may be undesirable). In cases when the structural deflections are caused by inertia forces, like in a revolute robot or a coordinate measuring machine (CMM) arm, use of a stiffer but heavier material can be of no benefit or even be counter-productive if its weight increases as much or more than its stiffness. Very frequently, stiffer materials are used to increase natural frequencies of the system. This case can be illustrated on the example of two single-degree-of-freedom dynamic systems in Fig. 1.4.1. In these sketches, γ, A_1, l_1, are density, cross-sectional area, and length, respectively, of the inertia element (mass m); A_2, l_2, h, b are cross-sectional area, length, thickness, width, respectively, of elastic elements (stiffness k). For the system in Fig. 1.4.1a (tension-compression elastic element), the natural frequency is

$$\sqrt{\frac{k}{m}} = \sqrt{\frac{EA_2/l_2}{\gamma A_1 l_1}} = \sqrt{\frac{E}{\gamma}} \sqrt{\frac{A_2}{\gamma A_1 l_1 l_2}}$$ (1.4.2)

For the system in Fig. 1.4.1b (elastic element loaded in bending)

$$m = \gamma A_1 l_1, \quad k = \frac{3EI}{l_2^3} = \frac{1}{4}\frac{Ebh^3}{l_2^3},$$ (1.4.3)

thus, the natural frequency is

$$\omega = \sqrt{\frac{k}{m}} = \sqrt{\frac{E}{\gamma}}\sqrt{\frac{1}{4}\frac{bh^3}{l_2^3 A_1 l_1}}$$ (1.4.4)

In both cases, the natural frequency depends on the criterion E/γ (*specific stiffness*) and on the system geometry. A similar criterion can be used for selecting structural materials for many non-vibratory applications. Some ways of enhancing effective values of E and E/γ are discussed in Chapters 2 and 7.

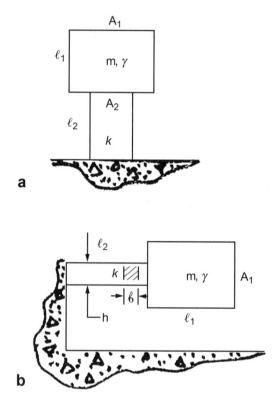

Fig. 1.4.1. Tension-compression (a) and bending (b) of single degree of freedom vibratory systems.

Use of materials characterized by a greater criterion E/γ allows to design a lighter structure having an adequate stiffness. Lighter structures require less driving power to accelerate/decelerate fast-moving components and to accurately follow non-straight trajectories with high rates of speed (e.g., feed drives in machine tools). Such structures also have higher natural frequencies and are easier to thermally stabilize.

To provide a comprehensive information, Table 1.1 lists data on E, γ, E/γ, ν for various structural materials. It has to be noted that most of the data in Table 1.1 (as well as in Table 1.2 below) are approximate, since stiffness, density, Poisson's ratio (and damping) depend also on small fluctuations of the constitutive elements of alloy-based materials. It is interesting to note that for the most widely used structural materials (steel, titanium, aluminum, and magnesium), values of E/γ are very close to each other, although the values of E and γ are very different. In such cases, selection

Table 1.1. Young's Modulus and Density of Structural Materials.

Material	E 10^5 MPa	γ 10^3 kg/m^3	E/γ 10^7 m^2/s^2	ν
Homogeneous Materials				
Carbon nanotubes, diameter < 8 nm	10–18	1.3–2.2	45–138	
Graphite	7.5	2.25	33.4	
Diamond	18.0	5.6	32	
Boron carbide, BC	4.50	2.4	19	
Silicon Carbide, SiC	3.9–5.6	3.1–3.2	17.5	
Carbon, C	3.6	2.25	16.0	
Silicon nitride	3.0–3.5	3.2–3.3	9.4–16.0	
Beryllium, Be	2.9	1.9	15.3	0.05
Boron, B	3.8	2.5	15.2	
Silicon carbide	3.9	3.1	12.6	0.19
Silicon nitride	3.5	3.3	10.6	
Sapphire	4.75	4.5	10.1	
Alumina, Al$_2$O$_3$ (99.9%)	3.9	4.0	9.8	0.22
Alumina (99.5%)	3.7	3.9	9.6	0.22
Alumina (96%)	3.0	3.7	8.2	0.21
Lockalloy (62% Be + 38% Al)	1.90	2.1	9.1	
Kevlar 49	1.3	1.44	9.0	
Titanium carbide, TiC	4.0–4.5	5.7–6.0	7.0–9.1	
Carbon nanotubes, diameter > 30 nm	2.0	2.25	8.9	
Silicon Si	1.1	2.3	4.8	
Tungsten carbide, WC	5.50	16.0	3.4	
Molybdenum	3.3	10.2	3.2	0.31
Zirconia	1.7	5.6	3.1	0.28
Alumin./Lithium (97% Al + 3% Li)	0.82	2.75	3.0	
Molybdenum, Mo	3.20	10.2	3.0	
Glass	0.7	2.5	2.8	
Steel 1010	1.90	7.8	2.4	
Steel 1020	2.0	7.8	2.6	
Steel 1040	2.1	7.8	2.7	
Steel 1045	2.0	7.8	2.6	0.29
Steel 1050	2.2	7.8	2.8	
Titanium, Ti	1.16	4.4	2.6	
Aluminum, Al (wrought)	0.72	2.8	2.6	0.33
Steel, stainless 440C	2.0	7.8	2.6	0.3
Nickel aluminide Ni$_3$Al (alloy IC-221M)	2.0	7.9	2.5	
Nitralloy 135M	2.0	8.0	2.5	0.29
Aluminum, Al (cast)	0.65	2.6	2.5	0.33
Steel, stainless 303	1.93	8.0	2.4	0.3
Steel stainl. (0.08–0.2%C; 17%Cr; 7%Ni)	1.83	7.7	2.4	
Magnesium Mg	0.45	1.9	2.4	
Tungsten (not alloyed)	4.2	19.3	2.2	0.3
Wood (along fiber)	0.11–0.15	0.41–0.82	2.6–1.8	
Marble	0.55	2.8	2.0	

Table 1.1. (continued)

Material	E 10^5 MPa	γ 10^3 kg/m³	E/γ 10^7 m²/s²	v
Tungsten (W + 2 to 4% Ni, Cu)	3.50	18.0	1.9	
Invar	1.5	8.0	1.9	0.3
Polymer concrete	0.45	2.45	1.8	0.23–0.3
Granite	0.48	2.7	1.8	
Iron (cast)	1.2	7.3	1.6	0.25
Beryllium copper	1.3	8.2	1.6	
Copper	1.2	8.9	1.3	0.34
Brass	1.1	8.5	1.3	0.34
Niobium	1.06	8.6	1.2	0.39
Zerodur	0.9	2.5	0.9	0.24
Polypropylene	0.08	0.9	0.9	
Granite	0.2	2.6	0.7	0.1
Nylon	0.04	1.1	0.36	
Paper	0.01–0.02	0.5	0.2–0.4	
Lead	0.15–0.18	11.3	0.16	
Composite Materials				
HTS graphite/5208 epoxy	1.72	1.55	11.1	
Boron/5505 epoxy	2.07	1.99	10.4	
Boron/6601 Al	2.14	2.6	8.2	
Lanxide NX - 6201 (Al + SiC)	2.0	2.95	6.8	
40% Al + 60% SiC powder (DuPont)	1.9	2.9	6.6	
T50 graphite/2011 Al	1.6	2.58	6.2	
Kevlar 49/resin	0.76	1.38	5.5	
50% Al + 50% Al₂O₃ powder (DuPont)	1.5	3.2	4.7	
80% Al + 20% Al₂O₃ powder	0.97	2.93	3.3	
Melram (80% Mg, 6.5% Zn, 12% SiC)	0.64	2.02	3.2	
E glass/1002 epoxy	0.39	1.8	2.2	

of a material is often determined by cost of the material itself as well as by requirements to its machining and other processing. It is important to note that both E and γ may vary, especially for derivative materials (such as carbides, nitrides, etc.) and for composite materials. When available, the ranges of parameter variations are given in Table 1.1. It is also important to note that thermal and mechanical treatments of metals (e.g., hardening of steel) as well as alloying by small percentage of alloying metals are not changing noticeably the Young's modulus. One design technique for enhancement of effective stiffness is described in Chapter 7.

Beryllium (Be) has the highest E/γ ratio for any structural metal. However, it is very expensive, has a limited formability (is rather brittle), and requires special precautions to protect operators from poisonous dust during machining. Lockalloy-like Be-Al alloys with 35% to 65% Be have a combination of high stiffness and low density, but with lower material cost and improved formability. The latter allows forming of near-net and net shapes, thus reducing or eliminating the need for machining and further improving the economics.

Diamond has the highest Young's modulus for conventional materials. Recently, there was a developed material comprising tin with embedded particles of barium titanate, which is claimed to be ten times stiffer than diamond at a narrow temperature range between 57°C and 59°C [22].

While graphite has a very high Young's modulus and the highest ratio E/γ (after carbon nanotubes) in Table 1.1, it does not necessarily mean that the graphite fiber-based composites can realize such high performance characteristics. First of all, the fibers in a composite material are held together by a relatively low modulus matrix (epoxy resin or a low E metal such as magnesium or aluminum). Second of all, the fibers realize their superior elastic properties only in one direction (in tension). Since mechanical structures are frequently used in a three-dimensional stress-strain environment, the fibers have to be placed in several directions, and this weakens the overall performance characteristics of the composite structures.

Figure 1.4.2 illustrates this statement on an example of a propeller shaft for a surface vehicle [23]. While a steel shaft, Fig. 1.4.2a, resists loads in all directions, in a shaft made of carbon fiber-reinforced plastic (CFRP), Fig. 1.4.2b, there is a need to place several layers of fiber at different winding angles. Fig. 1.4.2c,d show how bending and torsional rigidity of the composite shaft depend on the winding angles. While it is easy to design bending or torsional stiffness of the composite shaft to be much higher than these characteristics of the steel shaft, a combination of both stiffness values can only marginally be made superior to the steel shaft. In this case, the optimal winding angle is ~ 25 deg.

Another example of a stiffness-critical and natural-frequency-critical components are cones and diaphragms for loudspeakers [24]. Three important material properties for loudspeaker diaphragms are:

1. Large specific modulus E/γ (resulting in high natural frequencies) to get a wider frequency range of the speaker;
2. High flexural rigidity EI to reduce harmonic distortions;

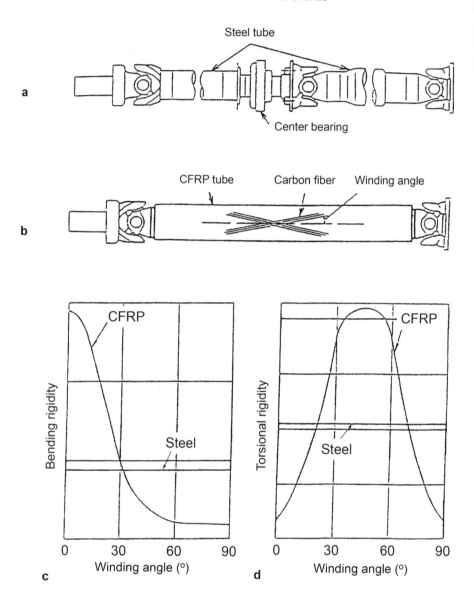

Fig. 1.4.2. Steel (a) and composite (b) propeller shafts for automotive transmissions and comparison of their bending (c) and torsional (d) rigidity

3. Significant internal energy dissipation (damping) characterized by the loss factor tan β (β is the "loss angle", see Appendix 1) to suppress breakups of the diaphragms at resonances.

While originally paper (a naturally fiber-reinforced composite material) and synthetic fiber-reinforced diaphragms had been used, their stiffness values were not adequate due to the softening influence of the matrix. It is demonstrated in [24] that use of beryllium and boronized titanium (25-μm thick titanium substrate coated on both sides with 5-μm thick boron layers) resulted in significant improvement of the frequency range for high frequency and mid-range speakers.

Table 1.2 lists available data on damping (loss factor η = tan β) and also "damping effectiveness product," $E\eta$ (e.g., see [25]), for some structural metals. The product $E\eta$ correlates with the frequently applicable stiffness/damping criterion $K\delta$, see Appendix 2.

Table 1.2. Young's Modulus E and "Damping Effectiveness Product" $E\eta$ for Structural Metals.

Material	E 10^5 MPa	$10^3\,\eta$	$E\eta$
Silentalloy (Fe, 12% Cr, 3% Al)	2.0	36˙	7,200
Nivco (72% Co, 23% Ni)	2.0	30	6,000
Incramute (Cu, 44% Mn, 1.8% Al)	1.2	31	3,700
Sonoston (Mn, 36% Cu, 4.5% Al, 3% Fe, 2% Ni)		40	
Nickel (Ni, pure)	2,0	18	3,600
Nitinol, optimally heat treated and prestressed	0.4	60–100	2,400–4,000
Iron (Fe, pure)	**2.0**	**16**	**3,200**
High carbon gray iron	**1.5**	**19**	**2,850**
Magnesium alloy (Mg, 0.5% Zr)	0.45	50	2,250
Magnesium (99.9% Mg, cast)	0.45	39	1,800
Nitinol (~ 50% Ni, ~50% Ti)	0.4	28	1,600
Steel stainless, martensitic	**2.0**	**8**	**1,600**
Steel, low carbon	**2.0**	**4**	**800**
Gray cast iron	**1.1**	**6**	**660**
Steel stainless, ferritic	**2.0**	**3**	**600**
Cast iron, nodular (malleable)	**1.3**	**2**	**260**
Steel stainless, austenitic	**2.0**	**1**	**200**
Steel, medium carbon	**2.0**	**1**	**200**
Superalloys, Ni-based	**1.1**	**0.2**	**< 22**
Aluminum, 1100	**0.7**	**0.3**	**21**
Aluminum, 2024-T4	**0.7**	**< 0.2**	**< 14**

1.5 GENERAL COMMENTS ON STIFFNESS IN MECHANICAL DESIGN

In most of the structures, their structural stiffness depends on the following factors:

1. Elastic moduli of structural material(s);
2. Geometry of the deforming segments (cross-sectional area A for tension/compression/shear; cross-sectional moment of inertia $I_{x,y}$ for bending; polar moment of inertia J_p for torsion);
3. Linear dimensions (e.g., length, L, width, B, height, H);
4. Character and magnitude of variation of the above parameters across the structure;
5. Character of loading and supporting conditions of the structural components;
6. In structures having slender/thin-walled segments, stiffness can depend on elastic stability of these segments;
7. Joints between substructures and/or components frequently are responsible for the dominant structural deformations (e.g., see data on the breakdown of tool-end deflections above in Section 1.3.6a).

While for most machine components a stiffness increase is desirable, there are many cases where stiffness values should be limited or even reduced. The following are some examples:

1. Perfectly rigid bodies are usually more brittle and cannot accommodate shock loads, e.g., see Section 1.2.2;
2. Many structures are designed as statically indeterminate systems, but if the connections in such a system are very rigid, the system would not function properly, since some connections might be overloaded. If the most highly loaded connection fails, others may also fail one after another;
3. Huge peak loads (stress concentrations) may develop in contacts between very rigid bodies due to presence of surface asperities;
4. Stiffness adjustment/tuning by preloading would not be possible for very rigid components;
5. High stiffness may result in undesirable values for the structural natural frequencies.

REFERENCES

[1] Tobias, S.A., 1965, "Machine Tool Vibration", Blackie, London.

[2] Rivin, E.I., 2003, "Passive Vibration Isolation", ASME Press, N.Y.

[3] Orlov, P.I., 1972, "Fundamentals of Machine Design", vol.1, Mashinostroenie Publ. House, Moscow [in Russian].

[4] Rivin, E.I., 2002, "Machine Tool Vibration", Ch. 40 in Harris' Shock and Vibration Handbook", 5th Edition, McGraw-Hill, N.Y.

[5] Davenport, A.G., Novak, M., 2002, "Vibration of Structures Induced by Wind", Ch. 29, part. II, in Harris' Shock and Vibration Handbook", 5th Edition, McGraw-Hill, N.Y.

[6] Blevins, R.D., 2002, "Vibration of Structures Induced by Fluid Flow", Ch. 29, part. I, in Harris' Shock and Vibration Handbook", 5th Edition, McGraw-Hill, N.Y.

[7] Timoshenko, S.P., Gere, J.M., 1961, "Theory of Elastic Stability", McGraw-Hill, N.Y., 541 pp.

[8] Jubb, J.E.M., Phillips, I.G., 1975, "Interrelation of Structural Stability, Stiffness, Residual Stress and Natural Frequency", *J. of Sound and Vibration*, 39 (1), pp. 121–134.

[9] Alabuzhev, P., Gritchin, A., Kim, L., Migirenko, G., Chon, V., Stepanov, P., 1989, "Vibration Protecting and Measuring Systems with Quasi-Zero Stiffness", Hemisphere Publishing Co., N.Y.

[10] Platus, D.L., "Vibration Isolation System", *U.S. Patent 5,178,357*.

[11] Dyrba, D.A., Lavendel, E.E., 1970, "Force-Deflection Characteristic for Hollow Rubber Seals", Voprosi dinamiki i prochnosti, No. 20, Zinatne Publish. House, Riga [in Russian].

[12] Rivin, E.I., Sayal, G. Johal, P., 2006, "Giant Superelasticity Effect" in NiTi Superelastic Materials and Its Applications", *J. of Materials in Civil Engineering*, 18(6), pp. 851–857.

[13] Rivin, E.I., 1988, "Mechanical Design of Robots", Mc Graw-Hill, N.Y.

[14] Vulfson, I.I., Kolovskii, M.Z., 1968, "Nonlinear Problems of Machine Dynamics", Mashinostroenie, Leningrad, 284 pp [in Russian].

[15] Rivin, E.I., 2007, "Dynamic Overloads and Negative Damping in Mechanical Linkage: Case Study of Catastrophic Failure of Extrusion Press", *Engineering Failure Analysis*, 14(7), pp. 1301–1312.

[16] Vinokur, R., 2003, "Vibroacoustic Effects in MEMS", *S)V Sound and Vibration*, No. 9, pp. 22–26.

[17] Rivin, E.I., 1980, "Role of Induction Driving Motor in Transmission Dynamics", ASME Paper 80-DET-96.

[18] Rivin, E.I., Kang, H., 1989, "Improvement of Machining Conditions for Slender Parts by Tuned Dynamic Stiffness of Tool," *Int. J. of Machine Tools and Manufacture*, 29(3), pp. 361–376.

[19] Tlusty, J., Heczko, O., 1980, "Improving Tests of Damping in the Cutting Process", Proceed. of the 8th NAMRAC Conference, *SME*, pp. 172–176

[20] Zhang, B., Wang, J., Yang, F., Zhu, Z., 1999, "The Effect of Machine Stiffness on Grinding of Silicon Nitride", *Int. J. of Machine Tools and Manufacture*, 39, pp. 1263–1283.

[21] Slocum, A. H., 1992, "Precision Machine Design", Prentice Hall, 750 pp.

[22] "Stiffer than Diamond", 2007, *Mechanical Engineering*, #9, pp. 22–23.

[23] Kawarada, K., et al, 1994, "Development of New Composite Propeller Shaft", *Toyota Technical Review*, 43(2), pp. 85–90.

[24] Yamamoto, T., Tsukagoshi, T., 1981, "New Materials for Loudspeaker Diaphragms and Cones. An Overview", *Presentation at the Annual Summer Meeting of Acoustical Society of America*, Ottawa, pp. 1–10.

[25] Ver, I.L., Beranek, L.L., (eds) 2005, "Noise and Vibration Control Engineering", Wiley Inc., N.Y.

CHAPTER

2

Modes of Loading and Stiffness of Structural Components

     tiffness and, to a lesser degree, damping characteristics of me-
chanical systems are, obviously, dependent on stiffness and
damping characteristics and modes of loading of its components. These issues are addressed in this chapter, mostly for components having linear load-deflection characteristis. The structural
elements with nonlinear characteristics are described in Chapter 3.
An important special case of components/subsystems having multiple load-carrying elements is addressed in Appendix 6. Both stiffness
and damping characteristics of mechanical systems are strongly dependent on connecting units (joints) between the components. This
issue is specially addressed in Chapter 4.

2.1 INFLUENCE OF MODE OF LOADING ON STIFFNESS

There are four principal types (modes) of structural loading: tension,
compression, bending, and torsion [1]. Parts experiencing tension-
compression loading demonstrate much smaller deflections for similar
loading intensities and, thus, usually are less stiffness-critical. Figure
2.1.1a shows a rod of length L having a uniform cross-sectional area A
along its length and loaded in tension by its own weight W and by an
external force P. Fig. 2.1.1b shows the same rod loaded in bending by the
force P or by distributed weight $w = W/L$ as a cantilever built-in beam,
and Fig. 2.1.1c shows the same rod as a double-supported beam.

Deflections of the rod in tension are

$$(f_t)_P = \frac{PL}{EA}; \qquad (f_t)_w = \frac{WL}{2EA} \qquad (2.1.1)$$

Bending deflections for cases of Fig. 2.1.1b,c, respectively, are

$$(f_{bb})_P = \frac{PL^3}{3EI}; \qquad (f_{bb})_w = \frac{WL^3}{8EI} \qquad (2.1.2)$$

$$(f_{bc})_P = \frac{PL^3}{48EI}; \qquad (f_{bc})_w = \frac{5WL^3}{384EI} \qquad (2.1.3)$$

where I = cross-sectional moment of inertia, and subscripts's meanings
are as follows: t = tension; b = bending, bb = bending per Fig. 2.1.1b, bc =
bending per Fig. 2.1.1c. For a round cross section (diameter d, $A = \pi d^2/4$,
$I = \pi d^4/64$, and $I/A = d^2/16$)

$$\frac{f_b}{f_t} = k\frac{L^2}{d^2} \qquad (2.1.4)$$

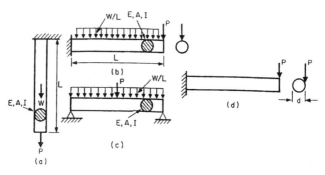

Fig. 2.1.1. Various modes of loading of a rod-like structure.
(a) Tension; (b) bending an a cantilever mode; (c) bending in a
double-supported mode; and (d) bending with an out-of-center load.

where coefficient, k, depends on loading and supporting conditions. For example, for a cantilever beam with $L/d = 20$, $(f_{bb})_P/(f_t)_P = 2,130$ and $(f_{bb})_W/(f_t)_W = 1600$; for a double-supported beam with $L/d = 20$, $(f_{bc})_P/(f_t)_P \cong 133$ and $(f_{bc})_W/(f_t)_W \cong 167$. Thus, bending deflections are exceeding tension-compression deflections by several decimal orders of magnitude.

Figure 2.1.1d shows the same rod whose supporting conditions are as in Fig. 2.1.1b, but which is loaded in bending by the same force P with an eccentricity, thus causing both bending as described by the first expression in Eq. (2.1.2) and torsion, with the translational deflection on the rod periphery (which is caused by only the torsional deformation) equal to

$$f_{to} = \frac{PLd^2}{4GJ_p} \tag{2.1.5}$$

where J_p is the polar moment of inertia and G - shear modulus of the material. Since $J_p = \pi d^4/32$ for a circular cross-section,

$$\frac{f_{to}}{f_t} = \frac{d^2}{4}\frac{EA}{GJ_p} = 2\frac{E}{G} \cong 5 \tag{2.1.6}$$

since for structural metals $E \cong 2.5G$. Thus, torsion of bars with solid cross-sections is also associated with deflections substantially larger than those under tension-compression.

These simple calculations help to explain why either bending or torsional compliances, or both are in many cases critical for the structural deformations. Many stiffness-critical mechanical components are loaded in bending, but bending is associated with much larger deformations than tension/compression of similar size structures under the same loads. Because of this, engineers were trying, since times immemorial, to replace bending with tension/compression. The most successful designs of this kind are trusses and arches.

Advantages of truss structures are illustrated by a simple case in Fig. 2.1.2 [2.2], where a cantilever truss having overhang l is compared with a cantilever beam of the same length and loaded by the same load P. If the beam has the same cross-section as links of the truss (case a), its weight is 0.35 of the truss weight, but its deflection is 9000 times larger, while stresses are 550 times higher. To achieve the same deflection (case c), diameter of the beam has to be increased by the factor of ten, thus the beam becomes 35 times heavier than the truss. The stresses are equalized (case b) if diameter of the beam is increased by 8.25 times; weight of such beam is 25 times that of the truss. Ratio of the beam deflection f_b to the truss deflection f_t is expressed as

$$\frac{f_b}{f_t} \approx 10.5 \left(\frac{l}{d}\right)^2 \sin^2 \alpha \cos \alpha \qquad (2.1.7)$$

Maximum stress ratio σ_b/σ_t and deflection ratio f_b/f_t for the truss loaded as in Fig. 2.1.3a are plotted in Fig. 2.1.3b,c as functions of l/d and α.

Similar effects are observed if a double-supported beam loaded in the middle of its span as shown in Fig. 2.1.4a, is replaced by a truss (Fig. 2.1.4b). In this case,

$$\frac{f_b}{f_t} \approx 1.3 \left(\frac{l}{d}\right)^3 \sin^2 \alpha \cos \alpha \qquad (2.1.8)$$

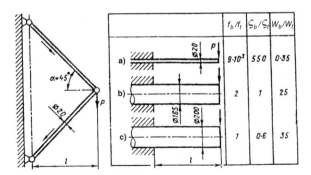

Fig. 2.1.2. Comparison of structural characteristics of a truss bracket and cantilever beams.

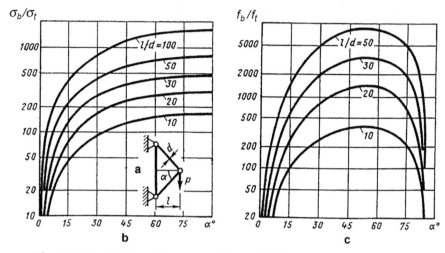

Fig. 2.1.3. Ratios of stresses (b) and deflections (c) between a cantilever beam (diameter, d, length, l) and a truss bracket (a).

Maximum stress ratio, σ_b/σ_t, and deflection ratio, f_b/f_t, for the beam in Fig. 2.1.4a and the truss in Fig. 2.1.4b are plotted in Fig. 2.1.5b,c as functions of l/d and α. A similar effect can be achieved if the truss in Fig. 2.1.4b were transformed into an arch (Fig. 2.1.4c).

These principles of transforming the bending mode of loading into the tension/compression mode of loading can be utilized in a somewhat "disguised" way in designs of basic mechanical components, such as brackets (Fig. 2.1.6). The bracket in Fig. 2.1.6aI is loaded in bending. An inclination of the lower wall of the bracket, as in Fig. 2.1.6aII, reduces deflection and stresses, but the upper wall does not contribute much to the load accommodation. Design in Fig. 2.1.6aIII provides a much more

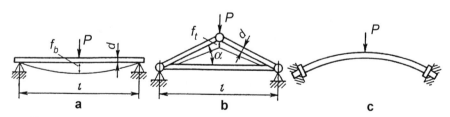

Fig. 2.1.4. Typical load-carrying structures: (a) double-supported beam; (b) truss bridge; (c) arch.

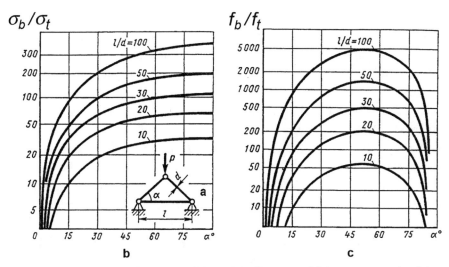

Fig. 2.1.5. Ratios of stresses (b) and deflections (c) between a double-supported beam in Fig. 2.1.4a and a truss bridge in Fig. 2.4b (a).

uniform loading of the upper and lower walls, which allows to significantly reduce size and weight of the bracket.

Even further modification of the "truss concept" is illustrated in Fig. 2.1.6b. Load P in case 2.1.6bI (cylindrical bracket) is largely accommodated by segments of the side walls, which are shown in the cross-section in solid black. Tapering the bracket, as in Fig. 2.1.6bII, results in a more even stress distribution. The face wall is an important feature of the

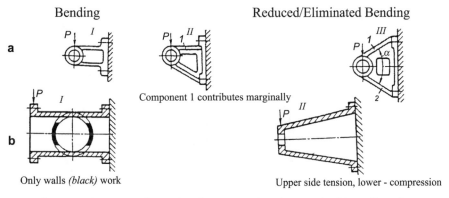

Fig. 2.1.6. Use of tension/compression instead of bending for structural components.

system, since it prevents distortion of the cross-section into an elliptical one, and it is necessary for achieving the optimal performance.

There are many other design techniques aimed at reduction or elimination of bending in favor of tension/compression. Some of them are illustrated in Fig. 2.1.7. Fig. 2.1.7aI shows a mounting foot of a machine bed. Horizontal forces on the bed cause bending of the wall and result in a reduced stiffness. "Pocketing" of the foot as in Fig. 2.1.7aII aligns the anchoring bolt with the wall and, thus, reduces the bending moment and also increases the effective cross-section of the foot area, which resists bending. The disc-like hub of a helical gear in Fig. 2.1.7bI bends under the axial force component of the gear mesh. Inclination of the hub as in Fig. 2.1.7bII enhances stiffness by introducing the "arch concept." Vertical load on the pillow block bearing in Fig. 2.1.7cI causes bending of its frame, while in Fig. 2.1.7cII, it is accommodated by compression of the added central support. Bending of the structural member under tension in Fig. 2.1.7dI is caused by its asymmetry. After slight modifications as shown in Fig. 2.1.7dII, its effective cross-section can be reduced due to total elimination of bending.

Majority of spring designs illustrated in Section 2.7 below use bending or torsion. Since bending- and torsion-induced stresses are non-uniformly distributed across cross-sections of the basic elements, the spring material is underutilized. The "frictional spring" uses tension/compression loading. Since in this spring the stresses are much more uniformly distributed, a better utilization of the material is achieved, thus reducing weight and size of the spring for a given rated load.

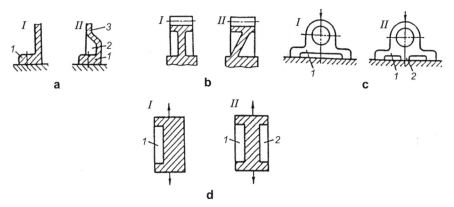

Fig. 2.1.7. Reduction of bending deformations in structural components.

Some structural materials, such as cast iron, are better suited to accommodate compressive rather than tensile stresses. While it is more important for strength, stiffness can also be influenced if some microcracks are present, which can open under tension. Fig. 2.1.8 gives some directions for modifying components loaded in bending so that the maximum stresses are compressive rather than tensile. While the maximum stresses in the horizontal beam whose cross-section is shown in Fig. 2.1.8aI are tensile (in the bottom section), turning this beam upside down, as in Fig. 2.1.8aII, brings the maximum stresses to the compressed side (top). The same is true for the beam in Fig 2.1.8b. A similar principle is used in transition from the bracket with the stiffening wall shown in Fig. 2.1.8cI to the identical but oppositely mounted bracket in Fig. 2.1.8cII.

After the design concept per the above discussion is selected, its optimization can be easily performed by application of the Finite Element Method (FEM).

Practical Case 1. *Tension/Compression Machine Tool Structures ("Parallel Kinematics Machines" or PKM).* While use of tension/compression mode of loading in stationary structures is achieved by using trusses and arches, there are also mechanisms which provide up to six degrees-of-freedom positioning and orientation of objects using only tension/compression actuators. A basic tension/compression mechanism is the so-called Stewart Platform [3]. First attempts to use the Stewart Platform for machine tools (machining centers) were made in the former Soviet Union in the mid-1980s [4], Fig. 2.1.9 shows the design schematic of the Russian machining center based on application of the Stewart Platform mechanism. Positioning and orientation of platform 1 holding spindle unit 2, which carries a tool machining workpiece 3, are achieved

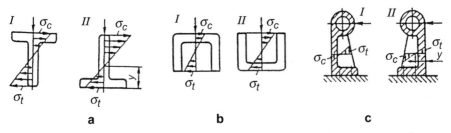

Fig. 2.1.8. Increasing compressive stresses at the expense of tensile stresses.

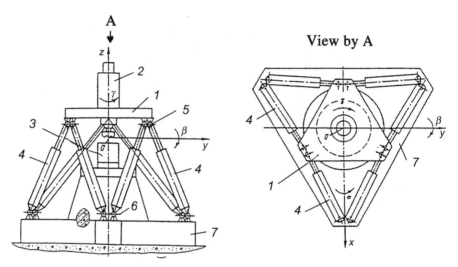

Fig. 2.1.9. Design schematic and coordinate axes of Russian machining center based on the Stewart platform kinematics.

by cooperative motions of six independent tension/compression struts/actuators 4, which are pivotably engaged via spherical (in some cases universal/Cardan) joints 5 and 6 with platform 1 and base plate 7, respectively.

Cooperation between the actuators is realized by using rather complex controlling software commanding each actuator to participate in the programmed motion of the platform. One shortcoming of such machining centers is a limited range of motion along each coordinate and numerous singularities constraining the usable work space. These factors result in a rather complex shape of the work zone as illustrated in Fig. 2.1.10. The work zone is limited by singularities of the link mechanism, e.g., see [5, 6]. In some PKM machines, the work zone may have numerous, up to 40 singular points.

Due to the large number of nonlinear parameters and boundary conditions, such as singularities, collisions, etc., design of a PKM is a complex optimization process. With the exception of few highly symmetrical designs, the optimization problem cannot be solved analytically, but is usually addressed by computer simulation [5].

However, there are many advantages which make such designs promising for many applications, such as machine tools, vibration isolators,

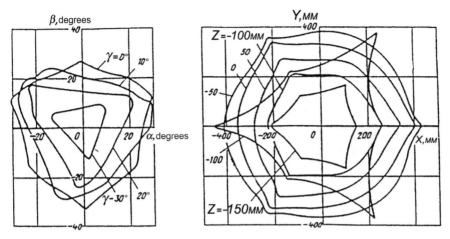

Fig. 2.1.10. Work zone of machining center in Fig. 2.1.9.

flight simulators, etc. Publication [4] claims that while stiffness along y-axis (k_y) is about the same as for conventional machining centers, stiffness k_z is about 1.7 times higher. The overall stiffness is largely determined by deformations in spherical joints 4 and 5, by platform deformations, and by spindle stiffness, and can be enhanced 50% to 80% by increasing platform stiffness in the x-y plane and by improving the spindle unit. The machine weighs significantly less than a conventional machining center and is much smaller (two-to-three times smaller footprint). While it is claimed in [4] that such PKM machine costs three to four times less due to use of standard identical and not very complex actuating units and has three to five times greater feed force, the recent results of extensive R&D on such machines are not as optimistic. Still, popularity of this concept and its modifications, especially for CNC machining centers and milling machines is increasing [5, 6]. The further advancement of this concept depends to a great degree on development of high stiffness, no backlash, preloaded spherical and universal joints, e.g., as described in [7, 8] (also see Chapter 8).

Practical Case 2. Tension/Compression Robot Manipulator. Tension/compression actuators also found application in robots [9]. Fig. 2.1.11 shows schematics and work zone of a manipulating robot from NEOS Robotics Co. While conventional robots are extremely heavy in relation to their rated payload (weight-to-payload ratios 15 to 25 [1]), the NEOS

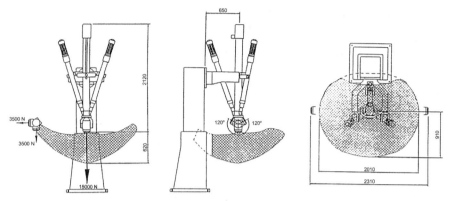

Fig. 2.1.11. Design schematic and work zone of NEOS Robotics robot utilizing tension/compression links.

robot has much better performance characteristics for its weight (about 300 kg), as listed in Table 2.1.

2.2 INFLUENCE OF BEAM DESIGN ON ITS BENDING STIFFNESS AND DAMPING

Significant gains in stiffness and/or weight of structural components loaded in bending can be achieved by a judicious selection of their cross-sectional shape and other design features, such as a formed helical groove on its outside surface and by separating load-carrying outer shelves of the beam by light spacers loaded in compression.

Table 2.1.   Specifications of NEOS Robot.

Load Capacity	Handling payload	150 kg
	Turning torque	200 Nm
	Pressing, max.	15,000 N
	Lifting, max.	500 kg
Accuracy	Repeatability (ISO 9283)	≤ ±0.02 mm
	Positioning	≤ ±0.20 mm
	Path following at 0.2 m/s	≤ ±0.10 mm
	Incremental motion	≤ 0.01 mm
Stiffness	Static bending deflection (ISO 9283.10)	
	- X and Y directions	
	- Z direction	0.0003 mm/N
		0.0001 mm/N

2.2.1 Round versus Rectangular Cross Section

One of the parameters which can be modified to comply better with these constraints is the shape of the cross-section. Importance of the cross-section optimization can be illustrated on the example of robotic links which have to comply with numerous, frequently contradictory, constraints. Some of the constraints are as follows:

1. The links should have an internal hollow area of a uniform cross-section to provide conduits for electric power and communication cables, hoses, power-transmitting components, control rods, etc.
2. At the same time, their external dimensions are limited to extend the usable workspace.
3. Links have to be as light as possible to reduce inertia forces and to allow for the largest payload per given size of motors and actuators.
4. For a given weight, links have to possess the highest possible bending (and in some cases torsional) stiffness.

The two basic cross-sections are hollow round (Fig. 2.2.1a) and hollow rectangular (Fig. 2.2.1b). There can be various approaches to the comparison of these cross-sections. Two cases are analyzed below [1]:

1. The wall thickness in both cross sections is the same, and $t \ll D, a$.
2. The cross-sectional areas (i.e., weight) of both links are the same.

Hollow Round   **Hollow Rectangular**

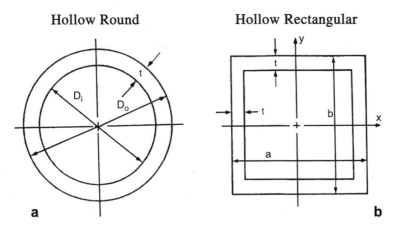

Fig. 2.2.1. Typical cross-sections of a manipulator link. (a) Hollow round (ring-like) and (b) hollow rectangular.

The bending stiffness of a beam is characterized by its cross-sectional moment of inertia, I, and its weight is characterized by the cross-sectional area, A. Subscripts: rd = round; sq = square; re = rectangular. For the round cross-section in Fig. 2.2.1a,

$$I_{rd} = \frac{\pi(D_o^4 - D_i^4)}{64} = \frac{\pi[D_o^4 - (D_o - 2t)^4]}{64} \cong \frac{\pi D_o^3 t}{8}\left(1 - \frac{3t}{D_o} + \frac{4t^2}{D_o^2}\right) \quad (2.2.1)$$

$$A_{rd} = \frac{\pi(D_o^2 - D_i^2)}{4} = \pi D_o t\left(1 - \frac{t}{D_o}\right). \quad (2.2.2)$$

For the rectangular cross-section in Fig. 2.2.1b, the value of I depends on the direction of the neutral axis in relation to which the moment of inertia is computed. Thus,

$$I_{rec,x} = \frac{ab^3}{12} - \frac{(a-2t)(b-2t)^3}{12}; \quad I_{rec,y} = \frac{a^3 b}{12} - \frac{(a-2t)^3(b-2t)}{12}. \quad (2.2.3a)$$

For the square cross-section,

$$I_{sq} = \frac{a^4}{12} - \frac{(a-2t)^4}{12} \approx \frac{2}{3}a^3 t\left(1 - \frac{3t}{a} + 4\frac{t^2}{a^2}\right). \quad (2.2.3b)$$

The cross-sectional areas for the rectangular and square cross-sections, respectively, are

$$A_{re} = ab - (a - 2t)(b - 2t) = 2t(a + b) - 4t^2; \quad A_{sq} = 4at(1 - t/a). \quad (2.2.4)$$

For case 1, $D_0 = a$, and t is the same for both cross-sections. Thus,

$$\frac{I_{sq}}{I_{rd}} = \frac{2/3}{\pi/8} = 1.7; \quad \frac{A_{sq}}{A_{rd}} = 1.27. \quad (2.2.5)$$

Thus, a square cross-section provides a 70% increase in rigidity with only 27% increase in weight; or 34% increase in rigidity for the same weight.

For case 2 ($D_o = a$, $A_{rd} = A_{sq}$, and $t_{rd} \neq t_{sq}$), if $t_{rd} = 0.2D_o$, then $t_{1sq} = 0.147D_o = 0.147a$ and

$$I_{rd} = 0.0405D_o^4; \quad I_{sq} = 0.0632a^4; \quad I_{sq}\Big/I_{rd} = 1.56. \quad (2.2.6a)$$

If $t_{2rd} = 0.1D_o$, then $t_{2sq} = 0.0765D_o = 0.0765a$, and

$$I_{rd} = 0.029D_o^4; \quad I_{sq} = 0.0404a^4; \quad I_{sq}\Big/I_{rd} = 1.40. \quad (2.2.6b)$$

Thus, for the same weight, a beam with the thin-walled square cross section would have 34% to 40% higher stiffness than a beam with the hollow round cross section. In addition, the internal cross-sectional area of the square beam is significantly larger than that for the round beam of the same weight (the thicker the wall, the more pronounced is the difference).

From the design standpoint, links of the square cross-section have also an advantage of being naturally suited for realizing prismatic joints and for using roller guideways. The round links have to be specially machined when used in prismatic joints. On the other hand, round links are easier to fit together (e.g., if telescopic links with sliding connections are used).

Both stiffness and strength of structural components loaded in bending (beams) can be significantly enhanced if a solid cross-section is replaced with the cross-sectional shape in which the material is concentrated farther from the neutral line of bending. Fig. 2.2.2 shows comparisons of both stiffness (cross-sectional moment of inertia I_0) and strength (cross-sectional modulus W) for solid round and square cross-sections versus round hollow (tubular) and standard I-beam profile, respectively, for the same cross-sectional area (weight).

2.2.2 Stiffness and Damping of Helically Patterned Tubular Beams

Helically patterned tubes are frequently used in heat exchangers and also in structural applications (e.g., in vibration-resistant smokestacks). An extensive study in [10] has shown that forming a helical groove on the external surface of a tube results in significant changes of its stiffness and damping characteristics. This modification might be useful for applications.

It was found that width of the helical groove as well as wall thickness of the tube have only minimal influences on stiffness and damping characteristics, while depth h and pitch S of the groove are important

Section	Ratios		
	d/D , h/h_0	I/I_0	W/W_0
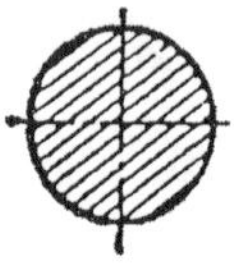	0	1	1
	0.6	2.1	1.7
	0.8	4.5	2.7
	0.9	10	4.1
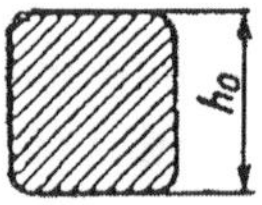	—	1	1
	1.5	4.3	2.7
	2.5	11.5	4.5
	3.0	21.5	7.0

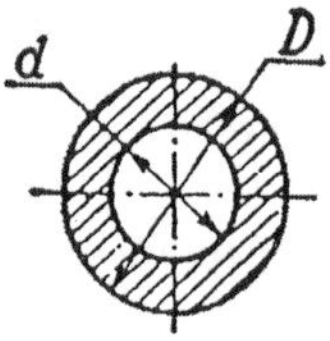

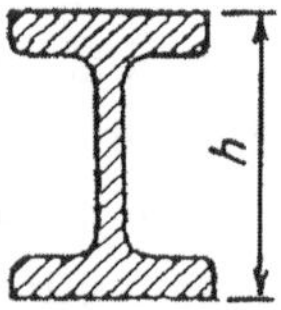

Fig. 2.2.2. Relative stiffness (cross-sectional moment of inertia, I) and strength (section modulus, W) of various cross-sections having the same weight (cross-sectional area A).

parameters. The studies had been performed for brass tubes with tube OD D = 16–24 mm, h = 0.4–1.5 mm, and S = 8.3–24 mm. Bending stiffness of the helically patterned tube $(EI)_{hp}$ is less than bending stiffness $(EI)_s$ of the original smooth tube. Test results are represented by the plot in Fig. 2.2.3 and can be described, within ±10%, by an empirical expression

$$(EI)_{hp} = (EI)_s \, e^{-4.8 \, (h/S)}. \tag{2.2.7}$$

Log decrement of the smooth brass tube was measured as δ_s = 0.0019. Tubes with the helical patterning have up to seven times higher damping,

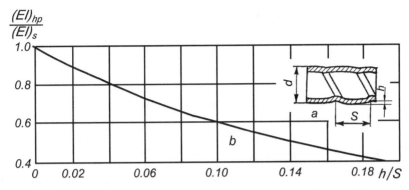

Fig. 2.2.3. Helically grooved tube (a) and its bending stiffness as function of the groove dimensions (b).

especially after the tube is heat treated (annealed) (Fig. 2.2.4), probably due to stress concentrations introduced by the helical grooves.

Sometimes it is important to know how the internal volume of the tube is changing after forming the helical groove. The test results on reduction of the internal volume V are shown in Fig. 2.2.5.

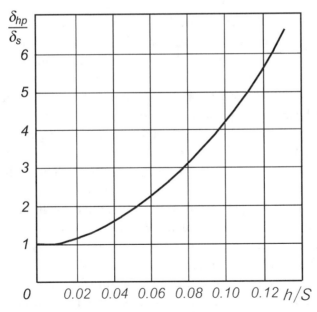

Fig. 2.2.4. Damping as function of helical groove dimensions.

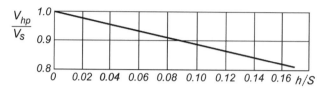

Fig. 2.2.5. Internal volume as function of helical groove dimensions.

2.2.3 Composite/Honeycomb Beams and Plates

Bending resistance of beams is largely determined by segments of their cross-sections, which are farthest removed from the neutral plane. Thus, enhancement of bending stiffness-to-weight ratio for a beam can be achieved by designing its cross-section to be of such shape that the load-bearing parts are relatively thin shelves (strips) on the upper and lower sides of the cross-section. However, there is a need for some structural members maintaining integrity and stability of the cross-section so that the positions of the load-bearing strips do not change noticeably by loading of the beam. Forming of an integral beam by rolling or casting (e.g., *I*-beams and channel beams in which an elongated wall holds the load-bearing strips) can achieve this. Another approach is by using composite beams in which the load-bearing strips are separated by an intermediate filler (core) made of a light material or by a honeycomb structure made from the same material as the load-bearing strips or from some lighter metal or synthetic material. Such composite beams can be lighter than the standard profiles such as *I*-beams or channels, and they are frequently more convenient for applications. For example, it is not difficult to make composite beams of any width (*composite plates*), to provide the working surfaces with smooth or threaded holes for attaching necessary components (such as *"breadboard" optical tables*), or to use high damping materials for the middle layer (or to use damping fillers for honeycomb structures).

It is important to realize that there are significant differences in the character of deformation between solid beams (plates) and composite beams (plates). Bending deformation of a beam comprises two components: moment-induced deformations and shear-induced deformations. For beams with solid cross-sections made from a uniform material, the shear deformation as a fraction of total bending deformation can be neglected for $L/h \geq 10$, where L = length of the beam, and h = height of its cross-section. For example, for a double-supported beam loaded with a uniformly distributed force with intensity q per unit length, deflection at the mid-span, considering both moment-induced and shear deformation is [11]

$$f_{ms} = \frac{5qL^4}{284EI}\left(1 + \frac{48\alpha_{sh}EI}{5GAL^2}\right), \tag{2.2.8a}$$

where E = Young's modulus, G = shear modulus, A = cross-sectional area, and α_{sh} is the so-called shear factor ($\alpha_{sh} \approx 1.2$ for rectangular cross-sections, α_{sh} = 1.1 for round cross-sections). If the material has E/G = 2.5 (e.g., steel), then for a rectangular cross-section ($I/A = h^2/12$),

$$f_{ms} = \frac{5qL^4}{384EI}\left(1 + 2.4\frac{h^2}{L^2}\right). \tag{2.2.8b}$$

For L/h = 10, the second (shear) term in brackets in Eq. (2.2.8) is 0.024, less than 2.5% of total term.

For a double-supported beam loaded with a concentrated force, P, in the middle, deformation under the force is

$$f_{ms} = \frac{PL^3}{48EI}\left(1 + \frac{12\alpha_{sh}EI}{GAL^2}\right). \tag{2.2.9a}$$

Again, the second term inside the brackets represents the influence of shear deformation. For rectangular cross-section and E/G = 2.5, then

$$f_{ms} = \frac{PL^3}{48EI}\left(1 + 3\frac{h^2}{L^2}\right) \tag{2.2.9b}$$

which shows a slightly greater influence of shear deformation than for the uniformly loaded beam. Deformation of a cantilever beam loaded at the free end by force, P, can be obtained from Eq. (2.2.9a) if P in the formula is substituted by $2P$, and L is substituted by $2L$. For I-beams, the shear effect is two-to-three times more pronounced, due to the smaller A than for the rectangular cross-section beams. However, for laminated beams in which the intermediate layer is made of a material with a low G or for honeycomb beams in which A and possibly G are reduced, the deformation increase (stiffness reduction) due to the shear effect can be as much as 50%, even for long beams, and must be considered.

However, even considering the shear deformations, deformations of laminated and honeycomb beams under their own weight are significantly less than those of solid beams (for steel skin and steel core honeycomb beams—about two times less). Stiffness-to-weight ratios (and natural frequencies) are significantly higher for composite and honeycomb beams than they are for solid beams.

Instead of honeycomb reinforcement, cellular (foamy) metals and plastics can be used. Analysis of such beams is given in [12], their application for machine tool structures is described in [13].

2.3 TORSIONAL STIFFNESS

The basic Strength of Materials expression for torsional stiffness, k_t, of a tubular member of length, l, whose cross-section is a circular ring with outer diameter D_o and inner diameter D_i (for a round cylindrical bar $D_i = 0$) is

$$k_t = \frac{T}{\theta} = \frac{GJ_p}{l} = \frac{G}{l}\frac{\pi}{32}\left(D_o^4 - D_i^4\right), \qquad (2.3.1)$$

where T is torque, θ = angle of twist, G = shear modulus of the material, and J_p = polar moment of inertia. However, if the cross-section is not round, has several cells, or is not solid (has a cut), the torsional behavior may change very significantly.

For a hollow solid (without cuts) cross-section of an arbitrary shape (but with a constant wall thickness, t), Fig. 2.3.1, torsional stiffness is (e.g., [14])

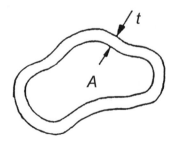

Fig. 2.3.1. Single-cell thin-walled torsion section.

$$k_t = \frac{4GA^2t}{Ll},$$
(2.3.2)

and the maximum stress is, approximately,

$$\tau_{max} = \frac{T}{2At},$$
(2.3.3)

where A = the area within the outside perimeter of the cross-section, and L = the peripheral length of the wall.

If Eq. 2.3.2 is applied to the round cross-section (cylindrical thin-walled tube) and $(D_o - D_i) = 2t << D_o$, then

$$J_p = \frac{\pi}{32}\left(D_o^4 - D_i^4\right) \approx \frac{\pi D_o^3}{8}\left(D_o - D_o^4\right) = \frac{4A^2t}{L}$$
(2.3.4)

Let this tube then be flattened out first into an elliptical tube and finally into a "double flat" plate. During this process of gradual flattening of the tube, t and L remain unchanged, but the area, A, is reduced from a maximum for the round cross-section to close to zero for the double flat. Thus, the double flat cannot transmit any torque of a significant magnitude for a given maximum stress (or the stress becomes very large even for a small transmitted torque). Accordingly, for a given peripheral length of the cross-section, a circular tube is the stiffest in torsion and develops the smallest stress for a given torque, since the circle of a given peripheral length, L, encloses the maximum area, A. One has to remember that the formula (2.3.4) is an approximate one, and the stiffness of the "double flat" is not zero. It can be calculated as an open thin-walled cross-section, see below.

Another case is represented by two cross-sections in Fig. 2.3.2a,b [14]. The square box-like thin-walled section in Fig. 2.3.2a is replaced by a similar section in Fig. 2.3.2b, which has the same overall dimensions, but also has two internal crimps (ribs). The wall thickness t is the same, and the cross-sectional area, A, is about the same for these cross-sections, but they have different peripheral lengths, L ($L = 4a$ for Fig. 2.3.2a, $L = 16a/3$ for Fig. 2.3.2b). Thus, the crimped section is 33% less stiff than the square box section, while being ~30% heavier and having greater maximum shear stress for a given torque.

A very important issue is torsional stiffness of elongated components whose cross-sections are not closed, such as ones shown in Fig. 2.3.3.

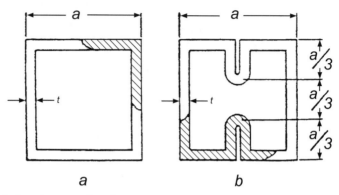

Fig. 2.3.2. Square (box) sections without (a) and with (b) crimps. In spite of the greater weight of the section (b), it has the same torsional shear stress as (a) and is less stiff than (a) by a factor of 4/3.

Torsional stiffness of such bars with the uniform section thickness, t, is [14]

$$k_t = \frac{Gbt^3}{3l} \qquad (2.3.5)$$

where $b >> t$ is the total aggregate length of wall in the section. If the sections have different wall thicknesses, then

$$k_t = \frac{G}{3l} \sum_i b_i t_i^3 \qquad (2.3.6)$$

where b_i is the length of the section having wall thickness t_i. It is very important to note that the stiffness in this case grows only as the first power

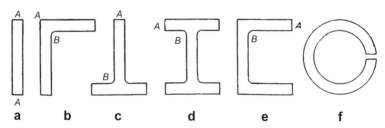

Fig. 2.3.3. Typical cross-sections to which Eqs. (2.3.5) and (2.3.6) for torsional stiffness apply. Corners A have zero stress and do not participate in torque transmission; corners B have large stress concentrations depending on the fillet radius.

of b. It is illustrative to compare stiffness of a bar having annular cross-section with the slit in Fig. 2.3.3f with stiffness of a bar having the solid annular cross-section in Fig. 2.2.1a of the same dimensions D_0, D_i with wall thickness $t = (D_0 - D_i)/2 = 0.05\,D_0$. The stiffness of the former is

$$k_{t1} = \frac{G}{3l}\pi\frac{D_o + D_i}{2}\left(\frac{D_o - D_i}{2}\right)^3 = 15.5\times10^{-6}\frac{GD_o^4}{l}, \qquad (2.3.7)$$

but the stiffness of the latter is

$$k_{t2} = \frac{GJ_p}{l} = \frac{G}{l}\frac{\pi}{32}\left(D_o^4 - D_i^4\right) = 3.4\times10^{-2}\frac{GD_o^4}{l}. \qquad (2.3.8)$$

Thus, torsional stiffness of the bar with the solid (uninterrupted) annular cross-section is about 2180 times (!) greater than torsional stiffness of the same bar whose annular cross-section is cut, so that shear stresses along this cut are not constrained by the ends.

Another interesting comparison of popular structural profiles is made in Fig. 2.3.4. The solid round profile in Fig. 2.3.4a has the same cross-sectional surface area as the standard I-beam in Fig. 2.3.4b (all dimensions are in cm). Bending stiffness of the I-beam about axis x is 41 times greater than bending stiffness of the round rod with the cross-section

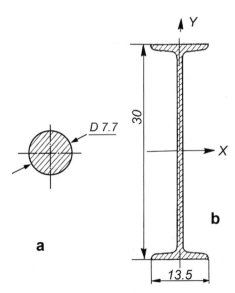

Fig. 2.3.4. Two structural profiles having the same cross-sectional areas.

shown in Fig. 2.3.4a; bending stiffness about axis y is two times higher than bending stiffness of the round rod, but torsional stiffness of the I-beam is 28.5 times lower than that of the round rod.

2.4 INFLUENCE OF STRESS CONCENTRATIONS

Stress concentrations (stress risers), e.g., caused by sharp changes in the cross-sectional area along the length of a component, are very detrimental to its strength, especially fatigue strength. However, much less attention is given to influence of local stress concentrations on stiffness and damping of the component. This influence can be very significant since local stress concentrations may induce local deformations, which significantly change the global deformation pattern of the component. If the component is dynamically loaded, microdisplacements also may develop in the stress concentration area. These microdisplacements result, on the one hand, in increasing damping (usually a desirable effect) and, on the other hand, in fretting corrosion, a highly undesirable effect. Reinforcing the stress concentration areas leads to increased stiffness and reduced or eliminated microdisplacements. The latter reduces damping but also reduces fretting, thus extending the life of the structural component.

Fig. 2.4.1 [2] compares performance of three round bars loaded in bending. The initial design, case 1, is a thin bar (diameter, d = 10 mm, length, l = 80 mm). Case 2 represents a much larger bar (diameter 1.8d) but which has two circular grooves required by the design specifications. While the solid bar of this diameter would have bending stiffness ten times higher than the bar 1, stress concentrations in the grooves result in only doubling the measured actual stiffness. The stress concentrations can be substantially reduced by using the initial thin (case 1) bar with reinforcement by tightly fit bushings (case 3). This results in 50% stiffness increase relative to case 2, as well as in strength increase (the ultimate load, P_1 = 8KN; P_2 = 2.1 P_1; P_3 = 3.6 P_1).

2.5 STIFFNESS OF FRAME/BED COMPONENTS

Presently, complex mechanical components such as beds, columns, plates, etc. are often analyzed for stresses and deformations by application of FEM techniques. However, the designer frequently needs some

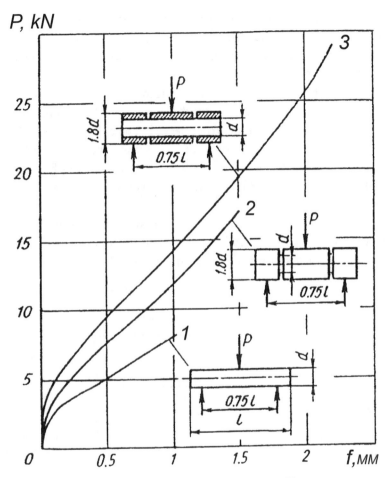

Fig. 2.4.1. Design influence on stiffness.

simple "quick and dirty" guidelines for initial design of these complex components. While this section addresses stiffness characteristics of integrated (cast or welded) structures, the current trend for flexibility of machine configurations, toward reconfigurable systems, led to increasing popularity of assembled basic (frame/bed) components. Such designs are presented in Chapter 3.

Machine beds, columns, etc., are typically made as two walls with connecting partitions or rectangular boxes with openings (holes), ribs, and partitions. While the nominal stiffness of these parts for bending and torsion is usually high, it is greatly reduced by local deformations of walls causing distortions of their shapes, and by openings (holes). The actual stiffness is about 0.25 to 0.4 of the stiffness of the same components but with ideally working partitions.

Fig. 2.5.1 shows influence of longitudinal ribs on bending (cross-sectional moment of inertia I_{ben}) and torsional (polar moment of inertia J_{tor}) stiffness of a box-like structure [2]. The table in Fig. 2.5.1 also compares weight (cross-sectional area, A) and weight-related stiffness. It is clear that diagonal ribs are very effective in increasing both bending and, especially, torsional stiffness for the given outside dimensions and weight.

Box-shaped beams in Fig. 2.5.2 have transversal ribs only (cases 2, 3) or the transversal ribs in combination with a longitudinal diagonal rib (case 4), harmonica-shaped ribs (case 5), or semi-diagonal ribs supporting

	Factors				
Profile	I_{ben}	I_{tors}	A	$\dfrac{I_{ben}}{A}$	$\dfrac{I_{tors}}{A}$
	1	1	1	1	1
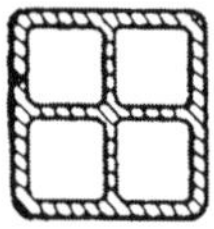	1.17	2.16	1.38	0.85	1.56
	1.55	3	1.26	1.23	2.4
	1.78	3.7	1.5	1.2	2.45

Fig. 2.5.1. Stiffening effect of reinforcing ribs.

Case #	Rib location	K_x %	K_t	W %
1	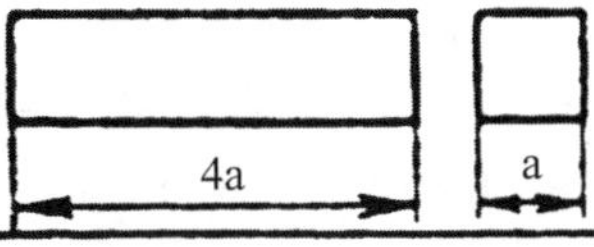 4a a	100	100	100
2		101	103	108
3		102	109	125
4	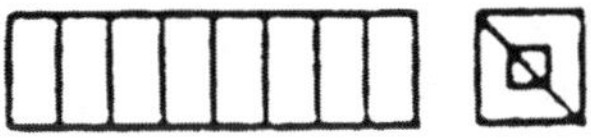	116	132	130
5	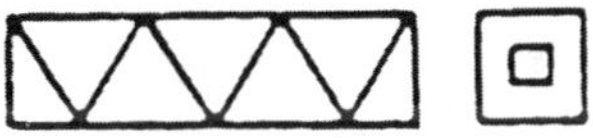	113	112	115
6	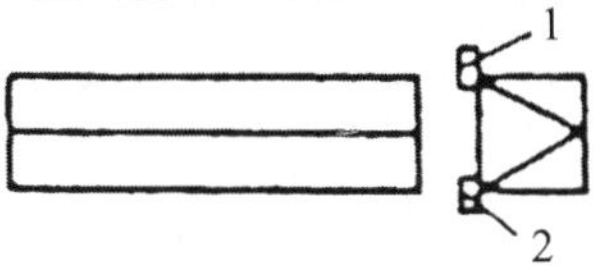	135	–	140

Fig. 2.5.2. Reinforcement of frame parts by ribs.

guideways 1, 2 (case 6). The table compares bending stiffness, k_x, torsional stiffness, k_t, and weight, W, of the structure. It can be concluded that:

1. With increasing number of ribs, weight W is increasing faster than stiffnesses, k_x, k_t;
2. Vertical transversal ribs are not effective; simple transversal partitions with diagonal ribs (case 4) or V-shaped longitudinal ribs supporting guideways 1, 2 (case 6) are better;

3. Ribs are not very effective for closed cross-sections, but are necessary for open cross-sections.

Machine frame components usually have numerous openings to allow for accessing mechanisms and other units located inside. These openings can significantly reduce stiffness (increase structural deformations), depending on their relative dimensions and positioning. Fig. 2.5.3 illustrates some of these influences; δ_x and δ_y are deformations caused by forces F_x and F_y, respectively; δ_t is angular twist caused by torque, T. Fig. 2.5.3 shows that:

1. Holes (windows) significantly reduce torsional stiffness;
2. When the part is loaded in bending, the holes should be located close to the neutral plane (case I);
3. Location of the holes in opposing walls in the same cross-sections should be avoided;
4. Holes having diameters exceeding 1/2 of the cross-sectional dimension ($D/a > 0.5$) should be avoided.

The negative influence of holes on stiffness can be reduced by embossments around the holes or by well-fit covers. If a cover is attached by bolts, then tightening of the bolts results in a reliable frictional connection of the cover to the base part. The loss of stiffness due to the presence of the hole would be compensated if the preload force of each bolt is [15]

$$Q \geq \frac{T(b_0 + l_0)}{Ffn}, \qquad (2.5.1)$$

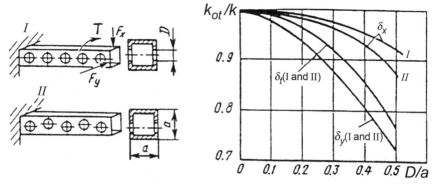

Fig. 2.5.3. Influence of holes in frame parts on stiffness.

where F = cross-sectional area of the beam undergoing torsion; T = torque applied to the beam; b_o, l_o = width, length of the holes; f = friction coefficient between the cover and the beam; n = number of bolts.

2.5.1 Local Deformations of Frame Parts

Local contour distortions due to torsional loading and/or local bending loading may increase elastic deformations up to a decimal order of magnitude in comparison with a part having a rigid reinforcement. The most effective way of reducing local deformations is by introducing tension/compression elements at the area of peak local deformations. Fig. 2.5.4a shows local distortion of a thin-walled beam in the cross section where an eccentrically applied load causes a torsional deformation. This distortion is drastically reduced by introduction of tension/compression diagonal ribs as in Fig. 2.5.4b.

Fig. 2.5.5 [2] shows distortion of a thin-walled beam under shear loading (a). Shear stiffness of the thin-walled structure is very low, since it is determined by bending stiffness of the walls and by angular stiffness of the joints (corners). The same schematic shows the deformed state of a planar frame. The corners (joints) can be reinforced by introducing

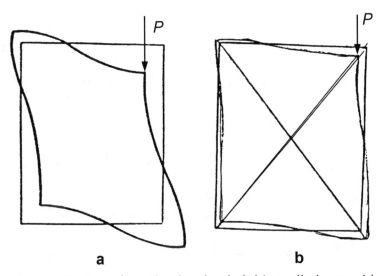

a   b

Fig. 2.5.4. Contour distortion in a loaded thin-walled part without (a) and with (b) reinforcing ribs.

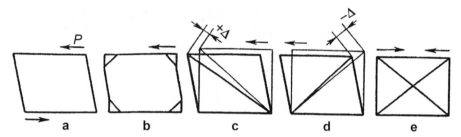

Fig. 2.5.5. Diagonal reinforcement for shear loading.

corner gussets holding the shape of the corners (Fig. 2.5.5b). The most effective technique is introduction of tensile, Fig. 2.5.5(c), or compressive,
Fig. 2.5.5(d), reinforcing diagonal members (diagonal ribs in the case
of a beam). Tilting of the cross-section is associated with stretching/
compression of the diagonal member by an increment Δ. Since tension/
compression stiffness of the diagonal member(s) is much greater than
bending stiffness of the wall, the overall shear stiffness significantly increases. Loading of the diagonal member in tension is preferable, since
the compressed diagonal member is prone to buckling at high force magnitudes. When the force direction is alternating, crossed diagonal members, as in Fig. 2.5.5e, can be used.

A different type of local deformation is shown in Fig. 2.5.6. In this
case, the local deformations of the walls are caused by internal pressure.
However, the solution is based on the same concept—introduction of a
tensile reinforcing member (lug bolt 2) in the axial direction and a reinforcing ring, 1 also loaded in tension, to prevent bulging of the side wall.

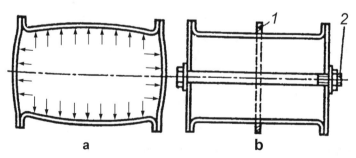

Fig. 2.5.6. Reduction of local deformations.

These reinforcing members not only reduce local deformations, but also reduce vibration and ringing of walls as diaphragms.

2.6 GENERAL COMMENTS ON STIFFNESS ENHANCEMENT OF STRUCTURAL COMPONENTS

The most effective design approaches for stiffness enhancement of a structural component without increasing its weight are:

1. Replacement of bending by tension/compression;
2. Optimization of load distribution and support conditions if the bending mode of loading of a component is inevitable;
3. Judicious distribution of the mass to achieve either the largest cross-sectional or polar moments of inertia, or both, for a given mass of a component;
4. Use of adjacent (connected) parts for reinforcement of the component; to achieve this effect, special attention has to be given to reinforcement of the areas where the component is joined with other components;
5. Reduction of stress concentrations; to achieve this, either sharp changes of cross-sectional shapes or areas, or both, have to be avoided or smoothed;
6. Use of stiffening ribs, preferably loaded in compression;
7. Reduction of local deformations by introduction of ties parallel or diagonal in relation to principal sides (walls) of the component;
8. Use of solid, non-interrupted cross-sections, especially for components loaded in torsion;
9. Optimized geometry of the component, which has a great influence on both stiffness values and stiffness models:
   a. For short beams (e.g., gear teeth), shear deformations are commensurate with bending deformations and may even exceed them; in machine tool spindles, *shear deformations may constitute up to 30% of total deformations*;
   b. For longer beams, their shear deformations can be neglected (bending deformations prevail); for example, for $L/h = 10$, where L = length and h = height of the beam, shear deformation is ~3% of the bending deformation for a solid cross-section, but increases to 6% to 9% for *I-beams*. Contribution from shear is even greater for multi-layered honeycomb and foam-filled beams;

10. If the cross-sectional dimensions of a beam are reduced relative to its length, the beam loses resistance to bending moments and torques, as well as to compression loads, and is ultimately becoming an elastic string;

11. Reduction of wall thickness of plates/shells transforms them into membranes/flexible shells, which are able to accommodate only tensile loads;

12. Cross-sectional shape modifications can enhance some stiffness values relative to the other;

13. Beams with open cross-sections, like in Fig. 2.6.1a, may have high bending stiffness but very low torsional stiffness;

14. Slotted tubular structures (Fig. 2.6.1b), may have high torsional but low bending stiffness;

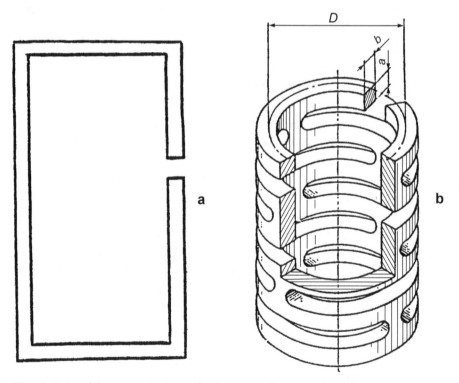

Fig. 2.6.1. (a) Cross-section of a beam stiff in bending but having very low torsional stiffness; (b) Compression (slotted) spring with high torsional stiffness.

15. Plates and shells can be designed to have anisotropic stiffness by a judicious system of ribs or other reinforcements.

2.7 STIFFNESS-CRITICAL METAL ELASTIC ELEMENTS (SPRINGS)

Very often mechanical systems comprise components having specified stiffness values in one or more directions, both translational and rotational. Such components may define performance of the system under static loading as well as under dynamic loads and in a vibratory environment. In some cases, these elements are required to have a minimum energy dissipation, e.g., in resonant machines typical for vibratory conveyance and vibratory processing equipment. In other cases, e.g., in vibration isolators, high damping is required for better performance [16]. Stiffness values in various directions for springs made from metal and from some hard plastics do not depend (or weakly depend) on amplitude and frequency of load application.

This section discusses elastic characteristics of some basic metal (or hard plastic) elastic elements with linear or quasi-linear load-deflection characteristics, which are not adequately addressed in general textbooks and handbooks on machine elements. Elements with specified nonlinear load-deflection characteristics are described in Chapter 3. Metal springs can be used for generating large static deformations (relative to their dimensions) and for accommodation of wide range of loads, from very small to very large. They usually exhibit very low creep, unless exposed to high temperatures, their dynamic stiffness is very close to static stiffness, and they are not sensitive to high or low temperatures (unless high damping shape memory alloys are used). Some frequently used metal spring designs are shown in Fig. 2.7.1, in which the springs are classified by the mode of stressing their material. Characteristic for many metal spring designs are low damping and relatively high costs. In cases where a higher damping is required, it can be achieved by using high damping metals (see Table 1.2), or other materials with high rigidity and high damping, such as shape memory materials and hard plastics; by incorporating friction interaction into spring designs (Fig. 2.7.1j,m), and by attaching dampers to springs. Basic parameters of the springs for mechanical design applications are stiffness, maximum stresses, and stability. Springs are usually made from appropriately heat-treated spring

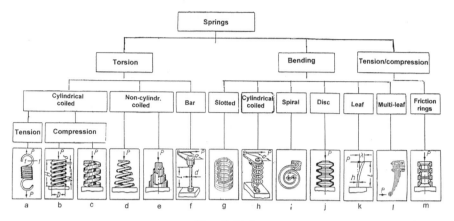

Fig. 2.7.1.

steel alloys having high yield strength, but can be also made from solid plastics.

In general design applications, special dampers are used only infrequently. Special damping devices are used mostly in vibration control devices, such as vibration isolators and dynamic vibration absorbers, which are beyond the scope of this book.

2.7.1 Coil Springs

Coil springs, both coiled from steel wire or bars and machined from tubular blanks are the most widely used types of metal elastic elements.

Cylindrical (linear) coil springs are often characterized by *spring index*, $C = D/d$. Axial (compression/tension) stiffness, k_a, of a cylindrical coil spring with the constant pitch of coils and constant round cross-section of the wire along its length is:

$$k_a = \frac{Gd}{8C^3 n} = \frac{Gd^4}{8D^3 n}. \tag{2.7.1}$$

Here D = diameter of a coil, measured by the center lines of the wire; d = wire diameter; G = shear modulus of spring material; n - number of coils. For springs with ground end surfaces, the number n of active coils can be taken as the total count of coils minus 1.75.

Shear stress in the coils is determined by torsion of the wire caused by the axial force (asymmetrical due to the coil curvature), shear stress due to direct action of the axial force, and by additional stresses due to inevitable eccentricity of the axial load, decreasing with increasing n. The maximum shear stress considering these factors is

$$\tau = K\,\tau_0 = K\frac{8PD}{\pi d^3} = K\frac{8PC}{\pi d^2}\,, \tag{2.7.2}$$

where τ_0 is the nominal stress, and K is *correction factor*

$$K = \frac{4C-1}{4C-4} + \frac{0.615}{C} \tag{2.7.3}$$

Variations of the coiling operation as well as of installation techniques may result in a significant scatter of spring parameters. To improve performance and consistency of the load-deflection characteristics, the end coils of compression springs are often flattened and restrained in the lateral directions, e.g., as shown in Fig. 2.7.2a. Still, scatter of a spring stiffness of the order of ±10% is rather typical.

Compression springs loaded by static or by short duration cyclical loads can be treated by "scragging" or "setting" treatment. This is performed by loading the spring for several hours by the load having the same sign as the payload but a significantly greater magnitude than the rated load of the spring, thus inducing stresses in the coils, which are 10% to 30% greater than the yield strength of the material. Residual stresses generated by this treatment are of the sign opposite to the stresses induced by the payload, thus reducing the stresses induced by the payload and increasing the rated load of the spring by at least 30%, and allowing reduction of the spring size for a given application and leaving more design space for stiffness optimization.

Extension springs have tightly wound coils in contact with each other when the spring is not loaded. While the spring iis being produced (wound), the wire is twisted, thus pressing coils together and creating a preload. When the spring is loaded by an axial force, the preload force must be overcome before the spring starts deforming. The preload force can be controlled by manufacturing and should be always considered when the spring is used as a stiffness-critical element.

Springs with rectangular coil cross-section, axb, have a better utilization of wire cross-section due to more uniform stress distribution. They can be fabricated either by coiling from rectangular cross-section wire,

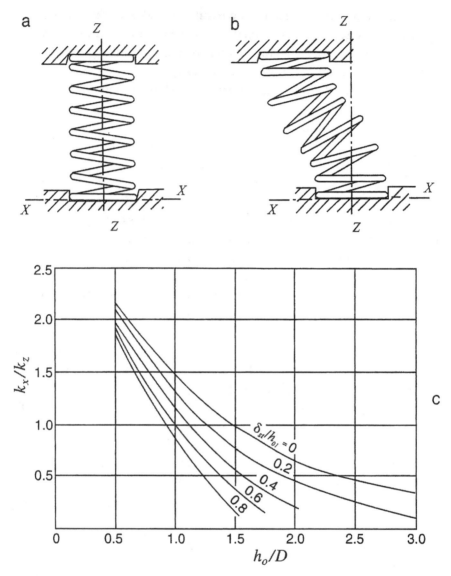

Fig. 2.7.2.

or by machining (turning or milling) coil springs with rectangular wire cross-section from tubular blanks (Fig. 2.7.3). All machined springs have usually a rectangular cross-section of the coil. The machined springs have better accuracy of dimensions and of stiffness, down to about ±5%, and with precision machining down to ±0.1% They also have a better

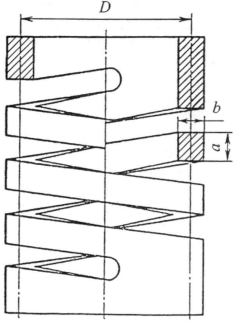

Fig. 2.7.3

assembly accuracy, since they can be directly fastened, e.g., by bolts, to the object and to the base. In addition, such springs can be made smaller. Stiffness of and stress in springs with rectangular wire cross-section are:

$$k_a = \frac{G\delta^4}{vD^3n}; \quad \tau_o = \frac{PD}{2\chi a^2 b}. \tag{2.7.4}$$

Here P = axial force; δ = the smallest value from a, b; χ and v are determined from Table 2.2.

While the axial deflection/stiffness of a coil spring is mostly due to torsion of the wire, the lateral deflection/stiffness is due to both torsion and bending of the wire and of the spring itself (Fig. 2.7.2b). Thus, it depends on the axial deflection of the spring and can be adjusted by varying axial compression (preload) of the spring. Fig. 2.7.2c gives the ratio, k_x/k_z, for the springs coiled from round cross-section steel wire for different axial loading (static deflection δ_{st} under the axial weight load), depending on the initial working height h_o of the spring. These plots are obtained for

Table 2.2. Coefficients for calculations of coil springs with rectangular cross section of wire.

b/a	χ	v	b/a	χ	v
1.0	2.40	5.57	3.0	0.625	1.0
1.5	1.44	2.67	4.0	0.44	0.7
1.75	1.2	2.09	6.0	0.28	0.44
2.0	1.02	1.71	10	0.25	0.16
2.5	0.78	1.26			

the conditions in which the ends of the spring are constrained to remain parallel in the process of the lateral loading, e.g., as in Fig. 2.7.2a,b.

Torsional stiffness of a coil spring about its longitudinal axis is:

$$K_t = \frac{Gd^4(1+v)}{32Dn} \tag{2.7.5}$$

where v = Poisson's ratio of the spring material. Coil springs are often used as torsional springs (with appropriate designs of their ends).

Long (initial length, $h_o > 2.5D$) coil springs can buckle at a large static deformation δ_{st}, unless at least one of the following two stability conditions are satisfied:

$$\frac{\delta_{st}}{h_o} \leq 0.81\left[1 - \sqrt{1 - 6.9(D/h_o)^2}\right] \tag{2.7.6}$$

or

$$\frac{k_x}{k_z} \geq 1.2\left(\frac{\delta_{st}}{h_s}\right), \tag{2.7.7}$$

where h_s is axial length of the deformed spring.

Since damping of metal coil springs is low (unless they are made from high damping alloys), high frequency resonance vibrations (*surge*) at natural frequencies of the spring considered as a distributed parameters body can be very intense. These natural frequencies can be calculated as

$$f_n = k\frac{2d}{\pi D^2 n}\sqrt{\frac{G}{32\gamma}}, \tag{2.7.8}$$

where γ = density of the spring material, $k = 1, 2, 3...$

Both axial and transverse stiffness values of coil springs can be adjusted. The axial stiffness can be adjusted by varying the active length of the spring by screwing a threaded plug fitted to the coil pitch into the spring, while the transverse stiffness can be adjusted by preloading the spring thus changing its effective length.

A coil spring can also be used under radial loading when the force is applied transversely to its axis (Fig. 2.7.4). This mode of loading allows accommodation of much higher loads (with much smaller deflections and higher stiffness values) and realizes a noticeable damping due to friction between the OD of the spring and the supporting surfaces during the deformation process of the spring. Both extension springs (tightly coiled, Fig. 2.7.1a) and other types of coil springs can be used in this mode. Since elastic stability of the coils limit the maximum loads, high load capacity of such elements is better realized if the axial movements of the coils are prevented, e.g., by mechanical constraining elements (like walls 1, 2 shown in Fig. 2.7.4) or by enhanced friction on the contact surfaces. Deflection δ_{st} (and stiffness $dP/d\delta_{st}$) and equivalent normal stress σ_{eq} for the properly constrained spring can be calculated as

$$\delta_{st} = \frac{3.74PD^3}{\pi^2 End^4} \frac{1 + 8.86\sin^2 \alpha}{\cos^3 \alpha}, \qquad (2.7.9)$$

$$\sigma_{eq} = \frac{4PD}{\pi^2 nd^3}\sqrt{\pi^4 \tan^2 \alpha + (4 + \tan^2 \alpha)^2} . \qquad (2.7.10)$$

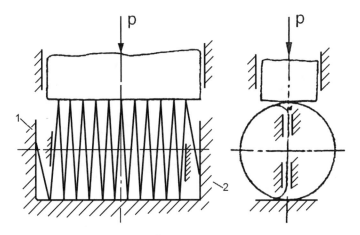

Fig. 2.7.4.

Here α = helix angle of the coils ($\alpha \cong 0$ can be assumed for tightly coiled springs); E = Young's modulus of the spring material.

The load-carrying capacity of such element can be increased (at the price of increased stiffness) by modifying the contact surfaces (increasing their curvature, making rectangular or gothic arch contact grooves, enhancing the friction coefficient, etc.).

2.7.2 Slotted Springs

Cylindrical slotted springs shown in Fig. 2.6.1b may have high or low axial stiffness, high lateral stiffness, and high load carrying capacity with high accuracy in a small outline. The spring is very stable and sturdy, since it has very high torsional stiffness around its axis and a high degree of symmetry. The number of slots in each transverse section of the springs is usually two, thus the spring is, essentially, a series connection of beams loaded at mid-span. Axial stiffness, k_a, and effective stress σ_{ef} are:

$$k_a = \frac{Eab^3}{D^3 n} \frac{1}{\alpha_n}; \quad \sigma_{ef} = k_\sigma \frac{PD}{ab^2} \beta \qquad (2.7.11)$$

Here E = Young's modulus of the spring material, α_n and β are determined from Table 2.3. The stress concentration factor, k_σ, for fillets between the beams is plotted in Fig. 2.7.5.

2.7.3 Friction Springs

Friction springs (Fig. 2.7.6a), comprise a stack of alternating outer and inner conforming double-tapered rings [17]. An interacting pair of one outer and one inner rings represents one element of the spring. An axial force P induces a wedge effect resulting in high radial forces between

Table 2.3.   Coefficients for calculations of slotted springs.

b/a	β	α_n	b/a	β	α_n
0.1	0.44	0.10	1.0	0.61	0.12
0.25	0.44	0.10	1.5	0.66	0.13
0.5	0.5	0.11	2.0	0.68	0.14
0.66	0.53	0.11	10	1.36	0.45

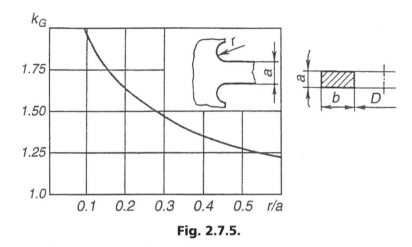

Fig. 2.7.5.

the rings, thus causing extension of the outer rings, compression of the inner rings, and axial sliding between the outer and inner rings (axial deformation s of the spring). The spring cannot be overloaded, since at the maximum allowable load/deformation, the adjacent outer and inner rings are touching each other thus rigidizing the spring (Fig. 2.7.6b). Since the material of rings (high grade spring steel) is loaded in tension/ compression, the stresses in the rings are relatively uniform, thus resulting in a good utilization of the material and small dimensions of springs for a given rated load. Sliding between the rings during the deformation process results in a significant energy dissipation (damping) (Fig. 2.7.7a). The load-deflection characteristics of such springs are quite linear as can

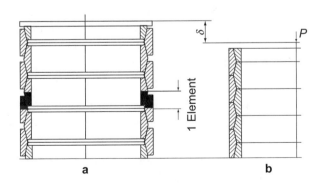

Fig. 2.7.6.

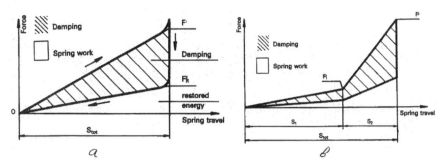

Fig. 2.7.7.

be seen in Fig. 2.7.7a. A nonlinear characteristic can be achieved by making axial slots in some of the inner rings. Compression of a slotted inner ring requires a very small force, thus resulting in a lower stiffness of the friction spring at low axial forces, $P < P_1$, first segment of the load-deflection characteristic in Fig. 2.7.7b. At $P = P_1$, the slots are closed, and the inner rings start behaving as solid, thus significantly increasing stiffness. The load-deflection characteristic can be tuned by changing the number of the slotted rings and width of the slots.

Friction springs are used as shock absorbers, since they may absorb and dissipate a significant amount of kinetic energy and as small high force springs for overload protection devices.

2.7.4 Miniature Tubular Elastic Elements Utilizing Giant Superelasticity Effect

Miniaturization of mechanical systems, especially fast advancement of Micro Electro-Mechanical Systems (MEMS) requires small elastic elements, which can accommodate relatively large forces and develop large elastic deformations relative to their dimensions [18]. Miniaturization of springs described in the previous section, e.g., the ones shown in Fig. 2.7.1, even if theoretically possible, is very costly and often not practical. Utilization of the recently discovered "Giant Superelasticity Effect" (GSE) opens a possibility to design inexpensive very small and versatile springs.

GSE is realized by using a special mode of loading of wires and tubes made from Nickel-Titanium (NiTi) alloys exhibiting the so-called "superelasticity effect" above the martencite transition temperature (e.g., [19]).

If wires or tubes made from NiTi materials exhibiting the superelastic effect are subjected to tension, they demonstrate a unique load-deflection characteristic like the one shown in Fig. 2.7.8 [18]. The initial phase of the tensile loading represents a quasi-linear load-deflection characteristic up to relative tension ε_{max} = ~1.5%. After this deformation is reached, the tensile deformation is increasing with a very small increase in the tensile force. Up to ε = 6% to 8%, the tensile deformation is fully recoverable, the specimen returns to its initial length after removal of the load. Fig. 2.7.8 depicts the load-deflection plot corresponding to ε_{max}. While the magnitudes of the recoverable deformations are many times greater than elastic deformations of other metals, the practical applications of this effect are rather limited, since the load magnitudes are quite small, the "plateau" representing the greatest deformation while the load is hardly increasing is not useful in typical mechanical design applications, and use of such elements in tension requires an elaborate end-clamping devices. Some known applications utilize high damping (large hysteresis loop) associated with this mode of loading.

It was found in [18] that if a round specimen (solid or hollow, wire or tube) made from a superelastic NiTi alloy is loaded in *radial compression* perpendicular to its longitudinal axis between flat parallel surfaces, like

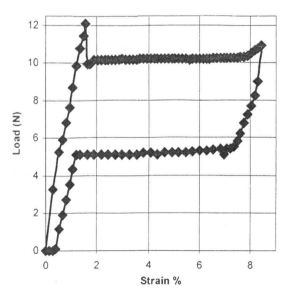

Fig. 2.7.8.

in Fig. 1.3.3, then its load-deflection characteristic is totally different (Fig. 2.7.9). It is quasi-linear, and the deformation is fully recoverable up to relative deformations ε_{max} = 10% to 15%. These relative compression deformations correspond to compression forces 850 N/mm to 1000 N/mm for radial loading of solid wire 1.2 mm diameter, 90 N/mm to 130 N/mm for solid wire 0.165 mm diameter, and 50 N/mm to 60 N/mm for radial loading of tube 0.41 mm OD and 0.2 mm ID. Depending on the needs of a specific design, solid wires or tubes can be used as spring elements, with the compression forces P_{max} at ε_{max} being about five to ten times greater for solid wires than for tubular specimens. The P_{max} can be "fine-tuned" by varying the diameter and wall thickness of the tube.

It was also shown in [18], both analytically and experimentally, that radial loading of solid and tubular elements made from common materials, such as steel, results in elastic deformations more than an order of magnitude exceeding their elastic tensile strain at the yield strength. A stainless steel wire has recoverable deformations under radial loading of $\varepsilon_{max} \approx 1.5$ % (at $P_{max} \approx 450$ N/mm for D = 1.6 mm). Stainless steel tube demonstrated recoverable deformation $\varepsilon_{max} \approx 3.4\%$ (at $P_{max} \approx 50$ N/mm for OD = 1.3 mm, ID = 1.0 mm). Although much smaller than for superelastic wires and tubes, these magnitudes far exceed $\varepsilon_{max} \approx 0.1\%$ to 0.2% for tensile loading of stainless steel specimens and are quite significant. Accordingly, tubular elements under radial loading can be used as in-

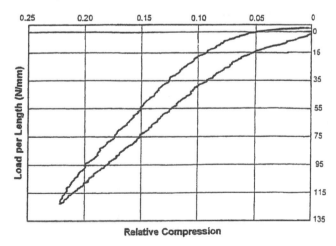

Relative Compression

Fig. 2.7.9.

expensive high quality springs, e.g., for miniature load cells and other MEMS applications. Use of steel greatly increases load-carrying capacity and temperature range of such springs.

The ε_{max} magnitudes can be further increased if thin-walled tubes are used. For example, radial compression of a NiTi superelastic tube with outer diameter D = 6.75 mm and wall thickness t = 0.2 mm was shown to be recoverable up to ε_{max} = 62% (at the compressive force P_{max} = 18.4 N/mm). These magnitudes of the recoverable deformation lead to naming this effect the "Giant Superelasticity Effect" (GSE).

One important application of the above elastic elements is as elastic compensators in mechanical assemblies [18, 20]. The attachment method for GSE tubing to supporting surfaces is proposed in [21].

2.8 STATIC DEFORMATION CHARACTERISTICS OF QUASI-LINEAR RUBBER ELEMENTS

Rubber-like or elastomeric materials have unique characteristics, both static and dynamic, suitable for designing extremely versatile elastic elements, which can be tailored for extremely broad range of applications. However, use of rubber structural flexible elements in mechanical design applications is still limited, largely due to low familiarity of mechanical design engineers with their potential. Basic characteristics of rubber-like materials are described in Appendix 4. A load-deflection characteristic in tension for rubber specimens is linear up to 10% to 15% deformation, up to somewhat lesser deformation in compression but much greater (up to 300%) in shear, unless the specimen was intentionally designed to have a nonlinear load-deflection characteristic. This section provides information on static stiffness characteristics of typical linear (or quasi-linear) rubber springs. Nonlinear rubber elements are described in Chapter 3. Close form expressions for the stiffness values are given for basic loading patterns (compression, torsion, shear, etc.). Due to rather complex elastic properties of rubber, these expressions represent only a first approximation but still are very useful in the first stages of designing rubber flexible elements. The next step is experimental verification and/or Finite Element Analysis. The special case of *rubber flexible elements for torque-transmission (couplings)* is addressed in Appendix 5, while another special case of vibration isolators with rubber flexible elements is addressed in [16].

For small deformations (relative deformations for each loading mode less than ~10%), the loading/deformation patterns are independent, but at greater deformations, this independence disappears, and loading in one direction may cause development of forces in other directions. Thus, loading in torsion of a rubber cylinder with axial dimension h (torque, T, angular deformation, ϕ) generates an axial compression force $P_c = T\phi/2h$. Shear by angle γ_{xy} of a rubber block bonded to stationary and parallel metal platens is accompanied not only by shear stress, $\tau_{xy} = G\gamma_{xy}$, but also by compression stresses, $\sigma_y = \sigma_z = -G\gamma_{xy}^2$. If a rubber element of a significant thickness h and bonded to rigid (metal) plates on both faces is subjected to shear deformation, some bending deformation also develops. It can be considered by using a bending correction term in calculating shear deformation x caused by shear force P,

$$x = \frac{Ph}{GA}\left(1 + \frac{h^2}{36i^2}\right), \tag{2.8.1}$$

where A = cross-sectional area of the loaded specimen, and i = bending radius of inertia of the cross-sectional area about its neutral axis.

Typical deformation modes of rubber elements are shown in Fig. 2.8.1 [22].

A rubber element configured as in Fig. 2.8.1c is characterized by an effective modulus of elasticity E_{eff} intermediate between magnitudes of K and M [23],

$$E_{\text{eff}} = \frac{E_o(1 + \beta S^2)}{1 + \frac{E_o}{K}(1 + \beta S^2)}. \tag{2.8.2}$$

Here S = the *shape factor*, the ratio of one loaded area of the element to the total of load-free surface areas, and β = numerical constant dependent on rubber hardness H, $\beta \approx 2.68 - 0.025H$ @ $H = 30 - 55$; $\beta \approx 1.49 - 0.006H$ @ $H = 60 - 75$.

The shape factor of a rubber cylinder (diameter D, height h) is $S_{\text{cyl}} = D/4h$; for a prismatic rubber block (sides, a, b, height, h), $S_{\text{rec}} = ab/h(a + b)$. For rubber elements with reasonably uniform dimensions (D/h, a/h, $b/h < \sim 10$), Eq. (2.8.2) can be written as:

$$E_{\text{eff}} \cong E(1 + \beta S^2) = 3G(1 + \beta S^2). \tag{2.8.3}$$

The elements with the uniform dimensions have linear load-deflection characteristics up to ~10% compression deformation and quasi-linear

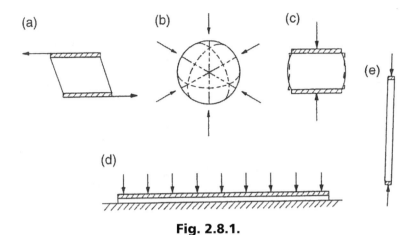

Fig. 2.8.1.

characteristics up to ~15% deformation. If $b \gg a$ for a rectangular shape element, expression (2.8.3) should be replaced by [23]:

$$E_{eff} = (2/3)\, E(2 + \beta S^2) = 2G(2 + \beta S^2), \qquad (2.8.4)$$

where $S = a/2h$.

While designers of metal parts are usually limited by allowable maximum stresses under specified loading conditions, it is customary to design rubber components using allowable strains, or relative deformations. For rubber components bonded to metal inserts or covers, a conservative strain limit for shear loading is relative shear, $\gamma = 0.5$ to 0.75, for the load applied and held for a long time and for a multicycle fatigue. Some blends can tolerate higher values of relative shear, up to $\gamma = 0.75$ to 1.0. For shorter times of load application, $\gamma = 1.0$ to 1.5 can be tolerated, up to $\gamma = 2.0$ for some rubber compositions and bonding techniques. A compression loading of a bonded rubber specimen is usually associated with the allowable relative compression, $\varepsilon = 0.1$ to 0.15, with a somewhat larger limit for softer rubber blends.

Fig. 2.8.2a illustrates compression deformation under force, P_z, of cylindrical rubber element 1 bonded to metal end plates 2 and 3. Due to incompressibility of rubber-like materials (see Appendix 4), compression deformation can develop only at the expense of the bulging of element 1 on its free surfaces. If an intermediate metal layer 4 is placed in the middle of and bonded to rubber element 1, as in Fig. 2.8.2b, thus dividing

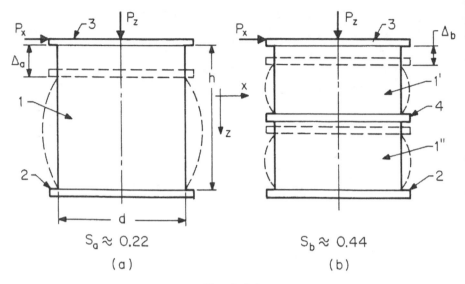

$$S_a \approx 0.22$$
(a)

$$S_b \approx 0.44$$
(b)

Fig. 2.8.2.

it into two layers 1´ and 1˝, the bulging becomes restricted, and compression deformation under the same force P_z is significantly reduced.

High damping is often a beneficial feature of rubber elastic elements. However, increase of material damping in elastomeric components may also have undesirable consequences. Usually (although not universally) high damping elastomeric compounds have higher creep rates than lower damping compounds. While a damping increase in the flexible element of a resonating oscillatory system results in a *reduction* of heat generation due to reduction of the resonance amplitude (see [16] and Section 1.1.2 above), heat generation increases with increasing damping in cases of forced motion, such as in flexible elements in vibratory machines or in misalignment compensating couplings. If there are stress concentrations in such loaded elastomeric elements made of high damping elastomers, they may overheat.

Expressions for shear stiffness of rubber springs can be derived using regular Strength of Materials approaches up to relative shear 60% or even larger in cases of low durometer rubber. For compression loading, four different cases (*a* to *d*) can be distinguished for quasi-linear rubber flexible elements bonded to metal, depending on influence of edge effects in the bonded area [24] and one case (*e*) of nonlinear rubber elements:

a. Axially loaded *bars* whose axial dimension (length or height h) is at least five times greater than the smallest transverse dimension. The allowable axial compression forces, P_c, are relatively small and limited by buckling, with the critical force

$$P_{cc} = \frac{12\pi^2 GI}{h^2},$$ (2.8.5a)

where G = the shear modulus, and I = cross-sectional moment of inertia. For a round cross-section (radius R),

$$P_{cc} = 3\pi^2 G \frac{R^4}{h^2}.$$ (2.8.5b)

Due to manufacturing errors resulting in asymmetry and eccentricity of load application, actual buckling may start at $P = 0.7$ to $0.8P_{cc}$. Small tensile deformations ($\Delta < 0.15h$) under force, P_t, can be calculated as

$$\Delta = \frac{P_t h}{EA},$$ (2.8.6)

where $E \approx 3G$ = tensile modulus, A = cross-sectional area. At larger deformations, the load-deflection characteristic becomes nonlinear.

b. *Regular size block* (pad), $5a > h > 0.1$ to $0.15a$, where a = the smallest cross-sectional dimension. For analysis of such elements, rubber can be considered as volumetrically incompressible; the load-deflection characteristic is determined to a large extent by shapes of metal elements bonded to rubber.

c. *Low block* (pad), $0.15a > h > 0.05a$. Compressibility of rubber, which is characterized by its volumetric modulus, K, or by deviation of the Poisson's ratio from $v = 0.5$, plays a significant role in determining load-deflection characteristic of the element.

d. *Thin-layered elements*, $h < 0.05a$, wherein the volumetric compressibility plays a dominant role, while conditions of the free surface are not important. Such elements have nonlinear load-deflection characteristic even at small deflections (see Chapter 3).

e. Elements whose either loaded or free surfaces, or both (and, consequently, shape factor S), are changing during the deformation process, thus resulting in a nonlinear load-deflection characteristic (see Chapter 3).

All close-form expressions for stiffness characteristics of rubber elements bonded to metal as given below are approximate, especially when compressibility of rubber is of any significance. In some cases, two different expressions from different sources can be provided, thus allowing to average them and thus narrow the deviation of the calculated values from actual values.

2.8.1 Stiffness of Bonded Rubber Blocks

Table 2.4 contains expressions for *compression* (along z-axis) *stiffness*, k_z = P_z/Δ_z, for cylindrical rubber elements (radius R) bonded to metal end plates (Fig. 2.8.3) [25]. The first line (small deformations) is characterized by linear load-deflection characteristic (with an exception of thin layer, $\rho > 20$), the second line (large deformations) is characterized by nonlinear load-deflection characteristics.

Shear stiffness, $k_s = P_s/\Delta_s$ between the top and bottom bonded plates

$$k_s = \frac{\pi G R^2}{h} = \pi G R \rho \qquad (2.8.7)$$

If the block is compressed (compression deformation Δ_z), relative compression $\lambda_z = \Delta_z/h$, then

$$k_s = \pi G' R^2 \rho \frac{\varsigma}{\tan \varsigma}; \quad G' = G\left[1 + 0.015 \frac{\Delta_z}{h}\right] = G(1 + 0.015\lambda_z) \qquad (2.8.8)$$

Table 2.4. Compression stiffness of cylindrical rubber blocks.

Δ_z/h	$\rho < 7$	$7 < \rho < 20$	$\rho < 20$
≤ 0.1	$k_{z1} = 3\pi G\rho(0.92 + 0.5\rho^2)R$	$k_{z2} = \dfrac{K\pi R\rho(0.92 + 0.5\rho^2)}{0.92 + 0.5\rho^2 + 0.33\theta}$	$k_{z3} = \pi K'\rho R$
≤ 0.5	$k_{z4} = \dfrac{\pi R^2 G}{\Delta_z}(1 + 0.413\rho^2)(\lambda_z^{-2} - \lambda_z)$	$k_{z5} = \dfrac{k_{z4}k_{z3}}{k_{z4} + k_{z3}}$	----

In Table 2.4, $\rho = R/h$; $\lambda_z = 1 - \Delta_z/h$; $\theta = K/G$; $K' = K[1 + 0.015 \, K \, (\Delta_z/h)]$.

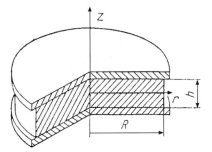

Fig. 2.8.3.

where ζ is determined from an equation

$$1 - \frac{\Delta_z}{h} = \frac{\sin 2\varsigma}{2\varsigma} \qquad (2.8.9)$$

Torsional stiffness about z-axis, $k_t = T_z / \phi$, where ϕ = angular deformation of the element, is

$$k_t = (\pi/2)GR^3\rho. \qquad (2.8.10)$$

If $\Delta_z = 0$, application of torque T_t around axis z results in development of a compressive force

$$P_t = T_z\phi/2h. \qquad (2.8.11)$$

Torsional stiffness of a pre-compressed block is

$$k_t = (\pi/2)GR^3\lambda_z^{1.5}\rho \qquad (2.8.12)$$

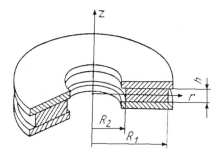

Fig. 2.8.4.

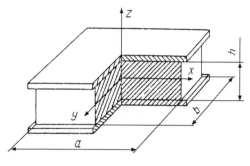

Fig. 2.8.5.

In case of a simultaneous compression and torsion of the element, compression force, P_z, torque, T_z, λ_z, and ϕ are interrelated by the following expressions

$$P_z = \pi GR^2 \left[(\lambda_z - \lambda_z^2) + \frac{\phi^2 \rho^2}{4\lambda_z^3}(1 - 2\lambda_z) \right]; \quad T_z = \frac{\pi}{2}GR^3 \rho \phi \lambda_z^{3/2}. \quad (2.8.13)$$

Angular (Cardan) stiffness $k_a = M_x/\psi$ (resistance to a moment M_x around one of horizontal coordinate axes in Fig. 2.8.3 causing angular deformation, ψ) at $\rho \le 5$; $\psi\rho \le 0.1$ is

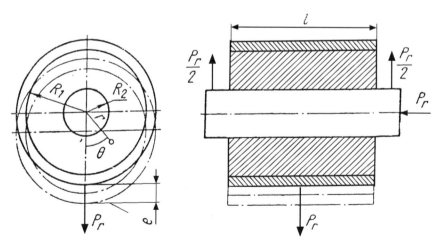

Fig. 2.8.6.

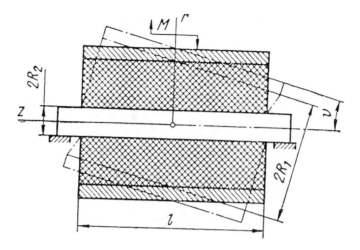

Fig. 2.8.7.

$$k_a = \frac{7GR^3}{3\rho^{-1} - \tanh(3\rho^{-1})}.$$ (2.8.14)

Expressions for compression stiffness, k_z, of ring-shaped rubber blocks bonded to metal plates (Fig. 2.8.4) are listed in Table 2.5.

Shear stiffness, $k_s = P_s/\Delta_s$ between the top and bottom bonded plates

$$k_s = \pi G'R_1\rho(1-\alpha^2).$$ (2.8.15)

If the block is pre-compressed (compression deformation Δ_z), then

$$k_s = \pi G'R_1(1-\alpha^2)\frac{\xi}{\tan\varsigma}$$ (2.8.16)

where ς is determined from Eq. (2.8.9).

Table 2.5. Compression stiffness of ring-shaped rubber blocks.

Δ/h		$(1-\alpha)\rho < 6$	$6 < (1-\alpha)\rho < 15$ $(1-\alpha)\rho > 15$
≤ 0.1		$k_{z1} \approx \pi R_1\rho G(1-\alpha^2)$ $\{1+0.413[\rho(1-\alpha)]^2\}$	$k_{z2} = \dfrac{k_{z1}k_{z3}}{k_{z1}+k_{z3}}$ $k_{z3} = \pi R_1\rho K'(1-\alpha^2)$
≤ 0.5	$k_{z4} = \dfrac{\pi R_1^2 G}{\Delta_z}(1+0.413\rho^2[1-\alpha^2])(\lambda_z^{-2}-\lambda_z)$	$k_{z2} = \dfrac{k_{z4}k_{z3}}{k_{z4}+k_{z3}}$	-----

In Table 2.5, $\rho = R_1/h$; $\lambda_z = 1 - \Delta_z/h$; $\alpha = R_2/R_1$; $K' = K[1 + 0.015 K (\Delta_z/h)]$.

Torsional stiffness about z-axis, $k_t = T_z/\phi$, is

$$k_t = (\pi/2)GR^3\rho\,(1 - \alpha^2). \qquad (2.8.17)$$

Angular (Cardan) stiffness, $k_a = M_x/\psi$ (resistance to a moment M_x around one of horizontal coordinate axes in Fig. 2.7.7) at $\rho \le 5$; $\psi\rho \le 0.1$ is

$$k_a = \frac{2.35(1-\alpha^4)NGR_1^3}{N\rho^{-1} - \tanh(N\rho^{-1})}; \quad N = 3\sqrt{\frac{1-\alpha^4}{1-\alpha^6}}. \qquad (2.8.18)$$

Table 2.6 contains expressions for *compression stiffness* for a prismatic bonded rubber block in Fig. 2.8.5
Shear stiffness, $k_s = P_s/\Delta_s$ between the top and bottom bonded plates

$$k_{sx} = k_{sy} = Gab/h. \qquad (2.8.19)$$

If the block is compressed (compression deformation Δ_z), then

$$k_{sx} = k_{sy} = \frac{Gab}{h}\frac{\zeta}{\tan\zeta}. \qquad (2.8.20)$$

Torsional stiffness about z-axis, $k_t = T_z/\phi$ at $\beta\phi \le 0.2$ is

$$k_t = k_t = \frac{Gab}{h}(a^2 + b^2). \qquad (2.8.21)$$

Rubber-metal bushings (Fig. 2.8.6) are widely used in industry, e.g., as link connectors in steering and suspension linkages in vehicles. The bushings can be molded or assembled.

Table 2.6. Compression stiffness of prismatic rubber blocks.

Δ/h	$\alpha, \beta > 1, \beta < 10$	$\alpha \gg \beta, \beta < 10$
≤ 0.1	$k_{1z} = Ga\beta h\dfrac{\pi^2}{2}\dfrac{36+\pi^2(\alpha^2+\beta^2)+\dfrac{\pi^4}{48}\alpha^2\beta^2}{48+\pi^2(\alpha^2+\beta^2)};$	$k_{2z} = Ga\beta h(3+\beta^2);$
≤ 0.5	$k_{4z} = \dfrac{ab}{\Delta_z}G(1+0.13\alpha^2)(\lambda^2 - \lambda^{-2}),\ a = b;$	$k_{5z} = \dfrac{ab}{\Delta_z}G\left[-4.8\ln\lambda + \dfrac{\beta^2}{2}(\lambda^{-2} - 1)\right]$

In Table 2.6, $\alpha = a/h$; $\beta = b/h$; $a \ge b$.

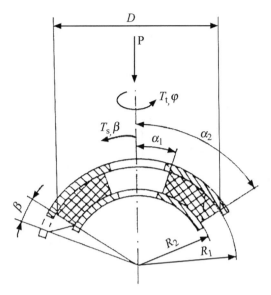

Fig. 2.8.8.

For a molded bushing *compression stiffness* under radial force, P_r, is

$$k_r = Gl\left[1.94\frac{1+\alpha}{1-\alpha}\left(3+n-\frac{(1+n)^2}{4+n+0.82\rho}\right)\right], \qquad (2.8.22)$$

where

$$\rho = \frac{l}{R_1-R_2}; \alpha = \frac{R_2}{R_1}; n = 1.57\left[3+n\frac{(1+\alpha)^2}{(1-\alpha)^2}\right]. \qquad (2.8.23)$$

For a bushing with relatively thin rubber layer, compressibility of rubber (deviation of Poisson's ratio, v, from 0.5) can be a significant factor for *radial stiffness* of the bushing. This factor is considered in [25]. Radial compliance, $e_r = 1/k_r$ is

$$e_r = \frac{1}{\pi l G}\left\{2(1-2\mu)\left(\frac{\alpha-1}{\alpha+1}\right)+\frac{2}{3}\left(\frac{\alpha-1}{\alpha+1}\right)^3\frac{\left[l^2/r^2+3\left(\frac{1+\alpha}{\alpha}\right)^2\right]}{\left[l^2/r^2+6\left(\frac{1-\alpha}{\alpha}\right)^2\right]}\right\} \qquad (2.24)$$

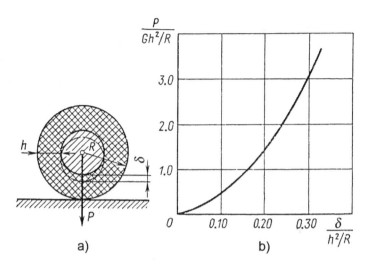

Fig. 2.8.9.

Axial shear stiffness under force P_z is, at $\dfrac{\Delta_z}{R_1(1-\alpha)} \leq 0.5$,

$$k_z = -\frac{2\pi Gl}{\ln \alpha}. \tag{2.8.25}$$

Torsional stiffness about z-axis, $k_t = T_z/\phi$, is

$$\text{at } \phi \leq 0.3(1-\alpha), k_t = \frac{4\pi GlR_2^2}{1-\alpha^2}, \tag{2.8.26}$$

$$\text{at } \phi \leq 1-\alpha, \quad \phi = \arccos\left(\frac{T_z}{4\pi GlR_1^2}\right) - \arccos\left(\frac{T_z}{4\pi GlR_2^2}\right). \tag{2.8.27}$$

Angular (Cardan) stiffness $k_c = M/v$ when the moment, M, is acting in the plane containing axis of the bushing (Fig. 2.8.7) is [25]:

$$k_c = \pi Gl^3[0.33C_2 - 4(C_3R_1^4 + C_4R_1^{-2})R_1^2l^{-2}], \tag{2.8.28}$$

where

$$C_2 = \Delta_2 D; \ C_3 = R_1^{-2}\Delta_3 D; \ C_4 = R_1^2\Delta_4 D; \ D = 0.5[(\alpha - \alpha^{-1})^2 + (\alpha^{-2} - \alpha^2)\ln\alpha]^{-1}; \ \Delta_2 = 2(\alpha^2 - \alpha^{-2}); \ \Delta_3 = \alpha^{-2} - 1; \ \Delta_4 = \alpha^2 - 1.$$

In assembled bushings, the rubber sleeve is made separately and is oversized by external and undersized by internal surfaces, its outer radius is $R'_1 = R_1 + \Delta_1$ and its inner radius is $R'_2 = R_2 - \Delta_2$. During assembly, this rubber sleeve is pressed (with the radial preload) into the smaller annular space between the outer and inner metal sleeves having the nominal dimensions, R_1 and R_2. Another technology is to make oversized/undersized metal sleeves and reduce the size of the assembled bushing to the required one by "swagging".

Torsional stiffness of the assembled (internally preloaded) bushing about z-axis, $k_t = T_z/\phi$, is

$$k_t = \frac{4\pi GlR_1R_2R'_1 R'_2}{R_1R'_1 - R_2R'_2}.$$

(2.8.29)

Axial shear stiffness is

$$k_s = Gl \frac{2\pi(1+6A)}{\ln \alpha_1^{-1} - 0.5\ln\left[\dfrac{1+6A-2B(\alpha\lambda_2)^{-2}}{1+6A-2B\lambda_1^{-2}}\right]},$$

(2.8.30)

where

$$\alpha_1 = \alpha\frac{\lambda_2}{\lambda_1}; \quad \lambda_1 = \frac{R_1}{R'_1}; \quad \lambda_2 = \frac{R_2}{R'_2}; \quad \alpha = \frac{R'_2}{R'_1};$$

$$A = \frac{\Delta_1}{R'_1} - B; \quad B = \frac{\alpha^2}{1-\alpha^2}(\lambda_2 - \lambda_1)$$

(2.8.31)

Shear stiffness of spherical elastomeric element bonded to metal shells (angular shear, β, in Fig. 2.8.8) is

$$k_\beta = \frac{T_s}{\beta} = \frac{1}{6\pi BG}\frac{R_1^3 - R_2^3}{R_1^3R_2^3}, \quad B = \cos\alpha_1 - \cos\alpha_2 - \frac{1}{3}(\cos^3 \alpha_2 - \cos^3 \alpha_1).$$

(2.8.32)

Twist stiffness (angle, ϕ, in Fig. 2.8.8) is

$$k_\phi = \frac{T_t}{\phi} = \frac{1}{6\pi AG}\frac{R_1^3 - R_2^3}{R_1^3R_2^3}, A = \cos\alpha_1 - \cos\alpha_2 + \frac{1}{3}(\cos^3 \alpha_2 - \cos^3 \alpha_1).$$

(2.8.33)

Compression stiffness (force, P, *deformation* Δ_c) for a not very thin rubber layer

$$k_c = \frac{P}{\Delta_c} = \frac{3(R_1 - R_2)}{2\pi R_1^2 D(\cos^5 \alpha_2 - \cos^5 \alpha_1)}.$$

(2.8.34)

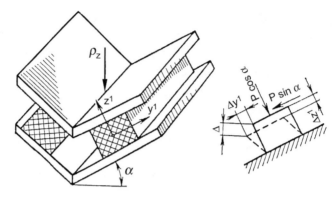

Fig. 2.8.10.

Stiffness, P/δ, of rubber coated roller in Fig. 2.8.9a acted upon by radial force, P, can be determined from the plot in Fig. 2.8.9b.

"Chevron"-shaped rubber-metal elements like the one in Fig. 2.8.10 are quite popular. Usually two rubber elements are identical, and each has compression stiffness, k_z', and shear stiffness, k_y'. Total deformation, Δ, under force, P_z, depends on compression, Δ_z', and shear, Δ_y', deformations of the rubber elements caused by the components, $P_z' = P\cos\alpha$ and $P_y' = P\sin\alpha$, respectively, of the force, $P = 0.5\,P_z$,

$$\Delta = \Delta_z'\cos\alpha + \Delta_y'\sin\alpha. \tag{2.8.35}$$

Since $\Delta_z' = P_z'/k_z'$, $\Delta_y' = P_y'/k_y'$, then

$$\Delta = \frac{P\cos^2\alpha}{k_z'} + \frac{P\sin^2\alpha}{k_y'} = P\left(\frac{\cos^2\alpha}{k_z'} + \frac{\sin^2\alpha}{k_y'}\right) = \frac{P_z}{2k_z}, \tag{2.8.36}$$

or

$$k_z = 2\left(\frac{\cos^2\alpha}{k_z'} + \frac{\sin^2\alpha}{k_y'}\right)^{-1}. \tag{2.8.37}$$

Table 2.7. Coefficients for calculating deformations of rubber sphere

x/D	α_s	β_s
0.225	0.0138	0.0007
0.293	0.024	0.002
0.367	0.039	0.004
0.452	0.063	0.010
0.553	0.101	0.025
0.684	0.175	0.083

REFERENCES

[1] Rivin, E.I., 1988, "Mechanical Design of Robots", McGraw-Hill, New York.

[2] Orlov, P.I., 1972, "Fundamentals of Machine Design", vol.1, Mashinostroenie Publ. House, Moscow, [in Russian].

[3] Stewart, D., 1965, "A Platform with Six Degrees of Freedom", Proceed. of the Institute of Mechanical Engineers, vol. 180, Part 1, No. 15, pp. 371–386.

[4] Astanin, V.O., Sergienko, V.M., 1993, "Study of Machine Tool of Non-Traditional Configuration", Stanki i instrument, No. 3, pp. 5–8 [in Russian].

[5] Liu, X.-J., Wang, J., Pritschow, G., 2006, "Kinematics, Singularity and Workspace of Planar 5R symmetrical Parallel Mechanisms", Mechanism and Machine Theory, vol. 41, No. 2, pp. 145–169.

[6] Weck, M., Staimer, D., 2002, "Parallel Kinematics Machine Tools — Current State and Future Potentials", Annals of the CIRP, vol. 51/2, pp. 1–13.

[7] Rivin, E.I., "Joint", U.S. Patent 6,588,967.

[8] Rivin, E.I., "Universal Cardan Joint with Elastomeric Bearings", U.S. Patent 6,926,611.

[9] Brumson, B., "Parallel Kinematics Robots", http://www.roboticsonline.com/public/articles/archivedetails.cfm?id=797.

[10] Plotnikov, P.N., Klimanov, V.I., Brodov, Yu.M., Kuptsov, V.K., 1983, "Stength and Vibratory Characteristics of Profiled Twisted Tubes", Teplotekhnika, No. 6, pp. 68–71 [in Russian].

[11] Timoshenko, S.P., Gere, J.M., 1972, "Mechanics of Materials", Van Nostrand Reinhold Co., N.Y.

[12] Gibson, L.J., Ashby, M.F., 1999, "Cellular Bodies – Structure and Properties", Cambridge University Press.

[13] Neugebauer, R., Hipke, T., "Machine Tool with Metal Foam", 2006, Advanced Engineering Materials, vol. 8, No. 9, pp. 858–863.

[14] DenHartog, J.P., "Advanced Strength of Materials", Dover Publ., 1987, 387 pp.

[15] Kaminskaya, V.V., "Load-Carrying Structures of Machine Tools", in Components and Mechanisms of Machine Tools, ed. By D.N.Reshetov, Mashinostroenie, Moscow, 1973, vol.1, pp. 439-562 [in Russian].

[16] Rivin, E.I., "Passive Vibration Isolation", 2003, ASME Press, 426 pp.

[17] Ringfeder® Friction Springs in Mechanical Engineering, Ringfeder Corp., Catalog R60.

[18] Rivin, E.I., Sayal, G., Johal, P.R.S., "Giant Superelasticity Effect in NiTi Superelastic Materials and Its Applications", *J. of Materials in Civil Engineering*, 2006, December, vol. 18, No. 6, pp. 851–857.

[19] Hodgson, D.E., "Shape Memory Alloys", 1988, at *www.sma-inc. com/SMAPaper.html*.

[20] Rivin, E.I., "Precision Compensators Using Giant Superelasticity Effect", 2007, Annals of the CIRP, vol. 56/1/2007.

[21] Rivin, E.I., "Mechanical Contact Connection", U.S. Patent 6,779,955.

[22] Snowdon, J.C., "Rubberlike Materials, Their Internal Damping and Role in Vibration Isolation", *J. of Sound and Vibration*, 1965, vol. 2, pp. 175–193.

[23] Freakley, P.K., Payne, A.R., "Theory and Practice of Engineering with Rubber", *Applied Science Publishers*, L., 1978.

[24] Lavendel, E.E., Elastic Elements of Vibratory Machinery, in "Vibrations in Engineering", 1981, Mashinostroenie Publish. House, vol. 4, pp. 187–222, [in Russian].

[25] Lavendel, E.E., "Computational Analysis of Industrial Rubber Products", 1976, Mashinostroenie Publish. House, Moscow, 232 pp [in Russian].

CHAPTER

3

Nonlinear and Variable Stiffness Systems; Preloading

3.1 DEFINITIONS

Since stiffness is the ratio of the force to the displacement caused by this force, the load-deflection plot (characteristic) allows to determine stiffness as a function of force or displacement. It is much easier to analyze both static and dynamic structural problems if the displacements are proportional to the forces that caused them, i.e., if the load-deflection characteristic is *linear*. However, most of the load-deflection characteristics of real-life mechanical systems are nonlinear. In many cases, the degree of nonlinearity is not very significant, and the system is considered as linear for the sake of simplicity. While a significant nonlinearity must be always considered in the analysis, even a weak nonlinearity may noticeably change behavior of the system, especially for analysis of dynamic processes in which nonlinearity may cause very specific important, and frequently undesirable, effects, e.g., see [1]. At the same time, there are many cases when the nonlinearity may play a useful role by allowing adjustment of stiffness parameters of mechanical systems.

There are two basic types of nonlinear load-deflection characteristics as presented in Fig. 3.1.1. Line 1 represents the case when the rate of increase of deflection x slows down with increasing force, P. If the local (differential) stiffness is defined as ratio between increments of force (ΔP) and deflection (Δx),

$$k = \frac{\Delta P}{\Delta x}, \tag{3.1.1}$$

then the stiffness along the line 1 is increasing with the increasing load,

$$k_1' = \frac{\Delta P_1'}{\Delta x_1'} < k_1'' = \frac{\Delta P_1''}{\Delta x_1''}. \tag{3.1.2}$$

Such characteristic is called *hardening load-deflection characteristic*.

The rate of increase of deflection x accelerates with increasing force P along line 2, thus the local stiffness along line 2 is decreasing with the increasing load,

$$k_2' = \frac{\Delta P_2'}{\Delta x_2'} > k_2'' = \frac{\Delta P_2''}{\Delta x_2''}. \tag{3.1.3}$$

Such characteristic is called *softening load-deflection characteristic*.

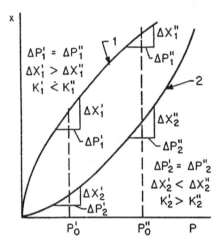

Fig. 3.1.1. Hardening (1) and softening (2) nonlinear load-deflection characteristics.

Consideration of nonlinearity is especially important for dynamic analysis of mechanical systems. If for linear systems natural frequencies do not depend on amplitudes of vibratory displacements, resonance frequencies of nonlinear hardening systems are increasing with increasing vibration amplitudes and for nonlinear softening systems are decreasing with increasing vibration amplitudes [1]. Such important dynamic effects as subharmonic and superharmonic resonances may be misunderstood if not correctly represented in the course of analysis of a nonlinear system.

Both types of nonlinear load-deflection characteristics allow for varying the actual stiffness by moving the *working point* along the characteristic. This can be achieved by applying a *preload force* to the system, which is independent from actual process forces (payload). If the preload force is not constant but changing with changing operational conditions of the system, then there is a potential for creating a system with *controlled, variable,* or *self-adaptive stiffness*. Such adjustability is very useful for assembling mechanical structures from components with preloaded connections (see Section 3.2.1 below).

3.1.1 "Constant Natural Frequency" Nonlinear Characteristic

A progressive hardening nonlinear characteristic with stiffness in the deforming force direction proportional to the magnitude of this force

["Constant Natural Frequency (CNF) Characteristic"] has special importance, especially for vibration isolators, since it is associated with their improved performance and reduced production costs [2]. If an object having weight, W, is mounted on flexible elements (springs), the natural frequency, f_n, of a mass, $(m = W/g)$ - spring (k), system is

$$f_n = \frac{1}{2\pi}\sqrt{\frac{k}{m}} = \frac{1}{2\pi}\sqrt{\frac{kg}{W}} \qquad (3.1.4)$$

where g = acceleration of gravity. To assure that f_n = *constant* for any weight, stiffness, k, must increase proportionally to the weight load W on the element. It represents a nonlinear elastic element with *a special hardening characteristic* for which

$$\Delta P/\Delta x = k = AW \qquad (3.1.5)$$

where A is a constant. The load range within which the Eq. (3.1.5) holds is described by ratio of the maximum W_{max} and minimum W_{min} weight loads of this range. Nonlinear load deflection characteristics of CNF flexible elements result in their low sensitivity to inevitable variations of their dimensions (as well as durometer, see Appendix 4, for rubber elements) within the batch. Low sensitivity to production tolerances is a unique feature of CNF isolators regardless of design features of their flexible elements, such as coil springs or rubber flexible elements, etc. The rated (nominal) natural frequency of a CNF flexible element is determined by its geometry. For small loads on the element, $W < W_{min}$, the flexible element usually can be assumed to be linear. The CNF characteristic starts at W_{min}, where deformation of the flexible element is Δ_{min}. Since at this point the flexible element can still be considered as linear, the natural frequency in the weight load application direction is:

$$f_{z_o} = \frac{1}{2\pi}\sqrt{\frac{k_z}{m}} = \frac{1}{2\pi}\sqrt{\frac{k_z g}{W_{min}}} = \frac{1}{2\pi}\sqrt{\frac{g}{\Delta_{min}}}, \qquad (3.1.6)$$

where k_z = stiffness of the flexible element at the linear (constant stiffness) segment of its load-deflection characteristic. When the weight load is increasing within the CNF load range $W_{min} - W_{max}$, the natural frequency is practically constant and is equal to

$$f_{z_{nom}} \approx f_{z_o}. \qquad (3.1.7)$$

Thus, the nominal natural frequency associated with the CNF isolator is determined by deformation Δ_{min} of the flexible element at which the load-deflection characteristic becomes nonlinear of the CNF type.

Elastic elements with CNF characteristic can be used as springs with hardening nonlinear load-deflection characteristic or as CNF mounting elements. In the latter case, an interesting property of such elements is their low sensitivity to production uncertainties [2]. If a CNF mount is manufactured without any deviations from the nominal design dimensions (coil diameters for a mount using coil springs, see Section 3.2.2.a below or shape of rubber elements as described in Section 3.3 below), the load-natural frequency characteristic is presented by line 1 in Fig. 3.1.2. At low weight loads, the spring isolator has a linear load deflection characteristic (e.g., no touching between the coils), and its natural frequency, f_z, is decreasing with the increasing load. As the deformation of the spring reaches Δ_{min} (at the load, P_{min}, when the large diameter coils start touching), the CNF region begins. If the spring is wound on a slightly larger mandrel, the spring would be softer, and its linear region is represented by line 2 indicating lower values of f_z for the same load than for the linear region of line 1. The deformation Δ_{min} occurs at a lower weight load P'_{min}; the CNF characteristic thus begins at a lower load but *at the same* Δ_{min} and $f_{z_0} = f_{z_{nom}}$ determined by the value of Δ_{min}. The

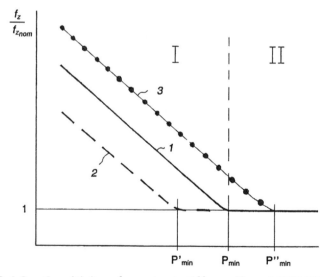

Fig. 3.1.2. Sensitivity of constant stiffness (I) and CNF (II) elastic elements to production variations.

deviation of the shape of the elastic element, e.g., generatrix of the spring from its required shape would, of course, result in deviations of the line, $f_z/f_{z_{nom}}$, from line 1. Similarly, for a smaller mandrel, the spring would be stiffer, its linear region represented by line 3 would indicate higher f_z values than line 1, and the CNF characteristic would begin at a higher weight load, P''_{min}, but again *at the same* Δ_{min} and $f_{z_0} = f_{z_{nom}}$ determined by the value of Δ_{min}.

For CNF isolators with rubber flexible elements, their dimensions are determined by the mold and usually are quite consistent. However, due to inevitable inconsistencies of the rubber blending components and their composition, as well as curing parameters (such as temperature and time), hardness (durometer) of the rubber part can vary within ±5 units of durometer scale (±17% in stiffness) (Appendix 4). Again, line 1 represents the isolator with a "perfect" (nominal) rubber durometer, line 2 — with the reduced durometer and line 3 — with the elevated durometer. For all these variations, the natural frequency remains the same while the weight load range shifts.

The shifts of the load range for the CNF isolators are not significant for most applications, since the load range is usually very broad ($P_{max}/P_{min} \approx$ 1.5:1 to 2:1 for spring CNF isolators, $P_{max}/P_{min} = $ 5:1 to 20:1 for CNF isolators with rubber flexible elements. However, P''_{min}/P'_{min} usually does not exceed ~1.35; where P_{max} is the maximum allowable weight load on the isolator. This conclusion about robustness of the natural frequency values provided by CNF vibration isolators was validated by monitoring mass-produced (~700,000 units a year) vibration isolators for 2 years of production. Variation of the natural frequency was not exceeding ±3%, while variation of the rubber durometer was observed within ±7 units of durometer (stiffness variation ±25%). The observed small variation of the natural frequency could also be explained by variation of K_{dyn} for the rubber flexible elements due to production tolerances.

3.2 EMBODIMENTS OF MECHANICAL ELEMENTS WITH NONLINEAR STIFFNESS

Nonlinear stiffness is specific for:

1. *Elastic deformations of parts whose material is not exactly described by the Hooke's Law.*

2. *Contact deformations.* Joints between mechanical components loaded perpendicularly to the contact surfaces are characterized by a hardening nonlinearity due to increase of effective contact area with the increasing load. Tangential contact deformations may exhibit a softening nonlinearity. Characteristics of contact deformations are described in Chapter 4.

3. Changing *part/system geometry* due to a special design to obtain variable stiffness or due to deformations. Typical examples of components having nonlinear stiffness due to changing geometry are:
   a. Coil springs with variable pitch and/or variable coil diameter and/or variable wire diameter;
   b. Belleville springs;
   c. Elastic elements with contact surfaces changing with load, e.g., compressed spherical and radially compressed cylindrical components;
   d. Rubber elements with built-in constraining;
   e. Thin-layered rubber-metal laminates loaded in compression.

3.2.1 Material-Related Nonlinearity

There are many materials which exhibit nonlinear deformation characteristics. Compression deformations of components made from cast iron or concrete are characterized by softening load-deflection characteristics. Although the basic rubber components described in Chapter 2 are quasi-linear, their stiffness is increasing with load (hardening) if the component is loaded in compression and is slightly decreasing with load (softening) if the component is loaded in shear (e.g., see Appendix 4).

Deformations or fibrous mesh-like materials are caused by slippages in contacts between the fibers and by bending of the fibers (natural or synthetic polymer fibers, steel, bronze, etc., wires). Static stiffness of the fibrous mesh components in compression is, approximately, proportional to the compression load, like in constant natural frequency vibration isolators described in Sections 3.1.1 and 3.3.1. Dynamic stiffness k_{dyn} is much higher than static stiffness k_{st},

$$k_{dyn} = K_{dyn} k_{st}, \qquad (3.2.1)$$

where K_{dyn} = 1 to ~10 is the dynamic stiffness factor. Slippages between the wires/fibers and, thus, support conditions of the bending fibers, depend on vibration amplitudes. The number and the intensity of slippages increase with increasing vibration amplitudes. Accordingly, both K_{dyn} and k_{dyn} strongly depend on amplitude of vibrations as shown in Fig. 3.2.1 [2]. In a vibratory system "mass, m - nonlinear spring, k," the effective stiffness (thus, the *natural frequency*) is increasing with increasing amplitudes of excitation for hardening nonlinearity of the spring and decreasing with increasing amplitudes for softening nonlinearity. But Fig. 3.2.1 shows that dynamic (vibratory) stiffness of mesh-like materials is decreasing with increasing amplitudes (*softening* nonlinearity), while for the static loading, the nonlinearity is of the *hardening* type. Thus, the fibrous mesh-like materials have *dual nonlinearity*. Fig. 3.2.1 also shows amplitude dependency of internal damping (log decrement) of the mesh-like materials.

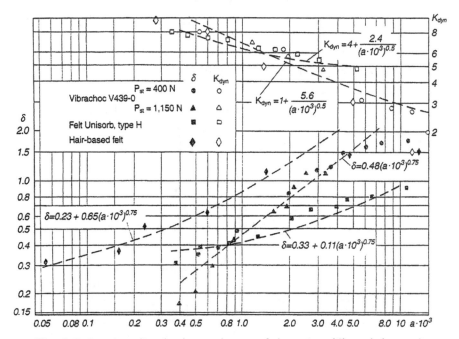

Fig. 3.2.1. Amplitude dependence of damping (δ) and dynamic stiffness coefficient (K_{dyn}) for mesh-like materials. ○ — wire mesh, low specific load; Δ — same, high specific load; □ — thin-fiber felt; ◊ — thick-fiber felt; clear symbols — K_{dyn}; solid symbols — δ.

A special case of materials with elastic characteristics not described by Hooke's Law are volumetrically incompressible elastomeric materials (Appendix 4). Compression deformation of thin elastomeric layers bonded to metal end plates is associated with high hydrostatic pressures resulting in changing of the effective volumetric (bulk) and shear moduli during the deformation process.

3.2.2 Geometry-Related Nonlinearity

3.2.2a Coil and Leaf Springs

Modified nonlinear coil springs are frequently used as statically nonlinear metal structural components. There are several modifications of nonlinear (compression) coil springs:

1. Springs coiled from constant cross-section wire but having non-cylindrical shapes, so that coil diameter is not constant along the length (e.g., Fig. 3.2.2a,b). Usually, such springs have a conical shape as in Fig. 3.2.2a, a "barrel" shape, or a "corset" shape as in Fig. 3.2.2b. The largest diameter coil is the softest, and at the predetermined load its deformation is so large that it touches one of the bases for a spring like the ones in Fig. 3.2.3a, b or another (adjacent) coil for a barrel-shaped spring, thus reducing the effective n and increasing stiffness. If the generatrix of the nonlinear cone is described by an exponential function, then the spring has the CNF characteristic, whereas its axial stiffness is proportional to the axial load in a specified load interval.

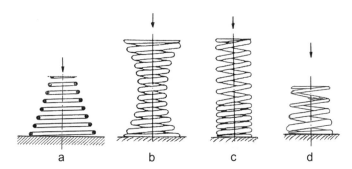

Fig. 3.2.2. Coil springs with nonlinear characteristics. a — conical spring with constant pitch; b — hourglass (corset) spring; c — cylindrical variable pitch spring; d — cylindrical spring coiled from variable diameter (tapered) wire.

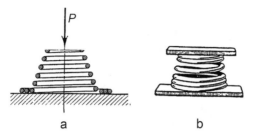

a   b

Fig. 3.2.3. Deformation patterns of conical springs.

2. Cylindrical springs coiled from constant cross-section wire but having variable pitch along their length (Fig. 3.2.2c). At a predetermined compressive force, the coils in the minimum pitch area touch each other, thus reducing the effective number n of coils and increasing stiffness of the spring.
3. Cylindrical springs with constant pitch coiled from variable cross-section ("tapered") wire (Fig. 3.2.2d). The coil made of the smallest cross-section wire segment is the softest, and at the predetermined load, it touches the adjacent coil thus increasing stiffness of the spring.
4. Combination of the above designs.

Cylindrical springs with variable pitch have different shear stresses for different coils, thus resulting in their larger size and weight. Coil springs made from tapered wire are also more bulky. Generally, variable stiffness cylindrical coil springs may develop noise and wear when the coils contact each other under intense vibration. Non-cylindrical coil springs designed in such a shape that the low stiffness (large diameter) coils are deformed until they contact not the adjacent coils but the supporting surface(s), like in Fig. 3.2.3a, usually have lower height, reduced weight, reduced contact pressures, and lower noise and wear.

Fig. 3.2.4 shows a nonlinear *torsional* coil spring 1. The applied torque causes twisting and reduction of its diameter, and variable diameter core 2 allows to change the number of active coils while the torque is increasing, thus creating a hardening characteristic "torque-twist angle". This is a typical example of a flexible element whose contact surfaces with other structural components are changing with increasing load. A similar concept is used in a nonlinear translational flat (leaf) spring shown in Fig. 3.2.5, whose effective length is decreasing (and, thus,

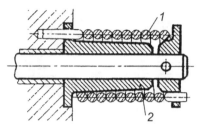

Fig. 3.2.4. Nonlinear hardening torsion spring.

stiffness is increasing) with increasing deformation. Another example of achieving nonlinear load-deflection characteristic by changing contact surfaces is presented by deformable bodies with curvilinear external surfaces. Frequently, rubber elastic elements are designed in such shapes, which result in changing their "footprint" with changing load (see Section 3.4).

3.2.2b Belleville (Disc) Springs

Belleville spring (Fig. 3.2.6) usually has $\alpha = 2$ deg. to 6 deg., $D/d = 2$ to 3. These springs are providing high load-carrying capacity in a small outline. Their load capacity, stiffness, and deformations in the axial direction can be easily adjusted by connecting two or several springs (the lateral stiffness, $k_{x,y}$ is very high). The series connection (Fig. 3.2.7a) provides reduced stiffness/increased deformation in proportion to the number of springs but without changing of the load rating. The parallel connection (Fig. 3.2.7b), results in increase of both rated load and stiffness with increasing number of stacked springs, but without changing the maximum deflection. A wide variety of stiffness/deflection/load characteristics can be obtained by combining the above stacking patterns (Fig. 3.2.7c). The Belleville springs provide significant damping, since their deformation s is accompanied by sliding between the spring and its support surface

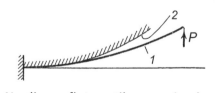

Fig. 3.2.5. Nonlinear flat cantilever spring having a shaped support surface.

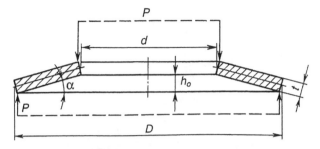

Fig. 3.2.6. Belleville spring.

and between the springs in the stacks as in Fig. 3.2.7, especially in the configurations of Fig. 3.2.7b,c. In the configuration of Fig. 3.2.7a, the amount of friction (and damping) is reduced to friction between the lowest spring and the supporting structure. For precision applications, influence of friction on load-deflection characteristics should be considered [3]. Belleville springs are produced in a huge variety of standardized dimensions and are easily available "off the shelf".

The load (P)–deflection (s) characteristic of Belleville springs is nonlinear,

$$P = \frac{\pi E t}{6(D-d)^2} s \left[\left(h_o - \frac{s}{2} \right)(h_o - s) + t^2 \right] \ln \frac{D}{d} \qquad (3.2.2)$$

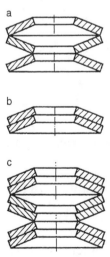

Fig. 3.2.7. Various assemblies of Belleville springs.

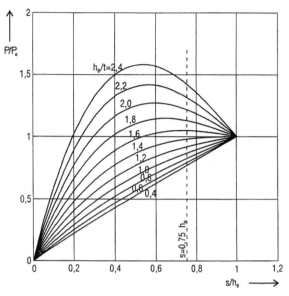

Fig. 3.2.8. Dimensionless load/deflection plots of Belleville springs.

As it can be seen in Fig. 3.2.8, where P_o = the load at which the spring is flattened ($s = h_o$), the character of nonlinearity is changing in a wide range depending on the spring dimensional parameters. It may have linear parts, "hardening" nonlinear parts (stiffness increasing with increasing deformation), and "softening" nonlinear parts (stiffness decreasing, while deformation is increasing). At $h_o/t >$ ~1.5, there even can be areas with quasi-zero stiffness.

The maximum normal stresses develop in the meridian cross-section of the tapered shell at its internal edge,

$$\sigma_{max} = \frac{4Es}{kD^2}(h_ok_o - sk_1 + t), \qquad (3.2.3)$$

where coefficients k, k_o, k_1 are taken from Fig. 3.2.9.

3.2.3 Nonlinear Spring Elements with Softening Nonlinear Characteristics

Examples described above mostly represent mechanical systems with hardening load-deflection characteristics. The only exception is Belleville spring whose load-deflection characteristic described by Eq. (3.2.1) may

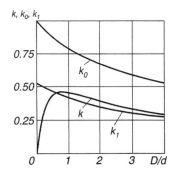

Fig. 3.2.9. Coefficients for computing characteristics of Belleville springs.

have some softening areas at certain combinations of the design parameters. It is more difficult to obtain a *softening load-deflection characteristic* in a mechanical system than to obtain a hardening characteristic. Usually, some ingenious design "tricks" are required. Fig. 3.2.10a shows such a specially designed device; its load-deflection characteristic is shown in Fig. 3.2.10b. The system consists of linear spring 1 and nonlinear spring 2 having a hardening load-deflection characteristic. These springs are precompressed by rod (drawbar) 3. Stiffness k of the set of springs acted upon by force P is the sum of stiffness values of spring 1 (k_1) and spring 2 (k_2), $k = k_1 + k_2$. With increasing load, P, deformation of nonlinear spring 2 is diminishing, and accordingly, its stiffness is decreasing. Since stiffness, k_1, is constant, the total stiffness is also

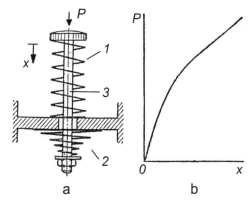

Fig. 3.2.10. Spring assembly with softening load-deflection characteristic.

decreasing as shown in Fig. 3.2.10b. This process continues until spring 2 is completely unloaded, after which event the system becomes linear, $k = k_1$.

Another design direction for realizing nonlinear systems with the softening load-deflection characteristic is by using thin-walled elastic systems that are usually capable of having two or several elastically stable configurations, i.e., capable of collapsing into the fundamentally stable configuration. Fig. 1.3.3b,c show stages of compression of a thin-walled cylinder in Fig. 1.3.3a. The first stable configuration (Fig. 1.3.3b) is characterized by the softening load-deflection characteristic 1 in Fig. 1.3.3d; the second stable configuration (Fig. 1.3.3c) has the hardening load-deflection characteristic 2, and there is a sizable segment of "negative stiffness" describing the collapsing process of the cylinder. The realizable load-deflection characteristic is shown by the solid line; the dimensionless coordinates in Fig. 1.3.3d are $\alpha = \Delta/R$ and $\beta = 12 \, PR^2/E\delta^3$.

The softening load-deflection characteristic is also exhibited by superelastic materials (Fig. 2.7.8 in Section 2.7.4).

Neutral (zero stiffness) load-deflection characteristics, wherein the deflection is increasing while the force is increasing very little, are beneficial for shock absorption systems, such as packaging systems. Such characteristics are often developing in elastically unstable (buckling) systems. Fig. 3.2.11a shows one design of a rubber packaging mat [4]. The load-deflection characteristics for two modifications of this design are shown in Fig. 3.2.11b. These characteristics are quasi-linear at low loads/ strains. At relative compression 0.15 to 0.2, there is an inflection of the plot, which becomes parallel to the strain axis (increase in strain occurs at the quasi-constant stress).

Some additional examples of structural elements with softening and/ or neutral (zero stiffness) load deflection characteristics and their practical applications are described in Sections 1.3.1 and 8.5.

Practical Case. Fig. 3.2.12 shows a somewhat different embodiment of a spring system with the softening load-deflection characteristic, which is used as a punch force simulator for evaluation dynamic and noise-radiation characteristics of stamping presses [5]. In this design, Belleville springs 2 are preloaded by calibrated (or instrumented) bolts 1 between cover plates 3 and 4 to the specified load P_s. The simulator is installed instead of the die on the press bolster. During the downward travel of the press ram, it contacts the head 6 of the simulator, and on its further

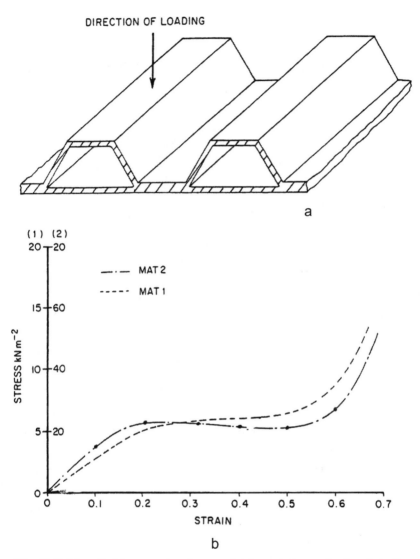

Fig. 3.2.11. Rubber packaging mat (a) and stress-strain plots for rubber packaging mats.

travel down, unloads the bolts from the spring-generated force. As the bolts are unloading, the ram is gradually loaded. This process of the press ram being exposed to the full spring load is accomplished when the ram travels a distance equal to the initial deformation of the bolts caused by their preloading to the force, P_s This initial deformation is very small (*high stiffness segment*) and can be adjusted by changing length/

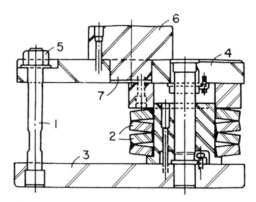

Fig. 3.2.12. Pulse load simulator.

cross-section of the bolts. After this process is complete, the ram is further compressing the springs (*low stiffness segment*). Thus, this device having the softening load-deflection characteristic provides simulation of the punching process consisting of a very intense loading during the "breakthrough" process and much less intense loading afterwards.

3.3 STATICALLY NONLINEAR RUBBER ELEMENTS

Although, as it is stated in Appendix 4, rubber elements comply with Hooke's Law in simple tension/compression tests up to much larger strains than metal elements, while quasi-linear deformations (up to 10% to 15% in tension and 3% to 5% in compression) are typical, there is nonlinearity in shear (Fig. A4.4 in Appendix 4), and in many cases, material- and geometry-related nonlinearities are intertwined. Load-deflection characteristics of rubber blocks bonded to metal plates are addressed in Section 2.7, they also have linear load-deflection characteristics up to significant deflection magnitudes. However, a judicious utilization of volumetric incompressibility of elastomers, with their Poisson's ratio being very close to 0.5 (see Appendix 4), allows to realize a great variety of desirable nonlinear load-deflection characteristics, both hardening and softening, in structural rubber components. The design techniques for designing nonlinear rubber components use shape sensitivity of the rubber deformation characteristics. While typical nonlinear rubber elements exhibit nonlinearity in the course of large deformations (relative to dimensions of the element), in some cases, a nonlinear load-deflection characteristic starts from small deflections is desirable.

The addressed in this section design techniques for nonlinear rubber elements include the following:

1. Use of constraints controlling bulging of the compressed elements;
2. Use of streamlined elements changing their contact conditions during the deformation process;
3. Using rubber elements with large shape factors (thin layer laminates), so that compression develops due to hydrostatic processes;
4. Using potentially unstable structures in order to obtain softening nonlinearities (see Section 3.3.3).

3.3.1 Compressed Elements with Controlled Bulging

This technique uses such designs of rubber elastic elements, in which the free surface is decreasing due to its gradual constraining in the process of compression by rigid components attached to the end metal plates thus resulting in increasing compression stiffness. Line I in Fig. 3.3.1 is an axial load-deflection characteristic of the cylindrical rubber element bonded to metal end plates and whose free side surface is maximum possible, like in Fig. 3.3.2a. This plot represents a lower stiffness limit for the synthesized nonlinear load-deflection characteristic. Line II in Fig. 3.3.1 represents the load-deflection characteristic of the same rubber element with all free side surface eliminated so that the elastic element is in the condition of hydrostatic compression under a compression force, the upper limit for the load-deflection characteristics is stiffer than plot I by up to 1.5 to 2 decimal orders of magnitude. The plot III is a desirable nonlinear load-deflection characteristic. Dimensions R and h of the rubber element with unrestricted free surface (Fig. 3.3.2a) are assigned so that the linear load-deflection plot of the element at small compression loads up to P_0 coincides with the first segment of the desirable plot, up to deflection Δ_0. These dimensions can be determined from expressions in Table 2.4.

The shape (bulging) of the free surface is expressed as [6]

$$u = 3\Delta\rho_0\left(\frac{1}{4} - \frac{z^2}{h_0^2}\right), \tag{3.3.1}$$

where z is vertical coordinate with the origin in the midpoint of the element z-axis. The second loading increment results in additional deformation

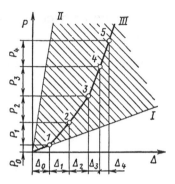

Fig. 3.3.1. Design sequence for synthesis of a nonlinear rubber flexible element.

Δ_1 under the force increment P_t. The higher stiffness is due to the reduced height, h_t, and increased radius, $R + u_t$, of the part of the element, which still has a free surface. Volume II in Fig. 3.3.2b can only deform due to its volumetric (hydrostatic) compression, but in the first approximation, the rubber can be considered as incompressible. This assumption allows for finding coordinates of point 1 on the lids of the end plates (Fig. 3.3.2b) by solving equation [6]

$$\rho_1(0.5\rho_1^2 + 0.92) = \frac{P_1}{3\Delta_1 \pi G R} \qquad (3.3.2)$$

for $\rho_1 = R/h_t$ to determine h_t. Point 1 is located at the distance $(h - h_t)/2$ from the flat surface of each end plate. In the horizontal direction, point

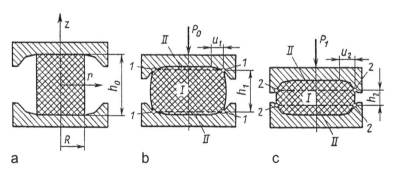

Fig. 3.3.2. Stages of deformation of the nonlinear rubber flexible element.

1 is at the distance u_1, from the rubber/metal corner of the element in Fig. 3.4.2,

$$u_1 = \frac{3\Delta_o \rho_o}{4}\left[1 - \left(\frac{h_1}{h_o}\right)^2\right]. \qquad (3.3.3)$$

By solving for ρ_2 in the equation,

$$\rho_2(0.5\rho_2^2 + 0.92) = \frac{P_2}{3\Delta_2 \pi G(R + u_1)}, \qquad (3.3.4)$$

h_2 is determined, and then from

$$u_2 = \frac{3\Delta_o \rho_o}{4}\left[1 - \left(\frac{h_1}{h_o}\right)^2\right] + \frac{3\Delta_1 \rho_1}{4}\left[1 - \left(\frac{h_2}{h_1}\right)^2\right] \qquad (3.3.5)$$

the second coordinate of the next point of the side barrier can be found, etc.

This algorithm was developed after a CNF vibration isolator for industrial machinery was empirically designed using this concept and successfully implemented in mass production (Fig. 3.3.3 [7], also see Catalog V105 at www.vibrationmounts.com). The flexible element is a monolithic rubber block molded in a relatively simple mold cavity, but the core of this element comprises two quasi-independent rubber rings 1 and 2.

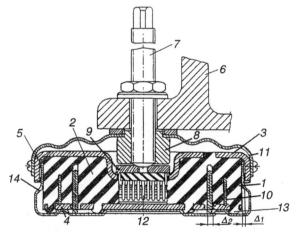

Fig. 3.3.3. CNF vibration isolator OB-31.

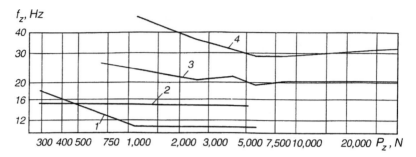

Fig. 3.3.4. Load-natural frequency characteristics of a line of CNF vibration isolators.

These rings are bonded to top 3 and bottom 4 metal covers. Top cover 3 has lid 5 so that there is a calibrated clearance (gap) Δ_1 between the inner surface of lid 5 and the outer surface of rubber ring 1. Another calibrated gap Δ_2 is between the inner surface of ring 1 and the outer surface of ring 2. The weight load from object 6 is transmitted by leveling nut 8 to leveling bolt 7, and by washer/flange 9 to top cover 3 and to the rubber block. At low weight loads, all four side surfaces of rings 1 and 2 are free to bulge thus resulting in the shape factor $S = {\sim}0.5$. With the increasing weight load, the bulge on the outer surface of ring 1 touches lid 5, and the bulge on the inner surface of ring 1 contacts the bulge on the outer surface of ring 2. With the increasing weight load, the gaps Δ_1 and Δ_2 are gradually filled with the bulging rubber, thus increasing S and stiffness

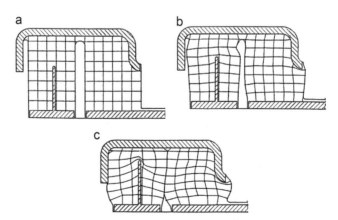

Fig. 3.3.5. Measured deformation patterns of CNF isolator in Fig. 3.3.4; a No weight applied; b Low weight load; c Intermediate weight load.

of the isolator. Plots in Fig. 3.4.4 illustrate load-natural frequency characteristics of several commercially realized CNF isolators of such design. It can be seen that the constant natural frequency (i.e., proportionality of stiffness to the weight load) was realized in the range of loads W_{maa} – W_{min}, with the ratio W_{maa}/W_{min} as great as 15 to 20:1 for some models.

Fig. 3.3.5 shows actually measured deformation patterns of the rubber flexible element of the CNF isolator in Fig. 3.3.4, similar but very different from the theoretical patterns in Fig. 3.4.2.

Bottom cover 4 in Fig. 3.3.4 has thin metal ring/rib 10 attached (welded) to it. Presence of ring 10 provides resistance of the flexible element to lateral forces, especially due to compression of rubber between ring 10 and lid 5. Presence of this rib results in a relatively constant and low ratio between the vertical (k_z) and horizontal ($k_{x,y}$) stiffness values, $\eta_{x,y} = k_z/k_{x,y} = 2.5 \pm 20\%$ across the W_{max} – W_{min} load range.

3.3.2 Streamlined Nonlinear Rubber Flexible Elements

Bonded elastomeric elements loaded in compression often exhibit intense stress concentrations, which limit their durability under intense dynamic loads, limit maximum static deformations, increase creep rates. For example, Fig. 3.3.6 [8] shows stress distribution for an axially compressed rubber cylinder whose diameter D is equal to length L for relative compressions $\varepsilon = 15\%$ (Fig. 3.3.6a) and $\varepsilon = 30\%$ (Fig. 3.3.6b). Performance characteristics of flexible rubber structural elements can be dramatically improved and their desirable nonlinear load-deflection characteristics realized by using "streamlined" or "ideal shape" elastomeric elements, such as radially loaded cylinders, spheres, toruses (O-rings), ellipsoids loaded in compression. The nonlinear load-deflection characteristic develops, since an incremental compression causes increasing loaded area, reduced bulging area, and reduced thickness (e.g., Fig. 3.3.7). These factors result in increasing shape factor (see Appendix 4). Radial compression of the streamlined elements is associated with very uniform patterns of stress (Fig. 3.3.8) and strain distribution. There are no areas with high stress, typical for loading of the bonded rubber elements, as in Fig. 3.3.6. Figs. 3.3.6 and 3.3.8 [8] are illustrating deformation of the same cylindrical specimen. However, the maximum tensile stresses are about five times lower for 15% compression and four times lower at 30% compression for the radially loaded specimen. Due

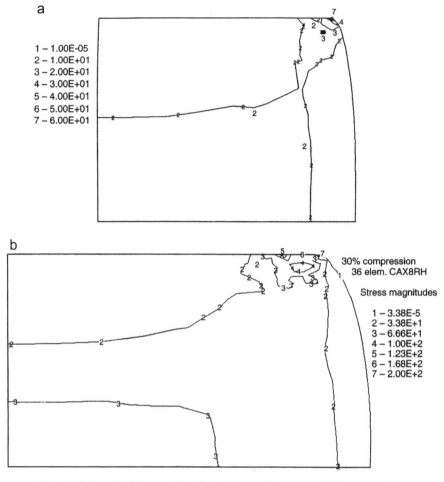

Fig. 3.3.6. Rubber roller (*D* = 60 mm) with grid lines on the face surface marked before loading; a-Undeformed; b-Loaded to 50% compression.

to the combination of the lower maximum stresses and the more uniform stress distribution, much higher relative compression values, up to and exceeding 40% can be tolerated for the radially loaded elements versus 10% to 15% for identical but bonded and/or axially loaded elements. Tests demonstrated a good durability under repeated compression of a rubber sphere cyclically loaded to 50% compression. Tests in [9] demonstrated much lower heat generation in rubber cylinders cyclically radially compressed by dynamic load (frequency 10 to 20 Hz) compared

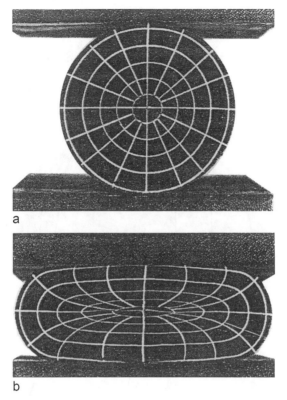

Fig. 3.3.7. Stress distribution in cylindrical rubber element ($D = L$) loaded in axial compression between flat loading surfaces and bonded to the loading surfaces; (a) – 15% compression; (b) – 30% compression.

with axially compressed cylinders with the same dynamic loading and with simulated bonding conditions.

It was demonstrated experimentally and analytically [8, 10] that creep rates of two identical rubber cylinders, one loaded axially and another radially, are different, with the axial loading of a bonded specimen having the creep rates 20% to 40% higher than the radial loading. In addition to reduced maximum stresses, the stress distribution in the case of radial compression is much more uniform than in axial compression of the same element bonded to the end plates. The surface layer of the cylinder circumference suffers practically no stress, while the internal circle undergoes high tension. The area subjected to the highest stress in tension is enveloped by layers, like bandages, which are less and less stressed

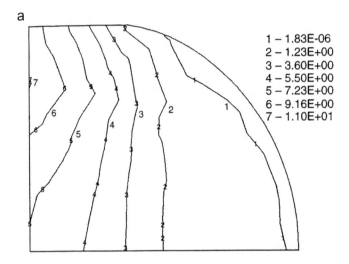

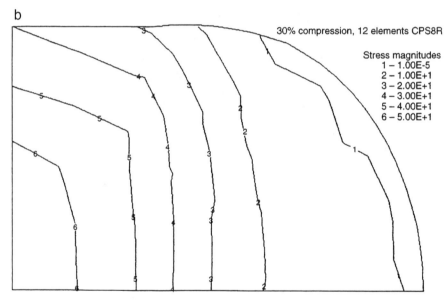

Fig. 3.3.8. Stress distribution in cylindrical rubber element ($D = L$) loaded in radial compression between flat loading surfaces; a 15% compression; b 30% compression.

the closer they are to the periphery (e.g., Figs. 3.3.7, 3.3.8). Thus, thanks to the practically non-existent tension on the surface, the streamlined element is insensitive to dirt, less sensitive to oil and ozone, etc., thus explaining its high robustness. Introduction of surface interruptions in a streamlined rubber element or its bonding to rigid components, disturbs the "natural" deformation process.

The ultimate advantages of using ideal shape flexible structural elements materialize when these elements are not bonded to rigid supporting or constraining components, since the bonding is associated with creating stress concentrations. The packaging problem was solved in [11] by embedding all the components in a foam matrix, e.g., as shown in Fig. 3.3.9. The foam maintains the relative positions of all components but does not change noticeably their load-deflection characteristics. It also makes interaction between the ideal shape rubber elements and the rigid components more consistent, since it eliminates friction between the rubber elements and the rigid components and replaces it by elastic deformation of the foam particles. The external shape of the foam matrix can be easily tailored to conform with the surrounding structural systems, such as a suspended/supported object, a foundation, etc. The foam can also provide protection of the rubber elements from aggressive environments, e.g., from oil in case of natural rubber components installed in an oil-mist environment.

The nonlinearity of the streamlined elastomeric elements due to continuous increase of the shape factor with increasing compression load results in the quasi "constant natural frequency" characteristic.

Fig. 3.3.10 shows load-natural frequency characteristic for a rubber torus with cross sectional diameter 12.7 mm (0.5 in.). The natural frequency is 12 Hz ±6% in the broad range of weight load, $W_{max}/W_{min} \cong 4.4$,

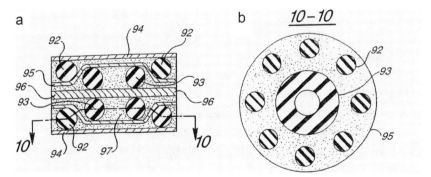

Fig. 3.3.9. Packaging of streamlined rubber elements 92, 93 and rigid inserts 95, 96 by foam matrix 97 without bonding.

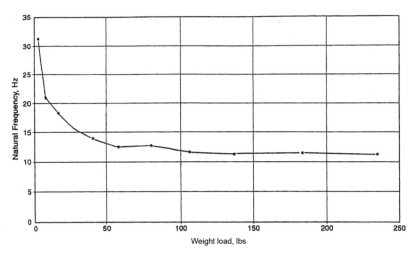

Fig. 3.3.10. Natural frequency versus axial compressive load for rubber torus ($H = 30$, $D = 50.8$ mm, $d = 12.7$ mm).

with compression deformation $\Delta_{max} = 30\%$ at W_{max}. While this weight load range is significantly narrower than the load ranges realized in specially designed CNF vibration isolators in Fig. 3.3.3, it would increase to $W_{max}/W_{min} \cong 6$ to 8 if used to 40% to 45% deformation. Due to larger allowable compression deformations (40% to 45% versus 10% to 15% for conventional bonded elements loaded in compression), use of ideal shape unbonded elastomeric elements allows for much more compact designs. The natural frequency in the CNF range can be changed by changing cross-sectional diameter of the cord/cylinder or of the O-ring. The square of the natural frequency for a given weight load (i.e., stiffness) is proportional to the cross-sectional diameter. Spherical and ellipsoidal rubber elements have similar load-deflection characteristics and can be used for lightly loaded flexible elements.

It was demonstrated that the radial loading of cylinders results in about 15% higher damping than for conventionally (axially) loaded elements made from the same rubber blend [9]. This effect can be explained by a greater role of shear deformations/stresses in deforming an unbonded cylinder under the radial load than in compression of a bonded element, since the bulk (hydrostatic) deformation characterized by volumetric (bulk) modulus K and playing a large role in compression of bonded elements, is not associated with a significant energy dissipation.

Deformation characteristic of a *radially loaded cylinder* or a *torus* is described reasonably well by the following analytically derived expression [12]:

$$\frac{P}{2LdG} = \frac{\lambda^{-3}}{16}(\Psi\lambda^{3/2} + 1.5\Psi\lambda^{1/2} + 1.5\sin^{-1}\Psi) + \frac{\lambda^{-2}-1}{16}\left(\frac{\pi}{2} - \Psi\lambda^{1/2} - \sin^{-1}\Psi\right)$$

$$-\frac{\lambda}{6}\sin^{-1}\Psi - \Psi\frac{\lambda^{1/2}}{16} \tag{3.3.6}$$

or an empirical expression [14]:

$$\frac{P}{Ld} = E_o\left[1.25\left(\frac{x}{d}\right)^{1.5} + 50\left(\frac{x}{d}\right)^6\right] \tag{3.3.7}$$

where P = compression force; L = length of the element (πD_o for torus, where D_o is its mean diameter); d = cross-sectional diameter of the cylinder; x = compression deformation; $\Psi^2 = x/d$, $\lambda = 1 - \Psi^2$; G = shear modulus, E_o is Young's modulus.

Stiffness of the radially loaded cylinder can be obtained by differentiating the load-deflection expressions (3.3.6) or (3.3.7). Differentiation of Eq. (3.3.7) results in:

$$k = \frac{dP}{dx} = LE_o\left[1.88\left(\frac{x}{d}\right)^{0.5} + 300\left(\frac{x}{d}\right)^5\right] \tag{3.3.8}$$

This expression leads to an interesting and somewhat unexpected conclusion that the radial stiffness of a rubber cylinder of length L made from a rubber with certain E_o does not depend on its diameter, d, but only on the relative radial compression, x/d. Of course, one has to remember that the expression (3.3.7) is an approximate one.

Deformation characteristic of a compressed rubber sphere (diameter, d) is described for small deformations (up to ~20%) by expression [4]:

$$\frac{P}{d^2} = 0.44E_o\left(\frac{x}{d}\right)^{1.5}; \quad k = \frac{dP}{dx} = 0.66D\left(\frac{x}{d}\right)^{0.5} \tag{3.3.9}$$

and for large deformations (20% to 65%) by expression:

$$P = E_o\,\pi D^2\,(\alpha_s + k\beta_s), \tag{3.3.10}$$

where $\qquad \alpha_s = (v - \ln v - 1)/8$; $\beta_s = (v^{-1} + 2\ln v - v)/64$.

Here $1 - v^{1/2} = x/D$ and α_s and β_s are given in Table 3.1.

Table 3.1 Coefficients for calculating deformations of a rubber sphere.

x/D	α_s	β_s
0.225	0.0138	0.0007
0.293	0.024	0.002
0.367	0.039	0.004
0.452	0.063	0.010
0.553	0.101	0.025
0.684	0.175	0.083

3.3.3 Thin-Layered Rubber-Metal Laminates

Very interesting and important for practical applications, nonlinear properties have thin-layered rubber-metal laminates [13]. Further splitting and laminating of the block in Fig. 2.8.2 leads to ever higher stiffness. When the layers become very thin, on the order of 0.05 mm to 1 mm (0.002 in. to 0.04 in.), compression stiffness becomes extremely high and highly nonlinear. The stress/strain condition of the thin rubber layer in compression is governed by the bulk modulus K of the rubber and by Young's modulus of the metal layers. Deformation in the perpendicular shear direction is governed by the much smaller shear modulus G. As a result, such element has highly anisotropic stiffness characteristics that may be useful for various structural applications. Usually, thin rubber layers are alternating with thin rigid layers made of strong metal or fabric woven from high strength fibers, with rubber and rigid layers connected by bonding. The multi-layered structure is needed to allow for larger shear displacements. The laminates for various applications may have different shapes — prismatic (Fig. 3.3.11a), spherical (Fig. 3.3.11b), cylindrical (Fig. 3.3.11c). The laminate usually consists of n layers of metal or other

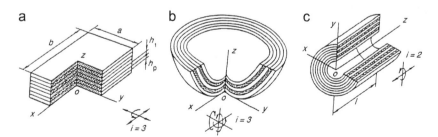

Fig. 3.3.11. Basic types of structural thin-layered rubber-metal laminates; a - prismatic, b - spherical; c - cylindrical.

rigid material and n-1 layers of rubber; for packaging purposes, the end metal layers are often made much thicker than the intermediate layers.

The bulk modulus K for different blends of rubber varies significantly. This variation is correlated to deviation of Poisson's ratio from the theoretical value for an incompressible material, $v = 0.5$ (see Fig. A.4.1). The usual requirement for the laminated elements for applications is high compression stiffness, which is determined by four major factors: deviation of v from $v = 0.5$ (the more the deviation, the lower the compression stiffness); shape factor, S (the larger the value, the greater the compression stiffness); deformations in the bonding (adhesive) layer between the rubber and the rigid layers, which effectively reduce compression stiffness; and deformations within the rigid (metal) layers under the hydrostatic pressure in the rubber layers (determined by the Young's modulus of the rigid layers and/or by their thickness). Usually, v is closer to 0.5 for softer (lower durometer) rubber blends, having less fillers as illustrated by Fig. A4.1. For example, a neoprene rubber with durometer $H = 40$ had $v = 0.4997$ versus $v = 0.4990$ for $H = 75$. As a result, using a higher durometer rubber for thin-layered laminates results in a much lesser increase in compression stiffness than the increase in shear stiffness. The highest compression stiffness and the greatest ratio between compression and shear stiffness values for a laminate can be achieved by using soft (low-filled) natural rubber blends. The low oil and ozone resistances of the natural rubber blends are not as critical for thin-layered laminates as they are for rubber elements with more uniform dimensions, since penetration of these aggressive substances into the element is limited by the small thickness of the rubber layer. In addition, the exposed edges of the rubber layers can be protected by appropriate coatings or by foam strips.

Effective compression modulus, E_{eff}, versus specific compression load, $p_z = P_z/A$, where A = the surface area of the laminate, is plotted in Fig. 3.3.12. E_{eff} is related to the total thickness of rubber, h_r, and was calculated from load-deflection plots as:

$$E_{eff} = \frac{\Delta p_z h_r}{\Delta z}, \qquad (3.3.11)$$

where Δz is compression deformation caused by increment, Δp_z, of the specific compression force (pressure). While compression of rubber elements of more uniform dimensions is characterized by a linear load-deflection characteristic up to relative deformations, $\varepsilon = 0.1$ to 0.15 (10%

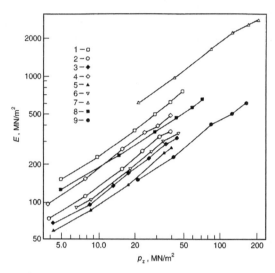

Fig. 3.3.12. Effective compression modulus for thin-layered rubber-metal laminates as function of compressive load (pressure, p_z); 1 to 4, 7 to 9, $H = 42$; 5, $H = 58$; 6, $H = 75$; 1 – 6, 0.05 mm brass metal layers; 7 to 9, 0.1 mm steel metal layers; $1 - A = 21.3$ cm^2, $t_r = 0.16$ mm, $n = 33$; $2 - A = 26.4$ cm^2, $t_r = 0.33$ mm, $n = 33$; $3 - A = 23.7$ cm^2, $t_r = 0.39$ mm, $n = 17$; $4 - A = 25.9$ cm^2, $t_r = 0.25$ mm, $n = 9$; $5 - A = 23.5$ cm^2, $t_r = 0.53$ mm, $n = 17$; $6 - A = 23.1$ cm^2, $t_r = 0.58$ mm, $n = 11$; $7 - A = 12.3$ cm^2, $t_r = 0.106$ mm, $n = 15$; $8 - A = 36$ cm^2, $t_r = 0.28$ mm, $n = 15$; $9 - A = 12.3$ cm^2, $t_r = 0.44$ mm, n = 14. Here H = rubber hardness, A = loaded surface area, t_r = thickness of one rubber layer, n = number of rubber layers.

to 15%), compression of the thin-layered laminates can be highly nonlinear (progressive increase of E_{eff} in Fig. 3.3.12 starting from the smallest deformations, down to $\varepsilon = \Delta z / h_r = 0.001$). In the highly constrained environment of a thin rubber layer bonded to the rigid surfaces, the major factor is hydrostatic pressure. Both shear modulus, G, and bulk modulus K are increasing with the increasing hydrostatic pressure in thin rubber layers. Increase of K is much more pronounced than increase of G. Consequently, increase of modulus, E_{eff}, is much more significant than increase of G, about one decimal order of magnitude per 1.5 decimal orders of magnitude increase of compression load. The thin-layered laminates can accommodate very high static compression loads, at least up to 500 MPa to 600 MPa (75 psi to 100,000 psi). The degree of nonlinearity is weakening, and the maximum compression loads are decreasing for thicker ($\geq$ ~1 mm) rubber layers. The ultimate static load of thin-layered laminates is limited by strength of the rigid (metal) layers, since the mode of failure at static loading was observed to be disintegration of the rigid (metal) layers. Thus, strength enhancement of a laminate for compression loading can be achieved by using a stronger metal or thicker metal layers. Absolute values of compression stiffness are quite high. Thus, for sample 7 in Fig. 3.3.12, $E_{eff} = 1{,}800$ MPa at $p_z = 100$ MPa, which is equivalent to 1 μm deflection per 1 MPa increase in compression pressure, about the same as contact deformation between two pressed together ground flat steel surfaces (see Chapter 4).

Effective shear modulus, $G_{eff} = \Delta p_x h_r / \Delta x$ is plotted in Fig. 3.3.13 versus shear force, $P_x = p_x A$, where Δx is shear deformation caused by increment $\Delta p_x = \Delta P_x / A$. The precompressed laminates loaded in shear demonstrate a slightly softening nonlinearity. For elements with thickness of one layer, $h_r < $ ~1.0 mm, there was observed only a relatively weak correlation between the magnitude of compression pressure on the specimen (in the range of $p_z = 0.5$ MPa to 150 MPa) and its effective shear modulus (Fig. 3.3.13 shows G_{eff} in the range of $p_z = 0.5$ MPa to 4.2 MPa). For $h_r = 2 - 4$ mm, G_{eff} increases as ~ $p_z^{0.5}$.

It was shown in [14] that stiffness of cylindrical (diameter d) or prismatic ($a*b$, $a \leq b$) thin-layered rubber-metal laminates with $0.01 \leq \alpha \leq 0.1$ ($\alpha = h_r/d$ or h_r/a) can be, in the first approximation, determined as

$$k = \frac{A}{h} \frac{(1+\Psi)G}{(\beta + c\alpha^2 + \gamma)}, \qquad (3.3.12)$$

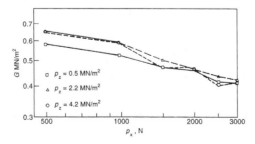

Fig. 3.3.13. Effective shear modulus of a compressed thin-layered (rubber layer 1 mm thick) rubber-metal laminate.

where $\gamma = \dfrac{\chi G h_r}{E_m h_m}$, $\chi = 1 + \dfrac{h_m}{h_r} \dfrac{v_m}{1-v_m}$, $\Psi = \dfrac{h_m + 2h_a}{h_r}$. Here h_r , h_m , h_a is thickness of rubber, metal, adhesive layer, respectively; $h = n\, h_r\,(1 + \Psi) =$ total thickness of the element; A = surface area of the element; $\beta = G/K$ for the rubber layers; E_m, v_m are Young's modulus, Poisson's ratio for the metal layers; c is a numerical constant ($c = 2.57$ for cylindrical elements, $c = 1.95$ for prismatic elements). This expression gives stiffness at small deformations when the moduli, G and K, are not yet distorted by high hydrostatic pressures. Some nonlinearity is reflected by the term, $c\alpha^2$, in the denominator.

Since compression deformation of the thin-layered laminates is due to so-called "hydrostatic compressibility" with, practically, no shear deformation involved, compression damping of the laminates is very low, even when the laminates are made from high-damping rubber blends. Damping associated with shear deformation is the damping characteristic for the used rubber blend. This unique material changes its stiffness by a factor of 10 to 50 during compressive deformation of only 10 μm to 20 μm. Another special feature of the rubber-metal laminates is anisotropy of their stiffness characteristics. Since the shear deformation is not associated with a volume change, shear stiffness does not depend on the design of the rubber block, only on its height and cross-sectional area. As a result of this fact, shear stiffness of the laminates stays very low, while the compression stiffness is increasing with the thinning of the rubber layers. Ratios of stiffness in compression and shear exceeding 3 to 5,000 are not difficult to achieve.

3.4 STIFFNESS MANAGEMENT BY PRELOADING (STRENGTH-TO-STIFFNESS TRANSFORMATION)

Contribution of stiffness to performance characteristics of various mechanical systems is diverse. In some cases, increase of stiffness is beneficial; in other cases, a certain optimal value of stiffness has to be attained. To achieve these goals, means for adjustment of stiffness are needed. In many instances, the stiffness adjustment can be achieved by *preloading* the components or their connections (joints) responsible for the stiffness parameters of the system. Preloading involves intentional application of internal forces to the responsible components, over and above the payloads. The proper preloading increases the stiffness but may reduce strength or useful service life of the system due to application of the additional forces. Thus, some fraction of the overall *strength is transformed into stiffness*.

3.4.1 Embodiments of the Preloading Concept

Preloading is most effective for nonlinear systems, but it can regulate stiffness both in linear and in nonlinear systems. Preloading of linear systems can be illustrated in the example of Fig. 3.4.1 [15].

In a belt drive in Fig. 3.4.1a, torque T applied to driving pulley 1 transforms into a tangential force, P_t, applied to driven pulley 2 by tensioning the leading (active) branch A of the belt. This tensioning leads to stretching of the branch A by an increment Δ. The idle branch, B,

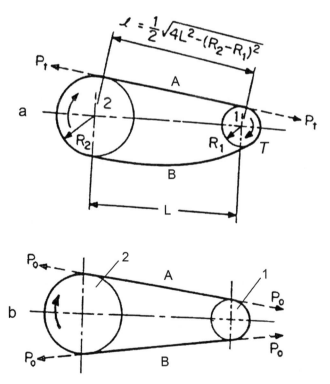

Fig. 3.4.1. Belt drive without (a) and with (b) preload. A — active branch, B — slack branch.

accommodates this elongation Δ by reducing slack. Stiffness k_{bd} of the belt drive is a ratio of the driving torque T to the angular deflection, α, of the driving pulley if the driven pulley is kept stationary (locked). In this case, it is more convenient to derive an expression for compliance, $e_{bd} = 1/k_{bd}$, rather than for stiffness, k_{bd}. Obviously,

$$e_{bd} = \frac{\alpha}{T} = \frac{\Delta/R_1}{P_t R_1} = \frac{\Delta}{P_t R_1^2} = \frac{P_t l / EA}{P_t R_1^2} = \frac{l}{R_1^2 EA} \qquad (3.4.1)$$

In this drive, only the branch A is transmitting the payload (tangential force), the other branch B is loose. The drive can be preloaded by pulling the pulleys apart during assembly so that each branch of the belt (both the active branch A and the previously loose branch B) is subjected to the *preload force*, P_o, even if no payload is transmitted. In this case, application of a torque T to the driving pulley will increase tension of the active branch A of the belt by the amount P_A and will reduce tension of the previously passive (loose) branch, B, by the amount, P_B. These forces are:

$$P_A = P_B = \frac{1}{2}\frac{T}{R_1} \qquad (3.4.2)$$

Total elongations of the branches A and B are, respectively,

$$\Delta_a = \frac{(P_o + P_A)l}{EA}; \qquad \Delta_b = \frac{(P_o - P_B)l}{EA} \qquad (3.4.3)$$

Incremental elongations of the branches, A and B, caused by the payload are, respectively,

$$\Delta_1 = \frac{P_A l}{EA}, \qquad \Delta_2 = \frac{P_B l}{EA} \qquad (3.4.4)$$

and the angular compliance of the preloaded drive is:

$$e'_{bd} = \frac{l}{2R_1^2 EA} \qquad (3.4.5)$$

or one half of the compliance for the belt drive without preload as in Eq. (3.4.1). In other words, preload has doubled stiffness of the belt drive, which is a system with linear load-deflection characteristic.

It is important to note that the active branch in the preloaded drive is loaded with a larger force $(P_0 + P_t/2)$ than in a nonpreloaded drive $(P_t < P_0 + P_t/2)$. Thus, only belts with upgraded strength limits can be preloaded, and accordingly, the described effect can be viewed as an example of *strength-to-stiffness transformation*.

While belts have approximately linear load-deflection characteristics, *chains* are highly nonlinear, both roller chains (Fig. 3.4.2a) and silent chains (Fig. 3.4.2b). Experimental data show that with the load increasing from $0.1P_r$ to P_r (P_r = rated load), the tensile stiffness of a silent chain increases about three-fold (Fig. 3.4.2c).

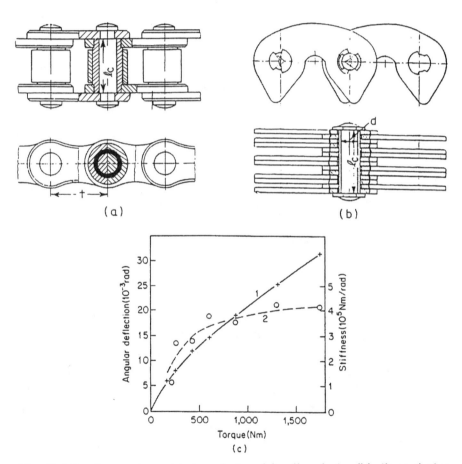

(a)   (b)

(c)

Fig. 3.4.2. Power transmission chains. (a) roller chain; (b) silent chain; (c) torque-angular deflection characteristic (1) and stiffness (2) of a silent chain drive.

The statics of a chain drive is very similar to that of a belt drive. If the chain is installed without initial preload, like the belt in Fig.3.4.1a, only the active branch will be stretched by the full magnitude of the tangential force, $P_t = T/R_1$. The tensile deformation is described by Eq. (3.4.1) in which now $A = l_c d$ is the effective cross-sectional area (l_c = the total load-carrying width of the chain, and d = the diameter of the roller-supporting axles for the roller chains or the laminae — supporting axles for silent chains (Fig. 3.4.2a and b), and $E = t/k_{ch}$ is the effective elastic modulus (t is the chain pitch, and k_{ch} is the compliance factor of the chain, which is a function of the tensile load, P_t). Thus

$$e_{ch} = \frac{k_{ch}(P_t)l}{R_1^2 A t} \qquad (3.4.6)$$

If a preload is applied, the chain drive stiffness would immediately double,which is described by an expression similar to Eq. (3.4.5)

$$e_{ch} = \frac{k_{ch}(P_{to})l}{2R_1^2 A t} \qquad (3.4.7)$$

However, in addition to this direct effect of the preload, with increasing preload, P_o, stiffness of each branch increases because of the hardening nonlinear load-deflection curve (Figs. 3.1.1 and 3.4.2c) as reflected in the expression describing dependence of k_{ch} on P_o. Thus, an increase in the preload would additionally enhance the stiffness, up to three times for a typical roller chain. The total stiffness increase, considering both effects, can be up to five to six times.

Another advantage of the preloaded chain drive is stability of its stiffness. Since only one branch transmits payload in a nonpreloaded chain, and this branch is characterized by a nonlinear hardening characteristic, the stiffness at low payloads is low, while the stiffness at high payloads is high. Thus, the accuracy of a device driven by a chain drive would be rather poor at relatively low torques. In a preloaded drive, stiffness for any payload force, $P_t < 2P_o$ will have the same (high) value, since the "working point" on the load-deflection plot in Fig. 3.1.1 (or torque-angular deflection in Fig. 3.4.2c) would always correspond to load P_o.

This statement is reinforced by the data in Section 4.4, Eqs. (4.4.7') and (4.4.7"), on preloading of *flat joints* to increase their stiffness in case of moment loading. Although contact deformations are characterized by nonlinear load-deflection curves, preloaded joints are essentially linear or having a constant stiffness when a moment is applied.

Of course, preloading of flat joints leads to increased stiffness not only for the moment loading but for the force loading as well. Indeed, loading with an incremental force $\pm P_1$ of a joint having a hardening characteristic and preloaded with a compressive force, P_o, $P_1 < P_o$, would result in stiffness roughly associated with the working point, P_o which is greater than the stiffness for the working point, $P_1 < P_o$, (Fig. 3.1.1). In addition, preloaded joints can work even in tension if the magnitude of the tensile force does not exceed P_o.

Similar, but numerically much more pronounced effects occur in *rubber-metal laminates* described above in Section 3.3.3 and in [13].

The concept of internal preload is very important for *antifriction bearings*, especially in cases when their accuracy of rotation and/or their stiffness have to be enhanced. The accuracy of rotation under load depends on elimination of backlash and, also, on the number of the rolling bodies participating in the shaft support (see Appendix 6). The deflections of the shaft (spindle) are also influenced by deformations of bearings.

Elastic displacements of shafts in antifriction bearings consist of elastic deformations of rolling bodies and races calculated by using Hertz formulae (see Table 4.1), as well as of deformations of joints between the outer race and the housing and between the inner race and the shaft. Since contact pressures between the external surfaces of both the races and their counterpart surfaces (shaft, housing bore) are not very high, contact deformations have linear load-deflection characteristics (see Section 4.5.1). The total compliance in these joints which are external to the bearing is:

$$\delta'' = \frac{4}{\pi} \frac{P k_2}{db} \left(1 + \frac{d}{D} \right), \text{ m} \qquad (3.4.8)$$

where P = radial force, N; d and D = the inner and outer diameters of the bearing, m; b = width of the bearing, m; k_2 = 5 to 25 x 10^{-11} m^3/N. Lower values of k_2 are representative for high-precision light interference cylindrical or preloaded tapered fits and/or high loads, and higher values are representative for regular precision tight or stressed fits, with reamed holes and finish ground shaft journals and/or light loads.

Hertzian deformations of rolling bodies in *ball bearings* are nonlinear and can be expressed as:

$$\delta_0 = (0.15 - 0.44d) \times 10^{-6} \, P^{2/3}, \, \text{m} \qquad (3.4.9)$$

For *roller bearings*

$$\delta r'' = \sim k_1 P, \, \text{m} \qquad (3.5.10)$$

where $k_1 = 0.66 \times 10^{-10}/d$ for *narrow roller bearings*; $0.44 \times 10^{-10}/d$ for *wide roller bearings*; $0.41 \times 10^{-10}/d$ for *normal width tapered roller bearings*; and $0.34 \times 10^{-10}/d$ for *wide tapered roller bearings* [15].

As a general rule, δ'' is responsible for 2% to 40% of the overall deformation at low loads (precision systems) and for 10% to 20% at high loads.

Accordingly, the internal preload of roller bearings (which can be achieved by expanding the inner race, e.g., by using a tapered fit between the bearing and the shaft, Fig. 3.4.3e) brings results similar to those

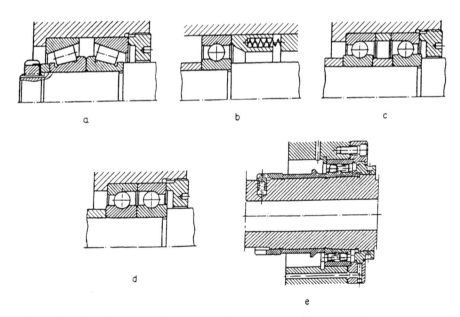

Fig. 3.4.3. Some practical techniques for preloading rolling friction bearings. (a) preload by tightening a nut; (b) preload maintained by a spring; (c) preload caused by unequal spacers for inner and outer races; (d) preload by reducing width of outer races; (e) preload by squeezing inner race along a shallow cone.

achieved in belt drives, namely, doubling their radial stiffness (reducing δ_r' in half) because of elimination of backlash and "bringing to work" idle rollers on the side opposite to the area compressed by the radial force (see also Appendix 6). In ball bearings, the effect of the internal preload on δ_b' is rather similar to that in chain drives and consists of (1) doubling the number of active balls and (2) shifting the working point along the nonlinear load-deflection curve.

The positive effects of internal preload of antifriction bearings are, besides the stiffness increase, also elimination of backlash and enhancement of rotational accuracy because of a more uniform loading of the rolling bodies and better averaging of the inevitable inaccuracies of the races and the rolling bodies. The negative effects are higher loading of the bearing components and associated with it higher working temperatures, as well as faster wear and higher energy losses (it can be expressed as "part of the strength had been used to enhance stiffness").

Since (1) hertzian deformations are responsible only for 60% to 90% of the total bearing compliance, the balance being caused by external contact deformations described by Eq. (3.4.8), and (2) the latter are not affected by the internal preload, in stiffness-critical cases the bearing preload has to be accompanied by more stringent requirements to the assembly of the bearings with the shafts and housings. The resultmg reduction of k_2 in Eq. (3.4.8) would assure the maximum stiffness-enhancement effect.

Fig. 3.4.3 shows some techniques used to create preload in the bearings. Fig. 3.4.3a shows preload application by tightening a threaded connection; Fig. 3.4.3b shows preload by springs; and Figs. 3.4.3c,d show preload by a prescribed shift between the rolling bodies and the races achieved through the use of sleeves of nonequal lengths between the outer and inner races (c) or by machining (grinding) the ends of one set of races (d). In the case shown, the width of each of the outer races is reduced by grinding. A technique similar to the one shown in Fig. 3.4.3d is used in "four-point contact" bearings. For preloading cylindrical roller bearings in stiffness-critical applications, their inner races are fabricated with slightly tapered holes (Fig. 3.4.3e). Threaded preload means are used frequently for tapered roller bearings, which are more sturdy and less sensitive to overloading than ball bearings. Since some misalignment is always induced by the threaded connections, the technique is generally used for nonprecision units. In cases where a high precision is required, the load from threaded load application means is transmitted to the bearing through a tight-fitted sleeve with squared ends as it is shown in Fig. 3.4.3e (or very accurate ground threads are used). Springs (or an adjustable hydraulic pressure) are used for high- and ultrahigh-speed ball bearings, which have accelerated wear that would cause a reduction of the preload if it were applied by other means.

The necessary preload force, P_o, is specified on the premise that after the bearing is loaded with a maximum payload, all the rolling bodies would still carry some load (no clearance is developing).

Similar preload techniques, and with similar results, are used in many other cases, such as in antifriction guideways, ball screws, traction drives, etc. (e.g., see [15]).

3.4.2 Antagonist Actuators

A special case of preload, especially important for manipulators, is used in so-called "antagonist actuators" (Fig. 3.4.4a). In such actuators (modeled from human limbs, in which muscles apply only pulling forces developing during their contraction), two tensile forces, F_1 and F_2, are applied at opposite sides of a joint. A general schematic of such an actuator is given in Fig. 3.4.4b. Link L is driven by applying parallel tension forces, F_1 and F_2, to driving arms 1 and 2 respectively, which are symmetrical and positioned at angles 90 deg. + α relative to the link.

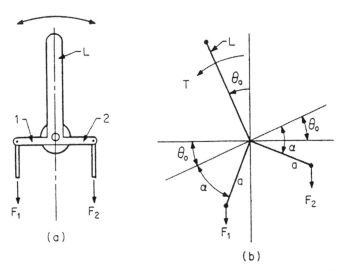

Fig. 3.4.4. Antagonist actuator (a) and its schematic (b).

Torque T applied to the link is:

$$T = -F_1 a \cos(\theta + \alpha) + F_2 a \cos(\theta - \alpha) \qquad (3.4.11)$$

or

$$T/F_1 = F_2/F_1 a \cos(\theta - \alpha) - a \cos(\theta + \alpha) \qquad (3.4.12)$$

It follows from Eq. (3.4.12) that one tension (e.g., F_1) can be chosen arbitrarily and then the required second tension (F_2) can be determined from the ratio

$$F_2/F_1 = [(T/F_1 + a \cos(\theta - \alpha)] / a \cos(\theta - \alpha) \qquad (3.4.13)$$

The stiffness of the actuator is:

$$
\begin{aligned}
dT/d\theta = k &= F_1 a \sin(\theta + \alpha) - F_2 a \sin(\theta - \alpha) \\
&= F_1 a[\sin(\theta + \alpha) - F_2/F_1 \sin(\theta + \alpha)]
\end{aligned}
\qquad (3.4.14)
$$

It follows from Eq. (3.4.14) that for a given F_2/F_1, stiffness is directly proportional to the absolute value of the tensile force F_1. Since, usually, a higher stiffness value is preferable, the highest tensile forces allowable by the strength of arms 1 and 2, by the joint bearings, by the available hydraulic pressures (for hydraulically driven actuators), etc., have to be assigned. Of course, the overall stiffness of the actuator system would be determined, in addition to Eq. (3.4.14), by the stiffness (compliance) of the other components, such as hydraulic cylinders and piping, etc. (see Chapter 6). As a result, an increase of tensions, F_1 and F_2 is justified only to the point at which an increase of stiffness k per Eq. (3.4.14) would still result in a meaningful increase in the overall stiffness.

3.4.3 Preloaded Flexible Elements with Variable Stiffness

If *nonlinear flexible elements* are used directly, their stiffness varies in accordance with the applied load. However, nonlinear flexible elements also can be used with preload. If variable preloading means are employed, the preloaded nonlinear flexible elements become structural elements with variable stiffness. Fig. 3.4.5a [2] shows an adjustable mount for an internal combustion engine 4 (only one mounting foot is shown). The adjustable mount comprises two nonlinear flexible elements, 1, 2, e.g. of Fig. 3.3.3 design, preloaded by bolt 5.

It is obvious that preloading of one flexible element (e.g., 2 in Fig. 3.4.5a) would result in a parallel connection of bolt 5 and isolator 2, thus resilience of isolator 2 would be lost. When two flexible elements 1 and 2 are used, mounting foot 4 of the engine is connected with foundation structure 3 via resilient elements of isolators 1 and 2, and stiffness of bolt 5 plays an insignificant role in the overall stiffness breakdown for the combined mount. However, tightening of the bolt increases loading of both flexible elements and thus is moving the "working points" along their load-deflection characteristics. If the load-deflection characteristic is of a hardening type (the most frequent case), then tightening of bolt 5 would result in the increasing stiffness. A typical case is shown in Fig. 3.4.5b. It is important to note that a preloaded combination of two identical nonlinear resilient elements, as in Fig. 3.4.5a, becomes a *linear* compound element of adjustable stiffness. When foot 4 in Fig. 3.4.5a is displaced downward by an increment Δx, load on the lower flexible element 2 is increasing by ΔP_2, while load on the upper flexible element 1 is decreasing by ΔP_1. If flexible elements 1 and 2 have identical hardening load-deflection characteristics, then stiffness of the upper flexible element is decreasing by Δk_1. Since the initial loading P_o and the initial

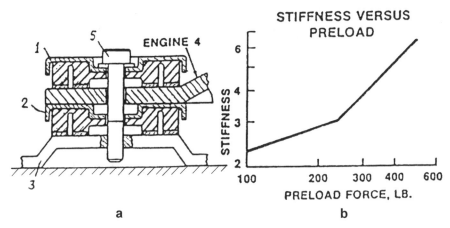

Fig. 3.4.5. Variable stiffness vibration isolator comprising two nonlinear (Constant Natural Frequency) isolators (a) and its stiffness versus load characteristic (b).

stiffness k_o of both flexible elements are the same, $\Delta P_1 = \sim\Delta P_2$ and $\Delta k_1 = \sim\Delta k_2$, thus the total stiffness,

$$\kappa_1 - \Delta k_1 + k_2 + \Delta k_2 = 2k_o - \Delta k_1 + \Delta k_2 = \sim 2k_o = \text{const} \quad (3.4.15)$$

Thus, preloading of the properly packaged pair of nonlinear flexible elements allows constructing of an element whose stiffness can be varied in a broad range. Such an element can serve as an output element of an active stiffness control system or a vibration control system. Fig. 3.4.6 shows test results of a similar system a in which radially loaded rubber cylinders were used as the nonlinear elements. Just two turns of the preloading bolt changed the natural frequency from 20 Hz to 60 Hz, which corresponds to a nine-fold increase in stiffness.

Fig. 3.4.7 shows application of nonlinear rubber elements (O-rings) for tuning a dynamic vibration absorber (DVA) for boring bar [16]. Fig. 3.4.7a shows the front end of a boring bar, housing DVA with a sleeve-like inertia mass made from "heavy metal" — machinable tungsten alloy. The inertia mass is fitted on the adjusting bolt. Turning the bolt results in

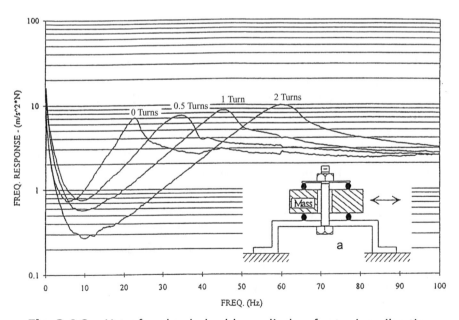

Fig. 3.4.6. Use of preloaded rubber cylinders for tuning vibration control devices.

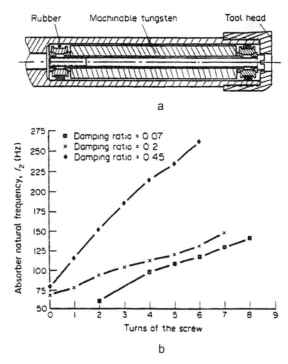

Fig. 3.4.7. Dynamic vibration absorber for boring bar (a) and its tuning range (b).

axial squeezing of two O-rings on which the bolt and the inertia mass are suspended in the radial direction. Since the axial squeezing of O-rings is changing their stiffness also in the radial direction, the natural frequency of the inertia mass is changing thus changing tuning of the DVA. Fig. 3.4.7b shows that the range of change of the natural frequency is different for O-rings made from different blends of rubber (distinguished by their damping). This range can be as great as 3.5:1, corresponding to 10:1 range of stiffness change.

Recently, a substantial interest was developed for using mechanical elements with variable stiffness for vibration control (e.g., [17]). This vibration control technique in application to a single-degree-of-freedom system "mass m–stiffness element k" can be described as changing the stiffness parameter, while the system is experiencing the resonance condition. Since reaching the resonance amplitude takes more than one cycle of vibration, changing the stiffness parameter during the process of reaching the resonance disrupts this process thus resulting in a significant reduction in vibration amplitudes. Varying stiffness according to a specified control law results in a reduction of the vibration response not only for resonance vibrations but for other vibratory processes such as pulse vibrations, random vibration, etc. This process is equivalent to adding damping but without energy dissipation.

Some widely used techniques for varying stiffness are using a magnetic field on flexible elements made from a magnetorheological elastomer and piezoelectric cirquitry [17]. These techniques are used when small deformations and thus high frequencies, 200 Hz to 300 Hz, are present. Mechanical means are used for discrete ("bang-bang") changes of stiffness. Use of nonlinear elastomeric elements with internal preload like in Figs. 3.4.5a and 3.4.6 can easily realize lower frequencies in the 5 Hz to 50 Hz range. Preload can be applied by judiciously selected actuators optimized for the required range of deformations/frequencies, e.g., by hydraulic actuators.

Use of internal preload (i.e., the *strength-to-stiffness transformation* concept) for enhancement of bending stiffness is addressed in Chapter 7.

3.4.4 Some Dynamic Effects Caused by Variable Stiffness

While nonlinear systems are characterized by stiffness varying with the loading conditions, in some mechanical systems, stiffness is changing

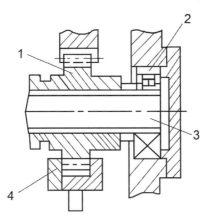

Fig. 3.4.8. System with variable stiffness (bearing 2 is fit on splined shaft 3).

with the changing geometry. While this effect is typical for linkage mechanisms wherein the links are not very stiff, e.g., see Section 1.3.4, it can also be present and can create undesirable effects in many other basic mechanical systems.

Fig. 3.4.8 shows a segment of a gearbox in which ball bearing 2 is mounted on a splined shaft 3. Since races (rings) of antifriction bearings are usually relatively thin, they easily conform with the profile of the supporting surface. Thus, the inner race of bearing 2 has different local effective stiffness in the areas supported by the splines and in the areas corresponding to spaces (valleys) between the splines. Even minute stiffness variations of support conditions of the thin inner race under balls during shaft rotation may cause undesirable, even severe, parametric vibrations.

Similar effects may be generated by a gear or a pulley having a non-uniform stiffness around its circumference and tightly fit on the shaft. Fig. 3.4.9 shows an oscillogram of vibration of a gear reducer housing caused bv a five-spoke gear, which was interference-fit on its shaft [18]. Effective stiffness of the gear rim is very different in the areas supported by the spokes and in the areas between the spokes. This resulted in a

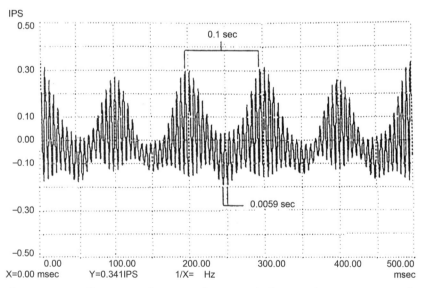

Fig. 3.4.9. Vibration of gear reducer with five-spoked gear press-fit on its shaft.

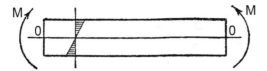

Fig. 3.5.1. Beam in a pure bending condition.

beat-like pattern of vibration amplitudes varying in the range of three-to-one. Such amplitude variation can cause intensive and annoying noise and other damaging effects.

3.5 ASSEMBLED FRAME-LIKE AND BEAM-LIKE STRUCTURES

Frequently, base- or frame-like structural parts are assembled from several components, mostly by bolts creating preload between the connected parts. The main requirements to such assembled frames are integrity (meaning that the assembled parts do not separate under foreseeable external forces and/or moments), and stiffness and damping of the assembled structures. A typical example of an assembled structure is engine block of an internal combustion engine. It is usually made as a separate cylinder block and an engine head, which are bolted together.

3.5.1 Integrity of Assembled Structures

3.5.1a Integrity of an Assembled Beam

If two moments, M, are acting on the ends of a straight beam with rectangular cross-section in Fig. 3.5.1, there are no stresses in the neutral plane O-O, thus no interaction between the upper and lower halves of the beam along this plane [19]. Obviously, splitting the beam into two thin ones along this plane should not change its stiffness. However, these two thin

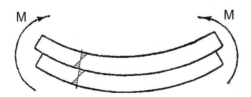

Fig. 3.5.2. Beam in Fig. 3.5.1 split along the neutral plane O-O.

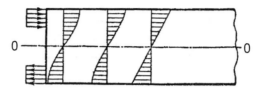

Fig. 3.5.3. Transformation of end stresses in beam loaded at the ends by bending moments.

beams assembled together and subjected to moments M would bend as shown in Fig. 3.5.2, with slippage along the contact surface and with a very different stress distribution in the cross-section. The key to resolving this contradiction is consideration of the patterns of forces/moments application at the beam ends. When the applied moments at the beam ends are shown in Fig. 3.5.1, it only means that the resultant moments are M each, with no specific information on what stress distributions on the beam ends results in moments, M. For a solid beam of a reasonable length shown in Fig. 3.5.1, it is not important, since any stress distribution at the ends would gradually become a linear stress distribution as shown in Fig. 3.5.1. Dynamics of the change of the stress distribution is shown in Fig. 3.5.3 for a special step-like stresses at the beam end. The transformation of the stress distribution occurs on a very short segment of the beam, but relatively strong tangential stresses are acting along this segment within the "neutral" plane O-O.

For the split beam, it is very important to consider stress distribution at the ends. If the end stresses are distributed identically for the initial

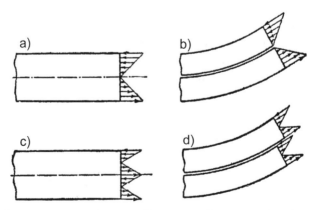

Fig. 3.5.4. Various realizations of moment loading.

Fig. 3.5.5. Split beam with clamped ends.

and the split beam, with zero stresses at the O-O plane, as shown in Fig. 3.5.4 a, c, then both beams perform identically. If the end stresses are distributed as in Fig. 3.5.4b, d, then the slippage along O-O would develop. The situation of Fig. 3.5.4a,c can be realized by clamping the split beam at the ends, as shown in Fig. 3.5.5.

3.5.1b Integrity of an assembled preloaded structure

In many cases, the only requirement to the assembled frame is its structural integrity under external forces and moments. A typical assembled structure is shown in Fig. 3.5.6 [20]. Its components 1 and 2 are identical, joined by a bolt, which is preloaded (tightened), thus creating tensile force in the bolt shank and compressive stresses at the joint surfaces

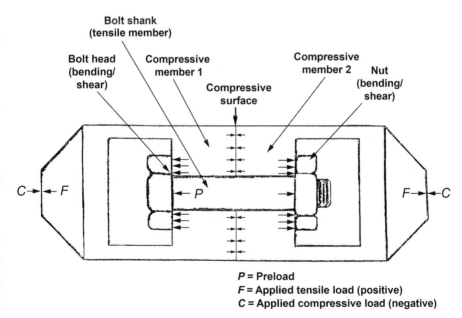

P = Preload
F = Applied tensile load (positive)
C = Applied compressive load (negative)

Fig. 3.5.6. Loads in a prestressed joint.

and in the areas of components 1 and 2 adjacent to the joint. Compression stiffness (a combination of structural and contact stiffness) of the compressed areas is k_c, tensile stiffness of the bolt is k_t. Obviously, there may be several bolts with the combined stiffness k_t. The compressed components and the stretched bolt can be modeled as a series connection of elastic elements with spring constants, k_c and k_t, so that the resulting stiffness of the connection is:

$$K = \frac{k_c k_t}{k_c + k_t},$$ (3.5.1)

and its total deformation under tightening (preloading) force P of the bolt is

$$\Delta = \frac{P}{K}.$$ (3.5.2)

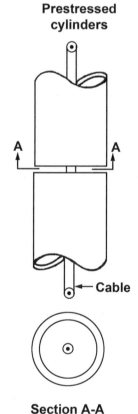

Prestressed cylinders

Cable

Section A-A

Fig. 3.5.7. Contact area between joined cylinders.

INITIAL GAPPING

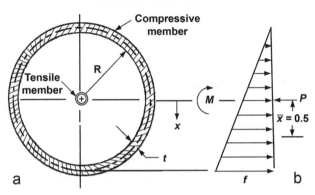

Fig. 3.5.8. Joint in bending.

"Gapping" is separation of the components bolted together into a joint and pulled apart by a large enough tensile force. "Tension release" results when an external force compresses the frame components, causing the bolts to become slack. Either condition can cause loss of stiffness, premature fatigue failure, part misalignment (possibly reducing preload), chatter, or other joint instability. The magnitude of the bolt preload should be set such as to avoid gapping or tension release. Since in most cases k_t << k_c, tension release is a relatively rare phenomenon.

The typical external loading pattern is a combination of an external force, P, and an external moment, M. Fig. 3.5.7 [20] shows connection of two relatively thin-walled cylinders 1 and 2 assembled by internal cable under tension. Fig. 3.5.8b shows stress p distribution on the contact surface in Fig. 3.5.8a under simultaneous action of compressive force, P, caused by the prestressed cable (compression stresses, p_c) and by external moment, M (tensile bending stresses, p_b) in the beginning of gapping ($|p_c| = |p_b|$).

The surface area, A, and the cross-sectional moment of inertia, I, of the contact surface are, respectively,

$$A = \pi R^2 - \pi (R-t)^2 = 2\pi Rt - t^2 \approx 2\pi Rt, \qquad (3.5.3)$$

$$I = (\pi/4)[R^4 - (R-t)^4] = (\pi/4)(4R^3t - 6R^2t^2 + 4Rt^3 - t^4) \approx \pi R^3 t. \qquad (3.5.4)$$

Compression stress, p_c, and the maximum magnitude of tensile stress, p_b, caused by bending are, respectively,

$$p_c = \frac{P}{A} = \frac{P}{2\pi R t}; \quad p_b = \frac{MR}{\pi R^3 t} = \frac{M}{\pi R^2 t}. \qquad (3.5.5)$$

At the start of gapping, $p_c = p_b$, or

$$\frac{P}{2\pi R t} = \frac{M}{\pi R^2 t}, \quad M_o = \frac{PR}{2} \text{ or } P_o = \frac{2M_o}{R}. \qquad (3.5.6)$$

Here M_o is the bending moment magnitude at which the joint preloaded by force, P_o, "opens up", and P_o is the minimum force still preventing gapping under moment, M_o. The maximum compressive stress, p_m, at the contact area in the beginning of gapping is twice as large as p_c,

$$p_m = p_c + p_b = \frac{M}{\pi R^2 t} + \frac{2M/R}{2\pi R t} = \frac{P}{\pi R t}. \qquad (3.5.7)$$

The compressive stresses at the beginning of gapping have a resultant at distance, x_r, from the axis of the connection.

3.5.2 Stiffness of Assembled Structures

Stiffness of structural connections in mechanical systems, such as in machine tool frames, is determined to a large degree by contact defor-

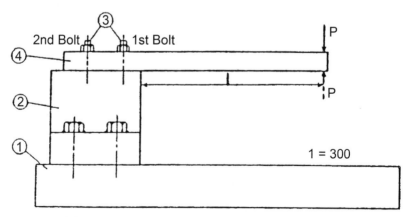

Fig. 3.5.9. Test setup for study of bolted joints.

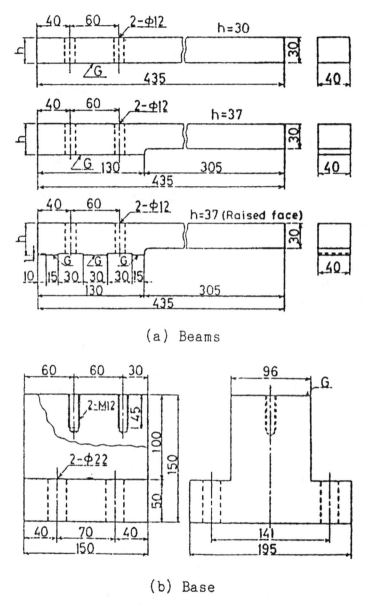

(a) Beams

(b) Base

Fig. 3.5.10. Components of the tested bolted joint.

mations between the connected surfaces and, thus, by contact pressures (see also Chapter 4). Frequently, the contact pressures are generated by tightening (preloading) bolts that pass through smooth holes in one part and engage with the threaded holes in the other part (base), thus creating

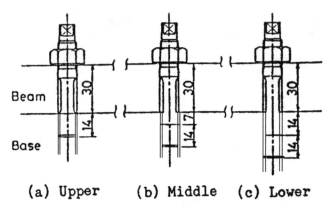

(a) Upper   (b) Middle   (c) Lower

Fig. 3.5.11. Tested position variants of bolt thread relative to base.

the joint-tightening force between the bolt heads and the thread. There are many factors to be considered in designing high-stiffness bolted connections. Some of these factors were studied experimentally in [21].

The test rig is shown in Fig. 3.5.9. The studied joint between steel cantilever beam 4 and base 2 is tightened by two bolts 3. Base 2 is bolted to foundation 1 by several large bolts. External load, P, at the beam end creates moment loading of the joint. Deflection, y, at the beam end is

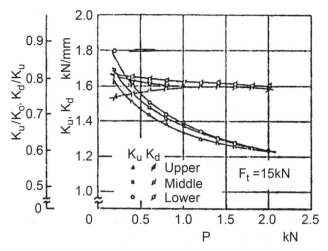

Fig. 3.5.12. Effect of bolt positioning on joint stiffness.

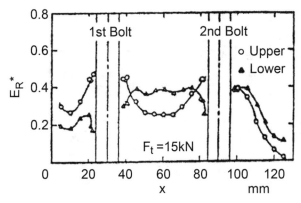

Fig. 3.5.13. Effect of thread positioning on joint contact stress distribution.

measured as the stiffness indicator. Force, P, can be applied downward (stiffness $K_d = P/y$) or upward (stiffness K_u).

Three configurations of the beam were tested as shown in Fig. 3.5.10a-c: beam 4 with uniform thickness, $h = 30$ mm, along its length (Fig. 3.5.10a); beam with a thicker ($h = 37$ mm) cross-section at the joint area (Fig. 3.5.10b); and beam modified from Fig. 3.5.10b by recessing areas at the bolt holes to reduce the contact area (Fig. 3.5.10c). The base is shown in Fig. 3.5.10d. Both surface roughness (R_a) and waviness were within

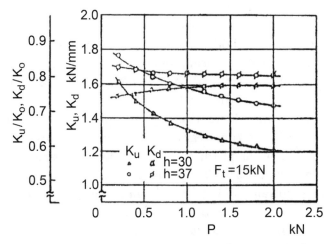

Fig. 3.5.14. Effect of moment loading of joint on bolt preload.

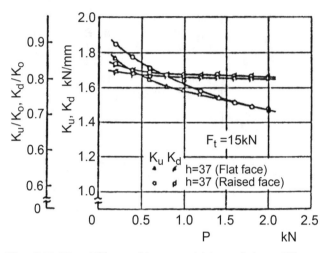

Fig. 3.5.15. Effect of beam width on joint stiffness.

2 μm (ground surfaces). Three sets of bolts had been tested: short (length of shank l = 30 mm), medium (l = 37 mm), and long (l = 44 mm), but with the same length (l = 14 mm) of the threaded segment. Accordingly, the bolts are engaging with upper, middle, and lower segments of the threaded holes of the base (Fig. 3.5.11). Preload force, F_t, of the bolts was measured by strain gages, and pressure distribution along the joint area was measured by an ultrasonic probe.

Fig. 3.5.12 shows stiffness values, K_d and K_u, as functions of the bolt preload force for three lengths of bolts. The much smaller values of K_u compared with K_d are to be expected, since the downward loading is associated with increasing contact pressures. However, the stiffness increase for longer bolts (both K_u and K_d) is not an obvious effect. It can be explained by a more uniform pressure distribution in the joint that is fastened by more resilient longer bolts, which also apply load to the base further from its surface (Fig. 3.5.13).

Fig. 3.5.14 illustrates incremental changes of the preload tensile force, ΔF_t, in the front ("first") and rear ("second") bolts as a result of applying upward (Fig. 3.5.14 a and b) and downward (Fig. 3.5.14c and d) forces to the beam (short bolts). The data in Fig. 3.5.14, especially in Fig. 3.5.14a, emphasizes importance of high initial tightening force, F_t, in the bolts to prevent opening of the joint. Due to nonlinearly of the joint deformation, increments ΔF, are much smaller (for the same P) when the initial F_t is larger.

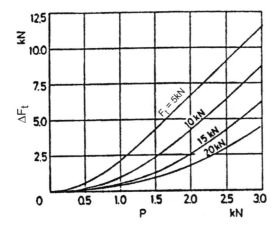

(a) Upward bending (the first bolt)

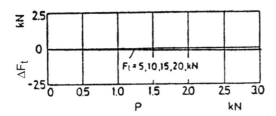

(b) Upward bending (the second bolt)

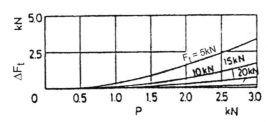

(c) Downward bending (the first bolt)

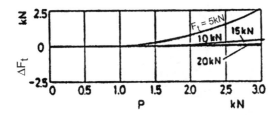

(d) Downward bending (the second bolt)

Fig. 3.5.16. Effect of contact surface on joint stiffness.

Thickness of the beam in the joint area is a very important factor for the joint stiffness. Fig. 3.5.15 shows significant improvements, especially for the most critical parameter K_u, achieved by changing thickness of the beam in the joint area from $h = 30$ mm to $h = 37$ mm. Reduction of the joint area as in Fig. 3.5.10c results in higher contact pressures for the same preload force, F_t. This results in higher values of the joint stiffness, especially for low external forces, P, as shown in Fig. 3.5.16.

3.5.3 Dynamics of Assembled Structures

Interesting results, somewhat correlated with the above described case, are obtained in [22]. Two types of bolted connections of three aluminum parallelepiped-shaped components simulating cylinder head, cylinder

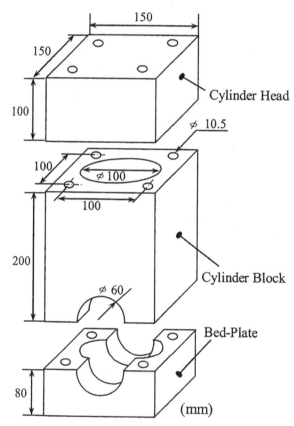

Fig. 3.5.17. Components of test model of one-cylinder engine.

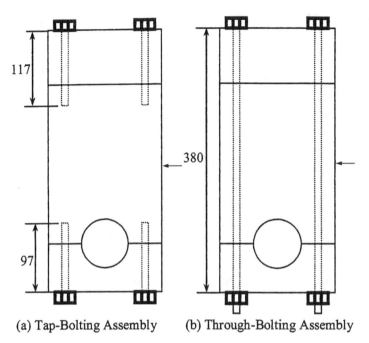

(a) Tap-Bolting Assembly   (b) Through-Bolting Assembly

Fig. 3.5.18. Tested bolt patterns.

block, and bed plate of a single-cylinder internal combustion engine (Fig. 3.5.17), had been tested and compared. All bolts have M10x1.5 thread. The bolts are made from steel with 310 MPa proof strength and tightened to tensile force 16.2 kN. In one design, the connection was realized by two sets of four "tap bolts" as shown in Fig. 3.5.18a. In the other design, the connection was realized by four "through bolts" (Fig. 3.5.18b). Dynamic tests of both assembled designs were performed by impact excitation at the center of the side surface of the assembly as shown by arrows in Fig. 3.5.18.

Fig. 3.5.19 shows a typical frequency response function (FRF) for the impact excitation of the models in Fig. 3.5.18. While FRF for the "tap bolt" model has 13 resonance peaks, FRF for the "through bolt" model has only 6 peaks, and the lowest natural frequency for the former case is 3143 Hz, 9.7% lower than the first natural frequency for the latter case (3450 Hz) in the frequency range 0 Hz to 6,500 Hz. This comparison indicates higher effective stiffness of the block assembled with the "through bolts". The overall level of surface vibration of the "tap bolt" connected

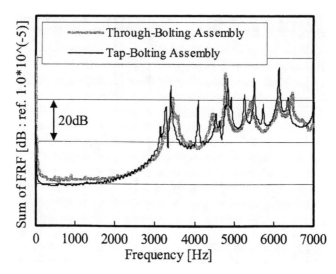

Fig. 3.5.19. Frequency response functions for tap- and through-bolt assemblies.

block was measured to be 10.6 dB higher than for the "through bolt" assembled block, thus indicating also a higher damping in the latter case. Similar results (10 dB lower surface vibration level, 5 dBA lower sound pressure level) were reported on comparison of a "real-life" four-cylinder engine assembled with "through bolts" with the similar engine assembled with the "tap bolts". No explanation for this effect is suggested, and no mention of contact stiffness/damping is present in [22]. It is possible that in the "tap bolts" case the system is held together by eight rigid connectors thus creating a highly over-constrained system (see also Chapter 5). Inevitable distortions in this system negatively affect both stiffness and damping. In the "through bolts" case, four long compliant bolts create much more uniformly stressed system, with a lesser degree of over-constraint.

REFERENCES

[1] DenHartog, J.P., 1952, "Mechanical Vibrations", McGraw-Hill, New York, 436pp.

[2] Rivin, E.I., 2003, "Passive Vibration Isolation", ASME Press, N.Y.

[3] Gurti, G., Montanini, R., 1999, "On the Influence of Friction in the Calculation of Conical Disc Springs", *ASME J. of Mechanical Design*, vol. 121, pp. 622–630.

[4] Freakley, P. K., Payne, A.R., 1978, "Theory and Practice of Engineering with Rubber", Applied Science Publishers, London, 666 pp.

[5] Rivin, E.I., 1983, "Cost-Effective Noise Abatement in Manufacturing Plants", *Noise Control Engineering Journal*, Nov.-Dec., pp. 103–117.

[6] Lavendel, E.E., 1980, "Elastic Elements of Vibratory Machinery", in *Vibration in Engineering*, vol. 4, pp. 211–213, Mashinostroenie, Moscow [in Russian].

[7] Rivin, E.I., "Resilient Anti-Vibration Support", *U.S. Patent 3,442,475*.

[8] Lee, B.-S., Rivin, E.I., 1996, "Finite Element Analysis of Load-Deflection and Creep Characteristics of Compressed Rubber Components for Vibration Control Devices", *ASME J. of Mechanical Design*, vol.148, No. 3, pp. 328– 336.

[9] Rivin, E.I., 1999, "Shaped Elastomeric Components for Vibration Control Devices", *S)V Sound and Vibration*, vol. 33, No. 7, pp. 18–23.

[10] Rivin, E.I., Lee, B.-S., 1994, "Experimental Study of Load-Deflection and Creep Characteristics of Compressed Rubber Components for Vibration Control Devices", *ASME J. of Mechanical Design*, vol. 116, No. 2, pp. 539 – 549.

[11] Rivin, E.I., "Nonlinear Flexible Connectors with Streamlined Resilient Elements", *U.S. Patent 5,954,653*.

[12] Dymnikov, S.I., 1972, "Stiffness Computation for Rubber Rings and Cords", Voprosi Dinamiki i Prochnosti [Issues of Dynamics and Strength], No. 24, Zinatne Publish. House, Riga, pp. 163–173 [in Russian].

[13] Rivin, E.I., 1983, "Properties and Prospective Applications of Ultra-Thin-Layered Rubber-Metal Laminates for Limited Travel Bearings", *Tribology International*, vol. 16, No. 1, pp. 47– 26.

[14] Gorelik, B.M., Kolosova, V.I., Tikhonov, V.A., Sshegolev, V.A., 1980, "Influence of Mechanical and Geometric Parameters of Thin-Layered Rubber-Metal Elements on Their Stiffness", *Kauchuk i resina*, No. 8, pp. 40–44 [in Russian].

[15] Rivin, E.I., 1988, "Mechanical Design of Robots", McGraw-Hill, New York, 320 pp.

[16] Rivin, E.I., Kang, H., 1992, "Enhancement of Dynamic Stability of Cantilever Tooling Structures", *Int. J. of Machine Tools and Manufacture*, Vol. 32, No. 4, pp. 539–562.

[17] Winthrop, M.F., Baker, W.P., Cobb, R.G., 2005, "A Variable Stiffness Device Selection and Design Tool for Lightly Damped Structures", Journal of Sound and Vibration, vol. 287, No. 4-5, pp. 667–682.

[18] Taylor, J., 1995, "Improvement in Reliability and Maintainability Requires Accurate Diagnosis of Problems", *P/PM Technology*, No. 12, pp. 38–41.

[19] Feodosiev, V.I., 2005, "Advanced Stress and Stability Analysis", Springer, 421 pp.

[20] Baumann, Th.R., 1991, "Designing Safer Prestressed Joints", *Machine Design*, April 25, pp. 39–43.

[21] Kobayashi, T., Mitsubayashi, T., 1986, "Considerations on the Improvement of Stiffness of Bolted Joints in Machne Tools," Bulletin of JSME, vol. 29, No. 257, pp. 3934–3937.

[22] Okamura, H., Yamashita, K., Shimizu, S., Sakamoto, H., 2005, "NVH Experiments and Analysis for a Single Cylinder Engine Model Assembled with Tap-Bolts and with Through-Bolts", SAE Paper 2005-01-2531.

CHAPTER

4

Contact (Joint) Stiffness and Damping

4.1 INTRODUCTION

When two solid blocks are in contact, their actual area of contact is generally very small. When *non-conforming* solid blocks with substantially different local curvature radii in the contact area are compressed together (such as spheres, cylinders, toruses contacting with flat surfaces, etc.), the initial contact area is small (ideally a point or a line) because of their geometry (Fig. 4.1.1*a, b*). Analysis of contact deformations between streamlined parts with initial point or line contact area can be performed using analytical formulas derived with rather broad assumptions. Some important issues related to contacts between non-conforming surfaces, including analytical expressions for contact deformations for some important cases, are addressed in Section 4.2.

When the contact surfaces are *conforming* (both are flat or shaped with identical or slightly different curvature radii), the actual contact area is small because of the roughness and waviness of the contacting surfaces (Fig. 4.1.1*c*). Since only the outstanding areas of surface asperities or waves are in contact, under small loads, only a small fraction (much less than 1%) of the nominal contact surface areas are in actual contact. This fact causes also reduction of thermal and electrical conductivity between the contacting bodies. While the contact deformations between the conforming surfaces could be very substantial — commensurate with or even exceeding the structural deformations —

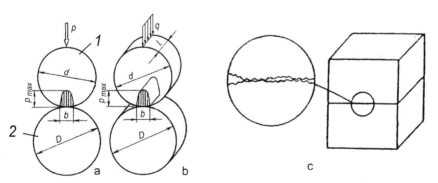

Fig. 4.1.1. (a), (b) Compression and pressure distribution in the contact areas of surfaces with substantially different curvature radii; (a) Two different size spheres in contact, (b) Two different size cylinders in contact; (c) Compression of surfaces with nominally conforming curvature radii.

purely analytical expressions for these cases are not always available, and a vast empirical database should be used. These contact deformations are especially important for high-precision mechanical systems, as well as for general-purpose machines, which have numerous joints. The largest contact deformations are associated with joints providing for the relative motion of the components, especially while they are experiencing low loads (e.g., at the final positioning stages of moving components in machine tools and robots when the highest accuracy and the highest stiffness are required). The role of contact deformations in machine tools can be illustrated by an example: contact deformation in typical sliding guideways of a machine tool is about 10^{-11} m^3/N (10 μm per 1 MPa contact pressure or 400 μin. per 150 psi contact pressure). It is equivalent to a compression deformation of one meter (40 in.) long cast-iron bar under the same axial pressure. Contact deformations between conforming surfaces are described in Section 4.2.

An intermediate case is contacting between quasi-conforming surfaces (usually, cylindrical and conical assembled surfaces with slightly different curvature radia of the assembled components). In such cases, the total deformations are determined by a combination of structural deformations of the contacting bodies and of deformations of the surface asperities. Often, these two deformation components are commensurate. This case (cylindrical connections with clearances) is addressed in Section 4.4.1a.

Due to nonlinearity of contact deformations, Fast Fourier transform (FFT)-based dynamic experimental techniques for mechanical systems, in which contact deformations play a significant role, should be used with caution. In many cases, more reliable results can be obtained by static testing (e.g., see Section 7.1).

4.2 CONTACT DEFORMATIONS BETWEEN NON-CONFORMING SURFACES

Contact between non-conforming surfaces is characterized by forces acting on small localized areas, thus resulting in high local stresses.

For contacting spheres 1 and 2 (Fig. 4.1.1a), the contact area is a circle of diameter

$$b = 1.4\sqrt[3]{\frac{PD_{\text{ef}}}{E}}, \tag{4.2.1}$$

where effective diameter, $D_{ef} = dD/(D \pm d)$ (plus for convex-convex contacts, minus for convex-concave contacts), and effective Young's modulus, $E = 2E_{1f}E_{2f}/(E_1 + E_2)$. The maximum contact pressure, p_{max}, is at the center of the contact area and is 1.5 times greater than the average contact pressure, p_{av}.

For contacting parallel axes cylinders (Fig. 4.1.1b), width of the contact area is

$$b = 1.5\sqrt{q\frac{D_{ef}}{E}}, \qquad (4.2.2)$$

where D_{ef}, E are effective diameter and effective Young's modulus, as above, and q = the load per unit length of the contacting cylinders. The contact pressure is the greatest along the midline of the contact rectangle and is 1.27 times greater than p_{av}

$$p_{max} = 1.27\frac{q}{b}. \qquad (4.2.3)$$

In the maximum stress zone, the material is in a condition of omnidirectional compression under mutually perpendicular compression stresses, σ_x, σ_y, σ_z, and shear stresses, $0.5(\sigma_z - \sigma_y)$, $0.5(\sigma_z - \sigma_x)$, $0.5(\sigma_y - \sigma_x)$ acting at 45 deg. to compression stresses. Distribution of these stresses in one contacting cylinder (as ratios to the maximum pressure, p_{max}, in the contact area) perpendicular to the contact surface with the distance into the body of the block is shown in Fig. 4.2.1 [1]. The distance into the block is measured as multiples of width b of the contact area. The nor-

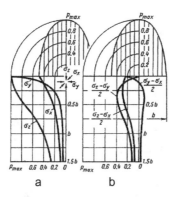

Fig. 4.2.1. Normal (a) and shear (b) stresses in the contact area between two cylinders with parallel axes.

Table 4.1. Displacements in Point and Line Contacts.

Sketch of contact	Auxiliary coefficients		Approach of contacting bodies
	(A)	**(B)**	
Sphere and plane	$(1/2R)$	$(1/2R)$	$178\sqrt[3]{\dfrac{P^2}{R}\left(\dfrac{1-v_1^2}{E_1}+\dfrac{1-v_2^2}{E_2}\right)^2}$
Sphere and cylinder	$1/2R_1$	$\tfrac{1}{2}(1/2R_1+1/2R_2)$	$1.41n_\Delta\sqrt[3]{P^2\,\dfrac{2R_2-R_1}{R_1R_2}\left(\dfrac{1-v_1^2}{E_1}+\dfrac{1-v_2^2}{E_2}\right)^2}$
Sphere and cylin. groove	$\tfrac{1}{2}(1/R_1-1/R_2)$	$1/2R_1$	$1.41n_\Delta\sqrt[3]{P^2\,\dfrac{2R_2-R_1}{R_1R_2}\left(\dfrac{1-v_1^2}{E_1}+\dfrac{1-v_2^2}{E_2}\right)^2}$
Sphere and thor. groove	$\tfrac{1}{2}(1/R_1-1/R_2)$	$\tfrac{1}{2}(1/R_1+1/R_3)$	$1.41n_\Delta\sqrt[3]{P^2\left(\dfrac{2}{R_1}-\dfrac{1}{R_2}+\dfrac{1}{R_3}\right)\left(\dfrac{1-v_1^2}{E_1}+\dfrac{1-v_2^2}{E_2}\right)^2}$
Roller bearing	$\tfrac{1}{2}(1/R_2-1/R_4)$	$\tfrac{1}{2}(1/R_1+1$	$1.41n_\Delta\sqrt[3]{P^2\left(\dfrac{2}{R_1}+\dfrac{1}{R_2}+\dfrac{1}{R_3}-\dfrac{1}{R_4}\right)\left(\dfrac{1-v_1^2}{E_1}+\dfrac{1-v_2^2}{E_2}\right)^2}$
Perpendicular cylinders	$1/2R_2$	$1/2R_1$	$1.41n_\Delta\sqrt[3]{P^2\,\dfrac{2R_2-R_1}{R_1R_2}\left(\dfrac{1-v_1^2}{E_1}+\dfrac{1-v_2^2}{E_2}\right)^2}$
Parallel cylinders	—	$\tfrac{1}{2}(1/R_1+1/R_{23})$	$\dfrac{2P}{\pi l}\left[\dfrac{1-v_1^2}{E_1}\left(\log_e\dfrac{2R_1}{b}+0.41\right)\right.$ $\left.+\dfrac{1-v_2^2}{E_2}\left(\log_e\dfrac{2R_2}{b}+0.41\right)\right]$
Roller between plates Δr = shrinking of roller Δ_{pl} = denting of each plate			$\Delta_r=4\dfrac{P}{l}\dfrac{1-v_1^2}{\pi E}\left(\log_e\dfrac{2D}{b}+0.41\right)$ $\Delta_{pl}=\dfrac{P}{l}\dfrac{1-v_2^2}{\pi E_2}\left(\log_e\dfrac{16d^2}{b^2}-1\right)$ $b=1.6\sqrt{\dfrac{P}{l}D\left(\dfrac{1-v_1^2}{E_1}+\dfrac{1-v_2^2}{E_2}\right)}$

A/B	1.00	0.404	0.250	0.160	0.085	0.067	0.044	0.032	0.020	0.015	0.003
n_Δ	1.00	0.957	0.905	0.845	0.751	0.716	0.655	0.607	0.546	0.510	0.358

mal stresses are the greatest on the surface ($\sigma_z = \sigma_y = p_{max}$; $\sigma_z = 0.5\ p_{max}$), but the shear stresses are the greatest at $0.25b$ to $0.4b$ depth.

It is important to note that in the condition of the omnidirectional compression the yield strength is changing and is four-to-five greater than the yield strength for the same material in the condition of uniaxial compression. This fact may explain very large elastic strains realized for both NiTi superelastic material and for steel cylinders as presented in Section 2.7.4.

Formulas for contact stiffness for some important cases are given in Table 4.1 [2]. Formulas in Table 4.1 were derived with the following assumptions: stresses in the contact zone are below the yield strength, contact areas are small relative to dimensions of the contacting bodies, contact pressures are perpendicular to the contact areas. These formulas also consider structural deformations of the contacting bodies.

The above listed assumptions presume static loads. If the contacting bodies are subjected to variable loads, or if they are in a relative motion relative to each other (e.g., in bearings — approximately pure rolling or in meshing gear teeth — rolling with sliding), then assuming the theoretical ("Hertzian") stress distribution in the contact area is not correct. If sliding is present, then friction forces would develop, thus changing the angle between the acting force and the perpendicular to the contact surface and violating one of the assumptions. Accordingly, the expressions in Table 4.1, derived for a static situation, represent only the first approximation.

4.3 CONTACT DEFORMATIONS BETWEEN CONFORMING AND QUASI-CONFORMING SURFACES

Contact deformations between conforming surfaces influence to a significant degree:

1. Vibration and dynamic loads;
2. Load concentration and pressure distribution in contact areas;
3. Relative and absolute positioning accuracy of the units.

The negative effect of contact deformations is obvious and is due to increased compliances caused by the structural joints. Role of contact deformations in the overall breakdown of deformations in mechanical systems can be illustrated by the following examples. Effective torsional compliance of power transmission systems (gearboxes) is comprised of three major sources: torsion of the shafts (~30%); bending of the shafts (~30%); contact deformations in the connections, such as keys, splines, and bearings (~40%) [3] (see also Chapter 6). Contact deformations in spindle units of machine tools are responsible for 30% to 40% of the total deformations at the spindle end. Contact deformations in carriages, cantilever tables, etc., constitute up to 80% to 90% of the total and in moving rams — 40% to 70% [4].

However, the positive effects of contact deformations also have to be considered. They include more uniform pressure distribution in joints and dissipation of vibratory energy (damping). Damping in joints between conforming surfaces is an important design feature in mechanical systems subjected to vibrations (see below, Section 4.7).

If the contact surfaces are perfectly flat (e.g., like contact surfaces of gage blocks) or if the curvature of one surface perfectly conforms with the curvature of its counterpart surface, and both surfaces have a high degree of surface finish, the magnitudes of contact deformations are insignificant. In a contact between two surfaces, which are neither perfectly flat nor perfectly conforming, and in which the average height of micro-asperities is commensurate with amplitudes of waviness, contact deformations are caused by Hertzian deformations between the apexes of the contacting waves and by similar deformations of micro-asperities representing roughness, as well as by squeezing lubrication oil from between the asperities.

There has recently been some progress in the computational analysis of interactions between non-ideal contacting surfaces. However, because of the extremely complex nature of the surface geometry, the reliable design information is still based on empirical data. For accurately machined and carefully matched flat contact surfaces of a relatively small nominal joint area (100 cm^2 to 150 cm^2 or 15 in.^{2} to 23 in.^{2}), contact deformations between cast iron and/or steel parts are nonlinear (Fig. 4.3.1) and can be expressed as:

$$\delta = c\sigma^m \qquad (4.3.1)$$

where σ = average contact pressure, MPa, and δ = contact deformation, μm (10^{-6} m; 1 μm = 40 μin). This expression describes a nonlinear hardening load-deflection characteristic with a degree of nonlinearity represented by the exponent m. For contacting steel and cast-iron surfaces, $m \approx 0.5$. The contact deformation coefficient, $c = 5.0$ to 6.0 for deep scraping, 3.0 to 4.0

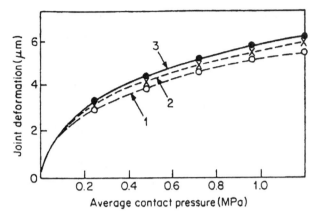

Fig. 4.3.1. Load-deflection characteristics for flat deeply scraped surfaces (overall contact area 80 cm^2). (1) No lubrication, (2) lightly lubricated (oil content 0.8×10^{-3} g/cm^2), and (3) richly lubricated (oil content 1.8×10^{-3} g/cm^2).

for medium-deep scraping, 0.6 to 1.0 for fine scraping, 0.5 to 1.0 for fine turning and grinding, and 0.2 for lapping [4]. As can be seen from Fig. 4.3.1, lubrication results in larger deformations. Since the formula (4.3.1) is empirical, it is approximate and parameters m and c may fluctuate.

Contacts between cast-iron or steel and plastic materials are characterized by $m \approx 0.33$ in Eq. (4.3.1) and by larger deformations, $c = 10$ to 18 for average scraped surfaces.

For very small joints (contact area 2 cm^2 to 3 cm^2 or ~ 0.3 in.^{2} to 0.5 in.^{2}), the waviness is hardly detectable, and as a result, the contact stiffness is about five times higher (or deformations for the same pressure σ are approximately five times smaller) [4]. Such small joints are typical for key connections, index pins, dead stops, etc.

For larger contacting surfaces, deformations are larger because of difficulties in matching the surfaces and of surface waviness effects. For large contacting surfaces, the exponent in Eq. (4.3.1) is closer to 1.0 (an almost linear load-deflection curve) because of higher local pressures and of the more pronounced role of structural deformations of the contacting components. Out-of-flatness of the order of 10 μm to 15 μm leads to 2 to 2.5 times higher c values in Eq. (4.3.1).

Contact stiffness is very sensitive to the method of surface finishing. There is some data showing that ball burnishing can increase contact stiffness by a factor of 1.5 to 2, and additional diamond smoothing can additionally increase it by about the same factor. Orthogonal positioning of machining traces (lay lines) on contacting surfaces increases contact stiffness by about 40% compared with a joint between the same surfaces with parallel machining traces.

Since the contact deformations are usually characterized by nonlinear hardening load-deflection characteristics, the contact stiffness can also be enhanced by increasing contact pressures. This can be achieved by *increasing* the load on the contact and/or by *reducing* the surface area of the joint. The incremental contact deformations of reasonably well-machined flat surfaces ($R_a \leq 4 \; \mu$m) can be neglected for contact pressures above 100 MPa.

Contact stiffness for vibratory loads (dynamic stiffness) is about the same as static stiffness if the joint is not lubricated. However, for lubricated joints, dynamic stiffness is about 5% higher than the static stiffness because of viscous resistance to squeezing oil from the contact area.

There were several studies on influence of thin film gaskets in the joints on their stiffness (e.g., [6]). Study of gaskets 0.1 mm thick made of various materials (PVC, polyethylene, aluminum and lead foils, rubber, etc.) has shown that use of the gaskets results in higher stiffness and damping. For example, use of a neoprene gasket in joints of a knee-type milling machine resulted in increase of dynamic stiffness from 1.5×10^7 to 2.6×10^7 N/m (83×10^3 lbs/in. to 140×10^3 lbs/in.), and increase of damping (as a percentage of the critical damping) from 2.5% to 3.2%. It has been found also that in the studied range of joint pressures, 2.5 MPa to 6.0 MPa (420 psi to 900 psi), stiffness does not significantly depend on the load, i.e., the joints with gaskets exhibit an approximately linear load-deflection characteristic. There are indications that a similar stiffening effect develops when the contacting surfaces are coated with a soft metal such as tin, lead, indium, silver, gold, etc. This effect of the thin, relatively soft, gaskets and coatings can be explained by their equalizing action, since they compensate effects of microasperities and, to some degree, also of waviness.

4.4 CONTACT STIFFNESS IN STRUCTURAL ANALYSIS

For design computations, small increments of contact deflections can be considered as proportional to the contact pressure

$$\delta = k(\sigma)\sigma \tag{4.4.1}$$

with stiffness $k(\sigma) = cm\sigma^{m-1}$ being dependent on the magnitude of pressure, σ.

Displacements of rigid parts such as brackets, massive carriages, etc., can be computed by considering their rotation and linear displacement relative to the supporting components due to contact deformations in their connections.

An example in Fig. 4.4.1a represents a rigid ram (weight, W), which is sliding in a prismatic joint comprising two support areas 1 and 2. External force, F, = applied to the end with an offset, h, and actuating force is $Q_1 = F_x + (R_1 + R_2)f$, where f = the friction coefficient. Reaction forces, $R_{1,2}$, acting on support areas can be found from equations of static equilibrium as

$$R_1 = \frac{W(1_c - a) + F_z(1 - a) - F_x h}{2a} \tag{4.4.2}$$

$$R_2 = \frac{Wl_c + F_z(1 + a) - F_x h}{2a} \tag{4.4.3}$$

The contact pressures and deformations in the support areas are:

$$\sigma_1 = \frac{R_1}{aB} \qquad \sigma_2 = \frac{R_2}{aB} \tag{4.4.4}$$

$$\delta_1 = c\sigma_1^m \qquad \delta_2 = c\sigma_2^m \tag{4.4.5}$$

$$\delta_F = \delta_1 - \phi(l + 2a) = \delta_1 - \left(\frac{\delta_1 + \delta_2}{2a}\right)(l + 2a) \tag{4.4.6}$$

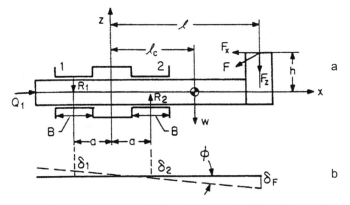

Fig. 4.4.1. (a) Rigid ram in a sliding joint and (b) deflected position of its axis.

Here B = width of the ram, δ_F = deflection at the point of application of force, F, σ_1 and σ_2 are contact pressures in support areas 1 and 2, δ_1 and δ_2 are contact deformations in the areas 1 and 2, and ϕ = the angular deflection of the ram caused by contact deformations (Fig. 4.4.1b).

A general case of a joint between two rigid parts involves are loading with a central force, P, and loading with a moment, M, (Fig. 4.4.2a) [4]. Equilibrium conditions can be written as:

$$B\int_{-a/2}^{a/2}\sigma_x dx = P \tag{4.4.7'}$$

$$B\int_{-a/2}^{a/2}\sigma_x x dx = M \tag{4.4.7''}$$

where B = width of the joint and σ_x = contact pressure at a distance x from the line of action of force, P. From Eq. (4.3.1), $\sigma_x = (\delta_x / c)^{1/m}$, where $\delta_x = \delta_0 + \phi x$ is the elastic displacement between the contacting surfaces in the cross-section, x, δ_0 = the elastic displacement under the force, P, and ϕ = the angular displacement in the joint.

With $\sigma_o = P/aB.= P/A$, and $\delta_0 \cong c\,\sigma_o^m$, it was shown in [4] that resolving Eqs. (4.4.7) gives for $m = 0.5$

$$\frac{\phi a}{\delta_0} \cong 6\frac{M}{Pa}, \tag{4.4.8}$$

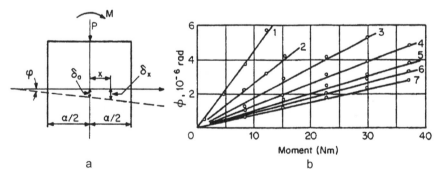

Fig. 4.4.2. (a) Flat joint loaded with compressive force, P, and moment, M, and (b) its torque versus angular deflection characteristics at various P. Contact area 75 cm^2; surface scraped; average pressure: (1) σ = 0.055 MPa, (2) 0.11 MPa, (3) 0.22 MPa, (4) 0.42 MPa, (5) 0.68 MPa, (6) 1.0 MPa, and (7) 1.35 MPa.

and for $m = 0.33$

$$\frac{\phi a}{\delta_0} \cong 4\frac{M}{Pa}. \tag{4.4.9}$$

The maximum and minimum values of the contact pressure for $m = 0.5$ are

$$\sigma_{max} = \frac{\delta_0^2}{c^2}\left(1 + \frac{\phi a}{2\delta_0}\right)^2 \cong \sigma_o\left(1 + \frac{3M}{Pa}\right)^2$$

$$\sigma_{min} = \frac{\delta_0^2}{c^2}\left(1 - \frac{\phi a}{2\delta_0}\right)^2 \cong \sigma_o\left(1 - \frac{3M}{Pa}\right)^2. \tag{4.4.10}$$

For $m = 0.33$

$$\sigma_{max} = \frac{\delta_0^2}{c^2}\left(1 + \frac{\phi a}{2\delta_0}\right)^2 \cong \sigma_o\left(1 + \frac{2M}{Pa}\right)^2$$

$$\sigma_{min} = \frac{\delta_0^2}{c^2}\left(1 - \frac{\phi a}{2\delta_0}\right)^2 \cong \sigma_o\left(1 - \frac{2M}{Pa}\right)^2. \tag{4.4.11}$$

From the expressions for σ_{min} in Eq. (4.4.10) and (4.4.11), it can be concluded that the joint "opens up" if $M_{max} > Pa/3$ for $m = 0.5$ or if $M_{max} > Pa/2$ for $m = 0.33$.

Expressions (4.4.8) and (4.4.9) show that for a constant average pressure, σ, angular deflection, ϕ is proportional to moment, M, or a *nonlinear joint preloaded with a force, P, has linear load-deflection characteristics for the moment loading up to M_{max}.*

An expression for the overall angular deflection for a rectangular joint can be easily derived from Eqs. (4.4.8) and (4.4.9) as

$$\phi \cong 12m\frac{\delta_o M}{Pa^2} \tag{4.4.12}$$

Substituting expressions for the cross-sectional moment of inertia and for the surface area of the joint, $I = Ba^3/12$ and $A = aB$ into Eq. (4.4.12)

$$\phi \cong m\frac{A\delta_o}{P}\frac{M}{I} \tag{4.4.13}$$

Using the expressions for δ_0 from Eqs. (4.4.9) and (4.4.10), the following expression for ϕ can be written:

$$\phi = k\frac{M}{I} \tag{4.4.14}$$

where

$$k = \mathrm{cm}\sigma_0^{m-1} \tag{4.4.15}$$

It was proven and confirmed by experiments in [4] that Eqs. (4.4.14) and (4.4.15) hold for $m = 1.0$, 0.5, and 0.33 not only for rectangular but also for hollow rectangular, ring-shaped, and round joints. For $m < 1$, k is diminishing with increasing σ (or P). Thus, preloaded joints behave as *linear angular springs stiffening with increasing preload force P*. Figure 4.4.2*b* shows the correlation between computations using Eq. (4.4.1), solid lines, and experiments, dots [4].

Effect of the preload on angular (moment) stiffness of a flat joint can be illustrated on a simple model in Fig. 4.4.3. The system in Fig. 4.4.3a comprises bar 1 supported by two identical nonlinear springs 2 and 3, $k_2 = k_3 = k$. Fig. 4.4.3b shows the (hardening) load-deflection characteristic of springs 2, 3. When bar 1 is loaded by a vertical force, $2P_a$, at the midpoint, stiffness of each spring is $k_2 (P_a) = k_3 (P_a) = k(P_a)$, and vertical stiffness of the system is $k_2 + k_3 = 2k(P_a)$. Thus, the system has a nonlinear stiffness in the vertical direction.

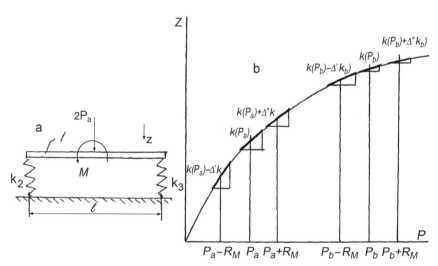

Fig. 4.4.3. Model of a nonlinear joint loaded with a moment: (a) Bar supported by nonlinear springs; (b) Load–deflection characteristic of the springs.

If the bar is loaded also with a moment M in addition to the force $2P_a$, this moment will be counterbalanced by two additional reaction forces, R_{2M} and R_{3M} having the same magnitudes but opposite directions,

$$R_{2M} = -R_{3M} = R_M = M/l.$$

Due to these additional forces, stiffness of spring 2 increases from $k(P_a)$ to $k(P_a + R_M) = k(P_a) + \Delta'k$, and stiffness of spring 3 decreases from $k(P_a)$ to $k(P_a - R_M) = k(P_a) - \Delta''k$ (see Fig. 4.4.3b). The incremental moment-induced deformation of spring 2 is $\Delta z_2 = R_M/k_2 = R_M/[k(P_a) + \Delta'k]$, and of spring 3, $\Delta z_3 = -R_M/k_3 = -R_M/[k(P_a) - R_M \Delta''k]$.

The angular stiffness of the system is $k_\alpha = M/\alpha$, where α is the angular deflection (tilt) of bar 1 caused by moment, M,

$$\begin{aligned} \alpha &= \frac{\Delta z_2 - \Delta z_3}{l} = \frac{1}{l}\left[\frac{R_M}{k(P_a) + \Delta'k} + \frac{R_M}{k(P_a) - \Delta'k}\right] \\ &= \frac{R_M}{l}\frac{k(P_a) + \Delta'k + k(P_a) - \Delta''k}{[k(P_a) + \Delta'k][k(P_a) - \Delta''k]} \cong \frac{2R_M}{lk(P_a)} \end{aligned} \tag{4.4.16}$$

since $\Delta'k \approx \Delta''k$ if $R_M \ll P_a$ (e.g., in the case of angular vibrations). Consequently,

$$k_{\alpha a} = \frac{M}{\alpha} = \frac{R_M l}{2R_M \Big/ lk(P_a)} \approx \frac{1}{2}k(P_a)l^2. \tag{4.4.16'}$$

If the normal force is increasing to $P_b > P_a$ (Fig. 4.4.3b), then, analogously,

$$k_{\alpha b} = \sim \tfrac{1}{2}k(P_b)l^2 \tag{4.4.16''}$$

Equations (4.4.16') and (4.4.16") show that the angular stiffness is increasing with increasing P but does not depend on M, or is *linear relative to M*.

If structural and contact stiffness values are commensurate, the computational procedure is more complex. Since, usually, one of the joined components is much more rigid than another, such computations can be performed with the assumption that the compliant component is a beam or a plate on a continuous elastic foundation (bed). Even for discrete contact points (like in a case of a sliding ram supported by rollers), substitution of continuous elastic foundation for discrete resilient mounts is fully justified. Computations were performed of the deflection diagrams for a beam with $EI = 3.6 \times 10^3$ Nm^2 and a concentrated load in the middle for two support conditions: (1) eight discrete resilient supports, each having stiffness $k = 1.37 \times 10^8$ N/m (corresponds to stiffness of a steel roller with 0.01 m diameter) and installed $a = 0.02$ m apart and (2) a distributed elastic foundation with stiffness $K = k/a = 0.68 \times 10^{10}$ N/m^2. The difference between the computed deflections of the two beams did not exceed 0.9%.

Figure 4.4.4 [4] shows a substantial effect of the joint compliance on a beam deflection as well as a good correlation between the experimental data and calculations based on considering the contact surface as an elastic foundation (bed). Such computations are important for sliding slender beams, such as robotic links in prismatic joints, sliding bars and sleeves in machine tools, etc. Elastic deflections of such components computed without considering joint deformations can be up to three times smaller than actual deflections, which can be determined using the elastic foundation technique.

Contact deformations in guideways are closely associated with local structural deflections of parts. Contact deformations can rise substan-

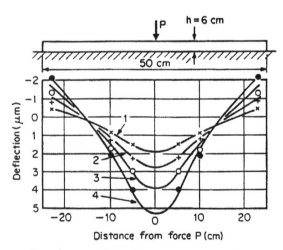

Fig. 4.4.4. Deflection profile of a beam caused by joint compliance (lines — computation, points — experiment); (1) average pressure $\sigma = 0.04$ MPa, $k(\sigma) - 1.8 \times 10^{-11}$ m^3/N; (2) $\sigma = 0.1$ MPa, $k = 1.0 \times 10^{-11}$ m^3/N; (3) $\sigma = 0.2$ MPa, $k = 0.65 \times 10^{-11}$ m^3/N; (4) $\sigma = 0.4$ MPa, $k = 0.45 \times 10^{-11}$ m^3/N.

tially (up to several times) because of distortions in guideways caused by local structural deformations.

From the information on contact stiffness given above, some recommendations on its enhancement can be made:

1. Better accuracy and surface finish to reduce values of c;
2. Proper orientation of machining traces on the contacting surfaces;
3. Adding thin plastic or soft metal gaskets or soft metal plating of the contacting surfaces;
4. Using smaller contact areas and/or higher specific loads (pressures) to take advantage of the hardening nonlinearity;
5. Using preload to take advantage of the hardening nonlinearity;
6. Optimization of the shape of contacting surfaces to increase their cross-sectional moments of inertia I (for cases of moment loading).

4.5 QUASI-CONFORMING CONTACT DEFORMATIONS IN CYLINDRICAL/CONICAL CONNECTIONS

Cylindrical and conical connections have special importance for design, since they are widely used both for sliding connections, when the male and female parts are fit with a clearance (bearings, guideways), and for holding (non-sliding) connections when there is an interference fit. Since frequently only a small fraction of an elongated part is interacting with a cylindrical or conical (tapered) hole, both radial and angular deformations in the connection are important. The angular deformations are projected to the end of the part and usually determine the effective stiffness of the system. Contact deformations in cylindrical and tapered connections cannot be analyzed by Hertz formulas, since those were derived for cases when the contact area is small in relation to curvature radii of the contacting bodies. These deformations cannot be directly analyzed using the basic notions of contact stiffness approach as described above in Section 4.3, since the degree of conformity and thus the nominal surface area of contact are dependent on tolerances and clearance in the connection.

Contact deflections in cylindrical joints depend on a degree of non-conformity, i.e., on magnitude of the allowance (clearance or interference). An increased clearance means a greater difference in curvature radii and, as a result, a reduction in the contact arc and a steep increase in contact deformations.

4.5.1 Cylindrical Connections

4.5.1a Connections with Clearance Fits

Cylindrical connections with clearance fits are illustrated in Fig. 4.5.1a [4]. Radii, r_1 and r_2, of the hole and the pin surfaces, respectively, are very close to each other, the clearance per diameter $\Delta = 2(r_1 - r_2)$ is several orders of magnitude smaller than r_1, r_2. Thus, the length of the contact arc is commensurate with the r_1, r_2. While Hertz formulas are not applicable in such cases, they can be used as an approximation when loads are small, and clearances are relatively large (thus, the contact area is small), so that

$$\frac{q}{\Delta} \leq 500 - 1000 \ \frac{N}{m * \mu m} = 0.5 \text{ to } 1.0 * 10^3 \ \frac{MN}{m^2} \left(70 \text{ to } 140 \frac{lb}{in * mil} \right),$$

$$(4.5.1)$$

where q, N/m (lb/in) is load per unit length of the connection, and Δ, μm (mil) is clearance (1 mil = 10^{-3} in).

For a more general case of higher loads and/or smaller clearances, deformations of the housing in which the hole is made are important.

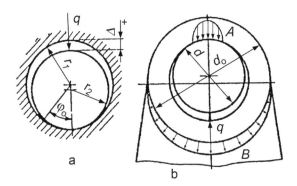

Fig. 4.5.1. (a) Cylindrical connection with a clearance fit loaded with axially distributed radial force q; (b) Assumed load distribution (A — acting load, B — reacting load).

Analysis of this problem was performed by D.N. Reshetov by cutting a cylinder from the housing and assuming a cosinusoidal distribution of the contact pressure between the shaft and the hole shown as A, and interaction of the cut out cylinder with the housing shown as B (cosinusoidal load distribution along 180 deg. arc; Fig. 4.5.1b). Fig. 4.5.2a shows ratio of maximum p_{max} to mean p_{av} contact pressures and contact angle, φ_o, as functions of q/Δ and d_o/d (d_o is external diameter of the cylinder cut from the housing, d = nominal diameter of the connection), and Fig. 4.5.2b shows shaft displacements, u. The housing is made from cast iron, the shaft (pin) is made of steel. The chain line represents an infinite size housing; it is obvious that comparison of lines 1, 2, 3 with the chain line shows that influence of thickness of the housing is very substantial.

Analysis illustrated by Fig. 4.5.2 assumed that both the pin and the hole are perfectly smooth, but real parts have micro-asperities and waviness whose deformations may add significantly to the overall deformations. Contact compliance of the surfaces with micro-asperities/waviness can be considered separately assuming that the associated deformations are much larger than deformations of the shaft/housing system which, for this phase of the analysis, is considered as rigid.

Two correlations between the contact pressures, p, and deformations, δ, are considered:

$$\delta = kp \quad \text{(a)}; \qquad \delta = cp^{0.5} \quad \text{(b)} \qquad\qquad (4.5.2)$$

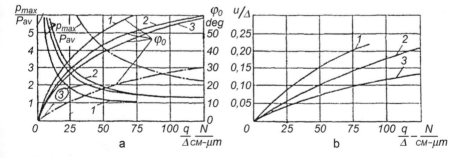

Fig. 4.5.2. Deformation parameters of cylindrical connection due to deformations of housing and shaft; (a) Length of contact arc φ_o, and contact pressure concentration, P_{max}/P_{av}; (b) Displacements u/Δ (u — shaft displacement, Δ — clearance); 1 — the bearing is loaded upwards, d_o/d =2; 2 — same, d_o/d = 3; 3 — downward loading, d_o/d =2.

In contact of two cylindrical surfaces with slightly differing r_1 and r_2, the radial clearance at angle, φ (Fig. 4.5.3a), is:

$$\Delta_\phi = 0.5\Delta\,(1 - \cos\phi). \tag{4.5.3}$$

Elastic radial displacement at angle φ to the direction of load, q (Fig. 4.5.3b),

$$\delta_\varphi = \delta\cos\varphi - \Delta_\varphi. \tag{4.5.4}$$

where $\delta =$ the displacement in the direction of the acting load, q. For linear contact compliance (4.5.2a), $\delta_\varphi = kp$, where k is contact compliance coefficient, and the vertical component of contact pressure at angle φ

$$p_v = p\cos\varphi = \delta_\varphi \cos\varphi/k \tag{4.5.5}$$

The total vertical load can be obtained by integrating p_v along the contact arc $2\varphi_0$,

$$q = (d/4k)\,(2\delta + \Delta)\,(\varphi_0 - \sin\varphi_0\cos\varphi_0). \tag{4.5.6}$$

Half-angle of contact arc φ_0 is determined from the condition $\delta_{\varphi 0} = 0$, and from Eqs. (4.5.3) and (4.5.4)

$$\cos\varphi_0 = \Delta/(2\delta + \Delta) \tag{4.5.7}$$

Thus, two unknowns, δ and φ_0, can be determined from Eqs. (4.5.6) and (4.5.7). Maximum pressure in the direction of force is:

$$p_{\max} = p_{av}[2(1 - \cos\varphi_0)]/(\varphi_0 - \sin\varphi_0\cos\varphi_0) \tag{4.5.8}$$

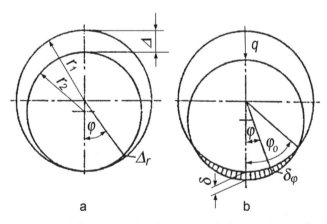

Fig. 4.5.3. Model for analysis of contact deformations of cylindrical connections with clearance; (a) Initial contact; (b) Contact under load.

where $p_{av} = q/d$. If there is no clearance, $\Delta = 0$, and

$$\varphi_o = \pi/2; \quad \delta = 4kq/\pi d; \quad p_{max} = (4/\pi)\, p_{av} \qquad (4.5.9)$$

Thus, $\pi\, d/4$ can be called an "effective width" of the connection.

Fig. 4.5.4 gives φ_o, p_{max}/p_{av}, and δ/Δ as functions of dimensionless parameter, $s_1 = qk/d\Delta$, where q is in 10^3 N/m, d in 10^{-2} m, Δ in $10^{-6}\, m$ (μm), k in 10^{-2} m/MPa.

For $\varphi_o \leq 10^{\circ}$ $\quad \sin\varphi_o \approx \varphi_o; \quad 1 - \cos\varphi_o \approx \varphi_o^2/2$, and after integration

$$\delta/\Delta = 0.83(qk/d\Delta)^{2/3}. \qquad (4.5.10)$$

If the contact deformation is nonlinear, $\delta = cp^{0.5}$, then analogously,

$$p_{av} = \frac{(\delta\cos\phi - \Delta_\phi)^2 \cos\phi}{c^2} \qquad (4.5.11)$$

$$q = \frac{d(2\delta + \Delta)^2}{4c^2}\left(\sin\phi_o - \frac{1}{3}\sin^3\phi_o - \phi_o\cos\phi_o\right) \qquad (4.5.12)$$

Considering also Eq. (4.5.7),

$$p_{max} = \frac{\delta^2}{c^2} = \frac{q}{d}\frac{(1 - \cos\phi_o)^2}{\sin\phi_o - \frac{1}{3}\sin^3\phi_o - \phi_o\cos\phi_o} \qquad (4.5.13)$$

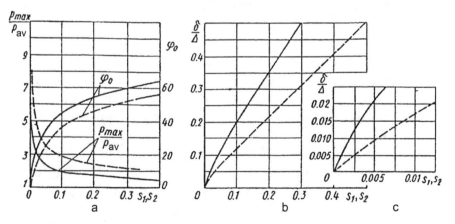

Fig. 4.5.4. Parameters of contact deformation of cylindrical connection.

For $\Delta = 0$

$$\delta = C\sqrt{\tfrac{3q}{2d}}, \; p_{max} = 1.5 \, p_{av}. \tag{4.5.14}$$

For a small contact arc (low load and/or large clearance)

$$p_{max} \approx \frac{15q}{8d}\sqrt{\frac{\Delta}{\delta}}; \; \frac{\delta}{\Delta} \approx \sqrt[5]{\frac{q^2 c^4}{d^2 \Delta^4}}; \; \varphi_o = 2\sqrt{\frac{\delta}{\Delta}} \tag{4.5.15}$$

Values δ/Δ, φ_o, and p_{max}/p_{av} are shown in Fig. 4.5.4 by broken lines as functions of dimensionless $s_2 = \frac{c}{d}\sqrt{\frac{2q}{d}}$.

Coefficients k and c were measured for connections shaft/ring with various clearances. Special attention was given to reduction of solid body deformations of the rings (such as making the outer diameter equal to three-times shaft diameter; applying load by a massive elastic semi-ring to distribute the load along a wide area). Both cast iron (HB180) and hardened steel (HRC 42) rings had been tested with hardened steel (HRC 45) shafts. Contact surfaces were machined to $R_a = 0.2$ μm; non-roundness was much less than the clearance magnitudes; tests were performed with $q = 60 \times 10^3$ N/m to 400×10^3 N/m (340 lb/in to 2260 lb/in). The test results are shown in Fig. 4.5.5a in comparison with analytical results. Plots in Fig. 4.5.4 were used to determine dimensionless parameters, s_1, s_2, for each δ/Δ and then k and c were determined.

Tests for high intensity loading (Fig. 4.5.5b), were performed with shafts having $R_a = 0.4$ μm and holes having $R_a = 0.4$ μm to 1.6 μm.

At high loading intensity, $q > 200 \times 10^3$ N/m (1130 lb/in) and contact arc $2\varphi_o > 80$ deg. to 90 deg., the best correlation between computational and test results is for the nonlinear dependence Eq. (4.5.2b) with values of coefficient c, close to its values for the flat joints having similar surface finish. For cases in Fig. 4.5.5a, $c = 0.18$; for cases in Fig. 4.5.5b, $c = 0.35$ to 0.45.

At low loads, $q < 200 \times 10^3$ N/m (1130 lb/in), $2\varphi_o < 90$ deg., values of c are increasing with increasing load, from $c = 0.08$ to 0.12 to 0.18 due to reduced influence of surface waviness for small contact surfaces, less than 5 cm^2 to 6 cm^2 (~1 in.^{2}) for the tested specimens. At low loads, better correlation was observed when the linear dependence Eq. (4.5.2a) was used. For the steel/cast iron pair coefficient k was 0.25 μm/MPa to 0.29

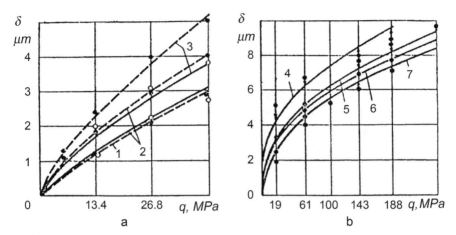

Fig. 4.5.5. Elastic contact displacements in cylindrical connections shaft/ring with clearance; (a) Low specific loads; (b) high specific loads; dots — test data, lines — computation; solid lines — steel-to-steel, broken lines - steel-to-cast iron. Clearance, Δ, per diameter: 1 — 3 μm; 2 — 13 μm; 3 — 29 μm; 4 – 7 – 20 — 30 μm. Dimensions $d \times b$, mm: 1–3 – 55x25; 4–75x8; 5–110x8; 6–80x8; 7–90x8.

μm/MPa for small clearance 3 μm to 10 μm and 0.3 to 0.32 for larger clearances (15 μm to 30 μm), while for the steel/steel pair, $k = 0.17$ to 0.2 for 10 μm to 15 μm clearance, and $k = 0.21$ to 0.24 for 29 μm clearance. These values also correlate well with values of $k = 0.3$ μm/MPa for flat ground joints hardened steel/cast iron, and hardened steel/hardened steel, $k = 0.2$ μm/MPa. Computed values of δ in Fig. 4.5.5a were determined using the linear dependence for low loads ($k = 0.3$ μm/MPa for steel/cast iron, $k = 0.2$ μm/MPa for steel/steel), and using nonlinear dependence for high loads ($c = 0.7$ μm/MPa$^{0.5}$ to 0.9 μm/MPa$^{0.5}$).

Combined influence of contact and solid body deformations can be obtained assuming their independence. Relative importance of the component deformations is determined by criterion

$$\gamma = \frac{k\pi^2 E}{4(1-v^2)10^4 d} \qquad (4.5.16)$$

where k = contact compliance coefficient, μm/MPa; E = Young's modulus of the housing, MPa; v = Poisson's ratio; d = connection diameter, cm.

When $\gamma \le 0.5$, displacements in the connection are determined largely by deformations of the shaft and the housing, and plots in Fig. 4.5.2 can be used. When $\gamma \geq 10$, the determining factor is contact deformations, and plots in Fig. 4.5.3 can be used. At $0.5 < \gamma < 10$, both components are commensurate, and the total deformation

$$u = \xi \delta, \tag{4.5.17}$$

where ξ = a correction factor given in Fig. 4.5.6.

Real shafts and holes have some deviations from the round shape, which may be commensurate with the clearance magnitude (Fig. 4.5.7). The influence of these deviations on contact deformations in the connection can be relatively easy to analyze if a sinusoidal pattern of non-roundness is assumed,

$$\Delta_{c\varphi} = \Delta_c \cos n\varphi, \tag{4.5.18}$$

where Δ_c = the maximum deviation from the cylindrical shape. This assumption is in many cases very close to actual shape of the surface. The surface radius at angle φ is

$$\rho = r\Delta_c \cos n\varphi, \tag{4.5.19}$$

and by changing n, typical basic patterns in Fig. 4.5.7 can be represented. For connections in which one surface is assumed to be perfectly round and the other having deviations as per Eq. (4.5.19), the initial clearance at angle φ is (Fig. 4.5.8 a, b),

$$\Delta\varphi - 0.5\Delta(1 - \cos\varphi) - \Delta_c(\cos n\varphi - \cos\varphi) \tag{4.5.20}$$

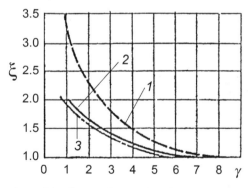

Fig. 4.5.6. Correction factor for computing total deformations in cylindrical connections with clearances; 1 — the bearing is loaded upwards, $d_o/d = 2$; 2 – same, $d_o/d = 3$; 3 — downward loading, $d_o/d = 2$.

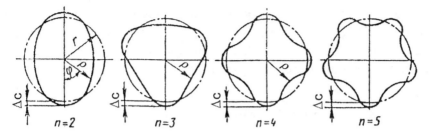

Fig. 4.5.7. Basic patterns of non-roundness.

and the correlations between δ, q, and φ_o are as follows:

$$qk/d\Delta - (\varphi_o - \sin\varphi_o\cos\varphi_o)/4\cos\varphi_o +$$
$$+(\Delta_c/2\Delta)\{[\sin(n-1)\varphi_o]/(n-1) + [\sin(n+1)\varphi_o]/(n+1) -$$
$$\cos n\varphi_o/\cos\varphi_o(\varphi_o + \sin\varphi_o\cos\varphi_o)\}; \qquad (4.5.21)$$

$$\delta/\Delta = 1/2\cos\varphi_o - 1/2 + (\Delta_c/2\Delta)(1 - \cos n\varphi_o/\cos\varphi_o) \qquad (4.5.22)$$

Plots in Fig. 4.5.8c give correction coefficient, c_c, for calculating an increase of elastic deformation due to deviations from roundness of one part of the connection for two basic patterns of deviations along axis, z_1. It can be seen that deformations may increase 1.4 to 1.7 times for small loads and 1.1 to 1.5 times for large loads. Loading in z_2 direction gives different deformations from loading along z_1. since in most cases the initial contact will be in two points. Thus, usually, displacements in z_1

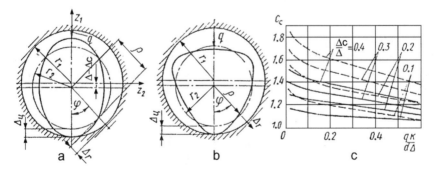

Fig. 4.5.8. Contacting between out-of-round surfaces; (a), (b) Typical shapes; (c) Correction coefficient for deformations of cylinders with deviations from roundness; solid lines — oval shape (case a), broken line — triangular shape (case b).

direction are larger than at $\Delta_c = 0$, but in z_2 direction, they are smaller. If there is a relative rotation between the parts, such stiffness variation can create undesirable effects, such as escalation of parametric vibrations or scatter of machining/measuring errors.

4.5.1b Interference-Fit Cylindrical Connections

Usually, displacements and pressures caused by interference fits are much larger than displacements and pressures caused by external forces. The initial interference-fit pressures create a preloaded system. Under external loading of the connection, total pressures on one side are increasing, while on the opposite side — decreasing, and magnitudes of incremental elastic deformations in two diametrically - opposed points are equal. While the maximum displacements caused by the external force are along the direction of force, distribution of the displacements along the circumference can be assumed to be cosine.

For large magnitudes of the interference fit pressures, p_i, approximately

$$\Delta = \left(\frac{d\delta}{dp}\right)_{p=p_1} \quad p \approx kp, \tag{4.5.23}$$

where p_i is interference pressure between the connected components; $p = 2q/\pi d$ is maximum pressure from the external load, having approximately the same magnitude on the loaded and unloaded sides; and $k = 0.5cp_i^{-0.5}$ is the contact compliance coefficient.

Elastic displacement in the connection thus can be determined as

$$\delta = 2kq/\pi d. \tag{4.5.24}$$

Dependence between the magnitude of interference (diametral interference, Δ) and pressure in the interference fit is, considering contact displacements,

$$\Delta = 10^4 p_i d \left[\frac{1}{E_1}\left(\left(\frac{1+\varsigma_1^2}{1-\varsigma_1^2} - v_1\right)\right) + \frac{1}{E_2}\left(\left(\frac{1+\varsigma_2^2}{1-\varsigma_2^2} - v_2\right)\right) \right] + 2cp_i^{0.5}, \tag{4.5.25}$$

where subscripts 1, 2 relate to male, female parts, respectively, E_1, E_2 are Young's moduli, MPa; v_1, $v_2 =$ Poisson's ratios; $\varsigma_1 = d_h/d$; $\varsigma_2 = d/d_o$; $d =$

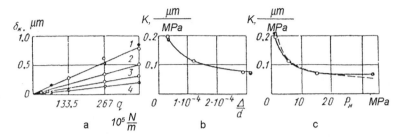

Fig. 4.5.9. Elastic displacements (a) and contact compliance coefficients (b, c) in cylindrical interference fit connections shaft/ring. 1 Diametral interference $\Delta = 2$ μm; $2\Delta = 7$ μm; $3\Delta = 13$ μm; $4\Delta = 16$ μm; surface finish $R_a = 0.2$ μm; • — hardened steel shaft, cast iron ring; 0 — hardened steel shaft and ring.

connection diameter, cm; d_h = diameter of a hole, if any, in the male part, cm; d_o = external diameter of the female part, cm.

Coefficients k were experimentally determined for shaft/ring connections similar to the ones used for the study of connections with clearances, with various interferences. Fig. 4.5.9a shows load-radial deflection characteristics, while Fig. 4.5.9b,c can be used to determine the value of k as a function of the relative interference, Δ/d, Fig. 4.5.9b, and interference pressure, Fig. 4.5.9c. The broken line in Fig. 4.5.9c represents correlation $k = 0.5cp_i^{0.5}$, computed for $c = 0.47$. This value of c is also typical for flat joints and for cylindrical connections with a clearance for the same surface finish.

4.5.2 Elastic Displacements in Conical (Tapered) Connections

Conical connections of mechanical components are very popular since they provide self-centering of the connection and also allow one to realize an easy to disassemble interference-fit connections. They are frequently used in machine tools and measuring instruments (e.g., Fig. 4.5.10), as well as in other precision devices. Deformations in such connections are due to large magnitudes of *bending moments* caused by long overhangs of cantilever tools, etc., connected by the taper, due to *bending deformations* of the cantilever tools, which cause very nonuniform pressure distributions inside the connection, and also due to inevitable differences in the taper angles of the male and female components of the connection. Angular deformation θ_o at the end of the connection due to contact

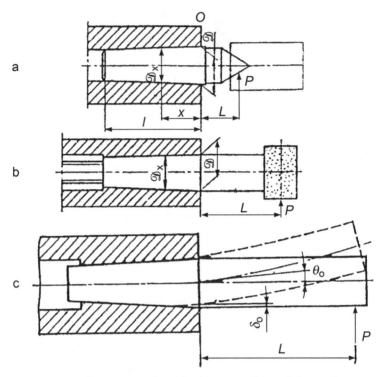

Fig. 4.5.10. Schematics of conical connections. (a) Dead center in lathe spindle; (b) Grinding mandrel in grinding machine spindle; (c) Deformations in conical connection.

deformations may cause large "projected" deformations at the tool end (Fig. 4.5.10c). Frequently, these projected displacements may even exceed bending deformations of the connected parts (tools, mandrels, etc.) on the overhang. Elastic deflection under the load is:

$$\delta = \delta_p + \delta_o + \theta_o L, \qquad (4.5.26)$$

where δ_p is deformation of the tool/mandrel of length, L, itself if it is considered "built-in" at the cross-section O (mouth of the female taper), $\delta_o =$ radial deformation in the connection, $\theta_o L =$ projected angular deformation.

4.5.2a Test Data

An experimental study in [4] was performed on spindles of internal grinders. The spindle was sturdily supported by bearings, especially under the front end, and mandrels with Morse #3 taper were inserted and axially preloaded by a drawbar. A load P was applied at the free end of the mandrel, and angular deflection at the mouth of the spindle hole was measured as shown in Fig. 4.5.11. Sometimes, the total displacement at the end of the mandrel was also measured. The testing was performed at small loads, typical for precision machines. Surface finish of the tapers was: for the mandrel, $R_a = 0.4\mu m$, for the spindle hole, $R_a = 0.2\mu m$. In all tested connections (with one exception), a "fitting dye" test has shown full conformity between the male and female tapers. The same mandrel was tested in several spindles to assess influence of manufacturing errors. The test results are shown in Fig. 4.5.12; the similar tests were performed with gage-quality tapers Morse #3 and #4, which were lapped together with the respective holes.

Tests in [7] were performed on machining center spindles, both on 7/24 taper #50 toolholders and on hollow HSK toolholders (taper 1/10) (Fig. 4.5.13), providing simultaneous taper and face contact due to radial deformation of the hollow taper when the axial (drawbar) force is applied.

All these tests have shown that:

1. Load-angular deflection characteristic is linear since the connection is preloaded;
2. A stronger axial preload results in increased stiffness (smaller angular deflection). This effect is more pronounced on imperfect

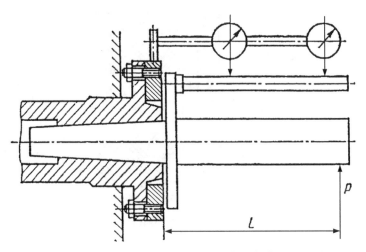

Fig. 4.5.11. Set-up for measuring angular deflections in tapered connection.

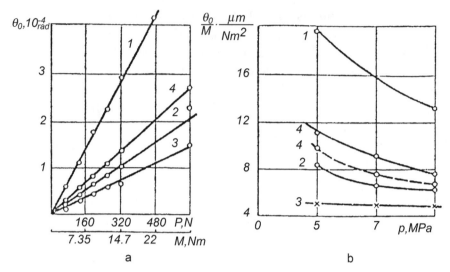

Fig. 4.5.12. Test results for four tapered connections 1 to 4 between mandrel and spindle (Morse #3); (a) Inclination angle at the mouth, interference pressure, 7 MPa (1000 psi); P — external force, M — bending moment at the mouth; (b) Angular compliance, θ_o/M, versus pressure p in the connection; broken line represents test results for mandrel rotated (90°) relative to spindle.

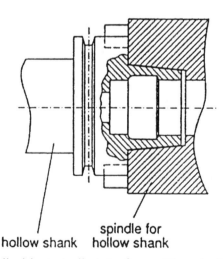

Fig. 4.5.13. Toolholder/spindle interface with a shallow hollow taper (HSK).

connections (low stiffness connections 1 and 4, Fig. 4.5.12), wherein increase of the drawbar force improves the contact conformity. For connections with a good initial fit (and high initial stiffness), elastic displacements do not depend significantly on the preload (connection 3). Such "saturation" of the stiffness values is confirmed by testing of steel 7/24 tapers #50 performed in [7] (Fig. 4.5.14). It can be seen that deformations of the tool end do not diminish noticeably at preloading (drawbar, entry) force exceeding 20,000 N for taper #50.

3. Manufacturing deviations are important, also see the following Section 4.5.2c. Deflections for several connections for the same test conditions and a good dye-tested conformity differ by 50% to 100% (lesser differences for high drawbar forces). Stiffness of the same connection after rotating of the mandrel may also change noticeably (25% to 30% for connection 4, Fig. 4.5.12). It is interesting to note that tests in [7] have shown, in some cases, a deterioration of stiffness for very precise connections, Fig. 4.5.15 (tolerances for taper grades AT3, AT4, AT5 per former ISO Standard 1947 "System of

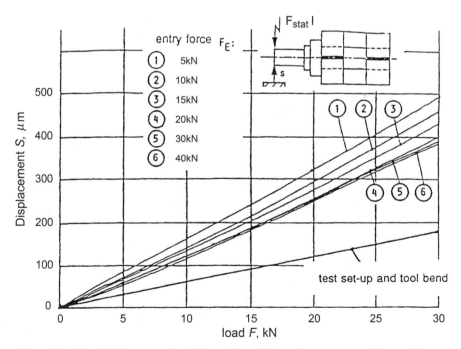

Fig. 4.5.14. Influence of axial preload (drawbar or entry force, F_E) on stiffness of 7/24 #50 taper connection.

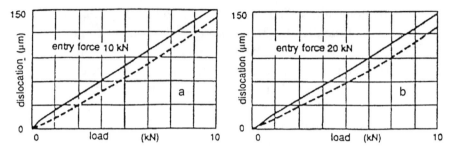

Fig. 4.5.15. Influence of fabrication accuracy of 7/24 taper connection on its stiffness. Solid lines — combination toolholder/spindle AT3/AT3; broken line — combination AT4/AT5.

Cone Tolerances" are equivalent to angular deviations 8, 13, 21 angular seconds, respectively, for both male and female tapers). These results can be explained by a possibility of some uncertainty of the contact area location for a very precise connection, since structural deformations may exceed the very small dimensional variations. The deformations caused by the radial force are due, to a large extent, to the angular mismatch between the male and female tapers. Accordingly, the radial stiffness can be significantly enhanced by bridging the clearance at one end of the connection resulting from the angular mismatch by elastic elements (see Section 7.2).

4. Radial stiffness of the lapped connections is higher than that of the ground connections, and the scatter between the 12 tested lapped connections was not exceeding 25% to 30%.

5. A typical load-deflection characteristic of a 7/24 taper #50 in the axial direction is shown in Fig. 4.5.16 [7]. Increase in the axial force causes a very significant axial deformation, which is equivalent to axial stiffness 0.6 – 0.8 μm/kN (the larger value at the linear region). The axial stiffness is not strongly dependent on accuracy of the connection, but is dependent on lubrication condition (lower stiffness for lubricated connections). Both axial stiffness and axial accuracy problems are solved by using the elastic taper providing a simultaneous taper/face contact (see Section 7.2).

6. The radial position accuracy (runout) of a component (such as a toolholder) connected to another component (e.g., spindle) by a tapered connection is uncertain due to tolerances on the taper angles and due to wear of the connection caused by repeated connections/

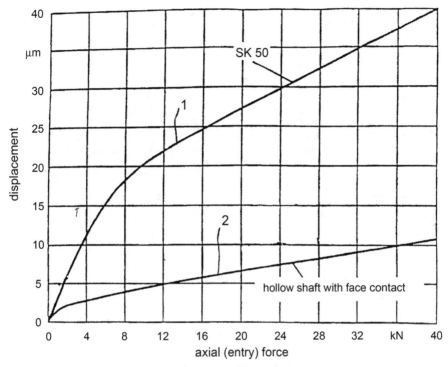

Fig. 4.5.16. Axial displacement of #50 taper (1) and HSK toolholder with face contact (2) with the spindle versus axial force.

disconnections. For a typical case when the taper angle of the male part is steeper than the taper angle of the female part (see below Section 4.5.2c), the wear pattern is the so-called "bell mouthing" of the spindle hole. This uncertainty can be significant as shown in Fig. 4.5.17 [7]. The uncertainty is decreasing with the increasing axial (drawbar) force (Fig. 4.5.17b), and is more pronounced for used (worn out) tapers Fig. 4.5.17c. The uncertainty is increasing with increasing angular mismatch. It can be significantly reduced by bridging the angular mismatch or by achieving a simultaneous taper/face contact (see Section 7.3).

7. A preloaded taper connection results in shrinking of the internal (male) part and expansion of the external (female) part of the connection. Such an expansion is undesirable, since it can change clearances in the bearings of the external part (spindle) and degrade their performance. A study in [7] demonstrated that expansion of

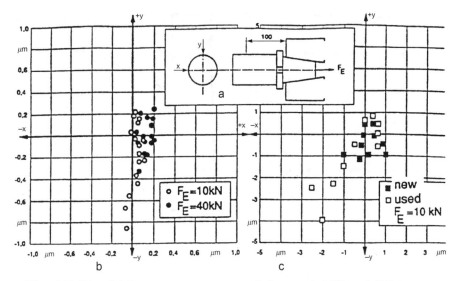

Fig. 4.5.17. (a) Setup to measure radial repeatability of #50 tapers measured at 100 mm from spindle face (a); (b) Repeatability as a function of axial (drawbar) force, F_E; (c) Repeatability as a function of wear.

the spindle by an axially preloaded solid steep 7/24 taper is very small and can be neglected. However, deformation of the spindle during insertion of a hollow taper with a small taper angle, as in Fig. 4.5.13, can be significant due to a greater mechanical advantage creating large radial forces necessary for radial contraction of the hollow taper to obtain the simultaneous taper and face contacts. The spindle expansion is shown in Fig. 4.5.18 for different modifications of the hollow taper/face interface [7]. Especially important is expansion under bearings, measuring point 4 in Fig. 4.5.18.

4.5.2b Computational Evaluation of Contact Deformations in Tapered Connections

A mandrel or a cantilever tool can be considered as a cantilever beam on elastic bed, which is provided by the surface contact between the male and female tapers. The connection diameter (i.e., width of the elastic bed, B_x) varies along its length, so

$$D_x = D\left(1 - \frac{2\alpha x}{D}\right); EI_x = \frac{E\pi D_x^4}{64}; B_x = \frac{\pi D_x}{2} \qquad (4.5.27)$$

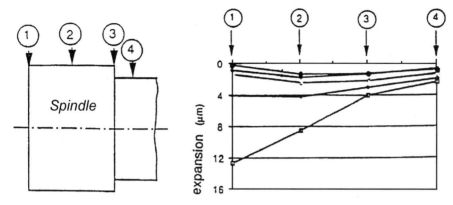

Fig. 4.5.18. Bulging of spindle caused by insertion of HSK-type toolholders at 15 kN axial force. (a) Positions of measuring points; (b) Test results.

where 2α = taper angle, D = larger (gage) diameter of the connection; D_x = diameter at distance x from the mouth. Since there is always an initial preload, it can be assumed that the contact arc is the whole circumference πD_x and that the pressure-deflection characteristic is linear. The deflected shape of such connection (deflection y in cross-section x) is described by differential equation:

$$\frac{d^2}{dx^2}\left(EI_x \frac{d^2y}{dx^2}\right) + \frac{B_x}{k} y = 0. \tag{4.5.28}$$

Its boundary conditions are: at $x = 0$, $d^2y/dx^2 = M/EI$, $d^3y/dx^3 = P/EI$; at $x = l$, $d^2y/dx^2 = 0$, $d^3y/dx^3 = 0$. Here $M = PL$, and P are bending moment and force at the mouth of the connection, respectively.

Elastic displacement δ_0 and angular displacement, θ_0, at the mouth are:

$$\delta_0 = \frac{0.2P\beta k}{B}(\beta L C_1 + C_2), \mu m \tag{4.5.29'}$$

$$\theta_0 = \frac{0.2P\beta^2 k}{B}(2\beta L C_3 + C_4), \mu m / cm \tag{4.5.29''}$$

where P = force, N; L = distance from the force to the mouth; $\beta = \sqrt[4]{\dfrac{Bx10^2}{4EIk}}$ = stiffness index of the connection, 1/cm; E = Young's modulus, MPa; $I = \pi D^4/$

Table 4.2. Coefficients C_1, C_2, C_3, C_4.

Morse Taper (1:10)					Steep (7/24) Taper			
$\lambda = \beta l$	C_1	C_2	C_3	C_4	$\lambda = \beta l$	C_1	C_2	C_3	C_4
2	1.23	1.2	1.2	1.23	1.5	2.34	2.06	1.70	2.34
3	1.10	1.08	1.015	1.10	2	2.16	1.94	1.35	2.16
4	1.06	1.04	1.01	1.06	2.5	1.65	1.64	1.17	1.65
5	1.05	1.03	1.0	1.05	3	1.45	1.48	1.07	1.45
6	1.04	1.03	1.0	1.04	3.5	1.34	1.36	1.05	1.34
					4	1.30	1.34	1.04	1.30

64 = cross sectional moment of inertia of the male taper at the mouth, cm^4; $B = 0.5\pi D$ = effective width of the connection; C_1, C_2, C_3, C_4 (Table 4.2) = correction coefficients considering influence of the variable diameter. For a cylindrical shrink-fit connection and for Morse tapers at $\beta l > 6$, C_1, C_2, C_3, $C_4 =$ ~1. Contact compliance coefficient, k, μm cm^2/N; was measured for connections of parts having $R_a = 0.4\mu$m and plotted in Fig. 4.5.19. For a good fitting of the connected tapers (gage-like accuracy, Fig. 4.5.19 (b)) and high pressures in the connection, $k = 0.1$ μm/MPa to 0.15 μm/MPa, for a poor fitting and low pressures, $k = 0.4$ μm/MPa to 1.0 μm/MPa.

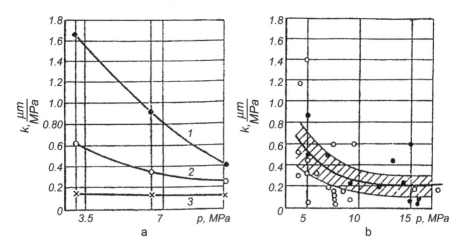

Fig. 4.5.19. Contact compliance coefficients for preloaded conical connections. (a) Mandrel/spindle (connections 1,2,3); (b) Gage tapers (hatched area represents 90% probability); ● — taper Morse #3; o — taper Morse #4; x — interference-fit cylindrical connection.

4.5.2c Influence of Manufacturing Errors

Inevitable manufacturing errors may cause a non-uniform contact or even a partial loss of contact in the connection. Typical errors are (Fig. 4.5.20):

1. Deviation from roundness of the cross-sections;
2. Deviations from straightness of the side surface;
3. Deviations from the nominal taper angle.

The most important parameter is angular difference between the male and female tapers, when the contact exists only on a partial length, l_e. Two cases may exist:

1. The hole has a steeper taper angle than the mandrel;
2. The hole has a more shallow angle than the mandrel.

In the former case, the contact area is at the smaller diameter, and the effective length of the mandrel overhang is increasing thus negatively influencing the bending stiffness of the mandrel. In the latter case, the contact area is at the larger diameter and stiffness of the connection is affected not very significantly when the loading is not too intense. However, at high loading (or dynamic loading, as during milling operations), the short frontal contact area behaves like a pivot. Small angular motions of the mandrel cause a fast wear of the contact area ("bell mouthing" of the spindle), fretting corrosion, and increased runout at the free end of the mandrel. These ill effects can be alleviated or eliminated by "bridging" the clearance at the back of the connection by using precision elastic elements (see Section 8.3).

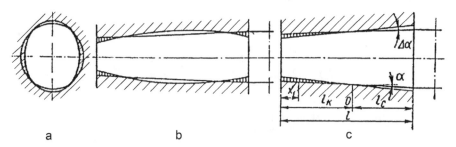

Fig. 4.5.20. Shape and fit errors of tapered connections. (a) Non-roundness; (b) Non-straightness; (c) Angle differential.

Preloading of the connection by axial pull of the mandrel by the drawbar results in radial displacements δ_r consisting of elastic deformation of the connected parts, δ_1, and contact deformations δ_2,

$$\delta_{rx} = \delta_1 + \delta_2, \tag{4.5.30}$$

where δ_{rx}, μm, is radial displacement at the coordinate x, cm. Neglecting change of diameter along the axis of the tapered connection,

$$\delta_1 = \frac{10^3 \, p D_{ef}}{E} \mu m; \qquad D_{ef} = \frac{D}{1 - \dfrac{D^2}{D_o^2}} \tag{4.5.31}$$

where p = contact pressure in the connection, MPa, D = nominal (mean) diameter of the connection cm; D_0 = external diameter of the housing, cm; D_{ef} is an effective diameter of the connection. Contact deformations, δ_2, are described by Eq. (4.3.1).

Computational analysis has shown that for realistic contact pressures and dimensions,

$$p = A \delta_{rx}^n. \tag{4.5.32}$$

It can be assumed that for ground and lapped parts in Eq. (3.3.1), $c = 0.3$; for this value of c, A and n are given in Table 4.3.

If the radial interference in the initial cross-section $x = 0$ is δ_{ro}, then in the cross-section with axial coordinate x, it is:

$$\delta_{rx} = \delta_{ro} - \Delta \alpha x, \tag{4.5.33}$$

where $\Delta \alpha$ is angle differential in μm/cm. Accordingly, contact pressure, p_x, and total axial preloading (drawbar) force, P_{db}, are

$$p_x = A \delta_x^n \tag{4.5.34}$$

$$P_{db} = \int_o^{\ell_k} p_x \pi D_x \tan(\alpha + \varphi) dx = \frac{\pi A \delta_{ro}^{n+1} D \tan(\alpha + \varphi)}{(n+1)\Delta \alpha}, \tag{4.5.35}$$

Table 4.3. Values of A and n.

D_{ef}, mm	20	30	40	50	70	100
A	3.9	3.1	2.6	2.1	1.7	12.5
n	1.5	1.35	1.33	1.25	1.2	1.18

where α = one-half of the taper angle, φ = friction angle (friction coefficient $f = \tan \varphi \approx 0.25$ for non-lubricated connections). Expression (4.5.35) is correct for $\ell_k \leq \ell$, where the length of contact zone ℓ_k is determined from the condition of disappearing contact pressure at some cross-section **o**,

$$\delta' = \delta_{ro} - \ell_k \Delta\alpha = 0; \tag{4.5.36}$$

$$\ell_k = \delta_{ro} / \Delta\alpha, \text{ cm} \tag{4.5.37}$$

or

$$\frac{\ell_k}{\ell} \approx \sqrt[n+1]{\frac{(n+1)p}{A(l\Delta\alpha)^n}} \tag{4.5.38}$$

Fig. 4.5.21a shows l_k / l for steel Morse tapers as a function of parameter at various D_{ef} and p, and as a function of Δ/δ_{av}, where $\delta_{av} = \sqrt[n]{p/A}$ (Fig 4.5.21b). Product $\Delta = \ell\Delta\alpha$ represents linear dimension deviation accumulated along the length of the connection, and δ_{av} = average contact deformation during preloading, and Δ/δ_{av} is the ratio of the linear error to the average contact deformation.

Thus, elastic deflection under load, P (Fig. 4.5.10c), for an ideal Morse taper connection (or a cylindrical "shrink fit" connection) can be calculated as

$$\delta = \frac{10^6 PL^3}{3EI} + \frac{0.02P\beta k(1 + 2\beta L + 2\beta^2 L^2)}{B}, \ \mu m \tag{4.5.39}$$

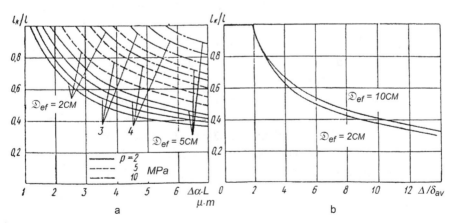

Fig. 4.5.21. Relative length of contact zone, ℓ_k / ℓ, as function of $\Delta\alpha$ (a) and $\Delta\delta_{av}$ (b).

where I is cross-sectional moment of inertia of the cantilever mandrel along its free length L; it is assumed that coefficients, C_1, C_2, C_3, $C_4 = 1$ in Eq. (4.5.39).

Influence of angular mismatch on stiffness can be approximately evaluated by substituting l_k as the length of the elastic bed and $(L + l_c)$ as the length of the overhang, where $l_c = L - l_k$. Then formula (4.5.39) becomes

$$\delta = \frac{10^2 PL^3}{3EI} + \frac{0.02P\beta'k[1 + 2\beta'(L + l_c) + 2(\beta')^2(L + l_c)^2]}{B'} \quad (4.5.39')$$

where

$$\beta^1 = \sqrt[4]{\frac{10^2 B^1}{4EI'k}}. \quad (4.5.40)$$

It is assumed in Eq. (4.5.39') that diameter of the cantilever part on the interval, l_c, is approximately equal to diameter of the cylindrical part. Effective parameters, B', and I, should be calculated for the initial cross-section at the distance l_c, from the mouth of the connection.

Example. Morse #3 taper connection ($l = 73$ mm, $D_{ef} = 3$ cm). Let's find out at what value of $\Delta\alpha$ and p the contact zone is along the whole length, $l_k /l = 1$. From Fig. 4.3.21b, this condition is satisfied if $\Delta/\delta_{av} < 1.9$. Then, at $\Delta\alpha = 5''$ (25 μrad or 0.24 μm/cm), it is required to create contact pressure, $p = 3$ MPa, at $\Delta\alpha = 10''$ (50 μrad or 0.48 μm/cm) $p = 7.5$ MPa, at $\Delta\alpha = 5''$ (75 μrad or 0.72 μm/cm) $- p = 12$ MPa.

Thus, for realistic values of $p = 3$ MPa to 8 MPa, only very small deviations of the angle are allowable. Actually, the allowable $\Delta\alpha$ are even less, since there is need to create a finite contact pressure at the mouth to prevent opening of the connection under load.

If at $p = 5$ MPa $\Delta\alpha = 10''$ (50 μrad), then $\Delta = l\Delta\alpha = 3.5$ μm. From Fig. 4.5.21a, $l_k /l = 0.82$ and $l_c = l(1 - l_k /l) = 1.3$ cm. At the mouth (gage) cross-section, $D = 2.38$ cm, $I = 1.58$ cm^4, $B = 3.74$ cm, $\beta = 0.62$ 1/cm. At the cross-section at the distance, $l_c = 1.3$ cm from the mouth, $D' = 2.38 - 0.05l_c = 2.31$ cm, $I' = 1.4$ cm^4, $B' = 3.62$ cm, $\beta = 0.64$ 1/cm. If the external force is applied at 7 cm from the mouth, then elastic deflection under the force, P, at $\Delta\alpha = 0$ is $\delta = 0.067P$, μm, but for $\Delta\alpha = 10''$ and $p = 5$ MPa, $\delta = 0.11P$ or 1.6 times greater. At $\Delta\alpha = 15''$ and $p = 5$MPa, elastic deflections are 2.5 times greater.

Even for very small angular deviations, for the positive angular difference between the hole and the mandrel, the contact zone can be only partial, causing much larger elastic deflections than for $\Delta\alpha = 0$. Experiments with Morse #4 taper have shown stiffness reduction by a factor 2 to 3 when $\Delta\alpha = 2''$ and by a factor 4 to 6 when $\Delta\alpha = 8''$. Experiments with 7/24 taper have shown that change of the angle α by $\Delta\alpha = 27''$ (for the same hole) increased elastic deflection at the mouth 9 times, and by $\Delta\alpha = 40''$ — 15 times.

For the angular difference in the opposite direction (the hole is more shallow than the taper on the mandrel), stiffness is not sensitive to the mismatch. The tests have demonstrated that deviations $\Delta\alpha$ up to $60''$ do not influence significantly the connection stiffness. However, as it was mentioned above, such deviations can cause a faster wear and increased radial runout at the end.

4.5.2d Finite Element Modeling of 7/24 Taper Connection

Results similar to the ones described above were obtained in an extensive computational (finite element) and experimental analysis of the 7/24 taper connection in [8]. The study in [8] addressed both axial displacement of the taper inside the hole under the axial (drawbar) force P_t and bending deformations at the spindle end under the radial force F. The design and manufacturing parameters whose influence was explored included: ratio of the outside diameter of the spindle, D, to the gage (maximum) diameter of the connection d_o; accuracy of the gage diameter; surface finish of the connected male and female tapers, R_a; friction coefficient f in the connection; magnitude of drawbar force, P_t. The study was performed for taper #40. It was found that the axial displacement for $D/d_o = 2$ for a very high friction ($f = 0.6$) and surface finish $R_a = 0.5$ μm (20 μin.) was 17 μm (0.0007 in.) for $P_t = 5$ kN (1,100 lbs), 21 μm (0.00085 in.) for $P_t = 10$ kN (2200 lbs.), and 23 μm (0.0009 in.) for $P_t = 15$ kN $= (3300$ lbs.). These deformations are mostly due to radial expansion of the spindle walls and, to a much smaller extent, contraction of the male taper. Deviation of d_o by ± 25 μm (0.001 in.), which represents a mismatch of the taper angles of the hole and the inserted in it taper, led to a further increase of the axial displacement by up to 20% to 30%. Increase of surface roughness to $R_a = 1.2$ μm (48 μin.) led to another 25% increase in the axial displacement. Reduction of f from 0.6 to 0.1 resulted in a 50% increase in the axial displacement.

The data from [8] on axial displacements in 7/24 tapered connections compares well with experimental data on axial displacements of various interfaces from [9] (Fig. 4.5.22). Fig. 4.5.22 indicates that not only flat joints and "curvic coupling" connections have much smaller axial deformation than the 7/24 tapered connection, but they return to their original non-deformed configurations when the axial force is removed. On the contrary, the 7/24 taper connection is held in its deformed condition after the axial force is removed, due to friction forces. The curvic coupling connection is an engagement of two identical flat spiral gears, which are held together by the axial force. One gear is fastened to the spindle flange, while another is fastened to the tool. Other characteristics of this connection are given in Section 4.8.3 below.

A change in the outside diameter of the spindle D has only a marginal effect on the axial displacement; at $D/d_0 = 4$, it is only 10% less than at $D/d_0 = 2$. However, bending deformation of the tapered insert (toolholder) as characterized by shear displacement between the male and female tapers (Fig. 4.5.23), is noticeably influenced by changes in D. The most dramatic increase of bending deformation is observed at $D/d_0 < 2$, due to increasing local deformations of spindle walls.

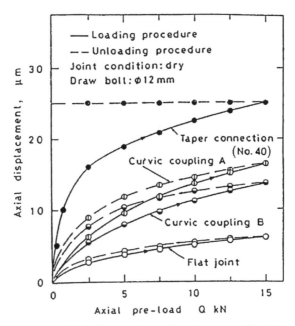

Fig. 4.5.22. Axial deformation of various toolholder/spindle interfaces.

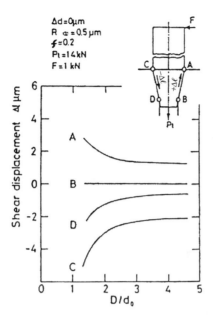

Fig. 4.5.23. Influence of spindle diameter on bending deformation of toolholder.

4.5.2e Short Taper Connections

Besides deformations in point or line contacts (Hertzian deformations, Section 3.1) and deformations in contacts between extended nominally conforming surfaces addressed above in Sections 4.2 to 4.4, stiffness of contact interfaces can also be influenced by misalignments and resulting interactions between the contacting bodies. This is especially, but not only, important for short taper fits when it is practically impossible to achieve perfect identity of the taper angles in the male and female components. The important embodiments of tapered contacts are interfaces between dead and live centers and center holes in the part ends for machining of oblong parts on turning and grinding machines. These interfaces deserve a special consideration due to much greater differences in taper angles between the male and female parts than are typical for precision connections considered above.

Axial contact stiffness between centers and center holes was studied in [10]; the studied system is shown in Fig. 4.5.24. Axial deformations of the connections have been analyzed using Finite Element Analysis approach with a variable mesh size model (Fig. 4.5.25), for various misalignments,

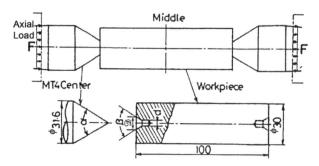

Fig. 4.5.24. Workpiece supported by two centers.

connection diameters, and friction coefficients. While the center angle in all cases was 60 deg., center hole was assumed to be one of three sizes: 59.5 deg., 60 deg. (perfect fit), and 60.5 deg.. For the connection having diameter 10 mm and length 4.5 mm, 0.5 deg. misalignment is equivalent to 20 μm clearance between the parts. Fig. 4.5.26 shows deformation patterns for these three cases and Fig. 4.5.27 gives load-deflection characteristics for the axial loading.

It is obvious from Fig. 4.5.27 that the connection with the perfect angular match is the stiffest. However, the connection in which the female taper has a *more shallow angle* (59.5 deg.) has the closest stiffness to the perfect match case, only 1.35 times lower stiffness for the friction coefficient $f = 0.7$, and about 2.0 times lower for a more realistic $f = 0.3$ (for the connection diameter 10 mm). The connection in which the female taper has a steeper angle (60.5 deg.) has a much lower stiffness, 2.6 times lower for $f = 0.7$ and 3.2 times lower for $f = 0.3$.

Qualitatively similar conclusions were derived in the study of larger tapers (Morse and "steep" 7/24 taper) used for toolholder/spindle interfaces in [4], see Section 4.5.2c.

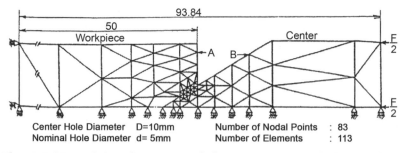

Center Hole Diameter D=10mm   Number of Nodal Points : 83
Nominal Hole Diameter d= 5mm   Number of Elements   : 113

Fig. 4.5.25. Finite element mesh for center-center hole interface.

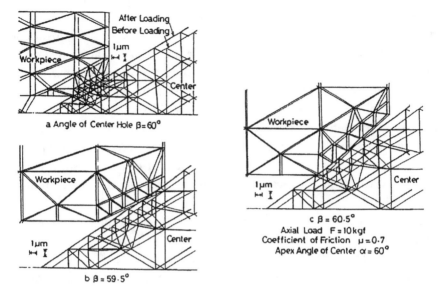

Fig. 4.5.26. Deformation patterns of center-center hole interface with different misalignments.

4.5.2f Some General Comments on Tapered Connections

The major advantage of the tapered connections is their self-centering while providing relatively easy disassemblable interference-fit-like joint. There are two major types of the taper connections. The connections with shallow tapers (e.g., Morse tapers, 1:10) are self-locking and do not require axial tightening. However the male Morse taper has to be axially

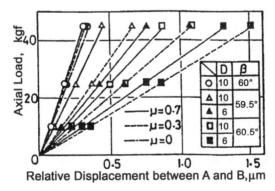

Fig. 4.5.27. Load-deflection characteristics of axial deformation of center-center hole interface.

pushed out of the hole with a significant force for disassembly. Another disadvantage of shallow taper connections is a very strong dependence of the relative axial position of the male and female parts on their diameter. For example, only a 10 μm diametral difference between two male tapers (e.g., at the gage diameter) results in about 100 μm difference in their axial positions. As a result, the Morse connections are not as popular presently as they were in the past when majority of machine tools were manually operated, extra seconds required for disassembling the connection were not critical, and the scatter of axial positioning between several toolholders could be compensated, if necessary, by manual adjustments.

Now one can see a resurgence of shallow taper connections for toolholders/spindle interfaces of high speed and/or high accuracy machine tools (Kennametal "KM," HSK, Sandvik "Capto," etc. [11]). One of the major requirements for the advanced interfaces is to provide an ultimate axial accuracy by assuring a simultaneous taper/face contact with the spindle. It is interesting to note that a drawback of the shallow taper connection — sensitivity of the relative axial position of the male and female tapers to minor variation of the gage diameter — has been turned to its advantage by making the male taper compliant. The high sensitivity of axial position to the diameter change allows for a significant adjustment of the axial position with relatively small deformations of the male taper. Still, designing even a shallow taper for a simultaneous taper/face connection requires extremely tight tolerances and creates many undesirable effects (e.g., [11]). The taper should have an enhanced radial compliance to be able to provide for the face contact with the spindle. To provide the compliance in KM and HSK systems, the tapered part must be hollow, as in Fig. 4.5.13. This requires to move the tool clamping device to the outside part of the toolholder (in front of the spindle), thus increasing its overhang and reducing its stiffness (Fig. 4.5.28) [10]; a special "kick-out" mechanism is required for disassembly of the connection; the connection is two to four times more expensive than the standard "steep taper" connection due to extremely tight tolerances, etc.

Connections with steep, most frequently 7/24 tapers, so-called CAT (from Caterpillar Co. which first introduced it for CNC machine tools with automatic tool changers) are not self-locking and are clamped by application of a significant axial force. This feature is very convenient for machine tools with automatic tool changers, since the assembly/dis-

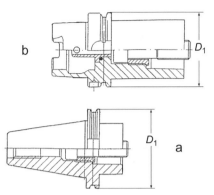

Fig. 4.5.28. Difference in tool overhang for toolholder tapers for connection with modular tools; (a) steep 7/24 taper; (b) HSK 1:10 taper.

assembly (tool changing) procedure is very easy to accomplish and no "kick-out" mechanism is required. The major shortcomings of this connection are indeterminacy of the axial position of the toolholder (see above, Sections 4.4.2b,d); practical impossibility to achieve the taper/face contact, even by making the taper body compliant due to a relatively low sensitivity of the axial position to diameter changes; more pronounced than in shallow taper connections micro-motions leading to fretting corrosion and fast wear at heavy dynamic lateral loading, typical for milling operations. These shortcomings have been overcome by introduction of external elastic elements ([12], also see Section 8.3).

4.6 TANGENTIAL CONTACT COMPLIANCE

If a flat joint is loaded with a force acting tangentially to the contacting surfaces, the connected parts will initially be elastically displaced so that, after removal of the force, their initial position is restored. After the force increases above the elastic limit of the joint, some plastic (non-restoring) displacements develop, and then the breakthrough occurs — continuous motion commences. Tangential compliance of joints, although much lower than the normal compliance addressed in the previous sections, is becoming more and more important with increasing stiffness requirements to precision machines and measuring apparatuses. It is also important for understanding mechanisms of functioning of the joints, their strength, fretting corrosion, and damping. Section 4.6.1 describes the results of an experimental study of tangential compliance based on [4], while Section 4.6.2 addresses analytical studies of mechanisms of dynamic behavior (stiffness and damping) of joints loaded in the normal direction and subjected to axial tangential dynamic loads.

4.6.1 Experimental Study of Tangential Compliance of Flat Joints

An extensive experimental study of tangential compliance was performed in [4] on flat annular joints with contact areas 51 cm^2 and 225 cm^2 loaded by a moment (torque) applied within the joint plane, while the joint was preloaded with a normal force. The connected cast-iron parts were very rigid so that their deformations could be considered negligible. The

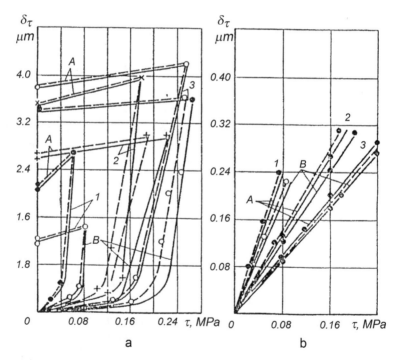

Fig. 4.6.1. Displacement δ_τ under tangential stress τ in a flat annular joint with surface area 51cm^2 (7.9 sq. in.); (a) First loading, (b) Repeated loadings; A — scraping, depth of dips 4 μm to 6 μm, number of spots 15/in.^{2} to 16/in.^{2}; B — grinding, surface finish $R_a = 0.4$ μm; solid lines — non-lubricated joint, broken lines — lubricated joint. Normal pressure: 1σ = 0.5 MPa (70 psi); 25 = 1.0 MPa (140 psi); 3σ = 1.5 MPa (210 psi).

Table 4.4. Coefficients f_e and f.

Surface condition	No lubrication		With lubrication	
	f_e	f	f_e	f
Fine turning, $R_a = 2\ \mu$m	0.13	0.25	0.13	0.25
Rough grinding	0.12	0.18	0.12	0.18
Grinding and lapping	0.17	0.35	0.14	0.30
Scraping, 8 – 10 μm dips	0.12	0.22	0.12	0.22
Fine scraping, 1 – 2 μm dips	0.14	0.28	0.1	0.24

contact surfaces were machined by fine turning, grinding, scraping with various depth of the dips, and lapping. Both carefully de-oiled surfaces and surfaces lubricated by light industrial mineral oil had been tested.

There is a big difference between the first and repeated tangential loading. At the first loading of the joint, the load-displacement characteristic is highly nonlinear (Fig. 4.6.1a). When the tangential force is removed, the unloading branch of the characteristic is parallel to the loading branch (the same stiffness), but there is a substantial hysteresis. At the following (repeated) force applications (Fig. 4.6.1b), the process is linear for loads not exceeding the loads in the first loading process, thus:

$$\delta_\tau = e_\tau \tau \tag{4.6.1}$$

where δ_τ = tangential displacement, μm, τ = specific tangential load, MPa, and e_τ = tangential compliance coefficient, μ/m MPa. For the plastic region, the tangential compliance coefficient is:

$$e_{\tau p} = (20 - 25)\, e_\tau \tag{4.6.2}$$

The magnitude of the tangential displacement before commencement of the continuous motion condition is increasing with increasing normal pressure and can reach 20 μm to 30 μm at the first loading. The important parameters for applications are maximum tangential stress τ_e of the elastic region and τ_{max} at the breakthrough event. The elastic range of the joint can be characterized by coefficient:

$$f_e = \tau_e / \sigma \tag{4.6.3}$$

where σ = the normal pressure. Table 4.4 gives f_e for annular specimen with the joint area 51 cm^2, as well as friction coefficient

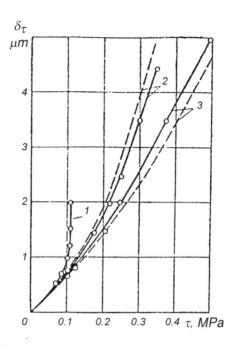

Fig. 4.6.2. Influence of holding pins on tangential compliance.
1 — joint without pins; 2 — two pins 12 mm diameter; 3 — two pins
16 mm diameter. Solid lines - experiments; broken lines - computation.

$$f = \tau_{max}/\sigma \qquad (4.6.4)$$

obtained from the same tests. It can be concluded that elastic displacements occur for the loads about one-half of the static friction forces. However, the maximum elastic displacement is only a small fraction of the total tangential displacement before the breakthrough. For lubricated surfaces, f_y and f are slightly smaller (for the light oil used in the tests), especially for high normal pressures (the lubricant is squeezed out). While tangential characteristics of the lubricated joints are not changing in time, f_e for dry joints increases 25% to 30% with increasing time of preloading from 5 minutes to 2.5 hours.

The magnitude of τ_{max} depends on the rate of application of the tangential loading. For a joint having 225 cm^2 (35 in.^{2}) area, increase in the rate of loading from 0.003 mm/second to 0.016 mm/second reduced τ_{max} from 0.07 MPa to 0.055 MPa, but the magnitude of the total displacement before the breakthrough did not change much.

If the contacting surfaces are also connected by pins or keys, tangential stiffness in the elastic region does not change much, since the tangential stiffness of the joint is usually much higher than stiffness of the pins/keys. However, in the plastic region, the joint stiffness is decreasing, and influence of the pins becomes noticeable. Fig. 4.6.2 shows results of testing of a 51-cm^2 joint with and without holding pins. At small tangential forces, the forces are fully accommodated by the joint itself, load-deflection characteristics with and without pins coincide. At the forces exceeding the elastic limit of the joint, its role in accommodating the external forces is gradually decreasing, while role of the pins is increasing. When the force exceeds the static friction force in the joint, it is fully accommodated by the pins (another linear section of the load-deflection curve).

The joint-pins system is statically indeterminate. Load sharing between the joint and the pins can be found by equating their displacements. Each pin can be considered as a beam on an elastic bed.

The total elastic displacements

$$\delta = \frac{\delta_o}{1 + \eta} \tag{4.6.5}$$

where δ_o = elastic displacement in the joint without pins and η = stiffness enhancement coefficient,

$$\eta = \frac{e_\tau}{e} \frac{z B_{ef}}{2 \beta F_j}. \tag{4.6.6}$$

Here z = number of pins; $B_{ef} = \pi d/2$ = effective width of the "elastic bed" of a pin; d = average pin diameter, cm; F_j - surface area of the joint, cm^2;

$$\beta = \sqrt[4]{\frac{10^4 B_{ef}}{4 E J e}} = \sqrt[4]{\frac{0.04}{d^3 e}} \quad \frac{1}{cm}; \tag{4.6.7}$$

J = cross-sectional moment of inertia of a pin, cm^4; E = Young's modulus of pin material, MPa; e = coefficient of normal contact compliance of the joint between pin and hole, $\mu m/MPa$.

This analysis is validated for the case of Fig. 4.6.2 (broken lines) when the assumed values of compliance coefficients were $e = 0.5$ $\mu m/MPa$; $e_\tau = 2.0$ $\mu m/MPa$; $e_{\tau p} = 40$ $\mu m/MPa$;

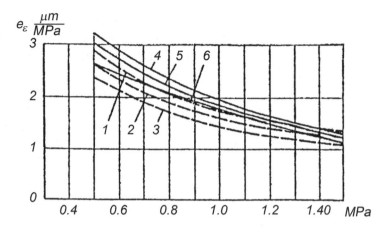

Fig. 4.6.3. Tangential contact compliance coefficients for specimens with joint area 51 cm^2; 1 — fine turning, R_a = 3.2 μm; 2 — grinding, R_a = 0.4 μm to 0.8 μm; 3 — grinding with lapping, R_a = 0.1 μm; 4,5,6 — scraping, depth of dip 8 μm to 10 μm, 4 μm to 6 μm, 1 μm to 2 μm, respectively.

In precision systems, it is important to realize such conditions when each joint is loaded in its elastic region. Usually these joints are designed from the condition that the external forces do not exceed static friction forces. In many cases, such approach is inadequate, since loading above f_e can lead to significant plastic deformations. It is especially important for dynamically loaded joints. To use this approach in design, it is important to know e_τ as a function of the surface finish (machining quality) and normal preload which is given in Fig. 4.6.3.

Use of the above data for actual machine tool units resulted in good correlation between calculated and tested data for τ_e and e_τ.

Value of e_τ = 3 μm/MPa measured for the joint guideways — spindle head of a jig borer, Fig. 4.6.4, compares well with e_τ = 3.5 μm/MPa measured on the scraped specimen in the lab.

For calculations of the total tangential displacement in a joint, the following formula considering both elastic and plastic deformations can be used:

$$\delta = \delta_e + e_{\text{тp}} \, (\tau - \tau_e), \tag{4.6.8}$$

Example. Find the allowable force on a dead stop (Fig. 4.6.5). Normal pressure in joints 1 (surface area, A_1 = 3.2 cm^2) and 2 (A_2 = 1.6 cm^2)

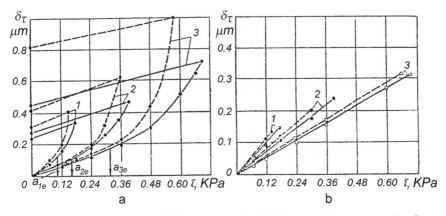

Fig. 4.6.4. Tangential displacements in sliding guideways at the first (a) and repeated (b) loading. $1\sigma = 0.09$ MPa; $2\sigma = 0.18$ MPa; $3\sigma = 0.35$ MPa. Solid lines — no lubrication; broken lines — lubricated.

is applied by two bolts M8 size (8 mm diameter), each preloaded to $P_o = 6000$ N), thus creating joint pressures $\sigma_1 = 30$ MPa (4200 psi) and $\sigma_2 = 60$ MPa (8400 psi). Tangential stresses in the joints from the external force, P, applied to the stop are:

$$\tau_1 = \tau_2 = \frac{P}{A_1 + A_2} \tag{4.6.9}$$

The allowable load (within the elastic region of the joints) at $f_e = 0.1$ is $P_{max} = 1.450$N. If P_{max} is exceeded, there is a danger of displacement of the stop due to non-restoring plastic deformations.

Tangential loading of joints preloaded by normal forces may generate a complex distribution of tangential stresses within the joint, a combination of elastic, plastic, and slippage zones (e.g., see Section 4.6.2 below).

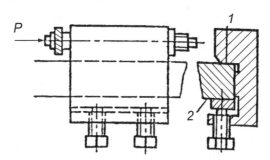

Fig. 4.6.5. Dead stop for a machine tool.

This can apply to both flat and cylindrical joints (interference fit between a shaft and a bushing).

The slippage zone can cause fretting corrosion. These zones can be identified if the tangential contact phenomena are considered. Depending on the load magnitude, the following cases are possible with the increasing load;

1. Elastic along the whole length ($\tau < \tau_e$);
2. Elasto-plastic with both elastic and plastic zones ($\tau > \tau_e$);
3. Plastic along the whole length ($\tau_e < \tau_{max}$); and
4. Plastic and slippage zones ($\tau_e = \tau_{max}$).

Fig. 4.6.6a shows computed distribution of contact tangential stresses in a cylindrical joint in Fig. 4.6.6b ($d = 25$ mm, $l = 20$ mm, ground surfaces with finish $R_a = 0.2$ μm; press fit with contact pressure $\sigma = 50$ MPa). The shaft is loaded by torque T at the end. Analysis without considering tangential contact compliance would conclude that the slippage starts from very low torque magnitudes, but in reality, it starts only at high torque magnitudes.

Energy dissipation (damping) can also be determined by computations (Section 4.6.2). Energy dissipation is mostly concentrated in the plastic and slippage zones.

If joints are loaded with dynamic forces outside their elastic regions,

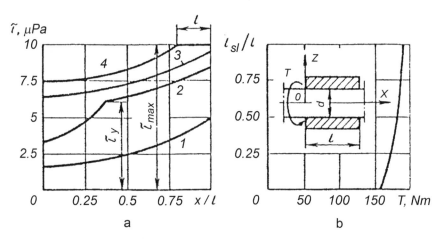

Fig. 4.6.6. (a) Tangential stress distribution in cylindrical press-fit joint. 1 — $T = 50$ Nm; 2 — $T = 120$ Nm, 3 — $T = 150$ Nm; 4 — $T = 170$ Nm; (b) Length of slippage zone.

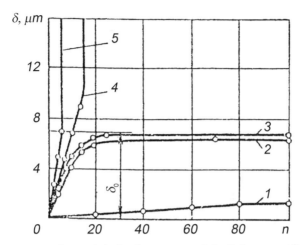

Fig. 4.6.7. Circumferential displacement, δ, in inference-fit cylindrical joint between steel parts with contact pressure $\sigma = 18$ MPa, $R_a = 0.2$ μm loaded by impact torque versus number of impacts n; $l/d = 3$; $E_{max} = 0.61$ Nm; $E_{min} = 0.21$ Nm. 1 — $E = 0.24$ Nm; 2 — $E = 0.27$ Nm; 3 — $E = 0.34$ Nm; 4 — $E = 0.42$ Nm; 5 — $E = 0.49$ Nm.

then residual displacements can develop. Depending on type and intensity of loading, these displacements can stabilize at a certain level or grow continuously, up to thousands of μm.

Study of behavior of joints under tangential impacts was performed for cylindrical interference-fit joints, $d = 25$ mm, loaded by an axial force or a torque, and on flat annular joints ($d_o = 62$ mm, $d_i = 42$ mm) loaded by torque [13]. Initial normal pressures in the flat joints were $\sigma = 13$ MPa to 16 MPa, in the cylindrical joints, $\sigma = 10$ MPa to 100 MPa. Some results are shown in Figs. 4.6.7 to 4.6.9.

When energy of each impact cycle is low, displacements in the joints are fully elastic and do not accumulate (grow). Beyond the elastic region, displacements always accumulate. If the energy of one impact E only slightly exceeds the energy of elastic deformation E_{min} then the process of accumulation of the residual displacement is slowing down. After a certain number of impact cycles, the process is stabilized, not reaching the magnitude of displacement, δ_o, at the breakthrough (lines 1 to 3 in Fig. 4.6.7). If the single impact energy is significantly greater than the elastic deformation energy E_{min} the displacement gradually grows up to δ_o, and after that, the surfaces abruptly commence a relative motion (lines 4 to 5 in Fig. 4.6.7). The motion can result in displacement measured in

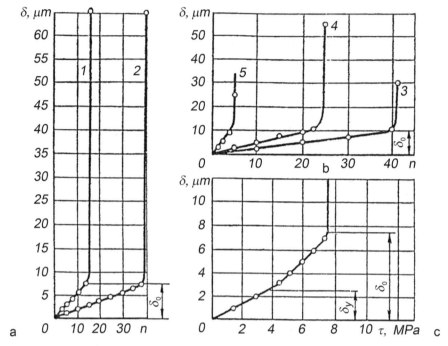

Fig. 4.6.8. Axial displacement, δ, in cylindrical interference-fit joints. (a), (b) impact loading; (c) Static loading. (a), (c) Contact pressure due to interference, $\sigma = 30$ MPa, $E_{min} = 0.25$ Nm, $E_{max} - 1.05$ Nm; (b) $\sigma = 50$ MPa, $E_{min} = 0.6$ Nm, $E_{max} = 2.4$ Nm; 1 — $E = 0.61$ Nm; 2 — $E = 0.415$ Nm; 3 — $E = 1.05$ Nm; 4 — E 1.35 Nm; 5 — $E = 1.65$ Nm.

hundreds of micrometers. Displacement magnitude δ is proportional to the number of impacts before occurrence of the breakthrough event (Fig. 4.6.8a, b). If the impact loading continues, displacements again accumulate until a new slip occurs, etc. (Fig. 4.6.9). At a certain level of impact energy, E_{max}, the initial slippage and resulting motion start after the first impact. For all tested joints, $E_{max}/E_{min} \approx 4$. These micro-slippage events are responsible for the fretting corrosion.

The results discussed above are recorded for a large time interval between the impacts, more than one second. For higher impact frequency, displacements always accumulate without a visible slippage. The total displacement may reach thousands of micrometers, much larger than displacement before the breakthrough event under static loading.

High frequency (10 Hz to 60Hz) *pulsating (0 to T_{max})* torque was applied to a cylindrical interference-fit connection. Interference pressure

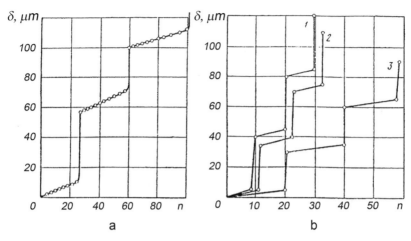

Fig. 4.6.9. Displacement, δ, under axial impact loading. (a) Cylindrical interference-fit joint of steel parts, $R_a = 0.2$ μm, $\sigma = 30$ MPa; (b) Flat annular cast iron preloaded joint, surfaces scraped, $\sigma = 16$ MPa, $E_{max} =$ 1.05 Nm, $E_{min} = 0.26$ Nm. 1 — $E = 0.82$ Nm; 2 — $E = 0.61$ Nm; 3 — $E =$ 0.35 Nm, $R_a = 0.2$ μm.

for the test specimens was $\sigma = 2.5$ MPa to 50 MPa. In these tests, accumulation of residual displacement is slowing down, and the displacement magnitude during each cycle is continuously diminishing (Fig. 4.6.10). The accumulated displacement converges to a certain limiting magnitude, and the slippage does not develop. The residual displacement δ_1 at the first torque pulse is 5 μm to 7 μm to 0.5 μm, depending on magnitudes of the normal pressure, σ, and amplitude of the loading torque, T. At σ = 30 MPa and $T = 32$ Nm $\delta_1 = 2.5$ μm; at $T = 26.5$ Nm $\delta = 1$ μm; at $T = 24$ Nm, $\delta_l = 0.5$ μm. As for impact loading, if the torque amplitude is increasing, at some amplitudes, the slippage develops at the first pulse; at small torque amplitudes, the displacements do not leave the elastic zone.

If the joint is again subjected to the pulsating torque after the first exposure to the pulsating torque (during which the slippage had not developed), then the residual displacement is reduced, and the torque causing continuous slippage becomes 1.25 to 1.5 times greater ("work hardening" of the joint). However, if during the first exposure there had been slippage, the work hardening does not develop.

Stabilization of the joint displacements under pulsating and impact loading with energy of one cycle not significantly exceeding E_{min} can be explained by increasing of the effective contact area (due to cyclical

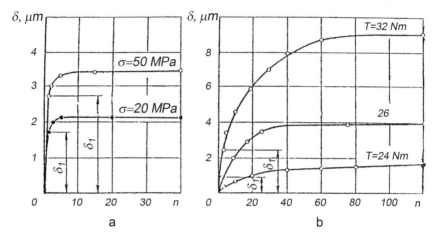

Fig. 4.6.10. Circumferential displacement, δ, in interference-fit cylindrical connection between steel parts with R_a = 0.2 μm loaded by pulsating torque (0 - T) versus number of cycles n. aT = 200 N-m, varying σ; bσ = 30 MPa, varying σ.

elasto-plastic deformation, squeezing out the lubricant, enhanced adhesion, etc.). During high intensity impact loading, repeated plastic deformation results, finally, in a breakdown of frictional connections. The process may repeat itself, since new frictional connections are developing. Another important factor is presence of high frequency harmonics in the impact spectrum. They can result in reduction of resistance forces.

4.6.2 Dynamic Model of Tangential Compliance and Damping

A basic model of a flat joint is shown in Fig. 4.6.11 [14]. Thin elastic strip 1 is pressed to rigid base 2 with pressure p. Strip 1 is axially loaded by a force αP cyclically varying from P_{min} ($\alpha = r$) to P ($\alpha = 1$). The maximum magnitude, P, of the force is assumed to be smaller than the friction force between strip 1 and base 2,

$$P < fpbl, \qquad (4.6.10)$$

where f = friction coefficient in the joint, b and l is width, length of strip 1, respectively. The friction is assumed to be dry (Coulomb) friction; the strip is assumed to have a linear load-deflection characteristic (comply-

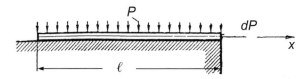

Fig. 4.6.11. Axial loading of elastic strip 1 pressed to rigid base 2.

ing with the Hooke's law). Condition (4.6.10) means that the strip cannot slide as a whole along the base. To analyze displacements between strip 1 and base 2, it is important to note that:

1. The *friction force is* equal to its *ultimate magnitude per unit length*

$$q = fpb \qquad (4.6.11)$$

in all areas within which strip 1 is slipping along base 2. These are the only areas where the strip is being deformed due to assumed dry friction.

2. The *friction force is zero* in the areas within which the strip is not being deformed. It follows from the fact that, in these areas, the strip is not axially loaded, while the friction force, if present, would load the strip.

Thus, friction force in any increment of the joint is either q or 0. The adopted Hooke's law for deformation and Coulomb's law for friction absolutely exclude a possibility that, in some area, there is a friction force which is non-zero and less than q.

Let's consider three basic phases of the load change:

1. The force, αP, is increasing from zero to maximum magnitude, P, or $0 \le \alpha \le 1$;
2. The force is decreasing from maximum magnitude P to minimum magnitude P_{min} or $r \le \alpha \le 1$;
3. The force is increasing from minimum magnitude, P_{min}, to maximum magnitude P, or $r \le \alpha \le 1$.

The last two phases are continuously repeating for cyclical variation of the force in $[P_{min}, P]$ interval.

First Phase of Loading. The length of the strip deformation area (slippage zone) is determined by the strip equilibrium condition and is

$$a_1 = \frac{\alpha P}{q},$$

(4.6.12)

Fig. 4.6.12a. With a gradual increase of force, αP, the length of this zone is increasing; according to Eq. (4.6.10):

$$a_{1max} = \frac{P}{q} < 1.$$

(4.6.13)

In accordance with Hooke's law, the relative elongation (strain) of an element of the deformed zone is:

$$\frac{du_1}{dx} = \frac{N}{EF},$$

(4.6.14)

where $u_1(x, \alpha)$ is displacement of the cross-section with coordinate, x, along x-axis; $N = N(x, \alpha)$ = tensile force in cross-section, x, of the strip; EF = tensile rigidity of the strip; and F = cross-sectional area of the strip. From equilibrium of an element, dx, it follows that:

$$\frac{dN}{dx} = q.$$

(4.6.15)

Substituting Eq. (4.6.14) into Eq. (4.6.15), arrive to:

$$\frac{d^2u_1}{dx^2} = \frac{q}{EF}$$

(4.6.16)

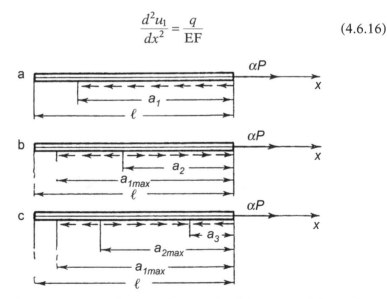

Fig. 4.6.12. Forces acting on elastic strip: (a) Loading (phase 1); (b) Unloading (phase 2); (c) Repeated loading (phase 3).

The integral of Eq. (4.6.16) is:

$$u_1 = A_1 + B_1 x + \frac{qx^2}{2EF} \tag{4.6.17}$$

The boundary conditions for the cross-section dividing the deformed and undeformed zones are:

$$u_1(l - a_1, \alpha) = 0; \quad \frac{du_1(l - a_1\alpha)}{dx} = 0. \tag{4.6.18}$$

These conditions reflect absence of displacement and axial force at this cross-section. From Eq. (4.6.18),

$$A_1 = \frac{(l - a_1)^2}{2EF}; \quad B_1 = -\frac{q(l - a_1)}{EF} \tag{4.6.19}$$

Substituting Eq. (4.6.19) into Eq. (4.6.17), arrive to:

$$u_1(x, \alpha) = \frac{q(l - a_1 - x)^2}{2EF} \tag{4.6.20}$$

where a_1 depends on a and is determined by Eq. (4.6.12). In the following, we will need these expressions:

$$u_1(x, l) = \frac{q(l - a_{1\max} - x)^2}{2EF} \tag{4.6.21}$$

$$u_1(l, \alpha) = \frac{\alpha^2 P^2}{2qEF} \tag{4.6.22}$$

Eq. (4.6.21) describes distribution of displacements, u_1, along the deformed zone in the end of the first phase of loading. Eq. (4.6.22) defines displacement of the end cross-section during the first phase of loading.

Second Phase of Loading. When the force, αP, starts to decrease, end elements of the strip start to shift in the negative x-direction; accordingly, friction forces in the positive x-direction appear. The equilibrium condition for the strip (Fig. 4.6.12b), is:

$$\alpha P + q a_2 - q(a_{1\max} - a_2) = 0. \tag{4.6.23}$$

This expression allows to find the length of the "back-shift zone" as

$$a_2 = \frac{P(1-\alpha)}{2q} \tag{4.6.24}$$

and

$$a_{2\max} = \frac{P(1-r)}{2q}. \tag{4.6.25}$$

To the left of the "back-shift zone," the displacements and friction forces existing at the end of the first phase of loading, remain frozen.

The equilibrium condition of some element within the "back-shift zone" is different from Eq. (4.6.15) and is:

$$\frac{dN}{dx} = -q. \tag{4.6.26}$$

Substituting Eq. (4.6.14) into Eq. (4.6.26), arrive to:

$$\frac{d^2 u_2}{dx^2} = -\frac{q}{EF}, \tag{4.6.27}$$

whose integral is

$$u_2 = A_2 + B_2 x - \frac{qx^2}{2EF}. \tag{4.6.28}$$

This solution should comply with the following boundary conditions:

$$u_2(l - a_2, \ \alpha) = u_1(l - a_2, \ 1); \quad du_2(l - a_2, \alpha) = \frac{du_1(l - a_2, 1)}{dx} \tag{4.6.29}$$

These conditions reflect the identity of displacements and axial forces at the cross-section $x = l - a_2$, which divides zones of "back-shift" and "direct" displacements. To construct the right-hand parts of expressions (4.6.29), Eq. (4.6.21) should be used. Thus,

$$A_2 = \frac{q}{EF}[(l - a_{1\max})^2 - 2(l - a_2)]; \quad B_2 = \frac{q}{EF}(l - 2a_2 + a_{1\max}). \tag{4.6.30}$$

Consequently,

$$u_2(x,\alpha) = \frac{q}{EF}[(a_2 - a_{1\max} - x)^2 - (l - a_2 - x)^2 - x^2 + 2(l - a_2)(a_2 - a_{1\max})],$$

$$\tag{4.6.31}$$

where a_2 is function of a and is determined by Eq. (4.6.24). The following expressions will be useful at the next phase:

$$u_2(x,r) = \frac{1+2r-r^2}{4qEF}P^2 - \frac{(l-x)rP}{EF} - \frac{q(l-x)^2}{2EF}; \qquad (4.6.32)$$

$$u_2(l,\alpha) = \frac{1+2\alpha-\alpha^2}{4qEF}P^2. \qquad (4.6.33)$$

Expression (4.6.32) describes distribution of displacements, u_2, within the "back-shift zone" at the end of the second phase of loading, while expression (4.6.33) represents shifting of the end cross-section during the whole second phase.

Third Phase of Loading. In the beginning of the third phase, the positive displacements are again developing in the end zone. Distribution of the friction forces is illustrated in Fig. 4.6.12c. The equilibrium condition for the strip is:

$$\alpha P - (a_{1\,max} - a_{2\,max})q - a_3 q + (a_{2\,max} - a_3)q = 0. \qquad (4.6.34)$$

From Eq. (4.6.34), the length of the end zone is determined as:

$$a_3 = \frac{\alpha - r}{2q}P. \qquad (4.6.35)$$

The differential equation again becomes as Eq. (4.6.16), and its integral is

$$u_3 = A_3 + B_3 x + qx^2/2EF. \qquad (4.6.36)$$

Boundary conditions in the cross-section with the coordinate $x = l - a_3$ are:

$$u_3(l - a_3, \alpha) = u_2(l - a_3, r); \quad du_3(l - a_3, \alpha)/dx = du_2(l - a_3, r)/dx, \qquad (4.6.37)$$

and these boundary conditions lead to:

$$A_3 = \frac{(1+2r-r^2)P^2}{4qEF} + \frac{q}{2EF}\left[(l-2a_3)^2 - 2a_3^2\right]; \quad B_3 = \frac{q}{EF}\left[\frac{rP}{q} + 2a_3 - l\right]. $$

$$(4.6.38)$$

Thus,

$$u_3(x,\alpha) = \frac{(1+2r-r^2)P^2}{4qEF} - \frac{(l-x)rP}{EF} - \frac{q(l-x)^2}{2EF} + \frac{q(l-a_3-x)^2}{EF}, \quad (4.6.39)$$

and displacement of the end cross-section, $x = l$, is expressed as:

$$u_3(l,\alpha) = \frac{(1-2\alpha r + 2r + \alpha^2)P^2}{4qEF}. \quad (4.6.40)$$

Fig. 4.6.13 illustrates functions, $u_1(l, \alpha)$, $u_2(l, \alpha)$, and $u_3(l, \alpha)$, for cases when $r = 0.5$ (small amplitude of the cyclical part of the load) and $r = 0$ (large amplitude). Effective stiffness of the connection corresponds to the ratio, $\Delta\alpha/\Delta u$. In the dimensionless parameters of plots in Fig. 4.6.13, for the hysteresis loop in Fig. 4.6.13a, $\Delta\alpha = 0.5$, $\Delta u = 0.1$, for the loop in Fig. 4.6.13b $\Delta\alpha = 1.0$, $\Delta u = 0.45$, or doubling the amplitude from Fig. 4.6.13a to Fig. 4.6.13b results in 2.25 times reduction in stiffness. The *static stiffness*, k_{st}, of the system in Fig. 4.6.11 (stiffness associated with the first phase of loading from $\alpha P = 0$ to $\alpha P = P$) is significantly smaller than *dynamic stiffness*, k_{dyn}, associated with the cyclical loading (the second and third phases of loading). For the static stiffness, $\Delta\alpha = 1.0$, $\Delta u = 1.0$, or the static stiffness is about two times lower than dynamic stiffness for the large cycle amplitude in Fig. 4.6.13a and five times lower than dynamic stiffness for the smaller cycle amplitude in Fig. 4.6.13b. The average value of static stiffness for the first phase of loading is the ratio of the maximum force, P, to maximum displacement of the end $u_1(l, 1)$. Using Eq. (4.6.22),

$$k_{st} = \frac{u_1(l,1)}{P} = \frac{2qEF}{P} \quad (4.6.41)$$

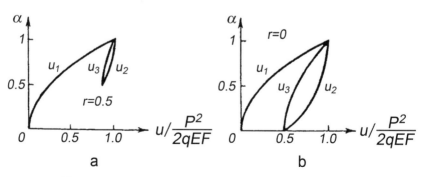

Fig. 4.6.13. Hysteresis loops: (a) $r = 0.5$; (b) $r = 0$.

or the system is softening nonlinear one (see Chapter 3) with the static stiffness decreasing inversely proportional to the applied force, P. The average value of dynamic stiffness is the ratio of the amplitude of cyclical force, P_v, to displacement amplitude u_v, or:

$$k_{dyn} = \frac{P_v}{u_v} = \frac{P_v}{P_v^2 / qEF} = \frac{qEF}{P_v} \qquad (4.6.42)$$

From Eq. (4.6.41) and Eq. (4.6.42), the dynamic-to-static stiffness ratio, K_{dyn} (see Chapter 3), can be derived as:

$$K_{dyn} = \frac{k_{dyn}}{k_{st}} = \frac{2P}{P_v} = \frac{4}{1-r} \qquad (4.6.43)$$

It follows from Eq. (4.6.43) that the dynamic-to-static stiffness ratio for the model in Fig. 4.6.11 is increasing with decreasing amplitude of the cyclical force (or, consequently, amplitude of vibratory displacement). This also is in a qualitative agreement with test data for the mesh-like and elastomeric materials in Fig. 3.2.1. It can be noted, that the mesh-like materials consist of wires or fibers pressed to each other by designed preload as well as by the performance-related forces.

The area inside the hysteretic loop represents the part, Ψ, of the work performed by the force, αP, which is lost (dissipated). This area can be computed as:

$$\Psi = \int_{P_{min}}^{P} [u_2(l,\alpha) - u_3(l,\alpha)]d(\alpha P) = P \int_{r}^{1} (u_2 - u_3)d\alpha. \qquad (4.6.44)$$

Substituting Eq. (4.6.33) and (4.6.40) into Eq. (4.6.44), arrive to

$$\Psi = \frac{P^3(1-r)^3}{12qEF}. \qquad (4.6.45)$$

This result is easier to understand if the force amplitude, αP, is designated as P_v, with $P(1-r) = 2P_v$. Then Eq. (4.6.45) becomes:

$$\Psi = \frac{2P_v^3}{3qEF}. \qquad (4.6.46)$$

This expression shows that energy dissipation depends only on the cyclical component, P_v, and does not depend on the median (d.c.) component of the force $P_m = (P_{min} + P)/2$. For practical calculations, it is convenient to express the energy dissipation in relative terms, as a relative energy dissipation, ψ, per one cycle, which is expressed as the ratio of Ψ to the maximum potential energy, V_{max}, of the system. For small ψ,

$$\psi \cong 2\delta, \qquad (4.6.47)$$

where δ = logarithmic decrement. From Fig. 4.6.11,

$$V_{max} = Pu_{max} - P\int_0^1 u_1(l,\alpha)d\alpha = \frac{2P^3}{3q\mathrm{EF}}, \qquad (4.6.48)$$

and

$$\psi = \frac{\Psi}{V_{max}} = \frac{P_v^3}{P^3} \approx 2\,\delta. \qquad (4.6.49)$$

Both this amplitude dependence of log decrement, δ, and addressed above effective stiffness of the system in Fig. 4.6.11 are qualitatively similar with test data for mesh-like and elastomeric materials in Fig. 3.2.1. For a symmetric cycle ($r = -1$, $P_v = P$), from Eq. (4.6.48), $\psi = 1$. Such large values of damping parameters are typical for the systems with hysteresis. However, they have to be treated carefully, since the model in Fig. 4.6.11 is rather simplistic. For example, if the strip in Fig. 4.6.11 continues further to the right, this extension would contribute to the value of V_{max} thus resulting in reduction of ψ in Eq. (4.6.48). A higher damping effect can be achieved (for the same displacement amplitude u_v) by increasing friction coefficient, f, and normal pressure, p [see Eq. (4.6.11)].

Similar analyses were performed in [14] for axial and torsional loading of cylindrical press-fit joints, for axial and torsional loading of riveted connections, for elasto-frictional connections (a model as in Fig. 4.6.11 in which the interaction between the strip and the base is not only frictional, but also elastic).

4.7 PRACTICAL CASE: STUDY OF A MODULAR TOOLING SYSTEM

A modular tooling system Varilock shown in Fig. 4.7.1, was found to have excessive runouts at the end of tool (up to 50 μm to 100 μm) if several spacers are used [15]. This prevents its use for long overhang tools without lengthy adjustments. The system uses round couplings with annular contact surfaces (D_o = 63 mm, D_i = 46.5 mm, surface area, F = 2.9 × 10^3 mm^2) interrupted by a deep key slot on one side of the annulus. The joints between the components are tightened by M19 (~3/4 in.) bolts with the manufacturer-recommended tightening torque, 150 lb ft (200 Nm). Axial force, P_a generated by this torque was calculated using an approximate formula:

$$T = 0.2 \, P_a D, \tag{4.7.1}$$

where T = tightening torque and D = bolt diameter. For T = 200 Nm and D = 0.019 m, P_a = 52,600 N, and the specific pressure in the joint is P_a/F = 18 MPa.

Inspection of the components has shown that due to the interrupted surface caused by the presence of the asymmetric key slot, the "springback" after grinding results in the radial edges of the key slot protruding by 2.5 μm to 5.0 μm.

Finite Element Analysis of compression of the spacer having a key slot has shown that due to lower stiffness of the key slot area as compared with stiffness of the un-interrupted area, compression at the side of the key slot (12 μm) was greater than at the diametrically opposing side (7 μm). Thus, the difference in compression deformations on the opposing sides of the contact surface of the spacer is 5 μm. Since both joined components have this deformation difference, the total asymmetry is 5 × 2 = 10 μm = 0.01 mm. This asymmetry results in the angular displacements between the two connected parts

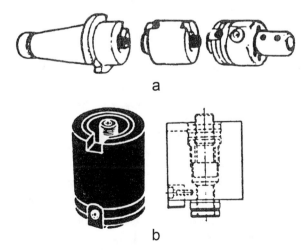

a

b

Fig. 4.7.1. Modular tooling system Varilock with axial tightening.

$$0.01/63 = 0.16 \times 10^{-3} \text{ rad.} \tag{4.7.2}$$

This angular deformation is due to the solid body compression of the contacting parts. In addition to this, there are also contact deformations of the surfaces. Since the surfaces are finely ground, $c = \sim 0.5$ (see Section 4.2 above). However, due to the presence of the after-grinding protrusions, whose role is similar to other asperities on the surface, the value of c would be different for different sides of the joint. For the side with the key slot/protrusions, it can be taken as for the fine-to-medium deep scraping (depth of dips 3 μm to 10 μm), $c \approx 2$. From Eq. (4.3.1), for this side

$$\delta = 2 \times 18^{0.5} = 8.5 \,\mu\text{m}, \tag{4.7.3}$$

while for the other side,

$$\delta = 0.5 \times 18^{0.5} = 2.1 \,\mu\text{m}. \tag{4.7.4}$$

Thus, the additional angular displacement due to the contact deformation is

$$(0.0085 - 0.0021)/63 = 0.1 \times 10^{-3} \text{ rad.} \tag{4.7.5}$$

The overall angular displacement in one joint per 200 Nm tightening torque is $(0.16 + 0.1) \ 10^{-3} = 0.26 \times 10^{-3}$ rad, or 0.13×10^{-5} rad/Nm. This angular deformation is approximately proportional to the specific pressure (tightening torque).

The initial position of the joint (with no tightening torque applied) is also inclined from the theoretical axis of the modular toolholder due to the presence of the protrusions. For the protrusions 5 μm high, the initial inclination is

$$0.005/63 = 7.9 \times 10^{-5} \text{ rad.} \tag{4.7.6}$$

This inclination is in the opposite direction relative to the torque-induced inclination. Thus, the total inclination at torque, T, is

$$\alpha = (0.13 \times 10^{-5} \, T - 7.9 \times 10^{-5}) \text{ rad.} \tag{4.7.7}$$

The minimum (zero) inclination would develop when

$$T = 7.9/0.13 = 61 \text{ Nm.} \tag{4.7.8}$$

Inclinations in the joints are projected to the end-of-tool thus creating eccentricity and runout. With one extension spacer 120 mm long, displacement at the end will be, $\Delta = 120\alpha$, with two assembled spacers, $\Delta = 240\alpha$, etc. The toolholder 80 mm long adds to this displacement, thus for one spacer, $\Delta = 200\alpha$, for two spacers, $\Delta = 320\alpha$, etc. The end runout is 2Δ.

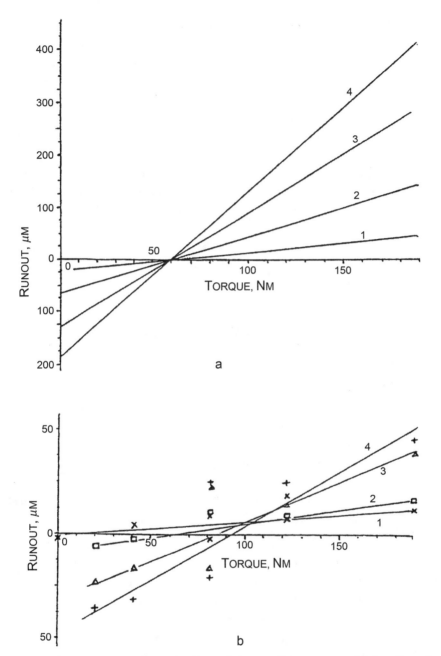

Fig. 4.7.2. Runout of Varilock modular tooling system; (a) Predicted, (b) Measured. 1 — taper and one spacer, x; 2 — same, but two spacers, □; 3 — same, three spacers, △; 4 — taper, three spacers, and tool head), +.

Fig. 4.7.2a shows the predicted runout of several assemblies. The run-out is fast increasing with increasing length of the assembly. At $T = 60$ Nm, the runout changes its direction (zero runout). Zero inclination does not necessarily mean zero runout due to other causes of runout, but it is associated with a much smaller runout.

These conclusions were tested by first measuring the assembly with one, two, three, four joints at $T = 20$ Nm, then all combinations were tested at 40 Nm, 80 Nm, 120 Nm, 190 Nm. The results are shown in Fig. 4.7.2b. The plots are similar to the prediction, but the measured runouts are much lower than calculated. Still, runouts at the manufacturer recommended $T = 200$ Nm are hardly acceptable. Also, it was known that runouts on the shop floor vary in a broad range and can become much larger.

The main conclusion of the analysis: *there is an optimal magnitude of the tightening torque (about 95 Nm from the test results, one-half of the manufacturer-recommended value).*

4.8 DAMPING OF MECHANICAL CONTACTS

There are four major processes contributing to loss of vibratory energy (damping) during vibratory processes in mechanical systems:

1. Loss of energy in material of the component (material damping);
2. Loss of energy in contact areas between the fitting parts (external friction during micro-motions, e.g., see Section 4.6, external and internal friction during deformation of micro-asperities on the contacting surfaces);
3. Loss of energy in layers of the lubricating oil, due to both viscous friction in the oil layer and viscous friction between micro-asperities.
4. Loss of energy in specially designed damping devices.

Usually, energy dissipation in contacts is much more intense than inside structural materials. The exceptions are materials with extremely high internal hysteresis, such as nickel-titanium alloys (see Section 3.2 and Section 8.3.1b), and elastomers as well as other specially designed polymers whose Young's modulus is too low for structural applications.

Results of a major study on energy dissipation in mechanical joints are presented in [4]. After numerous experiments, there were established damping parameter values and empirical correlations for evaluation of energy dissipation in joints considering materials of the components be-

Table 4.5. Stiffness, Damping, and Chatter-resistance Criterion for Various Mechanical Interfaces.

Q, kN	K, N/μm			δ			Kδ		
	5	10	15	5	10	15	5	10	15
Solid		18.5			0.006			0.11	
Flat joint	13	16.5	17.3	.075	.04	.03	.98	.66	.52
#45	12.5	13	14	.06	.04	.03	.75	.52	.42
Curvic coup. B	7.6	12.5	14	.11	.072	.045	.84	.9	.63
#40	10.5	12	12.5	.03	.02	.01	.32	.24	.125
Curvic coup. A	5.5	10.2	13	.13	.1	.075	.72	1.0	.98
#30	7.5	7.9	----	.02	.01	----	.15	.08	----

ing joined, surface finish, amount and viscosity of oil in the joint, frequency and amplitude of vibrations, shape and dimensions of the joint.

It is important to realize that character of damping in joints, as well as of material damping, is different from the classic "viscous damping" model (see Appendix 3). The damping is due, at least in the first approximation, to velocity-independent hysteresis of loading/unloading process, e.g., as discussed in Section 4.6.2, rather than due to velocity-dependent viscous friction in a viscous damper. Accordingly, such important dimensionless damping parameters as "damping ratio" c/c_{cr} and "log decrement" δ do not depend on the vibrating mass and can characterize damping of a joint or of a material. Another convenient damping parameter for hysteresis-based damping is "loss angle" β, or tan β.

It is also important to consider that, in both flat and cylindrical/tapered connections, the damping intensity is increasing with decreasing normal pressure in the connection, while stiffness is increasing with increasing normal pressure. There are many cases when performance of the system depends on a criterion including both stiffness and damping, thus such parameters should be considered simultaneously to optimize the system performance. This issue is addressed in Appendix 2. A comparison of stiffness, damping, and stiffness-damping criteria for various interfaces is given in Table 4.5

4.8.1 Damping in Flat Joints

Experimental findings described in [4] and elsewhere for damping of flat joints vibrating in the normal direction to the joint can be summarized as follows:

1. Damping (log decrement, δ) in joints between steel and cast iron parts has essentially the same magnitude, since the main source of energy dissipation is external friction.
2. In non-lubricated joints, damping does not depend on the pressure in the joint in the range of 0.1 MPa to 2 MPa (15 psi to 300 psi). For steel or cast iron joints with scraped or ground surfaces, $\delta = {\sim}0.075$, for cast iron–polymer joints, $\delta = {\sim}0.175$.
3. In lubricated joints, energy dissipation is growing with the increasing amount and viscosity of oil and decreasing with increasing mean pressure, σ_m (Fig. 4.8.1). The amount of oil in joints tested for Fig. 4.8.1 is typical for boundary friction condition.
4. Joints with scraped and ground surfaces *with the same* R_a exhibit practically the same damping.
5. Damping (log decrement, δ) does not depend strongly on frequency and amplitude of vibration (frequency and amplitude were varied in the range 1: 10). This is characteristic for the hysteresis-induced damping with $n = 1$ (see Appendix 1).
6. Damping does not depend significantly on the joint dimensions (in the experiments, the joint area was changing in the range of 2.2:1,

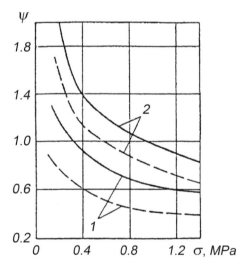

Fig. 4.8.1. Damping in flat joints cast iron–cast iron and steel–steel. 1 — rectangular joints with narrow contact surfaces; 2 — wide annular joints; *solid lines* — lubrication with medium viscosity industrial mineral oil; *broken lines* — low viscosity industrial mineral oil. The amount of oil in the joints 10/m^2.

and its cross-sectional moment of inertia in the range of 5:1). Damping is slightly increasing with increasing width of the contact area.

7. For joints with narrow contact surfaces:

$$\delta = \frac{A}{\sqrt[3]{\sigma_m}} \qquad (4.8.1)$$

where $A = 0.31$ for joints lubricated by industrial mineral oil of average viscosity and $A = 0.21$ for very thin oil, and σ_m is in MPa.

Damping for tangential vibration for joints preloaded with $\sigma_m = 50$ MPa to 80 MPa (750 psi to 1,200 psi) is $\delta = 0.04$ to 0.05 without lubrication and $\delta = 0.3$ to 0.37 for lubricated joints. Experimental studies on amplitude and frequency dependencies of tangential damping could not be found.

4.8.2 Damping in Cylindrical and Tapered Connections

Damping in cylindrical and tapered connections is, mostly, due to normal and tangential displacements and due to local slippages between the shaft and the sleeve.

Damping in non-lubricated preloaded tapered connections as well as in interference fit cylindrical connections is $\delta = 0.01$ to 0.05 for vibration perpendicular to the axis of the connection. It does not significantly depend on normal pressure. Damping in lubricated connections is about $\delta = 0.12$ to 0.35, and depends on the normal pressure as

$$\delta \approx 0.4\, \sigma^{-0.2}, \qquad (4.8.2)$$

where $\sigma =$ normal pressure in the connection, MPa. As for the flat joints, δ is amplitude-independent.

When a long sleeve (length, L) is fit on a relatively low stiffness shaft (diameter, D), energy dissipation may increase due to micro-slippages between the sleeve and the shaft caused by bending vibratory deformations of the shaft. For large ratios, L/D, and low normal pressures, $\sigma < 20$ MPa (3,000 psi), the energy loss caused by contact deformations is relatively small, up to 20% of the total, at $\sigma = 20$ MPa to 30 MPa (3000 psi to 4500 psi), it raises to 40% to 50%, and at $\sigma > 50$ MPa (7500 psi), the slippage zone is, practically, disappearing, and all energy dissipation is due to contact deformations. In more common connections with small L/D, the slippage disappears already at $\sigma = \sim 30$ MPa (4500 psi), and all energy dissipation is due to contact deformations. Optimal contact pressure, at which energy dissipation (damping) is the most intense usually is about 10 MPa to 40 MPa (1500 psi to 6000 psi).

4.8.3 Energy Dissipation in Power Transmission Components

Energy dissipation in key and spline connections is due to normal and tangential deformations and to slippages on the contact surfaces. In a key connection, angular deformations are associated with tangential displacements on the cylindrical contact surface between the shaft and the sleeve and with normal deformations between the key and the key slot. Radial deformations in the connection are associated with normal displacements in the cylindrical contact and tangential deformations along the key, i. Energy dissipation in a tight key or spline connection is relatively low and is characterized by $\delta = 0.05$ to 0.1. In a sliding spline connection, $\delta = 0.3$ to 0.4 due to larger slippage.

Energy dissipation in antifriction bearings during bending vibrations of the shafts is due to elastic deformations at contacts of the rolling bodies with the races, at contacts between the races and the shaft and the housing caused by radial and angular displacements of the shaft, and partly due to friction between the rolling bodies and the cage. In tapered roller bearings, an additional contributor is friction at the ends of the rollers.

It was observed during experiments that damping (δ) in antifriction bearings is amplitude- and frequency-independent. For bearings installed in single-bearing units with clearances not exceeding 10 μm, in average, $\delta = 0.1$ to 0.125 for ball bearings; $\delta = 0.15$ to 0.2 for single- and double-row roller bearings; $\delta = 0.15$ to 0.2 for tapered roller bearings at small angular vibrations. At low shaft stiffness resulting in larger angular vibrations, damping at tapered roller bearings is significantly higher. At high levels of preload in tapered roller bearings, $\delta = 0.3$ to 0.35. For double ball bearing units, $\delta = 0.1$ to 0.15; for a bearing unit consisting of one radial and one thrust ball bearings, $\delta = 0.25$ to 0.3. Energy dissipation at larger clearances in bearings (20 μm to 30 μm) and without constant preload is much more intense due to impact interactions (see Appendix 3) and can reach $\delta = 0.25$ to 0.35.

Use of this data in evaluating damping characteristic of a whole transmission system or other structures is described in Section 6.5.

REFERENCES

[1] Orlov, P.I., "Fundamentals of Machine Design", Vol. 1, 1972, Mashinostroenie Publishing House, Moscow, [in Russian].

[2] Pisarenko, G.S., Yakovlev, A.P., Matveev, V.V., 1975, "Strength of Materials Handbook", Naukova Dumka, Kiev.

[3] Rivin, E.I., 1988, "Mechanical Design of Robots", McGraw-Hill, N.Y.

[4] Levina, Z.M., 1971, Reshetov, D.N. "Contact Stiffness of Machines", Mashinostroenie, Moscow, 264 pp. [In Russian].

[5} Rivin, E.I., 1980, "Compilation and Compression of Mathematical Model for Machine Transmission," ASME Paper 80-DET-104, ASME.

[6] Eibelshauser, P., Kirchknopf, P., 1985, "Dynamic Stiffness of Joints", Industrie- Anzeiger, vol. 107, No. 63, pp. 40–41 [In German].

[7] "Valenite STS Quick Change System. Technical Information", 1993 (Translation of a Technical Report from RWTH Aahen).

[8] Tsutsumi, M., et al, "Study of Stiffness of Tapered Spindle Connections", Nihon Kikai gakkai rombunsu [Trans. Of Japan Society of Mechanical Engineers], 1985, C51 (467), pp. 1629–1637 [in Japanese].

[9] Hasem, S., et al, 1987, "A New Modular Tooling System of Curvic Coupling Type", Proceed. Of the 26th Intern. Machine Tool Design and Research Conference, MacMillan Publishing, pp. 261–267.

[10] Kato, M., et al, 1980, "Axial Contact Stiffness between Center and Workpiece", Bul. Japan Society of Precision Engineering, 1980, vol. 14, No. 1, pp. 13–18.

[11] Rivin, E., 2000, "Tooling Structure: Interface between Cutting Edge and Machine Tool", Annals of the CIRP, vol. 49/2, pp. 591–634.

[12] Agapiou, J., Rivin, E., Xie, C., 1995, "Toolholder/Spindle Interfaces for CNC Machine Tools", Annals of the CIRP, vol. 44/1, pp. 383–387.

[13] Reshetov, D.N., Kirsanova, V.A., 1970, "Tangential Contact Compliance of Machine Elements", *Mashinovedenie*, No. 3 [In Russian].

[14] Panovko, Ya.G., 1960, "Internal Friction in Vibrating Elastic Systems", Fizmatgiz, Moscow, 196 pp. [in Russian].

[15] Rivin, E.I., Xu, L., 1992, "Toolholder Structures — A Weak Link in CNC Machine Tools", *Proceed of the 2nd International Conference on Automation Technology*, Taipei, July 4-6, vol. 2, pp. 363–369.

CHAPTER

5

Supporting Systems/ Foundations

T he effective stiffness of mechanical systems is determined not
   only by stiffness characteristics of their components but also
   by designed interactions between the components. Influence
of interactions is both due to elastic deformations in the connections
between the interacting components (*contact deformations*; see Chapter 4) and due to the location, size, and shape of the connection areas
(*supporting conditions*). Relatively slight changes in the supporting
conditions may result in up to one to two decimal orders of magnitude changes of deformations under the specified forces, as well as in
very significant changes in stress conditions of the components.

The influence of parameters of the supporting elements on effective damping of the supported component or structure is more
difficult to assess. It depends not only on parameters (stiffness and
damping) of the supporting elements, but also on rather complex correlations between dynamic characteristics of the supported system/
component and the supporting system. In some cases, this influence
can be analyzed, at least approximately, in a close form. Such cases
are represented by an analysis of influence of the mounting (vibration isolation) system on dynamic stability (chatter resistance) of a
machine tool, e.g., given in [1] and in Appendix 2. However, in many
cases, influence of the supporting structure on the effective damping
has to be analyzed by testing and by estimation.

This chapter addresses effects of the supporting systems/founda-
tions on static stiffness/deformations of the supported systems. Some
important effects of the supporting conditions on damping are discussed in Section 5.3.

5.1 INFLUENCE OF SUPPORT CHARACTERISTICS

Fig. 5.1.1 [2] compares stiffness of a uniform beam of length l loaded by
a concentrated force P or by distributed forces of the same overall magnitude and with a uniform intensity, $q = P/l$. Maximum deformation, f,
and stiffness, k, are, respectively,

$$f = \frac{Pl^3}{aEI}, \quad k = \frac{P}{f} = \frac{aEI}{l^3}, \tag{5.1.1}$$

where E = Young's modulus of the beam material, I = cross-sectional moment of inertia, and a = coefficient determined by the supporting conditions. The cantilever beam (cases 5, 6) is the least stiff, while the double
built-in beam (cases 3, 4) is the stiffest, 64 times stiffer than the cantilever beam for the case of concentrated force loading and 48 times stiffer
for the case of distributed loading.

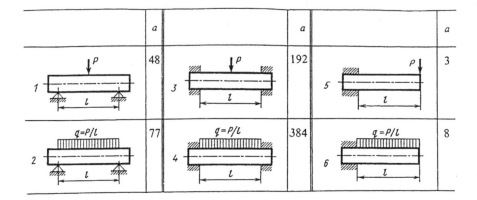

Fig. 5.1.1. Influence of load pattern and support conditions on beam stiffness.

Such simple cases are not always easy to identify in real life systems because of many complicating factors.

Fig. 5.1.2 [2] compares 14 designs of joint areas between piston, piston pin, and connecting rod of a large diesel engine. The numbers indicate values of deformation f of the piston pin under force P and of maximum stress σ in the pin as fractions of deformation $f_1 = Pl^3/48EI$ and *stress*, $\sigma_1 = Pl/4W$ in the pin considered as a double-supported beam (the first case), where W = the section modulus. If can be seen that seemingly minor changes of the piston and the connecting rod in the areas of interaction (essentially, design and positioning of stiffening ribs and/or bosses) may result in changes of deformation in the range of ~125:1 and of stresses in the range of ~12:1.

However, the models in Fig. 5.1.2 are oversimplified, since they do not consider clearances in the connections. Fig. 5.1.3 illustrates influence of the clearances. When the connection is assembled with tight fits (Fig. 5.1.3a), the loading in the pin model can be approximated as in a double built-in beam. A small clearance (Fig. 5.1.3b), may result in a very significant change of the loading conditions caused by deformations of the pin and by the subsequent nonuniformity of contacts between the pin and the piston walls and between the pin and the bore of the connecting rod. For a larger clearance (Fig. 5.1.3c), the contact areas are at the inner edges, and the effective loading pattern (thus, deformations and stresses in the system) is very different from the patterns in Fig. 5.1.3 a,b. Mean-

	σ	f		σ	f
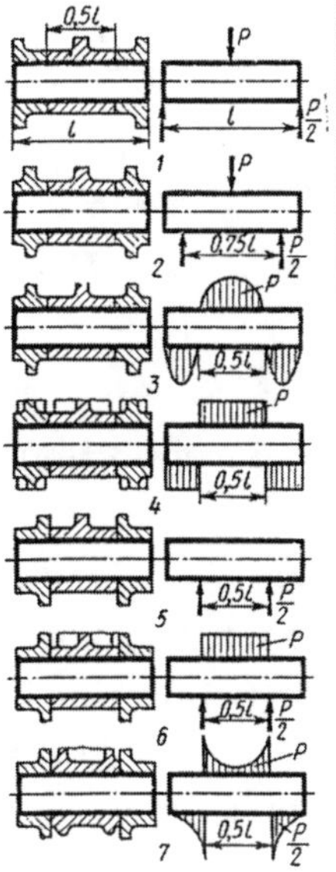 1	1	1	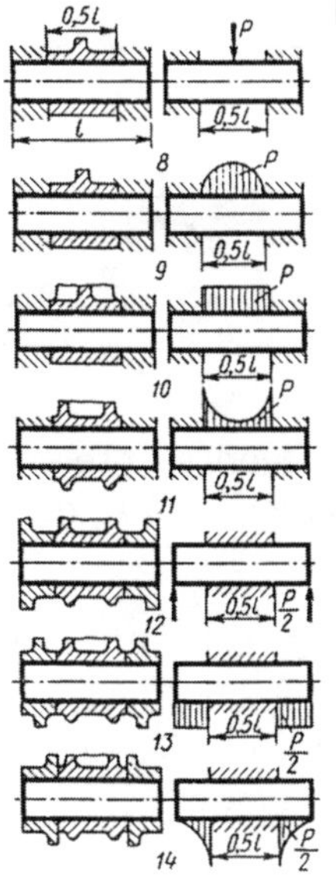 8	0.25	0.031
2	0.75	0.42		0.21	0.023
3	0.56	0.36	9	0.083	0.0156
4	0.5	0.25	10	0.04	0.008
5	0.5	0.125	11	0.5	0.125
6	0.25	0.078	12/13	0.25	0.047
7	0.12	0.04	14	0.12	0.021

Fig. 5.1.2. Influence of components' design on piston pin stiffness.

ing of the "small" and "large" clearance depends, in its turn, not only on the absolute magnitude of the clearance, but also on its correlation with deformation of the pin as well as of two other components.

The substantial influence of the supporting conditions on deformations in mechanical systems gives a designer a powerful tool for controlling the deformations. Fig. 5.1.4 compares three designs of a shaft carrying a power transmission gear. The system is loaded by force P

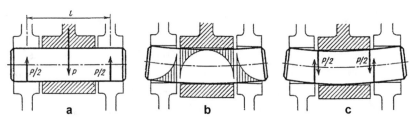

Fig. 5.1.3. Influence of fit between interacting parts on loading schematic.

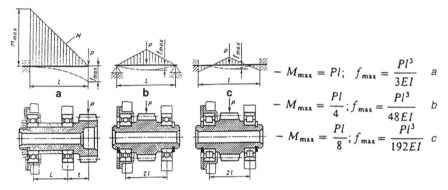

$$- M_{max} = Pl; \quad f_{max} = \frac{Pl^3}{3EI} \quad a$$

$$- M_{max} = \frac{Pl}{4}; f_{max} = \frac{Pl^3}{48EI} \quad b$$

$$- M_{max} = \frac{Pl}{8}; f_{max} = \frac{Pl^3}{192EI} \quad c$$

Fig. 5.1.4. Influence of selection and configuration of supporting elements on effective loads and on stiffness.

acting on the gear from its counterpart gear. In Fig. 5.1.4a,b, the shaft is supported by ball bearings. Since the ball bearings do not provide angular restraint, the shaft can be considered as a double-supported beam. The cantilever location of the gear in Fig. 5.1.4a results in an excessive magnitude of the bending moment M and deflection (f_{max}), as well as in angular deformation of the gear. Location of the gear between the supports as in Fig. 5.1.4b greatly reduces M and f_{max}. In Fig. 5.1.4c, the same shaft is supported by roller bearings in the same configuration as in Fig. 5.1.4b. The roller bearings resist angular deformations and, in the first approximation, can be modeled as a built-in support, thus further reducing deflection f_{max} by the factor of four, while reducing the maximum bending moment M_{max} (and thus maximum stresses), by the factor of two. In both cases of Fig. 5.1.4b,c, due to symmetrical positioning of the gear, the angular deflection or the gear is zero, thus the meshing process is not distorted. There would be some angular deflection and distortion of the meshing process if the gear were not symmetrical.

Large angular deflections, which develop if gears are mounted on cantilever shafts, result from deformations of the double supported shaft, but also due to contact deformations in the bearings (see Chapter 6). In Fig. 5.1.5a, the greatest deformation is in the front bearing 1, which is acted upon by a large reaction force, N_1. A very important feature of cantilever designs, as the one in Fig. 5.1.5a, is the fact that the reaction forces, especially N_1, can be of a substantially higher magnitude than the acting force P as illustrated by Fig. 5.1.6a. As can be seen from Fig. 5.1.6a, the reaction forces are especially high when the span between the supports

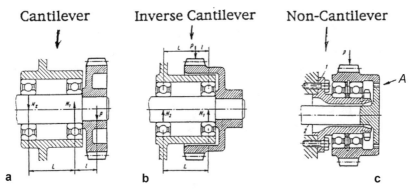

Fig. 5.1.5. Reduction of reaction forces in supporting elements by design.

(bearings) is less than two overhang lengths, l. In addition to high magnitudes, the reaction forces for the model in Fig. 5.1.6a have opposite directions, thus further increasing the angular deflection of the shaft in Fig. 5.1.5a. The reaction forces in cases when the external force is acting within the span are substantially lower, as shown in Fig. 5.1.6b.

To alleviate this problem in cases when the force-generating component, such as a gear, cannot be mounted within the span between the supports, such components can be reshaped so that the force is shifted to act within the span, even while the component is actually attached outside the span, like the gear in Fig. 5.1.5b (*"inverse cantilever"*).

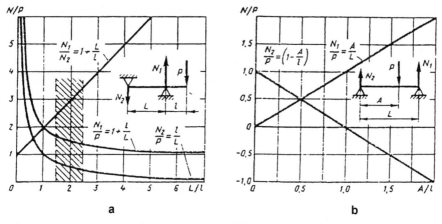

Fig. 5.1.6. Reaction forces on a double-supported beam with (a) out-of-span (cantilever) loading and (b) in-span loading.

Even better performance of power transmission gears and pulleys, as well as of precision shafts (e.g., the spindle of a machine tool) can be achieved by total separation of forces acting in the mesh, belt preloading forces, etc., and torques transmitted by the power transmission components. Such design is shown in Fig. 5.1.5c, where the gear is supported by its own bearings mounted on the special embossment of the shaft housing (gear-box, headstock). The shaft also has its own bearings, but the connection between the gear and the shaft is purely torsional via coupling A, and no force acting in the gear mesh is acting on the shaft. In this case, the gear is maintained in its optimal non-deformed condition, and the shaft is not subjected to bending forces.

The same basic rules related to influence of supporting conditions on deformations/deflections apply to piston-pin-connection rod in Fig. 5.1.2 and to power transmission gears on shafts in Figs. 5.1.4 and 5.1.5. They are true also for machine frames and beds mounted on a floor or on a foundation, a massive table mounted on a bed, etc. While these rules were discussed for beams, they are (at least, qualitatively) similar for plates. For example, for round plates, *"built-in"* support (Fig. 5.1.7b), reduces deformation 7.7 times versus. "simple edge" support (Fig. 5.1.7a).

It is very important to correctly model the supporting conditions of mechanical components to improve their performance. Incorrect modeling may significantly distort the actual force schematics and lead to very wrong estimations of the actual stiffness. Figure 5.1.8 [2] presents a case of a shaft supported by two sliding hydrodynamic bearings, A and B, and loaded in the middle by a radial force P also transmitted through a hydrodynamic bearing. This model describes, for example, a "floating" piston pin in an internal combustion engine. The actual pressure distribution along the length of the load-carrying oil film in the bearings is parabolic (left diagrams in Fig. 5.1.8c). The peak pressures are 2.5 to 3.0

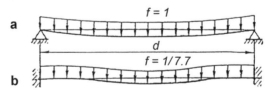

Fig. 5.1.7. Influence of support contour design on deformation of round plates under uniformly distributed load (weight); (a) Simple support contour; (b) Built-in contour.

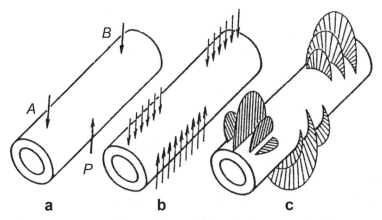

Fig. 5.1.8. Loading schematic for shaft supported by two hydrodynamic bearings.

times higher than the nominal (average) pressures. In transverse cross-sections, the pressure is distributed along a 90 deg. to 120 deg. arch (right and center diagrams in Fig. 5.1.8c).

Comparison of actual loading diagrams in Fig. 5.1.8c with simplified mathematical models in Figs. 5.1.8a and 5.1.8b shows that the schematic in Fig. 5.1.8a overstates the deformations and stresses, while the schematic in Fig. 5.1.8b understates them. None of these models considers transverse components of the loads and associated deformations and stresses. It is important to remember that a more realistic picture of the loading pattern in Fig 5.1.8c may change substantially in the real circumstances due to elastic deformations of both the shaft and the bearings, excessive wedge pressures, etc. Design of the front end of the diesel engine crankshaft in Fig. 5.1.9a [2] experienced such distortions. While

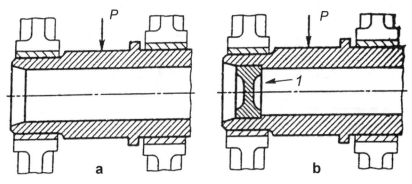

Fig. 5.1.9. Stiffness enhancement of front-end bearing of a crankshaft.

the nominal loading on the front end journal was relatively low, the bearing was frequently failing. It was discovered that the hollow journal was deforming and becoming elliptical under load. The elliptical shape of the journal resulted in reduction of the hydrodynamic wedge in the bearing and deterioration of its load-carrying capacity. The design was adequately improved by enhancing stiffness of the journal using a reinforcing plug 1 (Fig. 5.1.9b).

Fig. 5.1.10 illustrates typical support conditions for power transmission shafts supported by sliding bearings and antifriction ball bearings [3]. Although the bearings are usually considered as simple supports, it is reasonably correct only for the case of the single bearing (Fig. 5.1.10a). For the case of tandem ball bearings (Fig. 5.1.l0b), the bulk of the reaction force is accommodated by the bearing located on the side of the loaded span (the inside bearing). The outside bearing is loaded much

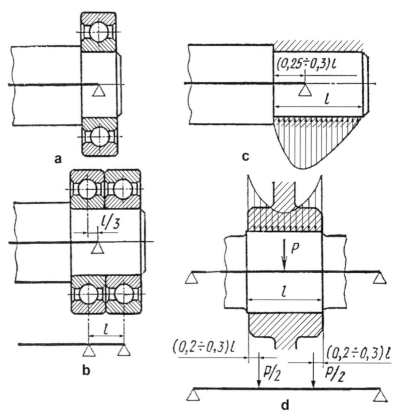

Fig. 5.1.10. Computational models of shafts in ball bearings.

less and might even be loaded by an oppositely directed reaction force if there is a distance between two bearings. Thus, it is advisable to place the support in the computational model at the center of the inside bearing or at one-third of the distance between the bearings in two bearing support, toward the inside bearing (Fig. 5.1.10b). Due to shaft deformations, pressure from sliding bearings onto the shaft is nonsymmetrical (Fig.5.1.10c), unless the bearing is self-aligning. Accordingly, the simple support in the computational model should be shifted off-center and located (0.25 to 0.3) l from the inside end of the bearing (Fig. 5.1.10c).

Radial loads transmitted to the shafts by gears, pulleys, sprockets, etc., are usually modeled by a single force in the middle of the component's hub (Fig. 5.1.10d). However, the actual loading is distributed along the length of the hub, and the hub is, essentially, integrated with the shaft. Thus, it is more appropriate to model the hub-shaft interaction by two forces as it is shown in Fig. 5.1.10d. Smaller shifts of the forces $P/2$ from the ends of the hub are taken for interference fits and/or rigid hubs; larger shifts are taken for loose fits and/or for not adequately rigid hubs.

5.2 RATIONAL LOCATION OF SUPPORTING/ MOUNTING ELEMENTS

The number and location of supporting elements have a great influence on deformations/stiffness of the supported system. Frequently a "flimsy" system may have a decent effective stiffness if it is judicially supported.

One of the optimization principles for locating the supports is *balancing* of deformations within the system so that the maximum deformations in various "peak points" are of about the same magnitude. For simple supported beams, the balancing effect is achieved by placing the supports at so-called Bessel's points. This approach is illustrated in Fig. 5.2.1 showing a beam loaded by a uniformly distributed load (e.g., by its weight load). Replacement of simple supports in Fig. 5.2.1a with built-in end fixtures (Fig. 5.2.1b), reduces the maximum deformation by a factor of five. However, placing just simple supports at the Bessel's points located about $0.23L$ from the ends of the beam, or $L/l = 1.86$ (L = length of the beam, l = distance between the supports) reduces the maximum deformation by 48 times (Fig. 5.2.1c). Instead of one point of the system where the maximum deformation is observed in cases of Fig. 5.2.1a, b (at the

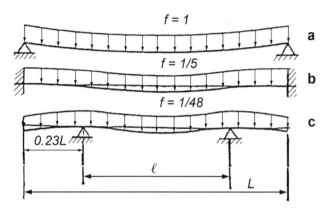

Fig. 5.2.1. Influence of type and location of supports on maximum deflection of a beam.

midspan), in case of Fig. 5.2.1c there are three points where the deformation is maximum — midspan and two ends.

The same principle can be applied in more complex cases of plates. Stiffness of plate-like round tables of vertical boring mills was studied in [4]. Since the round table is rotating in circular guideways, the critical performance factor is relative angular displacements in the radial direction between the guiding surfaces of the table and its supporting structure (base). If the base is assumed to be rigid (since it is attached to the machine foundation), then the radial angular deformation of the table along the guideways is the critical factor. There are several typical designs of the round table system, such as:

1. The round table is supported only by the circular guideways;
2. There is a central support (thrust bearing), which keeps the center of the table at the same level as the guideways regardless of loading; and
3. There is a central unidirectional support which prevents the center of the table from deflection downward but not upward.

For the first design embodiment, three optimal loading schematics were considered:

1. Load is uniformly distributed along the outer perimeter (D_o) of the table (total load, P) and creates the same angular deformation at the guideways (diameter D_g) as the same load, P, uniformly distributed along the circle having diameter $D_g/2$. Angular compliances, A, for

both loading cases are plotted in Fig. 5.2.2a as functions of $\beta = D_o/D_g$. The parameter

$$A = \frac{PD_g}{16\pi N},$$ (5.2.1)

where N = so-called cylindrical stiffness of the plate. The angular compliances for these two loading cases become equal at $\beta_a = 1.4$.

2. Plate is loaded with its uniformly distributed weight, Fig. 5.2.2b (the case similar to the case of Fig. 5.2.1c). The angular deformation at the guideways vanishes at $\beta_b = 1.44$.

3. Load P is uniformly distributed across the whole plate and creates the same angular deformation at the guideways as the same load distributed across the circle having diameter, $D_g/2$ (Fig. 5.2.2c). The angular compliances become equal at $\beta_c = 1.46$.

While $\beta_a \approx \beta_b \approx \beta_c \approx 1.43$, there might be cases when different loading conditions are of interest and/or loading inside and outside the supporting circle have different application frequency and/or importance. In such cases, the selected value of β can be modified.

The similar plots for the design embodiment with a bi-directional central support are given in Fig. 5.2.3a,b for two cases of loading:

1. Weight load, there is no angular deformation at the guideways at $\beta_a \approx 1.33$ (Fig. 5.2.3a), and

2. Uniformly distributed load, P, across the whole plate and uniformly distributed load across the circle of diameter, $D_g/2$ (Fig. 5.2.3b); $\beta_b \approx 1.15$.

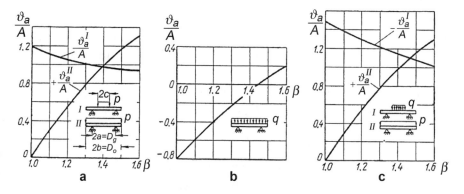

Fig. 5.2.2. Angular deformations at support contour for plates without central support.

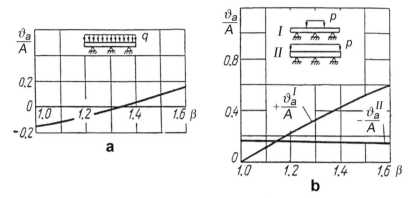

Fig. 5.2.3. Angular deformations at support contour of plate with bi-directional central support.

For the design with a unidirectional central support (Fig. 5.2.4), the optimal location of the guideways for the distributed loading is even closer to the periphery, $\beta \approx 1.06$.

Beds/frames of machines and other mechanical systems may have ribs and other reinforcements, may have box-like structures, etc. However, the same basic principles can be applied. Mounting of complex beds can reduce deformations due to moving loads (e.g., tables) even with a smaller number of mounts if their locations are judicially selected.

Example 5.1. This can be illustrated by a real-life case of installing a relatively large cylindrical (OD) grinder whose footprint is shown in Fig. 5.2.5. The 3.8 m (~ 13 ft) long bed exhibited excessive deformations causing errors in positioning the machined part when a very heavy table carrying the part was traveling along the bed. Also, excessive positioning errors develop when the heavy grinding head travels in the transverse direction. To reduce the deformations, the manufacturer (Schaudt Co., Germany) recommended to level the machine on 15 rigid leveling wedges as shown by **o** on Fig. 5.2.5. The leveling protocol required lifting the table and monitoring the leveling condition by measuring straightness of the guideways. However, due to substantial friction in the wedges after the leveling procedure is completed, the wedges remained in a strained condition after the leveling procedure had been completed. Vibrations caused by overhead cranes, by passing transport, by accidental impacts, etc., were relieving the strain, and the wedges were changing their height (e.g., see [5]). Since the re-leveling procedures are very expensive, the machine had to be isolated from vibrations. There are

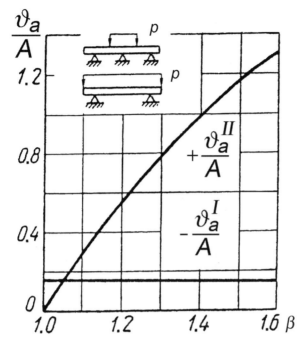

Fig. 5.2.4. Angular deformations at support contour of plate with uni-directional central support.

two major isolation techniques: (a) to install a massive and rigid foundation block on elastic elements, and then to mount the machine, using rigid wedges, on the foundation block; and (b) to mount the machine on the floor directly on vibration isolating mounts. While the second strategy is much less expensive and allows to quickly change the layout, it is suitable only for machines with rigid beds not requiring reinforcement from the foundation. The OD grinder in Fig. 5.2.5 definitely would not have an adequately rigid bed if 15 vibration isolators were placed instead of 15 rigid wedges. However, there was made an attempt to apply the load balancing, or the Bessel's points concept, to this machine. This was realized by replacing 15 mounting points with 7 as shown by **x** in Fig. 5.2.5. The effective stiffness of the machine installed on seven mounts increased dramatically, so much so that it could be installed on seven flexible mounts (vibration isolators), which eliminated effects of outside excitations on leveling, as well as on the machining accuracy. Table in Fig. 5.2.5 shows deformations (angular displacements) between the grinding wheel and the machined part during longitudinal travel of the

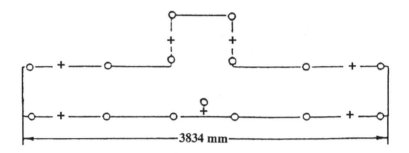

Type of Installation	Longitudinal Transverse Axis Rotation			Longitudinal Axis Rotation		
	Wheelhead position			Table position		
	Front	Rear	Diff.	Left	Right	Diff.
On 15 rigid wedge-type mounts	−3.0	−0.5	2.5	0	−1.0	1.0
On 7 isolators ($f_{vz} - 20$ Hz)	+3.3	+0.8	2.5	+0.8	+0.3	0.5

Fig. 5.2.5. Influence of mount locations on effective stiffness of cylindrical (OD) grinder.

table and during transverse travel of the grinding head. It is remarkable that the deformations are *smaller* when this large machine was installed on seven flexible mounts than when it was installed on 15 rigid wedges.

Influence of the number and locations of the mounting elements on relative deformations between the structural components (e.g., between the grinding wheel and the workpiece in the grinding machine in Fig. 5.2.5) is associated with changing reaction forces acting from the mounts on the structure and causing a change of its deformation pattern. There are two main reasons for the reaction-caused deformations:

1. The bed, or the frame, or another object is finish-machined, while resting on supports (mounts) which, in their turn, sit on a foundation. If the number of the supports is *three* (so-called "kinematic support"), then their reaction forces do not depend on variations of the foundation surface profile, variations of height and stiffness of the individual mounts, surface irregularities of the supporting surface or of the frame itself. The reaction forces are fully determined by the position of the center of gravity of the frame and by coordinates of the supports relative to the C.G.

If the frame is reinstalled on a different foundation, using different supports but at the same locations, the reaction forces would not change. As a result, relative positions of all components within the structure would be stable to a very high degree of accuracy.

However, in most cases there are more than three supports. In such cases, the system is a "statically indeterminate" one, and relocation, even without changes in distribution of mass within the system, would cause changes in the force distribution between the supports/mounts. The force distribution changes result in changes of the deformation map of the structure due to inevitable differences in relative height of the mounts and their stiffness values, minor variations of stiffness, and flatness of the foundation plates, etc.

If the object is finish-machined, while mounted on more than three supports, and then relocated (e.g., to the assembly station), the inevitable change in the mounting conditions would result in distortions of the machined surfaces, which are undesirable. This distortions can be corrected by leveling of the part (adjusting the height of each support until the distortions of the reference surface are within an acceptable tolerance). The same procedure must be repeated at each relocation of the part or of the assembly.

If a precision maintenance of the original/reference condition of the system is required, but the reference (precisely machined) surface is not accessible, e.g., covered with a table, then leveling is recommended, while the obstructing component is removed, like in the above example with the cylindrical grinder. This procedure is not very desirable (but often unavoidable), since not only is it very labor intensive and time consuming, but also the leveling procedure is usually performed for a system loaded differently than during its fabricating.

Although increasing the number of supports for a frame part of a given size reduces the distances l between the supports and reduces deformations, the system becomes more sensitive to small deviations of the foundation surface and condition and to height variations of the supports. These variations will significantly affect reaction forces, and since the system becomes more rigid with the increasing number of supports, any small deviation may result in large force (and thus deformation) changes or even may result in "switching off" of some supports.

Example 5.2. The table in Fig. 5.2.6 [5] shows reactions in the mounts (vibration isolators equipped with a reaction-measuring device) under a

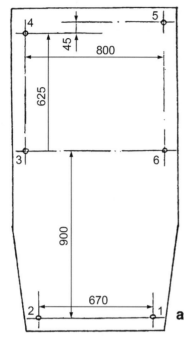
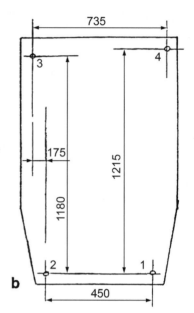

Vertical Milling Machine   **Horizontal Milling Machine**

Mounting Points

Mount Reactions, kN

Machine	Condition	1	2	3	4	5	6
a	Not levelled	4.0	20.9	5.0	3.7	17.7	5.9
	Levelled	9.25	9.1	11.5	10.1	8.9	9.9
b	Not levelled	9.1	12.2	9.8	6.6		
	Levelled	11.3	11.5	7.3	7.5		

Fig. 5.2.6. Influence of leveling on weight distribution between mounts for (a) vertical and (b) horizontal milling machines. Dimensions in mm.

vertical milling machine (Fig. 5.2.6a) and under a horizontal milling machine (Fig. 5.2.6b). Two cases for each installation were compared: first, the machine was installed without leveling and then leveled by adjusting the heights of the mounts until the table of the machine (the reference surface) becomes horizontal. One has to remember that the differences in the reaction forces on various mounts create bending moments acting on the structure and deforming it. It should also be noted that if these machines were installed on more rigid (e.g., wedge-type) mounts, the ranges of load variation would be much wider, and distortions in the structure, correspondingly, would be much more pronounced.

Another way of maintaining the same conditions of the frame and/or of the assembled object is by measuring actual forces on the supports at the machining station for the frame and/or at the assembly station for the object. At each relocation, the supports can be adjusted until the same reaction forces are reached. This approach is more accurate since stiffness values of the mounts would have a lesser influence. It was successfully used for precision installations of both the gear hobbing machine (weight 400 tons) for machining high precision large (5.0 m diameter) gears for quiet submarine transmissions and for the large gear blanks placed on the table of the machine [6]. The machine was installed on 70 load-sensing mounts, while the gear blank was installed on 12 load-sensing mounts. The loads on each mount were continuously monitored.

(B) The situation is somewhat different if mass/weight distribution within the structure is changing. Figure 5.2.7 shows variations of load distribution between rigid (wedge) mounts under a large turret lathe (weight 24,000 lb), while a heavy turret carriage (weight 2000 lb) is traveling along the bed. If the bed were absolutely rigid, the whole machine would tilt without any relative displacements of its constitutive components. Since the bed is not absolutely rigid, travel of the heavy unit would also cause structural deformations within and between the components. These structural deformations would be reduced if the mounts were rigid and provide reinforcement of the structure by reliably connecting it to the foundation. However, the rigid mounts (wedge mounts, screw mounts, shims, etc.) are prone to become strained in the course of the adjustment procedure. When the strain is relieved by vibration and shocks generated inside or outside the structure, their dimensions may change as in the above example of the OD grinder in Fig. 5.2.5. Similarly to group (A), leveling of these structures is easier and more reliable when

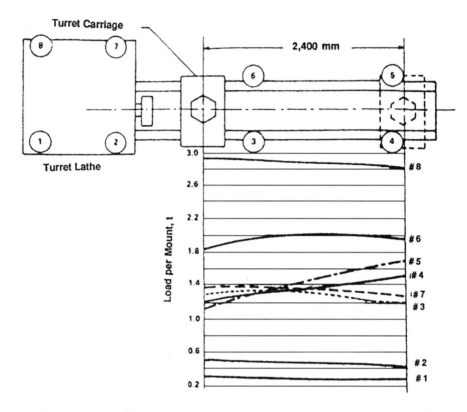

Fig. 5.2.7. Influence of traveling heavy turret carriage on weight distribution between mount under a turret lathe.

the weight load distribution between the mounts is recorded for a certain reference configuration. In this configuration, all traveling components should be placed at the specified positions, and specified additional components should be attached. For production systems, the workpiece or die weights should also be specified.

It is important to understand that the "kinematic support" condition in these circumstances *does not guarantee the perfect alignment* of the structural components, since relocation of a heavy component inside the structure is causing changes in the C.G. location and in mount reaction forces and thus generates changing bending moments that may cause structural deformations. Accordingly, the structures experiencing significant mass/weight distribution changes must be designed with an adequate stiffness, and the mounting point locations must be judiciously selected. This is illustrated by Fig. 5.2.8 [7], which shows the change of

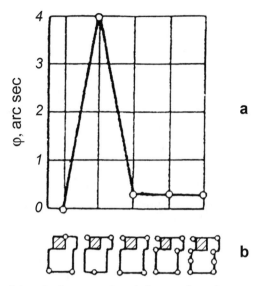

Fig. 5.2.8. (a) Relative angular deformations between table and spindle head of a precision jig borer for (b) different number and location of mounts.

angular deflection between the table and the spindle head of a precision jig borer between the extreme left and right positions of the moving table for different numbers and locations of mounts. While the minimal deflection corresponds to one kinematic support configuration, the maximum relative internal deflections correspond to another kinematic support configuration.

5.3 OVERCONSTRAINED (STATICALLY INDETERMINATE) SYSTEMS

The mounting systems discussed in the previous section are typical statically indeterminate systems if more than three mounting points are used. Statically indeterminate (overconstrained) systems are frequently used, both intentionally and unintentionally, in mechanical design. Static indeterminacy has a very strong influence on stiffness. Excessive connections, if properly applied, may serve as powerful stiffness enhancers and may significantly improve both accuracy and load carrying capacity. However, they also may play a very detrimental role and lead to a fast deterioration of a structure or a mechanism.

Effects of overconstraining depend on the design architecture, geometric dimensions of the structures, and performance regimes. For example, overconstraining of guideways 1 and 2 of a heavy machine tool (Fig. 5.3.1a) [8] plays a very positive role in increasing stiffness of the guideways. This is due to the fact that the structural stiffness of the guided part 3 is relatively low because of its large dimensions. Local deformations of part 3 accommodate uneven load distribution between the multiple guiding areas. On the other hand, guideways 1 and 2 for a lighter and relatively rigid carriage 3 in Fig. 5.3.1b are characterized by some uncertainty of load distribution that cannot be compensated by local deformations of carriage 3 due to its high rigidity and multiple constraints by multiple guideways.

It should not be forgotten that stiffness enhancement is not the real goal in many cases. In such cases, the real goal is reduction of deformations in the system. It can be achieved not only by increasing the stiff-

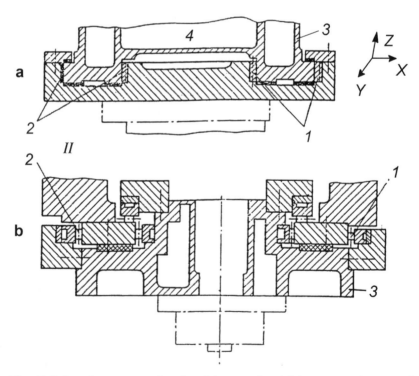

Fig. 5.3.1. Overconstrained guideways for (a) heavy carriage and (b) medium size carriage.

ness proper, but also by reduction of forces acting in the system (e.g., see Section 8.1). Frequently, such force reduction can be achieved by design means. For heavy parts moving in linear guideways, an example is carriage 3 in Fig. 5.3.1a. If the driving force along axis y is applied at the "center of gravity" 4 of the cross-section of the symmetrical part 3, then no moment about vertical axis z would be generated. However, if position of the driver (e.g., a power screw or a linear motor) is shifted from the C.G. in x-direction, the moment about z-axis would be generated, which would, in turn, generate reactions in the x-y plane causing increase of friction forces and of ensuing contact deformations. Rise in vibration amplitudes would also be possible.

Symmetrical driving of guided components was always one of the fetures characterizing a good design. This approach was recently re-emphasized by Mori Seiki Co in designs of vertical and horizontal CNC machining centers [9].

Excessive connections, if judiciously designed, may significantly improve dynamic behavior of the structure/mechanisms, such as to enhance chatter resistance of a machine tool. In the spindle unit with three bearings (Fig. 5.3.2a), presence of the intermediate bearing (the rear bearing is not shown) may increase chatter resistance as much as 50%. However, this beneficial effect would develop only if this bearing has a looser fit with the spindle and/or housing than the front and rear bearings. In such a case, the "third bearing" would not generate large extra loads in static conditions due to uncertainty caused by static indeterminacy, but would enhance stiffness and, especially, damping for vibratory motions due to presence of lubricating oil in the clearance. While the spindle is able to self-align in the clearance both under static loads and at relatively slow rpm-related variations, viscosity of the oil being squeezed from the clearance under high frequency chatter vibrations would effectively stiffen the connection and generate substantial friction losses (damping action).

A similar effect can be achieved by using an intermediate plain bushing that is fit on the spindle with a significant clearance (0.2 mm to 0.4 mm per diameter) and filled with oil (Fig. 5.3.2b) [8]. High viscous damping provided by the bushing dramatically reduced vibration amplitudes of the workpiece being machined (Fig. 5.3.2c), which represents an enhancement of dynamic stiffness of the spindle. The price for the higher stiffness and damping in both cases is higher frictional losses.

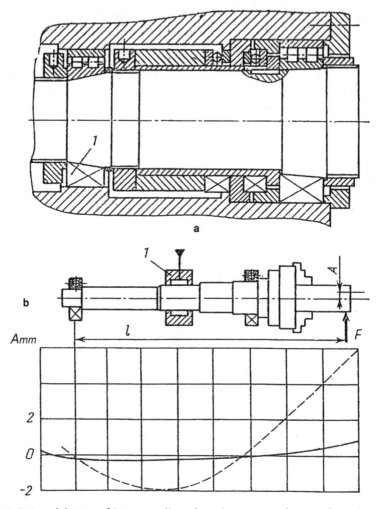

Fig. 5.3.2. (a) Use of intermediate bearing 1 to enhance damping and chatter resistance; (b) Influence of plain bushing fit with 0.2 mm to 0.4 mm clearance on fundamental vibration mode of the spindle (dashed line — no damping bushing; solid line — with damping bushing).

The same concept can also be used in linear guideways. Carriage 1 in Fig. 5.3.3 is supported by antifriction guideways 2. The (intermediate) supports 3 and 4 are flat plates having clearance $\delta = 0.20$ mm to 0.03 mm with the base plate 5. While the intermediate supports do not contribute to the static stiffness of the system, they enhance its dynamic stiffness and damping due to resistance of oil filling the clearance to "squeezing out" under relative vibrations between carriage 1 and base 3.

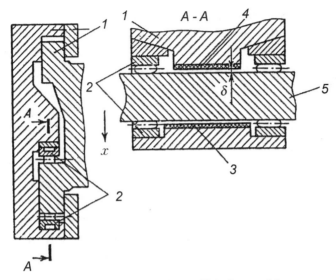

Fig. 5.3.3. Intermediate support in antifriction guideways to enhance dynamic stability.

Overconstraining (and underconstraining) of mechanical structures and mechanisms can be caused by thermal deformations, by inadequate precision of parts and assembly, and by design errors. Figure 5.3.4 shows round table 1 of vertical boring mill supported by hydrostatic circular guideway 2 and by central antifriction thrust bearing 3. At low rpm of the table, the system works adequately, but at high rpm (linear speed in the guideway 8 m/s to 10 m/s), heat generation in the hydrostatic guideway 2 causes thermal distortion of the round table as shown by the broken line in Fig. 5.3.4. This effect distorts the shape of the gap in the hydrostatic

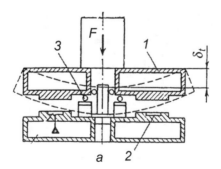

Fig. 5.3.4. Loss of linkage and loss of stiffness due to thermal deformations.

support, and the system fails to provide the required stiffness. The performance (overall stiffness) would *improve if thrust bearing 3 is eliminated* or if the system were initially distorted to create an inclined clearance in the guideway with the inclination directed opposite to the distortion caused by the thermal deformation of the round table.

Although thermal deformations in the system shown in Fig. 5.3.4 caused loss of linkage and resulted in an underconstrained system, frequently thermal defonnations cause overconstraining. For example, angular contact ball bearings 2 and 3 for shaft 1 in Fig. 5.3.5 may get jammed due to thermal expansion of shaft 1, especially if its length, $l \geq (8$ to 12) d. Similar conditions can develop for tapered roller bearings. Jamming in Fig. 5.3.5 can be prevented by replacing spacer 4 with a spring, which would relieve the overconstrained condition and the associated overloads.

A similar technique is used in the design of a ball screw drive for a heavy table in Fig. 5.3.6. The supporting bracket for nut 3 is engaged with lead screw 2, which propels table 1 supported by hydrostatic guideways 5 and 6. Bracket 4 contains membrane 7 connecting nut 3 with table 1. Such an intentional reduction of stiffness prevents overconstraining of the system due to uneven clearances in different pockets of hydrostatic guideways or to misalignment caused by imprecise assembly of the screw mechanism, among other things.

Frequently, the overconstrained condition develops due to design mistakes. Figure 5.3.7 shows a bearing unit in which roller bearing 1 accommodates the radial load on shaft 2, while angular contact ball bearing 3

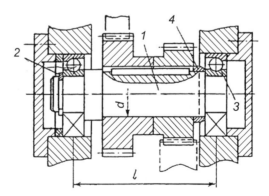

Fig. 5.3.5. Shaft supported by two oppositely directed angular contact bearings.

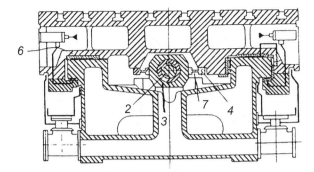

Fig. 5.3.6. Intentional reduction of stiffness (elastic membrane 7) to enhance performance of screw drive.

accommodates the thrust load. In the original design, both bearings are subjected to the radial loading, thus creating overconstraining. It would be better to remove the excessive constraints by dividing functions between the bearings as shown in Fig. 5.3.7 by a broken line. In the latter design, radial bearing 1 does not prevent small axial movements of shaft 2 thus making accommodation of the thrust load less uncertain.

Figure 5.3.8 shows a spindle unit in which the bearing system is underconstrained. Position of the inner race of the rear bearing 1 is not determined, since the position of the inner race is not restrained in the axial direction. Due to the tapered fit of the inner race, it can move, thus detrimentally affecting stiffness of the spindle. The design can be improved by placing an adapter ring between inner race 2 and pulley 3. The adapter ring can be precisely machined or, even better, be deformable to accommodate thermal expansion of the spindle.

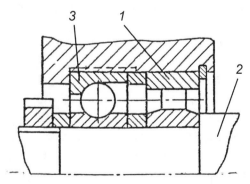

Fig. 5.3.7. Overconstrained bearing unit.

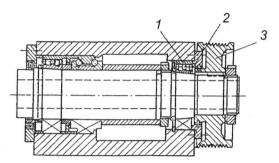

Fig. 5.3.8. Spindle unit with underconstrained rear bearing 3.

5.4 INFLUENCE OF FOUNDATION ON STRUCTURAL DEFORMATIONS

5.4.1 General Considerations

Deformations of a structure that is kinematically supported (on three mounting points) do not depend on stiffness of mounts and /or of the foundation block, but only on locations of the mounts. In many cases, there are more than three mounts, or the structure (e.g., the frame of a machine) is attached to the foundation block by a supporting area rather than by discrete mounts. In such cases, stiffness of the foundation as well as number and location of the supporting mounts may have profound influences on the effective stiffness of the structure.

For precision systems, location of the mounts must be assigned in such a way that influence of structural deformations on accuracy of measurements/processing is minimal. The supporting mounts and design of the foundation are especially critical for machine frames having large horizontal dimensions.

If a short bed is installed on discrete feet or mounts, the foundation block resists both linear and angular displacements of the feet. Influence of stiffness of the connection between the feet and the foundation on deformations of the bed depends on loads acting on the bed, and it is more pronounced for beds fastened to the foundation. In most cases, this influence does not exceed 10% to 20% and can be neglected.

However, stiffness of a long bed supported by an adequate number of mounts on the shop floor plate [monolithic concrete plate 0.15 m to 0.3 m (6 in to 12 in.) thick] is increasing by 30% to 40% even without anchoring and/or grouting. Anchoring or grouting may result in two to four times stiffness increase. Fastening of the bed to an individual foundation block [0.5 m to 1.5 m (20 in. to 60 in.) thick] may result in stiffness increase up to ten times [10]. Computational evaluation *of* deformations in the system: machine bed-foundation is performed to analyze their influence on accuracy of the system and to appreciate its stability in time. Some important modes of deformation of this system are torsion (e.g., twisting of the bed under the moving weight force of a heavy table moving in the transverse direction) and bending (e.g., in a long broaching machine due to reaction to the cutting forces).

The computational model of a machine bed fastened to an individual foundation is a beam on an elastic bed. To analyze deformations of machine bases mounted on the floor plate, the plate should be considered as a variable stiffness plate laying on an elastic bed. The higher stiffness of the plate would be in the area of the machine base installation due to the stiffening effect of the installed base on the foundation plate itself. A machine base installed on mounts that are not anchored/grouted to the plate should be considered as a beam on elastic supports whose deformations depend, in turn, on deformations of the plate on an elastic bed. Such models are extremely complex, but in many cases, a qualitative analysis is adequate for selecting parameters of the installation system. The approximate method [10] replaces the plate on an elastic bed with a beam having effective width, B_{ef}, and effective length, L_{ef}, which are determined so that the displacements under each mounting point are approximately the same as for the bed installed on the plate.

The effective width, B_{ef}, of an infinitely long beam can be calculated for one force acting on the plate and on the equivalent beam. For a more realistic case when two forces are acting, one on each side of the bed, then

$$B_{ef} = B + 13h, \qquad (5.4.1)$$

where B = width of the bed and h = thickness of the foundation plate. This simple expression results in the error not exceeding 20% between deformation of the plate and the beam in the area around the force application point.

The effective length, L_{ef}, of the beam is determined from the condition that the relative deflections of the finite and infinite length beams are close to each other within a certain span. As a first approximation, L_{ef} is determined in such a way that for a beam loaded with a concentrated force in the middle, the beam with length, L_{ef}, would have a deflection under the force relative to deflections at the beam's ends equal to deflection of the infinitely long beam under the force relative to deflections at the points at distances, $L_{ef}/2$ from the force. Under these conditions,

$$L_{ef} = 3.464 \sqrt{\frac{E_f I_{fy}(1 - v_b^2)}{E_b}}, \qquad (5.4.2)$$

where E_f = Young's modulus of the plate material; $I_{fy} = B_{ef}(h^3/12)$ = cross-sectional moment of inertia of the beam; and E_b and v_b = Young's modulus and Poisson's ratio, respectively, of the elastic bed (soil) on which the plate/beam is supported. Although expressions for B_{ef} and L_{ef} given above are derived for the case of the bending, they are also applicable for analyzing torsional deformations [10].

Analysis of the foundation's influence on stiffness of a machine base using the "effective beam" approach is performed with an assumption that the beam is supported by the elastic bed of the Winkler type (intensity of the distributed reaction force from the elastic bed at each point is proportional to the local deformation at this point). The foundation beams are classified in three groups depending on the stiffness index:

$$\lambda = \frac{L}{2} \sqrt[4]{\frac{kB}{4EI}}, \qquad (5.4.3)$$

where L and B = length and width of the beam, EI = its bending stiffness, and the elastic bed coefficient is:

$$k = \frac{0.65 E_b}{B_{ef}(1 - v_b^2)} \sqrt[12]{\frac{E_b B_{ef}^4}{E_f I_{fy}}} \qquad (5.4.4)$$

If $\lambda < 0.4$, then the beam is a *rigid beam*; for $0.6 < \lambda < 2$ it is a *short beam*; and for $\lambda > 3$ - a *long beam* [10]. An overwhelming majority of machine installations can be considered "short beams". The long beam classification is applicable to cases when a long machine base is installed on an individual foundation block without anchoring or grouting.

It has been suggested that one must consider influence on structural stiffness of the installed machine of the foundation block or plate reduced to the equivalent beam by using *stiffness enhancement coefficients*, R. Such coefficients can be defined for both bending and torsion. They are valid for analyzing deformations caused by so-called *balanced forces*, e.g., by cutting forces in a machine tool, which are contained within the structure. Weight-induced forces are not balanced within the structure. They are balanced by reaction forces on the interfaces between the machine base and the foundation and between the foundation and the soil or a substructure. Distribution of these reaction forces depends on stiffness of the bed and the foundation. If the supported beam (the machine base) has an infinite stiffness, $EI \rightarrow \infty$, or the elastic bed is very soft, $k \rightarrow 0$, then the reaction forces have a quasi-linear distribution.

Structural stiffening due to elastic bed coefficient, $k > 0$, can be estimated by first assuming uniform distribution of the reaction forces under the foundation block. Relative deformations of the bed are computed as for a beam acted upon by the main loading system (weight loads) and by the secondary loading system (uniformly distributed reactions), with the beam (bed) connected to the foundation block by rigid supports. Influence of deviation of actual distribution of the reaction forces from the assumed uniform distribution is considered by introduction of the stiffness enhancement coefficient.

Example 5.3. It is required to find deformation of the base of a boring mill or a large machining center from the weight of a heavy workpiece attached to the table located in the midspan of the bed. The weight of the workpiece is balanced by reaction forces uniformly distributed across the base of the foundation block (Fig. 5.4.1). Considering weight of the workpiece (part), W_p, as a concentrated force, deformation under the force relative to ends of the base (where rigid mounts are located) is:

$$f = \frac{W_p L}{128 EIR_b}\left[1 - \frac{4}{3}\left(\frac{L_f - L}{L_f}\right) + \frac{1}{3}\left(\frac{L_f - L}{L_f}\right)^4\right], \tag{5.4.5}$$

Here R_b is bending stiffness enhancement coefficient (discussed below).

Similarly, torsional deformation *of* the bed caused by weight, $W_t + W_p$, *of* the table with the part that travels transversely for a distance, l_1, can be determined. The torque, $T = (W_t + W_p) l_1$ is counterbalanced by the reaction torque uniformly distributed along the bottom *of* the foundation block (Fig. 5.4.1). The angle *of twist of* the loaded cross-section relative to ends of the base is then:

$$\phi = \frac{TL}{8GJR_t}\left(2 - \frac{L}{L_f}\right) \tag{5.4.6}$$

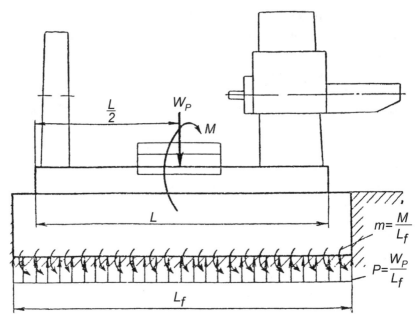

Fig. 5.4.1. Computational model for determining deformations of heavy boring mill/machining centers under weight loads.

Here J = polar moment *of* inertia *of* the cross-section *of* the base, and R_t = *torsional stiffness enhancement coefficient* (discussed below). It is possible to use these simplistic expressions for calculating (in the first approximation) *deformations of* the structures. However, use of the similar expressions for calculating *stresses* is absolutely unacceptable.

5.4.2 Machines Installed on Individual Foundations or on Floor Plate

If the machine base is anchored or grouted to the floor/foundation, deformations between the base and the floor can be neglected. It is assumed that the base and the foundation are deforming as one body relative to the axis passing through the effective center of gravity (C.G.) *of* the overall cross-section *of* the system "base-foundation." Position *of* the effective C.G. is determined considering different values of Young's moduli for the machine base (E) and for the foundation (E_f). For machines installed on the monolithic floor plate, the analysis replaces it with an "effective"/equivalent foundation beam whose width and length are given by

Eqs. (5.4.1) and (5.4.2). Distance from the upper surface of the base (e.g., guideways for machine tools) to the axis *of* the system: base-foundation is:

$$Z_{c.g.} = \frac{F_b Z_b + F_f Z_f \xi}{F_b + F_f \xi}.$$ (5.4.7)

Here $\xi = E_f/E$; F_b and F_f = cross-sectional areas *of* the base and the foundation, respectively; *and* Z_b *and* Z_f = vertical distances from the upper surface of the bases to cross-sectional centers *of* gravity *of* the base and the foundation, respectively.

Bending stiffness *of* the system in the vertical plane *X-Z* is dependent on the *effective rigidity*

$$(EI_y)_{ef} = EI + E_f I_f + E[F_b (Z_b - Z_{c.g.})^2 + F_f \xi (Z_f - Z_{c.g})^2].$$ (5.4.8)

Bending stiffness in the horizontal plane *X - Y* is

$$(EI_z)_{ef} = (EI_{bz} + E_f I_{fz}).$$ (5.4.9)

Bending deformations in the horizontal plane for this group *of* machines are very small and are usually neglected.

The torsional stiffness *of* the system machine base foundation is:

$$(GJ)_{ef} \approx GJ_b + G_f J_f + 0.83 \frac{G(z_f - z_b)}{\frac{\beta_b}{F_b} + \frac{\beta_f}{F_f \xi}},$$ (5.4.10)

where

$$\beta_b = 1 + \frac{l^2 G_b F_b}{14.4 EI_b}, \quad \beta_f = 1 + \frac{l^2 G_f F_f}{14.4 EI_b}$$ (5.4.11)

Here EI_b and EI = bending stiffness of the machine base in the horizontal and vertical planes, respectively; EJ = torsional stiffness of the base; I_{fy} and I_{fz} = cross-sectional moments of inertia of the foundation about the horizontal *y* and vertical *z* axes passing through the C.G. of the cross-section of the foundation block; J_f = polar moment of inertia of the cross-section of the foundation block; l = (average) distance between the anchor bolts fastening the base to the foundation; E_f and G_f = elastic moduli of the foundation material; F_b = cross-sectional area of walls of the base; and F_f = cross-sectional area of the foundation block.

Stiffness enhancement coefficients for the system machine bed-foundation due to supporting action of the soil for bending in the vertical plane, R_b, and for torsion, R_t, are

$$R_b = 1 + 0.123\gamma^4; \quad R_t = 1 + s_1\gamma_t^2. \tag{5.4.12}$$

Here

$$\lambda = \frac{L_f}{2}\sqrt[4]{\frac{K_f}{4(EI_x)_{ef.y}}}, \quad \lambda_t = L_f\sqrt[4]{\frac{K_{ft}}{4(GJ)_{ef}}} \tag{5.4.13}$$

For machines installed on a monolithic floor plate

$$L_f \approx 0.035\sqrt[4]{\frac{(EI_x)_{ef.y}(1 - v_s^2)}{E_s}} \tag{5.4.14}$$

Here v_s = coefficient of transverse deformation of soil (equivalent to Poisson's coefficient). It varies from $v_s = 0.28$ for sand to $v_s = 0.41$ for clay soil. Values of elastic modulus of soil E_s are given in Table 5.1. Coefficient $s_t = 0.02$ to 0.08; smaller values are to be used when the base is loaded with one concentrated torque in the midsection and with the counterbalancing torque that is uniformly distributed along the length; larger values are taken when the bed is loaded by two mutually balancing torques or by one torque at the end that is counterbalanced by a uniformly distributed torque along the length. K_f and K_b are stiffness coefficients of the interface between the foundation block and soil for bending and for torsion:

$$K_f \approx \frac{\pi}{2\log(4\alpha)}\frac{E_x}{B_f(1 - v_s^2)}, \quad \alpha = L_{ef}/\beta_f \tag{5.4.15}$$

$$K_{ft} = \frac{\pi}{16}\frac{E_s B_f^2}{(1 - v_s^2)} \tag{5.4.16}$$

For bending in the horizontal plane, $R \approx 1$.

Table 5.1. Elastic Modulus of Various Soils.

Grade of soil	E_s, MPa
Loose sand	150–300
Medium density sand	300–500
Gravel (not containing sand)	300–800
Hard clay	100–500
Medium hard clay	40–150
Plastic clay	30–80

Example 5.4. Table 5.2 contains computed data for deformations of the base of a horizontal boring mill (spindle diameter 90 mm) caused by weight of its table with part [$W = 39,000$ N (18,700 lb.)]. The system parameters are as follows: $l_1 = 1$ m; $EI = 7$ x 10^8 N m^2; GJ = 5 x 10^8 N m^2; $(EI_x)_{ef.y} = 25$ x 10^8 N m^2; $F_b = 0.085$ m^2; stiffness of each mount (when the machine is not tied to the foundation block by anchor bolts) $k_m = 5$ x 10^8 N/m; number of mounts, $n = 14$; surface area of the supporting (footprint) surface of the machine = 1.5 m^2; distance from the supporting surface to C.G. of the bed $c_b = 0.35$ m; $L = 5$ m; elastic modulus of soil $E_s = 120$ MPa; $v_s = 0.35$; and $s_t = 0.02$.

Table 5.2 shows the critical influence of thickness of the foundation block, especially when stiffness of the connection between the machine and the foundation is high (anchored). If the base is not anchored on a 1.4-m thick foundation, its deformation is about the same as deformation of the bed anchored to a 0.4-m thick foundation block. These results are illustrated in Fig. 5.4.2. This influence is not as great when the foundation block is not very rigid (small thickness), since deformations of the nonrigid foundation are comparable with deformations of the mounts.

While the bending deformations are smaller when the machine is attached to an individual foundation block, torsional deformations are substantially smaller on the floor plate. When the machine is anchored to an individual foundation block, soil stiffness does not significantly influence bending stiffness of the system. However, soil stiffness becomes a noticeable factor when the machine is installed on the floor plate. At $E_s = 20$ MPa, bending deformations are about two times greater than for $E_s = 120$ MPa. This effect can be explained as follows: on soils having low elastic modulus, a relatively large area of the floor plate responds to the

Table 5.2. Computed Deformations of Base of Horizontal Boring Mill (Spindle D = 90 mm).

Deformation of bed	Attachment to foundation	Thickness (height) of foundation, H_f (m)			
		1.4[a]	1.0[a]	0.4[a]	0.4[b]
Deflection	Anchored	0.0022	0.0043	0.015	0.017
f, mm	Not anchored	0.016	0.021	0.035	0.037
Twist	Anchored	0.0023	0.0046	0.02	0.014
φ mm/m	Not anchored	0.029	0.029	0.041	0.032

[a]Individual foundation: $L_f = 5.7$ m; $B_f = 2.1$ m.
[b]Floor plate foundation: $B_{ef} = 2.1$ m.

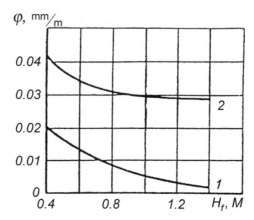

Fig. 5.4.2. Relative angular deformations between table and spindle of horizontal boring mill (1) anchored and (2) not anchored to foundation block as function of foundation block thickness.

external forces, since the soil reaction is weak. Thus, curvature of the plate under the machine is steeper. It can be concluded that individual foundations are especially effective with low stiffness soils.

5.5 DEFORMATIONS OF LONG MACHINE BASES

Beds having large longitudinal dimensions are relatively easily deformable, since height of the base is about the same for both short and long bases. Two factors are the most critical for effective stiffness of the long beds: deformations of the bed caused by weight of moving heavy components (e.g., a table carrying a part being machined or a gantry of a plano-milling machine, Fig. 5.5.1). These deformations can be somewhat reduced by increasing size (thickness) of the foundation block, but this direction for improvement is limited. Very thick and long foundation blocks are becoming counterproductive, since their construction cycle is very long, they are very expensive, they may have high magnitudes of residual stresses causing distortions of both the block and the machine base, and they are extremely sensitive to even very small temperature gradients. Monitoring of loads on the mounts under the base is not practical, since the loads depend on the weight of the part and are continuously changing in the process of moving the heavy units. The most effective way of enhancing effective stiffness of the long bases is by using an active

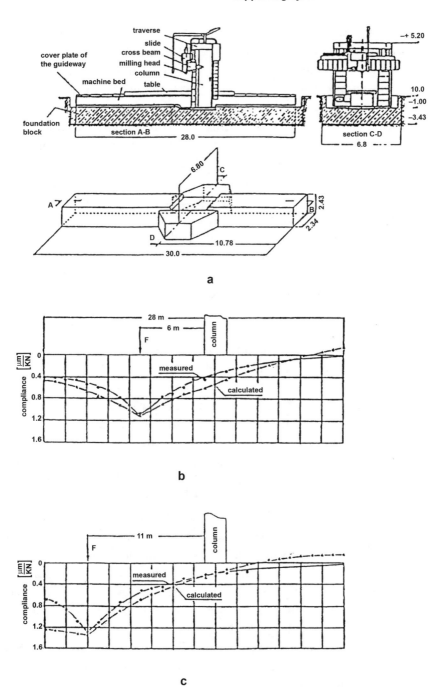

Fig. 5.5.1. Base deformations of a plano-milling machine tool under weight of moving gantry.

mounting system maintaining a constant distance between the supporting surface of the base and the reference frame that is not subjected to
loading and is maintained at a constant temperature (see Section 7.6).

Another important factor affecting performance of the long beds anchored to the foundation blocks, is nonuniform *sagging* of the foundation
in time along its length, thus causing distortion of the mounting surface
of the foundation. The sagging is due to densification of the soil under
the foundation and due to bending of the bi-material system caused by
temperature changes and temperature variation in the vertical direction
and different thermal expansion coefficients of the cast iron or steel bed
and the concrete foundation. Moving of heavy components, such as gan-
tries or tables also cause bending of the bed and the foundation and compression deformation of soil. The amount of sagging or *set* depends on
the specific load (pressure) on the soil, on the overall size of the foundation, and on compressibility of the soil. The rate of sagging/set depends
mostly on water permeability of the soil and on creep resistance of the
core structure of the soil.

Foundations for long machines are usually loaded nonuniformly; the
load under the gantry in a plano-milling machine in Fig. 5.5.1 is much
greater than at other locations. Accordingly, the sagging process is also
nonuniform, thus resulting in changing curvature of the foundation and
of the base in time. The maximum differentials between the level changes
along the foundation depend on properties of the soils, the rigidity of the
base-foundation system, and on the loading parameters. It is important
to assure that re-leveling of the machine is properly scheduled so that
during the time interval between the leveling procedures, the change in
curvature of the long bed does not exceed the tolerance. The rate of sagging is declining in time exponentially (initially, the rate of sagging is fast
and then gradually slows down). The fastest stabilization is observed for
soils with high water permeability, such as sands; the slowest stabilization is for the water-impermeable soils, such as clays. To utilize this exponential decrease in the sagging rate, heavy foundations are sometimes
artificially stabilized before installing the machine, by loading them with
"dead weights" whose total weight exceeds by 50% to 100% the weight of
the machine to be installed, and "aging" the loaded foundation for up to
4 months to 5 months.

The fully stabilized deformation due to the soil sagging of plano-
miling machine tool [$L = 7.7$ m, $B = 1.4$ m; $H_f = 1.0$ m; weight of the gantry

W = 120,000 N (26,500 lb.)] installed on a mixed sand + clay soil is 0.17 mm. Depending on the soil properties, this machine had to be re-leveled every 6 months to 12 months to achieve a relatively weak toler-ance for straightness of the guideways, 0.04 mm (0.00115 in.) [10]. The most effective technique to reduce deformations of long heavy structures caused by sagging of soil is by using an active mounting system with a reference frame isolated from the soil sagging effects. One such system is described in Section 7.6, another one is used for maintaining a stable level of foundation at the Kansai International Airport in Osaka, Japan [11]. The airport building is about 1.7 km long and is supported by a box-like foundation built on a relatively weak soil of an artificial island. The measuring system is continuously monitoring levels of the building col-umns (with 1 mm accuracy). When the nonunifonn sagging is detected, powerful hydraulic jacks (capacity of 300 tons each) are lifting the col-umns to allow for correcting shims to be placed under them.

Thermal expansion coefficients of cast iron/steel frames and of rein-forced concrete foundation blocks are, respectively, $\alpha_{is} = 8 \times 10^{-6}$ to 12×10^{-6} per 1°C and $\alpha_c \approx 14 \times 10^{-6}$ per 1°C. Heating of a bi-material strip, which is modeling the system "machine frame–foundation block" results in bending of the system so that the curvature radius is:

$$\rho = \frac{(E_1h_1^2 - E_2h_2^2)^2}{\dfrac{E_1E_2h_1h_2(h_1 + h_2)}{6\Delta T(\alpha_{is} - \alpha_{con})}}, \tag{5.5.1}$$

where $E_1 \approx 1.15 \times 10^5$ MPa for cast iron; $E_2 \approx 0.3 \times 10^5$ MPa for reinforced concrete; h_1, h_2 are effective heights of the machine frame, of the founda-tion block, respectively; ΔT = temperature change, °C. For length of the machine frame, $l = 20$ m, $h_1 = 1$ m; $h_2 = 2$ m, the temperature change $\Delta T = 1$°C would cause bending with $\rho = 0.5 \times 10^6$ m, corresponding to deforma-tion at the midspan:

$$\Delta = \frac{l^2}{4\rho} = 2 \times 10^{-4} \text{m} = 0.2 \text{ mm}, \tag{5.5.2}$$

and angular deformation at the bed ends $(2 \times 10^{-4})/10$ m/m or 0.02/1000 mm/mm, often an unacceptable amount.

This "bi-material effect" can be eliminated or greatly reduced if the machine bed and the foundation block had a relative mobility in the hori-zontal longitudinal direction.

REFERENCES

[1] Rivin, E.I., 2003, "Passive Vibration Isolation ", ASME Press, N.Y., 426 pp.

[2] Orlov, P.I., "Fundamentals of Machine Design", Vol. 1, 1972, *Mashinostroenie Publishing House*, Moscow, [in Russian].

[3] Reshetov, D.N., "Machine Elements, Mashinostroenie Publishing House, Moscow, [in Russian].

[4] Rivin, E.I., 1955, "Stiffness Analysis for Round Table-Bed System of Vertical Boring Mills," *Stanki i Instrument* No. 6, pp. 16–20 [in Russian].

[5] Rivin E.I., and Skvortzov, E. V., 1969, "Load-on-Each-Mount Measuring for Rubber-Metal Mountings", *Machines and Tooling*, #10, pp. 11–12.

[6] "Photo Briefs, 1966, " *Mechanical Engineering*, Vol. 88, No. 2.

[7] Polacek, M., 1965, "Detem1ination of Optimal Installation of Machine Tool Bed with Help of Modeling," Maschinenmarkt, No. 7, pp. 37–43 [in German].

[8] Bushuev, V.V., 1991, "Load Application Schematics in Design," *Stanki i Instrument*, No. 1, pp. 36–41 [in Russian].

[9] Mori Seiki Co., Ltd, Publication NV4000 DC6/0401 NAP 5000.

[10] Kaminskaya, V.V., 1972, 'Machine Tool Frames," Components and Mechanisms of Machine Tools, D.N. Reshetov, ed., *Mashinostroenie*, Moscow; Vol. I, pp. 459–562 [in Russian].

[11] Kanai, F., Saito, K., Kondo, S., Ishikawa, F., 1994, "Correction System for Nonuniform Sagging of a Floating Airport Foundation," Yuatsu to Kukiatsu [Journal of Japanese Hydraulic and Pneumatic Society], Vol. 25, No. 4, pp. 486–490 [in Japanese].

CHAPTER

6

Stiffness and Damping of Power Transmission Systems and Drives

6.1 BASIC NOTIONS

Power transmission systems and drives are extremely important units for many machines and other mechanical devices. Stiffness of these systems might be a critical parameter due to several factors depending on specifics of the device. Some of these factors are as follows:

1. Natural frequencies of power transmission systems and drives may play a significant role in vibratory behavior of the system, especially if some of them are close to other structural natural frequencies and/or to excitation frequencies from input (driving) devices such as internal combustion engines or from output (driven) devices, such as reciprocal compressions and pumps or milling cutters. Correct calculations of torsional natural frequencies depend on accurate information on inertia and stiffness of the components. While calculation of inertias is usually a straightforward procedure, calculation of stiffness values is more involved.

2. In precision devices, correct angular positioning between the driving and driven elements is necessary even under loaded condition. Inadequate stiffness of the drive mechanism connecting the driving and driven elements may disrupt proper functioning of the device.

3. Self-excited vibrations in positioning and production systems frequently develop due to inadequate stiffness of the drive mechanisms. Some examples include stick-slip vibrations of carriages and tables supported by guideways, chatter in metal-cutting machine tools, intense torsional vibrations in drives of mining machines, etc.

4. Low torsional stiffness and, especially, torsional backlash disrupt performance of widely used servo-controlled power transmission systems and require more advanced/expensive control systems.

A word combination "effective stiffness" or "effective compliance" is frequently used in analyses of power transmission systems and drives. This definition means a numerical expression of the response (deflection) of the structure at a certain important point (e.g., the arm end of a manipulator) to magnitudes of performance-induced forces (forces caused by interactions with environment, inertia forces, etc.) Such a response, i.e., effective stiffness and compliance is a result of five basic factors:

1. Direct structural deformations of load-transmitting components (e.g., torsional deformations of shafts, couplings, etc.). These can be

idealized for computational purposes as beams, rods, etc., also with idealized loading and support conditions;

2. Contact deformations in connections between the components or within the components (e.g., splines, bearings);

3. Reduced to torsional deformations bending deformations of shafts, gear teeth, etc., caused by forces in power transmission systems (gear meshes, belts, etc.);

4. Deformations in the energy-transforming devices (motors and actuators) caused by compressibility of a working (energy delivering) medium in hydraulic and pneumatic systems, deformations of an electromagnetic field in electric motors, etc.

5. Modifications of numerical stiffness values caused by kinematic transformations between the area in which the deformations originate and the point at which the effective stiffness is analyzed.

This chapter addresses basic issues, which have to be considered to perform the stiffness analysis of a power transmission or a drive system, and describes some cases of analytical determination of breakdown of deformations in complex real-life machines.

While there are substantial resources for enhancing effective stiffness of power transmissions and drives, utilization of these resources requires a clear understanding of sources of structural compliance, computing a breakdown of compliance, identifying weak links, and directing design efforts toward their improvement. Any efforts wrongly targeted to improve components and design features, which are not the major contributors to the overall compliance would result in waste. This can be substantiated by a simple example. If a component deformation is responsible for 50% of the overall deflection, making it two times stiffer would reduce the overall compliance (increase stiffness) about 30%. However, if a component deformation is responsible for only 5% of the overall deflection of the system, doubling its stiffness would reduce overall compliance only by an insignificant 2.5%. Costs of both treatments usually do not significantly depend on the component compliance and are comparable in both cases.

Transmission and drive systems usually have chain-like structures. Since the most important goal is reduction of deflections in structural chains caused by a force or torque applied at a certain point, the use of "compliance" values instead of "stiffness" values is natural in many cases since the "compliance breakdown" is equivalent to "deflection breakdown."

Stiffness of *bearings*, while important for analysis of power transmission systems and drives, has also other implications in designing precision mechanical devices. It is addressed in more detail in Section 6.5.

Another universally important component of power transmission systems and drives is *couplings*. Basic characteristics of couplings and principles of designing with couplings are described in Appendix 5.

Dynamic stability of transmission and drive systems is significantly influenced by damping. Damping contributions of various components in the compliance breakdown of the system are very different. Complexity of damping analysis is further increasing due to the necessity to consider vibratory modes of the transmission/drive system. These issues are addressed in Section 6.6.

6.2 COMPLIANCE OF MECHANICAL POWER TRANSMISSION AND DRIVE COMPONENTS

6.2.1 Basic Power Transmission Components

Compliance of mechanical elements employing joints and/or non-metallic (e.g., elastomeric) parts can be nonlinear. It is assumed that their nonlinearity can be approximated by the expression

$$x = x_o(P/P_r)^n \tag{6.2.1}$$

where P = the acting load (torque), P_r = the rated load (torque), x = the deflection (linear or angular) of the element under load, P_r, and n = the nonlinearity exponent. Usually, empirical values in expressions for compliance are given for a load magnitude $P = 0.5P_r$.

Data on *torsional compliance e_s, of shafts*, of various shapes and cross-sections are compiled in Table 6.1. In this Table, J_p is the cross-sectional polar moment of inertia.

The *equivalent torsional compliance e_k, of key and spline connections*, is caused by contact deformations in the connection and is described by expression [1]

$$e_k = \frac{k_k}{D^2 Lhz} \text{rad/Nm} \tag{6.2.2}$$

Table 6.1.   Torsional compliance of shafts.

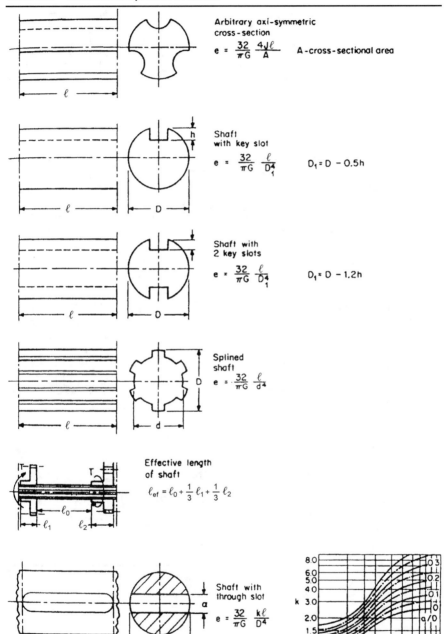

Arbitrary axi-symmetric cross-section

$$e = \frac{32}{\pi G} \frac{4 J \ell}{A}$$ A - cross-sectional area

Shaft with key slot

$$e = \frac{32}{\pi G} \frac{\ell}{D_1^4}$$ $D_1 = D - 0.5h$

Shaft with 2 key slots

$$e = \frac{32}{\pi G} \frac{\ell}{D_1^4}$$ $D_1 = D - 1.2h$

Splined shaft

$$e = \frac{32}{\pi G} \frac{\ell}{d^4}$$

Effective length of shaft

$$\ell_{ef} = \ell_0 + \frac{1}{3} \ell_1 + \frac{1}{3} \ell_2$$

Shaft with through slot

$$e = \frac{32}{\pi G} \frac{k\ell}{D^4}$$

Table 6.1. (*continued*)

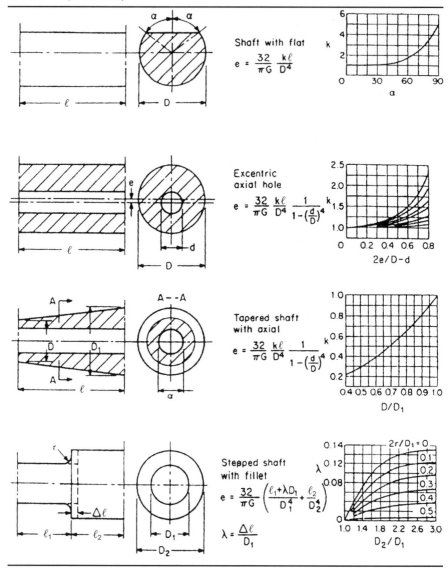

where D = the nominal diameter (for *spline* connections or *toothed clutches*, it is the mean diameter); L = the active length of the connection; h = the active height of a key, a spline, or of a clutch tooth; z = the number of keys, splines, or teeth. The joint compliance factor k_k is 6.4 × 10^{-12} m^3/N for a square key, 13 × 10^{-12} m^3/N for a Woodruff key, and 4 × 10^{-12} m^3/N for toothed clutches (even lower values for smaller z and better machining); nonlinearity exponent, $n \approx {}^2\!/_3$, for key connections and $n \approx \frac{1}{2}$ for spline connections.

Conventional key and spline connections are characterized by relatively large clearances ("backlash" or "play"), which are designed into the connection to facilitate the assembly procedure. This feature creates a highly nonlinear load-deflection characteristic of the connection as illustrated by line 1 in Fig. 6.2.1 [2]. The amount of play can be reduced, if critically required for functioning of the system, by more accurate machining of the components. Usually, it creates problems during assembly/disassembly operations, but the clearances are still developing during the life of the connection, especially if it is subjected to dynamic loads.

Many precision systems as well as servo-controlled systems benefit from using connections not having clearances and, preferably, preloaded connections.

One no-play key connection is shown in Fig. 6.2.2 [3]. The key consists of helical spring 5, which is made slightly larger in diameter than the inscribed circle of the combined opening of slots 3 and 4 in hub 1 and shaft 2, respectively. Before (or during) the insertion process, the helical spring is wound up between its ends (like a helical torsional spring), which causes a reduction of its outside diameter below the inscribed circle diameter. After insertion, the spring unwinds and fills the opening, developing friction forces in the contacts. Due to these friction forces, the "flimsy" spring is cemented into a stack of rings, which can accommodate very significant radial forces [4]. The unwinding process creates a preload thus completely eliminating the backlash. Since the spring key transmits the load by compression in contact with a concave surface of the slot, which has only a slightly larger curvature radius than the radius of the spring, stress concentrations in the connection are significantly reduced (about six to seven times reduction according to the test data).

Another type of no-play connection is a ball-spline connection, e.g., shown in Fig. 6.2.3. Ball splines are used for reducing friction and eliminating play in axially mobile connections. If the ball diameter is slightly

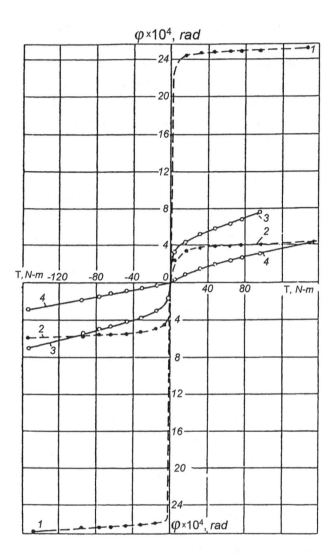

Fig. 6.2.1. Load-deflection characteristics of spline connections.
1 — sliding spline connection with play; 2 — same, with reduced play;
3 — ball spline connection with play; 4 — same as 3 but with
11 μm preload.

smaller that the inscribed circle of the cross-section of combined grooves (splines) in the shaft and the bushing, the connection becomes preloaded, and its play is eliminated (line 4 in Fig. 6.2.1), while the connection still retains its mobility. However, the torsional stiffness of a ball-spline con-

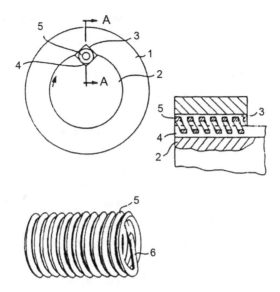

Fig. 6.2.2. Helical spring key connection. 1 — sleeve; 2 — shaft; 3 — key slot in sleeve; 4 — key slot in shaft, 5 — helical spring; 6 — optional tongue for twisting spring 5.

nection is lower than that of a similar size conventional (sliding) spline connection (see Fig. 6.2.1). Torsional compliance of the ball-spline connection [2] is

$$e_{bs} = \frac{4k\sqrt[3]{Q^2/d}}{D\sin\alpha}; \qquad Q = \frac{2T}{z_s z D \sin\alpha} \qquad (6.2.3)$$

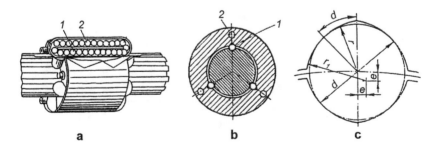

Fig. 6.2.3. Ball - spline connection. (a) General view; (b) Cross section; (c) Ball in « gothic arch » spline grooves. 1 — working balls; 2 — return channel.

where T = torque applied to connection; Q = force acting on one ball; z_s = number of splines; z = number of balls in one groove; r_1 = curvature radius of the groove cross-section; r = radius of the ball; D = diameter of connection; α = angle of contact; k = 0.021 $(m/N^2)^{1/3}$ for r_1/r = 1.03, and k = 0.027 for r_1/r = 1.10 (see Fig. 6.2.3c).

An important contributor to the equivalent torsional compliance of multishaft geared power-transmission systems is compliance e_b caused by *bending of the shafts, elastic displacements in the bearings,* and *bending and contact deflections of the gear teeth* [1].

Deformations of these elements lead to relative angular displacements between the meshing gears 1 and 2 and between 3 and 4 (Fig. 6.2.4a and b), thus the elastic member representing compliance e_b in the mathematical model should be inserted between the inertia members representing the meshing gears. More detailed modeling of the dynamic processes in the gear trains, in which both masses and moments of inertia of the gears in an intercoupled bending-torsional system are considered (e.g., [5]), gives a better description of high-frequency modes of vibration. However, it is not usually required for the compliance breakdown analysis (as well as for analysis of the lower modes of vibration which usually are the most important ones).

For calculation of e_b four steps can be followed:

1. Total vector deflection y_i of a shaft under the i-th gear caused by all forces acting on the shaft is calculated. Sleeves on the shaft reduce its bending deflection by factor, k_b (Fig. 6.2.5).
2. Vector displacement δ_i of the i-th gear caused by compliance of bearings is calculated (Fig. 6.2.6) as:

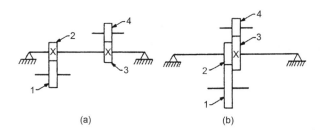

(a)   (b)

Fig. 6.2.4. Various configurations of a gear train.

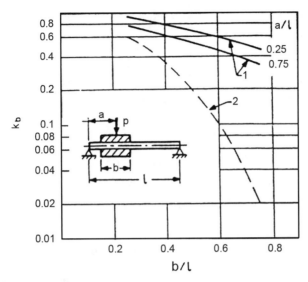

Fig. 6.2.5. Compliance reduction coefficient k_b of shaft bending due to bushings. (1) clearance fit and (2) interference fit.

$$\delta_i = (\delta_B - \delta_A)[a/(a + b)] + \delta_A$$
$$= \delta_B [a/(a + b)] + \delta_A [a/(a + b)], \tag{6.2.4}$$

where $\delta_A = e_A P_A$ and $\delta_B = e_B P_B$. P_A and P_B are vector reactions at the bearings A and B from all the forces acting on the shaft; e_A and e_B are compliances of the bearings A and B (see below).

3. Total (vector) linear displacement of the i-th gear is:

$$\Delta_i = y_i + \delta_i$$

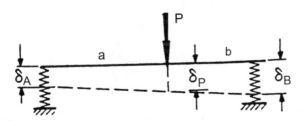

Fig. 6.2.6. Gear displacement caused by bearing deflections.

Relative displacement between the meshing i-th and $(i + 1)$-th gears is:

$$\Delta_{i,i+1} = \Delta_i - \Delta_{i+1}$$

As can be seen from Fig. 6.2.7, the resulting relative angular displacement of the two gears referred to the i-th gear is:

$$\alpha_i = [\Delta^T_{i,i+1} + \Delta^R_{i,i+1} \tan(\alpha + \rho)]/R_i \tag{6.2.5}$$

where $\Delta^T_{i,i+1}$ and $\Delta^R_{i,i+1}$ are, respectively, the tangential and radial components of the vector $\Delta_{i,i+1}$, R_i = the pitch radius of the i-th gear, α = the pressure angle, and ρ = the friction angle (friction coefficient $f = \tan \rho = {\sim}0.1$).

4. Equivalent torsional compliance:

$$e_b = \alpha/T_i = [\Delta^T_{i,i+1} + \Delta^R_{i,i+1} \tan(\alpha + \rho)] / R_i^2 P_i^T + e_m, \tag{6.2.6}$$

where T_i = the torque transmitted by the i-th gear, P_i^T = tangential force acting on the i-th gear, and

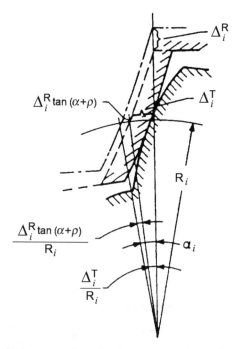

Fig. 6.2.7. Relative angular displacement of gears caused by their translational displacements.

$$e_m = k_m/b\, R_i^2 \cos^2 \alpha \qquad (6.2.7)$$

is compliance *of the mesh* caused by the bending and contact deformations of the engaging teeth [1]. In Eq. (6.2.7), b = tooth width and factor k_m = deflection of the engaged pair of teeth under the normal pressure that is acting on the unit of the tooth width (for steel spur gears, k_m = $\sim 6 \times 10^{-11}$ m^2/N, for steel helical and chevron gears, k_m = $\sim 4.4 \times 10^{-11}$, and for polyamide spur gears, k_m = ~ 5 to 10×10^{-10}). While Eq. (6.2.7) is an empirical formula, recently there were several publications describing computational approaches to determine gear tooth stiffness by using the Finite Element Method (FEM), e.g. [6]. Unfortunately, due to complexity of the problem, only spur gears are addressed. It should be noted that gear tooth deformations usually constitute only a small fraction of the total breakdown of torsional deformations in a transmission. However, calculation and/or testing of teeth stiffness variation during rotation of meshing gears gives information on cyclical stiffness change, which is important for analyzing parametric vibrations excited by the rotating gear. Fig. 6.3.7 below shows that using the above algorithm for computing compliances of mechanical components results in a good correlation with measured compliances of a gearbox.

The compliance of antifriction bearings and their fitting areas can be calculated from Eqs. (3.4.8), (3.4.9), and (3.4.10).

Compliance of multi-pad hydro-dynamic bearings (specifically, multi-pad bearings with screw adjustment of pads within the range of journal diameters $30 < d < 160$ mm) is:

$$e_{hd} \approx 5 \times 10^{-2}/d^{2.1}, \text{ mm/N}. \qquad (6.2.8)$$

The compliance of hydrostatic bearings without a special stiffness-enhancement servo-control system is:

$$e_{hs} \approx \Delta/3p_p F, \qquad (6.2.9)$$

where Δ = diametrical clearance in the bearing, p_p = pressure in the hydrostatic pocket, F = surface area of the pocket.

The compliance of a *belt* or *a steel band drive*, e_{bd}, is described by Eqs. (3.4.1) and (3.4.5) where the effective length is [1]

$$L_{ef} = L + 0.03v(R_1 + R_2) = \sqrt{L_c - (R_1 - R_2)^2} + 0.03v(R_1 + R_2) \qquad (6.2.10)$$

Here L = the actual length of a belt branch (the distance between its contact points with pulleys); L_c = the distance between the centers of pulleys 1 and 2; v = the linear speed of the belt, m/s; and $R_{1,2}$ are the radii of the pulleys, m. The second term in the expression for L_{ef} reflects the influence of centrifugal forces.

The tensile modulus E of V-belts with cotton cord increases gradually when the belt is installed in the drive and tensioned and could be up to 100% higher than the magnitude of E measured by tensioning a free belt. This effect is of a lesser importance for synthetic cords. With this effect considered, static modulus, E = 6 x 10^2 MPa to 8 x 10^2 MPa for V-belts with cotton cord; 2 x 10^2 MPa for V-belts with nylon cord; 1.4 x 10^2 MPa for flat leather belts and for knitted belts made of cotton; 2 x 10^2 MPa for knitted belts made of wool and for rubber-impregnated belts; and for high-speed thin polymer belts E = 23 x 10^2 MPa to 38 x 10^2 MPa. For laminated belts (e.g., polymeric load-carrying layer with modulus, E_1, thickness, h_1, and leather friction layer with E_2 and h_2), the effective modulus is

$$E = (E_1 h_1 + E_2 h_2)/(h_1 + h_2) \tag{6.2.11}$$

The compliance of *synchronous* (*timing*) *belts* is a combination of two parts: compliance of the belt branches and the teeth compliance,

$$e_{tb} = L/aR_1^2 EF + k_{tb}/b\, R_1^2 \tag{6.2.12}$$

where b = the belt width; F = the cross-sectional area (between the teeth); E = 6 to 40 x 10^3 MPa, — the effective tensile modulus of the belt cord, depending on its material and structure; k_{tb} = the factor of tooth deformation; $k_{tb} \approx 4.5$ x 10^{-10} m^3/N for ½ in. to ¾ in. pitch belt; and a = 1 for a belt drive without preload, and a = 2 for a preloaded drive. The compliance of timing belts can be reduced by using a high-modulus cord material (e.g., steel wire). Also, k_{tb} can be reduced by design means (modification of tooth thickness and profile).

The compliance of a *chain drive*, e_{cd} is expressed by Eqs. (3.13) and (3.14). When a chain is used without preload, k_{cd} = 0.8 to 1.0 x 10^{-12} m^3/N for roller chains and 20 to 25 x 10^{-12} m^3/N for silent chains.

Data for torsional and radial stiffness for some basic designs of torsionally flexible elastomeric couplings is provided in Appendix 5.

6.2.1a Stiffness of Ball Screws

Ball screws (Fig. 6.2.8) are widely used for actuating devices in mechanical systems. They have high efficiency, in the range of 0.8 to 0.95, as compared with 0.2 to 0.4 for power screws with sliding friction. Both efficiency and stiffness of ball screws can be adjusted by selection of the amount of preload. A judicious preloading allows to use ball screws both as a non-self-locking transmission (low preload) and as a self-locking transmission (high preload).

There are many thread profiles used for ball screws. Although the profiles, which are generated by straight lines in the axial cross-section (trapezoidal, square, etc.), are the easiest to manufacture, they have inferior strength and stiffness characteristics compared with the curvilinear cross-sections shown in Fig. 6.2.10a and b. For a profile generated by straight lines, curvature radius of the thread surfaces is $R_2 = \infty$ and $1/R_2 = 0$, thus both maximum contact (Hertzian) stress, σ_{max} (strength), and contact deformation, δ, (stiffness) are dependent only on the ball radius, R_1 (see Section 4). For the curvilinear profiles, both σ_{max} and δ are greatly reduced since usually $R_2 \approx (1.03 - 1.05)R_1$ and $(1/R_1 - 1/R_2) << 1/R_1$. The state-of-the-art ball screws use larger balls to enhance the load-carrying capacity, the stiffness, and the efficiency.

The two most frequently used cross-sectional profiles are semicircular, Fig. 6.2.9a, and gothic arch, Fig. 6.2.9b. The gothic arch profile is advantageous because of the small differential between R_1 and R_2 and, additionally, because of the doubling of the number of contact points (thus increasing the rated load capacity and stiffness). It also allows to

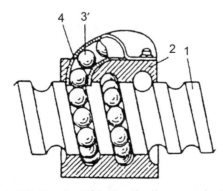

Fig. 6.2.8. Antifriction ball screw design.

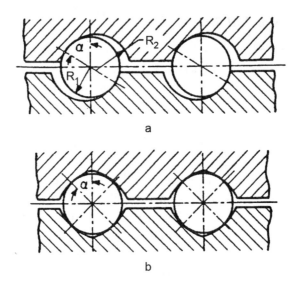

Fig. 6.2.9. Thread profiles for ball screws. (a) Semicircular and (b) gothic arch.

generate preload by using one nut and oversized balls. If a semicircular thread is used, the second nut is necessary for preloading. However, the gothic arch profile has higher manufacturing costs.

The *axial compliance* of ball screws can be a significant factor in the compliance breakdown of a drive. The determining factors for the ball screw compliance are contact deflections, tension-compression deformations of (long) screws, and deformations in the thrust bearings. Bending and shear of the thread coils can be neglected.

Contact deflections are very different for non-preloaded and preloaded mechanisms. For a *non-preloaded* screw, the axial displacement caused by an external axial force, Q, daN, is [7]:

$$\delta = 2CQ^{2/3}/Z_{ef}^{2/3} \sin^{5/3}a \, \cos^{5/3}\gamma, \text{ cm} \qquad (6.2.13)$$

and the compliance is:

$$e = \frac{d\delta}{dQ} = \frac{1.33C}{Z_{ef}^{2/3} \sin\alpha\cos^{5/3}\lambda \sqrt[3]{Q}}, \text{mm/N} \qquad (6.2.14)$$

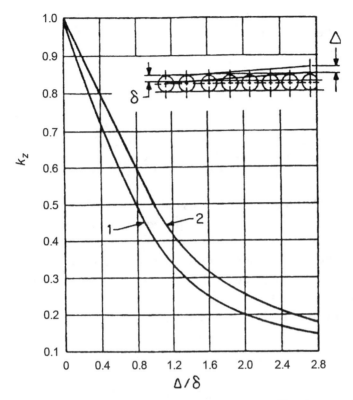

Fig. 6.2.10. Influence of ball-screw accuracy on effective number of active balls. 1 — without preload; 2 — with preload.

It can be seen from Eq. (6.2.14) that the compliance diminishes (or stiffness is increasing) with the increasing load (hardening nonlinearity). Here Z_{ef} is the effective number of balls (the actual number of balls participating in the force transmission);

$$C = 0.7753 \sqrt{\frac{E_1 + E_2}{E_1 E_2} \left(\frac{1}{R_1} - \frac{1}{R_2} \right)}$$

is elastic coefficient from the Hertzian formula for contact deformation; Section 4; E_1 and E_2 are Young's moduli for ball, screw, respectively; $\alpha =$ contact angle of the ball in the assembly (see Fig. 6.2.9); $\gamma =$ helix angle of the screw.

For a *preloaded* screw, the overall axial displacement caused by the contact deflection, δ, is

$$\delta = \frac{CQ}{1.5Z_{ef}\sin^2\alpha\cos^2\lambda\sqrt[3]{P_p}}\ \text{cm} \tag{6.2.15}$$

and

$$e = \frac{\delta}{Q} = \frac{C}{1.5Z_{ef}\sin^2\alpha\cos^2\lambda\sqrt[3]{P_p}},\text{N/mm} \tag{6.2.16}$$

where P_p = the preload force. Thus, with the initial preload, screw compliance does not depend on the applied load (in accordance with the conclusions in Chapters 3 and 4) and is much lower than it is for a non-preloaded ball screw. Since, with larger ball diameters, the initial efficiency is higher and greater preload forces can be used, a higher stiffness can also be achieved.

Example: α = 45 deg., λ = 4.5 deg., Z_{ef} = 50, P_p = 0 N or 300 N, and Q = 5000 N

1. For no preload, P_p = 0, compliance by Eq. (6.2.14) is:

   $e_a = 0.022C$ mm/N, stiffness $k_a = 45.4/C$ N/mm;

2. If P_p = 300 N, compliance by Eq. (6.2.16) is

   $e_b = 0.0086C$ mm/N, $k_b = 116.3/C$ N/mm,

or less than 40% of the compliance without preload.

The compliance depends significantly on the screw accuracy. If, because of variations in the ball diameters or inaccuracies of the threads, there is a difference Δ between the minimum and maximum clearance between the individual balls and thread surface and an axial force Q causes a displacement δ, not all the actual balls would take equal part in transmitting force, Q, but only the "effective number" of balls

$$Z_{ef} = K_z Z, \tag{6.2.17}$$

where Z = the actual number of balls in the system, and K_z = the coefficient depending on δ/Δ, which can be taken from Fig. 6.2.10. For a reasonably accurate ball screw, at high loads (close to the rated loads)

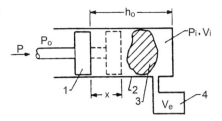

Fig. 6.2.11. Basic pneumatic actuator.

$$Z_{ef} = 0.7Z. \qquad (6.2.17')$$

6.2.2 Compliance of Pneumatic System Components

Pneumatic actuators are very popular because of their simplicity, their very beneficial economics, and the possibility of them being used in circumstances wherein electrical systems can create a safety hazard. One of the important disadvantages of pneumatic actuators is their reduced stiffness and natural frequencies caused by compressibility of air.

A generic pneumatic actuator — a linear cylinder-piston system — is sketched in Fig. 6.2.11. Piston 1 is moving inside cylinder 2 of any cross-sectional shape 3 (usually round or square) and having cross-sectional area A, instantaneous internal active volume V_i and attached external volume V_e, (e.g., associated with plumbing and with ancillary units 4 such as external tanks); instantaneous pressure p_i (excess over environmental pressure, p_e) is acting in both volumes, V_i and V_e; external force P is applied to the piston rod. The increment of work dW of force, P on an incremental piston displacement dx is equal to an incremental change in internal energy; thus:

$$dW + d(p_i V) = P\,dx + p_i\,dV + V\,dp_i = 0, \qquad (6.2.18)$$

where $p_i V$ = the internal energy of the system. Usually, pressure variations are quite small because of the system connection to the compressed air line and can be neglected, thus:

$$P = -p_i(dV/dx). \qquad (6.2.18')$$

For the most general polytropic process of the gas state variation:

$$(p_{i,0} + p_e)/(p_i + p_e) = p_{a,0}/p_a = (V/V_0)^{\gamma}. \qquad (6.2.19)$$

where p_a = the absolute pressure, γ = the polytrope exponent, and subscript 0 is assigned to the initial state.

Substituting p_i from Eq. (6.2.19) into Eq. (6.2.18') gives:

$$P = (dV/dx) [p_e - p_{a,0} (V_0/V)^\gamma]. \tag{6.2.20}$$

Substituting

$$V = V_0 - Ax \tag{6.2.21}$$

into Eq. (6.2.20) gives

$$P = A [p_{a,0} V_0^\gamma/(V_0 - Ax)^\gamma - p_e] \tag{6.2.22}$$

or

$$P = A \{p_{a,0}/[V_0 - (Ax/V_0)]^\gamma - p_e\} \tag{6.2.22'}$$

The value of γ depends on thermal exchange conditions between the working gas and the environment and is a function of the volume change rate in the system. Two limiting cases are:

1. Very low rates (static loading conditions, loading frequency below ~0.5 Hz) when there is an equilibrium between internal and external gas temperatures ("isothermal conditions," $\gamma = 1$), and
2. Very high rates (vibratory conditions at frequencies higher than ~3 Hz) when no energy exchange between internal and external gas can be assumed ("adiabatic conditions," $\gamma = 1.41$ for air).

The stiffness of the system can be easily derived by differentiating Eq. (6.2.22') as

$$k = dP/dx = (\gamma A^2 p_{a,0}/V_0)/(1 - Ax/V_0)^{\gamma+1} \tag{6.2.23}$$

or, if the vibratory variation of volume is small, $Ax \ll V_0$, and

$$k = \gamma A^2 p_{a,0}/V_0 = \gamma A (P_r + A p_e)/V_0 \tag{6.2.23'}$$

where $P_r = Ap_{io}$ is the rated load of the actuator.

Since P_r is specified for a given actuator, stiffness can be increased by increasing the cross-sectional area of the piston with a simultaneous reduction of the working pressure, p_{io}. Of course, a subsequent increase of the size and mass of the actuator should be considered. Another, and in many cases, more effective way of increasing stiffness is by reducing the total internal volume of the system V_o by means of shortening pipes, hoses, etc. Any effort to increase stiffness by design means would lead to some increase in the natural frequency and thus, to an increase in γ and, as a result, to an additional effective stiffness increase (since natural frequencies of pneumatic actuators are frequently located in a 0.5 Hz to 2.5 Hz range, which corresponds to a transition between isothermic and adiabatic conditions).

The wall compliance of flexible hoses and cylinders does not play a significant role in the overall compliance (provided that reinforced hoses are used) because of high compressibility of air.

The effective compliance of pneumatic actuators can be substantially reduced (or stiffness increased) by using additional mechanical reduction stages at the output (similarly to the modifications evaluated in Section 6.3).

It is worth noting that the compliance of a pneumatic system is not constant and varies along the stroke together with the changing volume in the pressurized part of the cylinder.

6.2.3 Compliance of Hydraulic System Components

Hydraulic actuators are characterized by the highest force (torque) - to - size ratios and thus, potentially, are the most responsive ones. Accordingly, they find a wide application for heavy payload actuators, although their applications for medium payloads are diminishing in favor of electromechanical systems. One of the substantial disadvantages of hydraulic actuators is their relatively low effective stiffness. Although the compressibility of oil is many orders of magnitude less than that of air, very high pressures in hydraulic systems lead to a significant absolute compression of oil as well as to deformations of containing walls (in cylinders, pipes, hoses, etc.).

The compliance of the compressed fluid in a cylinder or a pipe can be analyzed using the schematic in Fig. 6.2.12. By definition, compliance is the ratio between the incremental displacement Δx and the incremental force ΔP that causes the displacement.

Displacement Δx of the piston in Fig. 6.2.12 under the force increment ΔP applied to the piston is caused by compressibility of the fluid in the cylinder, pipe, or hose and is also caused by the incremental

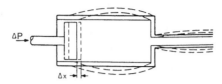

Fig. 6.2.12. Displacement of piston in hydraulic cylinder caused by oil compressibility and by cylinder and pipes expansion.

expansion of their walls. Compressibility of a liquid is characterized by its volumetric compressibility modulus $K_o = \Delta p/(\Delta V/V)$, where Δp is the increment of the internal pressure, V = the initial volume of liquid, and ΔV = the incremental change of the volume caused by application of Δp. For a typical hydraulic oil, $K_o \approx 1600$ MPa $= 1.6 \times 10^3$ N/mm^2.

The combined effects of fluid compressibility and wall expansion in pipes can be conveniently characterized by an effective modulus K_{ef}. If a section of the pipe is considered [length, l_p; internal diameter, d; wall thickness, t; Young's modulus and Poisson's ratio of the wall material, E and v, respectively; cross-sectional area, $A = \pi d^2/4$; internal volume, $V = (\pi d^2/4)l$], an increment of internal pressure would cause an increase in the internal diameter (e.g., see [7])

$$\Delta d = 2 \, \Delta p \, (d/E) \, d^2/[(d + 2t)^2 - d^2] \, [(1 + v) + (1 + v) \, (d + 2t/d)^2]$$
$$\approx (\Delta p/E)(d^2/t) \, [1 + 2(1 + v) \, (t/d)] \qquad (6.2.24)$$

since $t \ll d$. This increment of diameter is equivalent to a specific increment of pipe volume

$$\Delta V_p/V_p \approx 2(\Delta d/d) = 2(\Delta p/E)(d/t)[1 + 2(1 + v)(t/d)] \qquad (6.2.25)$$

An equivalent modulus associated with the volume change described by Eq. (6.2.25) is

$$K_p = \Delta p/(\Delta V_p/V_p) = Et/2d[1 + 2(1 + v)(t/d)] \qquad (6.2.25')$$

Thus, the total effective change of volume caused by compressibility of the liquid and by expansion of the pipe walls caused by a pressure increase, Δp, is

$$(\Delta V/V)_{ef} = \Delta p/K_o + 2 \, (\Delta p/E)(d/t)[1 + 2(1 + v)(t/d)] \qquad (6.2.26)$$

and the effective modulus of the pipe

$$K_{eff.p} = \Delta p/(\Delta V/V)_{ef} = K_o/\{1 + 2 (K_o/E)(d/t)[1 + 2(1 + v)(t/d)]$$
(6.2.26')

or

$$1/K_{eff.p} = 1/K_o + 1/K_p$$
(6.2.27)

The same formula is also applicable to hydraulic cylinders. For a steel pipe, $d = 25$ mm, $t = 2.0$ mm, $E = 2.1 \times 10^5$ MPa, $v = 0.23$, $K_0 = 1.6 \times 10^3$ MPa, and $K_{eff.p} = 0.81K_o$. For a copper pipe of the same dimensions, ($E = 1.2 \times 10^5$ MPa and $v = 0.35$), $K_{eff.p} = 0.71K_o$. For a steel cylinder ($d = 63.5$ mm and $t = 6.3$ mm), $K_{eff.p} = 0.84K_0$.

For flexible hoses, the values of E and v are hardly available, but in some cases, data on "percent volume expansion versus pressure" is available from the hose manufacturers, from which the equivalent modulus, K_h, can be calculated for any pressure, p. Figure 6.2.13 shows data provided by Rogan and Shanley, Inc., on the comparison of six hose types

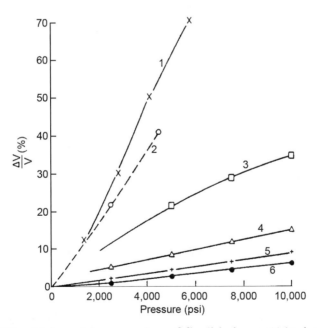

Fig. 6.2.13. Volumetric expansion of flexible hoses ¼ in. bore under pressure. (1) Aeroquip 100R2, 2781-4, (2) Synflex 3R80, (3) Synflex 3V10, (4) 4006 CSA (Kevlar R&S Polyflex 2006 St, and (6) R & S Polyflex 2006 StR.

of ¼-in. diameter. There is a very substantial, 10 to 20 times, scatter of the K_h value for various hose types. The best hoses are comparable with or even superior to metal pipes. For the pressure range, p = 3 MPa to 10 MPa, equivalent modulus, K_h = 0.1 x 10^3 MPa to 1.2 x 10^3 MPa. Since hose expansion is determined experimentally, it includes the effects of oil compressibility, thus $K_h = K_{eff,p}$. For the hoses in Fig. 6.2.13, $K_{eff,p}$ = 0.03 to 0.73K_0. Some hoses exhibit a very significant nonlinearity of the hardening type.

If a force, ΔP_c, is applied to a piston having an effective cross-sectional area A_c, it will result in a pressure increment, $\Delta p_c = \Delta P_c/A_c$. This pressure increment would cause volume changes: ΔV_c in the cylinder; ΔV_p in the rigid piping; and ΔV_h in the flexible hoses, whose magnitudes are determined by their respective volumes and effective moduli. The total volume change is:

$$\Delta V = \Delta V_c + \Delta V_p + \Delta V_h = \Delta X_c A_c \qquad (6.2.28)$$

or

$$\Delta X_c A_c = \Delta P_c \ (V_c/K_{eff,c} + V_p/K_{eff,p} + V_h/K_{eff,h}) \qquad (6.2.29)$$

or, finally,

$$\Delta X_c/\Delta P_c = (V_c/K_{eff,c} + V_p/K_{eff,p} + V_h/K_{eff,h})/A_c \qquad (6.2.30)$$

where ΔX_c is the incremental piston displacement caused by the volume changes.

One way to reduce the compliance originated in the hydraulic system is by the introduction of a mechanical reduction stage, as illustrated in Section 6.3.

The compliance of hydraulic systems changes along the stroke of the actuating cylinder(s) because of the changing volume of hydraulic fluid under compression.

6.2.4 Dynamic Parameters of Electric Motors (Actuators)

The electromagnetic field connecting the stator and rotor of a driving motor or actuator demonstrates quasi-elastic and/or damping properties [8]. Consideration of these properties could noticeably influence both the compliance breakdown and the dynamic characteristics of mechanical structures. A mathematical model of a motor can be, typically, approximated by a single-degree-of-freedom oscillator including rotor (moment of inertia, I_r) - compliance e_{em} - damping coefficient c_{em} - inertia of the power supply, where e_{em} and c_{em} are the effective compliance and damping coefficients of the electromagnetic field. While the inertia of the power supply can be considered as infinity in cases of general machinery drives, for servodrives it becomes a complex parameter representing the dynamic characteristics of the feedback and/or feedforward systems (e.g., [9]). For conventional motors, both for d.c. motors and for induction a.c. motors, the amplitude-frequency characteristic of a motor, while it is excited by a periodically varying torque on the shaft has a pronounced, although highly damped, resonance peak [8].

A specific feature of electric motors is their very high dynamic torsional compliance (low stiffness). Dynamic stiffness and damping coefficients associated with the electromagnetic field of industrial a.c. induction motors and d.c. motors are given in [8] as:

$$k_{em} = pT_{max}; \qquad c_{em} = s_{max}\omega_e I_r \qquad (6.2.31)$$

for induction motors, where p = the number of poles on the stator; f_e = the line voltage frequency, Hz; $\omega_e = 2\pi f_e$ is the angular frequency of the line voltage, rad/s; T_{max} is the maximum (breakdown) torque of the motor; s_{max} is slippage associated with T_{max}; and I_r is rotor moment of inertia. Eqs. (6.2.31) describe parameters of an induction motor when it runs on the stable (working) branch of its torque – speed characteristic (speed/slippage is decreasing with increasing torque) (e.g. see Fig. 1.3.12). During the start-up period, the motor initially is running along the unstable (starting) branch of its characteristic, on which increasing torque is associated with increasing rpm. As the result, induction motors may develop a very intense *negative damping*, which can produce self-exciting vibrations in the driven mechanical system and dangerous overloads (see [8] and Section 1.3). This peculiar feature of induction motors can be utilized in special cases when reduction of damping in the driven system or even negative damping is desirable.

For d.c. motors:

$$k_{em} = 1/\omega_o v \tau_e; \qquad c_{em} = \tau_e/I_r \qquad (6.2.32)$$

where $v = s/T$ is the slope of torque T versus slippage s characteristic; ω_0 is no-load speed of the motor, rad/s; $\tau_e = L_r/R_r$ is electromagnetic time constant, where L_r and R_r are inductance and resistance of motor windings. Similar properties of the electromagnetic field are characteristic also for linear motors.

Since natural frequencies of the electric motor systems are rather low, in the range of 1 Hz to 10 Hz for medium horsepower motors, their dynamic behavior can be greatly influenced by the control system. It is shown in [9] that effective damping of a *direct drive motor* can be substantially improved by using velocity feedback. Because of the close coupling between the motor and the control system, effective compliance of the motor is significantly influenced by parameters of the control system. It was shown that the effective linkage stiffness of a robot with direct drive motors is rather low [9] and corresponds to a fundamental natural frequency of 2.5 Hz to 7.5 Hz; the motor natural frequency was about 3.65 Hz.

For conventional (non-direct) drives, when a motor is connected to the driven component through a rpm-reducing transmission, effect of the electric motor compliance on the effective system compliance is usually small. According to the "reduction" procedure described in Section 6.3, the motor compliance has to be divided by the square of a usually high transmission ratio. However, because of the same reasons, influence of the electromagnetic stiffness (compliance) and damping of the driving motor on dynamics of high speed mechanical drives (whose rpm are equal or exceeding the motor rpm) can be quite significant. With a proper design approach (tuning), this influence can be used to noticeably improve effective damping of the drive system (section 6.6 and [8]).

6.3 PARAMETER REDUCTION IN MATHEMATICAL MODELS

In a complex mechanical system such as a robot, a machine tool, a vehicle, or even in a subsystem, such as a gearbox or a chain drive, there are usually compliant components present as well as inertias and energy-dissipating units. If all the compliance and/or inertia and/or damping values are calculated and known, the breakdown of their distribution throughout the system *cannot be done by simply adding up the numbers* and calculating the fractions (percentages) of each contributor's participation in the total sum. The reason for this is that contributions of partial compliances (as well as inertias and dampers) to the system's behavior depend on kinematic relationships between the design components whose parameters are considered. These contributions are reflected in roles, which the specific components play in the overall potential and kinetic energy expressions.

Accordingly, before the breakdown can be constructed and analyzed, all the partial values have to be *"reduced"* (*referred, reflected*) to a selected point (or to a selected part) of the system. If such a reduction is properly done, neither natural frequencies nor modes of vibration are affected. The overall compliance reduced (referred) to a certain component of the system would be the same as the compliance value measured by the application of force or torque to this component assembled in the system and then recording the resulting deflection. For the reduction algorithm to be correct, the magnitudes of both potential and kinetic energy should be the same for mathematical models of the original and the reduced systems.

Several typical and important examples of the reduction procedure are considered below. They are intended to serve as computational tools to be used in analyzing real systems as well as to illustrate general concepts. For generality, the reduction of both elastic constants and inertias is addressed.

Fig. 6.3.1 shows beam 1 pivoted to support structure 2 and also connected to it through revolute spring 3 having angular compliance e_ϕ (or angular stiffness $k_\phi = 1/e_\phi$). If force F is acting at distance l from the pivot, the angular deflection, ϕ, of the beam can be easily calculated by first calculating moment $M = Fl$ of force F relative to the center of rotation:

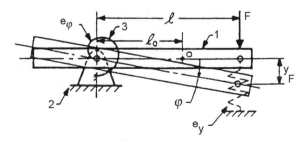

Fig. 6.3.1. Reduction of angular compliance to equivalent translational compliance.

$$\phi = e_\phi M = e_\phi Fl \tag{6.3.1}$$

However, in many cases, it would be beneficial to describe the compliance of this system in terms of linear compliance e_y at the force application point (e.g., the compliance at the end of arm for a robot). By definition, such a compliance is:

$$e_y = y_F/F \tag{6.3.2}$$

where y_F = the linear displacement at the force application point caused by force F. Using Eq. (6.3.1),

$$e_y = y_F/F = \phi\, l/F = e_\phi Fl^2/F = e_\phi\, l^2. \tag{6.3.3}$$

The potential energy of an elastic system is:

$$V = \tfrac{1}{2}\, k\Delta^2 = \tfrac{1}{2}\, \Delta^2/e \tag{6.3.4}$$

where k and e are *generalized stiffness and compliance*, and Δ is the *generalized deflection*. The equivalency of the "initial" compliance e_ϕ and "reduced" compliance e_y can be easily proven as:

$$V = \tfrac{1}{2}\, \Delta^2/e = \tfrac{1}{2}\, y_F^2/\, e_y = 1/2\, \phi^2\, l^2/e_\phi l^2 = \tfrac{1}{2}\, \phi^2/e_\phi \tag{6.3.5}$$

Of course, the reduction can be performed to an intermediate point as well, such as to the point o at a distance l_o from the center (Figure 6.3.1). Obviously, in this case,

$$e_o = e_\phi l_o^2. \tag{6.3.6}$$

Figure 6.3.2 shows the same beam as in Fig. 6.3.1 but instead of a revolute spring at the pivot, there is a linear spring 4 at the opposite side of the beam. Compliance of spring 4 is e_o (or, stiffness $k_o = 1/e_o$).

First, let us reduce compliance, e_o, to the force application point (reduced compliance, e_y). Force F is transformed by the leverage action of beam 1 into force $F_o = F(l/a)$ acting on spring 4. The deformation of spring 4 is:

$$y_0 = e_0\, F_0 = e_0 F(l/a). \tag{6.3.7}$$

This deformation is transformed by the leverage effect of beam 1 into deformation $y = y_0\,(l/a)$ at the force application point. Accordingly:

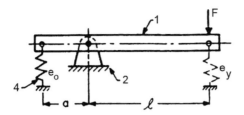

Fig. 6.3.2. Reduction of translational compliance to a different location.

$$e_y = y/F = y_o\,(l/a)/F = e_oF(L/a)(l/a)/Fl = e_o(l^2/a^2) \qquad (6.3.8)$$

Compliance e_0 can also be reduced to angular compliance e_ϕ at the center of beam rotation as follows:

$$e_\phi = \phi/M = (y_o/a)/Fl = e_oF(l/a)(l/a)/Fl = e_o/a^2 \qquad (6.3.9)$$

Naturally, Eq. (6.3.9) is identical to Eq. (6.3.3). It is easy to verify that the potential energy is invariant for both reduction procedures in Eqs. (6.3.8) and (6.3.9).

It is convenient to correlate the reduction formulae with transmission ratios in the respective cases. Transmission ratio $i_{b,c}$ between components or points b or c is defined as the ratio between the velocities of component or point b and component or point c,

$$i_{b,c} = v_b/v_c \qquad (6.3.10)$$

In case of Eq. (6.3.3), the ratio between the linear velocity of the force application point $v_F = l(d\phi/dt)$ and the angular velocity $d\phi/dt$ of the beam is:

$$I_{y,\phi} = l(d\phi/dt)/(d\phi/dt) = l. \qquad (6.3.11)$$

In the case of Eq. (6.3.8), the transmission ratio between the velocities of the force application point (v_F) and the spring attachment point (v_0) is:

$$i_{F,0} = l/a, \qquad (6.3.12)$$

in the case of Eq. (6.3.9), the transmission ratio is:

$$i_{F,0} = (v_0/a)/v_0 = 1/a \qquad (6.3.13)$$

Thus, in all of these cases, the reduction formula can be written as:

$$e_k = e_n/i^2_{n,k} = e_n\, i^2_{k,n} \qquad (6.3.14)$$

or

$$k_k = k_n\, i^2_{n,k} = k_n/i^2_{k,n} \qquad (6.3.15)$$

where e_n and k_n are the compliance and stiffness proper of component or point n, e_k is the compliance reduced to component or point k, $i_{n,k}$ is the transmission ratio between these components or points, and $i_{k,n}$ is the transmission ratio between the same points in the opposite direction. Expressions (6.3.9) and (6.3.15) cover reduction both between the same modes of motion (linear-linear, angular-angular) and between the different modes of motion (linear-angular, angular-linear).

If a system comprises several compliant components moving with different velocities, all the partial compliances can be reduced to a selected component (usually to the input or the output component). After the reduction procedure is performed, an analysis of the relative importance of various design components in the compliance breakdown could easily be done. This is illustrated below on the example of a system in Fig. 6.3.5.

Similar procedures can be developed for the transformation (reduction) of inertias. Figure 6.3.3 shows the same beam as Fig. 6.3.1, but in this case, the beam is assumed to be massless, and it carries a concentrated mass M at its end.

If a translational motion of mass M is considered (in association with the deformation of the linear spring e_y), the velocity is v_F mass is M and kinetic energy $T = \frac{1}{2}Mv_F^2$. If it is more convenient to deal with revolute motion (in association with the deformation of revolute spring 3), the angular velocity, $d\phi/dt = (1/l)\, v_F$, the moment of inertia

$$I = Ml^2 \qquad (6.3.16)$$

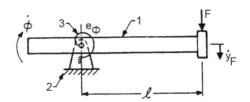

Fig. 6.3.3. Reduction of mass to a moment of inertia.

and kinetic energy remains the same,

$$T = \tfrac{1}{2}I\phi^2 = \tfrac{1}{2}Ml^2[(1/l)v_F]^2 = \tfrac{1}{2}Mv_F^2 \tag{6.3.17}$$

Analogously to Eq. (6.3.15). Eq. (6.3.17) can be written as:

$$I_y = I_\phi\, i^2{}_{\phi,y} \tag{6.3.18}$$

where $I_y = M$ and $I_\phi = I$ designate inertias associated with coordinates y and ϕ, and $i_{\phi,y}$ is the transmission ratio between these coordinates as defined in Eq. (6.3.10).

In the case of Fig. 6.3.4, it might be desirable to consider motion in the coordinate y_F associated with the right end of the beam, while both the actual spring, e_o, and the actual mass, M_o, are located at the left end. Reduction of the spring constant was discussed before; mass reduction can be performed from expressions for kinetic energy for the initial M_o and reduced M_F mass positions,

$$T_o = 1/2M_o v_o^2; \qquad T_1 = 1/2M_F v_F^2 = \tfrac{1}{2}[M_F\,(v_F/v_o)^2]v_o^2 \tag{6.3.19}$$

Thus,

$$M_F\,(v_F/v_o)^2 = M_o; \qquad M_F = M_o\,(v_o/v_F)^2 = M_o\, i_{o,F}^2 \tag{6.3.20}$$

In a simple gearbox in Fig. 6.3.5a, motion from the input pulley having rotational moment of inertia I_1 can be transmitted to the output pulley I_4 via the reducing pair of gears I_2/I_3 having numbers of teeth z_2, z_3 respectively. Obviously, multiplicating gears can also be used. The dynamic models for these two configurations are represented by the chain system in Fig. 6.3.5b. The compliance of the system is caused by torsional compliances of shafts 1 to 2 and 3 to 4 as well as by the bending compliances of these shafts. The bending compliance, according to Eqs. (6.2.5)

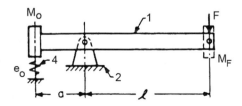

Fig. 6.3.4. Reduction of mass to a different location.

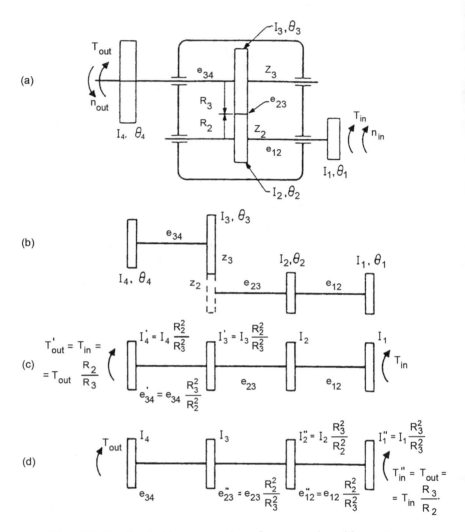

Fig. 6.3.5. Reduction procedure for a gearbox-like system.

and (6.2.6) is referred to one of the shafts depending on the gear whose radius is used as R_1 in Eqs. (6.2.6) and (6.2.7). Let us assume that in the case of Fig. 6.3.5, the bending compliance had been computed with reference to shaft 1 to 2. Accordingly, the first stage mathematical model of the transmission in Fig. 6.3.5a is shown in Fig. 6.3.5b. Gear Z_2 is shown in Fig 6.3.5b twice — once with solid lines, representing the moment of inertia of the actual gear, and also with dotted lines, representing a massless gear with number of teeth Z_2, engaged with the gear Z_3. Using the reduction technique described above, the model can be reduced either to the input shaft (Fig. 6.3.5c) or to the output shaft (Fig. 6.3.5d). In both cases, constructing a compliance and/or inertia breakdown is possible.

In the case of reduction to the input shaft, the output torque would be modified as:

$$T'_{out} = T_{out}\,(R_2/R_3) = T_{out}\,(Z_2/Z_3) = T_{out}\,i_{3,2} = T_{out}/i_{2,3}$$

All the compliances and inertias on the output shaft will be reduced to the input shaft (Fig. 6.3.5c) as:

$$e_{34}' = e_{34}/i_{2,3}^2 \qquad I'_4 = I_4'\,i_{3,2}^2$$

If the reduction is performed using the output shaft as a reference, then (Fig. 6.3.5d):

$$e''_{12} = e_{12}/i^2_{2,3} \qquad e''_{23} = e_{23}/i^2_{2,3}$$

$$I''_1 = I_1\,i^2_{2,3} \qquad I''_2 = I_2 i^2_{2,3}.$$

After the reduction is performed, a breakdown of compliance and/or inertia can be performed because all the components are now referred to the same velocity and thus, are comparable. The breakdown of the compliance is written as:

$$e_{in} = e_{12} + e_{23} + e'_{34} \tag{6.3.21a}$$

or

$$e_{out} = e''_{12} + e''_{23} + e_{34} \tag{6.3.21b}$$

The breakdown of inertia is written as:

$$I' = I_1 + I_2 + I'_3 + I'_4 \tag{6.3.22a}$$

or

$$I'' = I''_1 + I''_2 + I_3 + I_4 \tag{6.3.22b}$$

As it is clear from the reduction procedure, a change in the kinematic configuration (e.g., a gear shift for a different transmission ratio or a change in the transmission ratio of a variable transmission device or a change in a linkage configuration) would completely change the compliance and inertia breakdowns. Thus, to thoroughly understand the roles of various design components in the overall compliance and/or inertia breakdown, all the critical kinematic configurations have to be analyzed.

Case Study. The relative importance of various contributors to the compliance breakdown of a geared transmission as well as influence of a change in the kinematic configuration is illustrated in Fig. 6.3.6 [1]. Figure 6.3.6a shows the initial composition of the mathematical model for an actual gearbox (of a vertical knee milling machine) with the actual values of torsional, contact (keys and splines), and bending compliances as well as compliance of the rubber coupling (e_c) indicated in the model. All these parameters were computed in accordance to the formulas presented above in this chapter. Figures 6.3.6b, c, and d show changes in

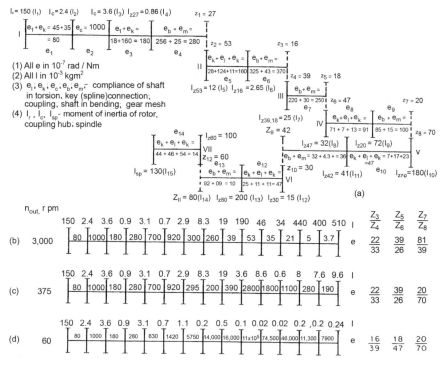

Fig. 6.3.6. Initial composition of mathematical model for a machine gearbox (a) and final mathematical models for high (b), medium (c) and low (d) output (spindle) rpm.

compliance and inertia breakdowns with all the components reduced to the motor shaft (n_m = 1460 rpm, N = 14 KW). Different transmission ratios in the three shifting gear stages lead to dramatic changes in the breakdowns. For a slow rotational speed of the output shaft (spindle), n_{out} = 60 rpm, the overall compliance is totally determined by the components close to the output shaft (slow-moving components whose reduction to the motor shaft involves very large multipliers), Fig. 6.3.6d. The overall inertia, on the other hand, is totally determined by the fast-moving components close to the motor, since reduction of inertias of the slow-moving components to the motor shaft involves very large dividers. Compliance, e_c, of the rubber coupling (the largest actual compliance of a single component in the diagram in Fig. 6.3.6a) does not have any noticeable effect. The totally reversed situation occurs for high rotational speeds of the output shaft (n_{out} = 3000 rpm, Fig. 6.3.6b). In this case, the overall compliance is about 100 times less and is totally determined by the relatively slow-moving components close to the motor (specifically, by the rubber coupling), while the overall inertia is largely determined by the fast-moving components close to and including the output shaft (spindle). An intermediate configuration (n_{out} = 375 rpm, Fig. 6.3.6c) shows a more uniform breakdown of both compliance and inertia. Influencing either the overall compliance or the overall inertia of the gearbox by design changes would, thus, require rather different design changes depending on which system configuration is considered. However, in each configuration, there are components dominating the breakdown (for n_{out} = 60 rpm, e_{10} represents 39% of the overall compliance, and I_9 is 17% of the overall inertia; and for n_{out} = 3000 rpm, e_1 is 26% of the overall compliance, and I_{15} is 30% of the overall inertia). Modification of these dominating components must be a starting point in any system improvement/modification process. Design approaches to such a modification can be developed after analyzing the physical origins of the dominating components by looking into an appropriate segment of the original model in Fig. 6.3.6a.

Fig. 6.3.7a presents total compliance values of the gearbox at all kinematic configurations (spindle speeds), computed using the algorithm described in Section 6.2, and reduced to the motor shaft (line 2). These values are compared with the measured values at two different torque settings. For the measurement, the motor shaft was clamped, and the torque was applied to the spindle whose angular displacements had been measured.

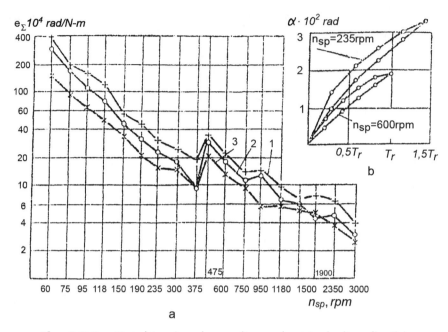

a

Fig. 6.3.7. Total torsional compliance (not including flexible coupling), referred to motor rotor of multispeed gearbox in Fig. 6.3.6 (a) and typical hysteresis loops (b). 1 — measured compliance at torque $T = 0\text{-}0.2\ T_r$; 3 — same at $T = 0.6$ to $0.8\ T_r$; 2 — computed values.

Fig. 6.3.7b presents two typical loading/unloading plots showing hysteresis loops characterizing damping of the gearbox, with $\psi = 2\delta = \sim 0.3$.

As a general rule, compliances of the slowest-moving components of a system tend to be the largest contributors to the compliance breakdown, while inertias of the fastest-moving components tend to be the largest contributors to the inertia breakdown. Some important practical conclusions from this general rule are:

1. To reduce the contribution of a physically very compliant component to the overall compliance, its relative speed in the system has to be increased (e.g., by using additional reducing stages after such compliant devices as harmonic drives or hydraulic actuators; see practical cases in Section 6.4). An important case is use of flexible shafts in stiffness-critical power transmission systems. Although the actual stiffness of flexible shafts is low, the reduced stiffness values might be acceptable if the shaft is used at the high speed stage of the system.

2. To reduce the contribution of a physically massive component to the overall inertia, its relative speed in the system has to be reduced (e.g., to reduce the inertia of the moving linkage as seen by the driving motor, the speed of the linkage relative to the motor has to be reduced, or for a given linkage speed, the speed of the motor shaft has to be increased, and the appropriate reduction means have to be introduced).

These practical conclusions should be judiciously balanced to avoid overall negative effects (e.g., an unwanted increase in the overall compliance caused by the introduction of excessive reducing means between the motor and the linkage, loss of cost-effectiveness because of the introduction of additional reduction stages, etc.).

Damping coefficients can be reduced to system components moving with different speeds by using the same algorithms as for the reduction of compliance, Eq. (6.3.14), and inertia, Eq. (6.3.17), namely [2],

$$c_i^k = c_i^n / i_{n,k}^2, \qquad (6.54)$$

where c_i^k = damping coefficient of unit i on shaft k and c_i^n is the same but reduced to shaft n.

6.4 PRACTICAL EXAMPLES OF STRUCTURAL COMPLIANCE BREAKDOWN

Compliance breakdown for a typical gearbox is presented in Fig. 6.3.6. It can be done also for more complex devices, such as a planetary transmissions [10]. Below, the breakdown of effective overall compliance for selected coordinate directions is presented for several robot manipulators (one hydraulic robot and three electromechanical robots of different structural designs [7]). Certain information about robots, such as types and dimensions of bearings, specifics of gears, etc., was not provided by the manufacturers but was taken (or assumed) from drawings and pictures available in the public domain. Compliance values of various design components were calculated using techniques described in Section 6.2 and in Chapter 3.

A special case of measured compliance breakdown for a precision OD grinder is described in Appendix 7.

6.4.1 A Hydraulically Driven Robot

The robot (Fig. 6.4.1) operates in spherical coordinates (radial arm extension with a maximum speed of the end effector of 0.9 m/s, rotation around the vertical axis with a maximum tangential speed of the end effector at its maximum radial extension of 2.6 m/s, and rotation in the vertical plane with a maximum tangential speed of the end effector at its maximum radial extension of 1.4 m/s). Since the rate of deceleration required for stopping within a time interval is greater and inertia forces are more intense the faster the motion, rotation around the vertical axis is the critical mode. Also, this is the most frequently used mode of motion (e.g., for all pick-and-place type applications). Since abrupt accelerations/decelerations usually excite vibrations of the end effector, which have to decay before its complete stop, it is obvious that the fundamental (lowest) natural frequency determines the upper limit of the rate of acceleration/deceleration and, thus, productivity of the manipulator.

Accordingly, this mode of motion is selected to be analyzed for overall compliance and fundamental (lowest) natural frequency. The sketch of the robot in Fig. 6.4.1 illustrates its design features that are relevant to this mode. Column 1 carries aluminum arm carriage 2 with extending

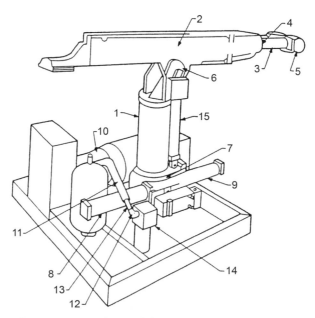

Fig. 6.4.1. Spherical frame robot manipulator.

steel rods 3 supported by bronze bushings 4 and carrying end block 5 with the wrist and/or end effector (not shown). Arm carriage 2 is connected to the top surface of column 1 through pin-hole connection 6.

Column 1 has rack-and-pinion unit 7 attached to its lower end with the rack driven by double cylinder systems 8 and 9. Pressurized oil is supplied to cylinders 8 and 9 from pump 10 through metal tubes 11 and 12, hose insert 13, and servovalve 14.

The overall deflection caused by payload inertia force, P, is composed out of the following components (reduced to the arm end):

1. Deflection of rods 3;
2. Contact deformation in bushings 4;
3. Deflection of carriage 2;
4. Contact deformation in pin-hole connection 6;
5. Twist of column 1;
6. Contact and bending deformations in rack-and-pinion unit 7;
7. Compression of oil in and deformation of the hydraulic cylinders;
8. Compression of oil in and deformation of the rigid plumbing;
9. Compression of oil in and deformation of the hose insert.

These components are evaluated below:

1. Deflection of the cantilever beam (Fig. 6.4.2) is $\delta_a = [Pl^2_1(l_1 + l_2)]/ 3EI$, where I = the cross-sectional moment of inertia of two rods, and $l_2 = 235$ mm is the distance between bushings 4; $l_1 = 800$ mm

$$\delta_a/P = 6.7 \times 10^{-4} \text{ mm/N}$$

2. One set of bushings 4, accommodating one rod, is acted upon by a force $P/2$.

Reaction forces between one rod 3 and its set of bushings 4 are (Fig. 6.4.2):

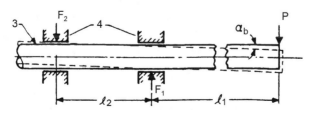

Fig. 6.4.2. Loading schematic of sliding rods.

$$F_1 = (P/2)[(l_1 + l_2)/l_2]; \qquad F_2 = Pl_1/2l_2.$$

Forces, F_1 and F_2, cause deformations between the rod and the bushings (caused by both structural deformation of the pin and of the bushings and by contact deformation of surface asperities) and, accordingly, an angular deflection of the rod. The contact deformations between a pin and a hole of slightly different diameters depend on the part materials, the diameter of the connection, the diametrical clearance (assumed in this case to be 20 μm or 0.0008 in.), and magnitude of the forces, since the contact deformation is non-linear (see Section 4.4.1). In this case, two magnitudes of the force, P, were considered. In the no-load case, the effective mass at the arm end consists only of the structural mass, which is equivalent to $m_1 = \sim 40$ kg, and inertia force with deceleration, $a = 2g = 20$ m/sec^2, and $P_1 \approx 800$ N. In the maximum load case, $m_2 = 40 + 50$ kg, $a = 2$ g, and $P_2 = 1800$ N. The contact deformations in the bushings under this force lead to an angular deflection α_b (Fig. 6.4.2), which in turn causes a linear displacement $\delta_b = \alpha_b l_1$ of the end effector. Contact deformations can be easily calculated by Eqs. (4.5.9), and (4.5.15); the results for deflection of masses, m_1 and m_2, respectively, are:

$$(\delta_b/P)_1 = 0.034 \times 10^{-4} \text{ mm/N}; \qquad (\delta_b/P)_2 = 0.033 \times 10^{-4} \text{ mm/N}$$

3. The arm carriage is a very massive cast aluminum part; its deformations are assumed to be negligibly small:

$$\delta_c/P \approx 0$$

4. The arm carriage is connected to the column via the pin-bushing connection shown in Fig. 6.4.3. In this case, the same computational approach is used as in paragraph 2 above; the diametral clearance was again assumed to be 20 μm, and the distance between the center line of the pin and the force, P, is equal to the full arm extension, $l_3 = 1.64$ m. The force, P_d, in each pin-bushing connection is $P_d = Pl_3/l_d$; contact deformation δ_d in each connection is calculated according to Eqs. (4.5.9), and (4.5.15). After reduction to the end effector, the result is:

$$(\delta_d/P)_1 = 3.5 \times 10^{-4} \text{ mm/N}; \qquad (\delta_d/P)_2 = 1.6 \times 10^{-4} \text{ mm/N}$$

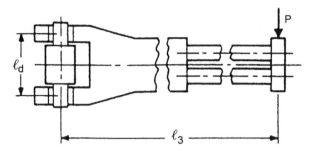

Fig. 6.4.3. Loading schematic of extended arm.

where subscripts 1 and 2 have the same meaning as they do in paragraph 2 above.

5. The torsional compliance of the steel column that is 1.0 m high and 0.23 m diameter with 13 mm thick walls is $(\phi/T)_e = 8.3 \times 10^{-11}$ rad/N mm, where T = the torque acting on the column; accordingly

$$\delta_e/P = (\phi/T)_e l^2{}_3 = 8.3 \times 10^{-11} \times 1640^2 = 2.2 \times 10^{-4} \text{ mm/N.}$$

6. The compliance of the rack-and-pinion mesh (pinion diameter 254 mm, mesh width 76 mm) is calculated using Eq. (6.2.7). The angular compliance reduced to the pinion is $e_{rp} = 6.2 \times 10^{-11}$ rad/Nmm. This corresponds to a translational compliance reduced to the end of arm equal to:

$$\delta_e/P = e_{rp} l^2{}_1 = 1.6 \times 10^{-4} \text{ mm/N,}$$

7, 8, and 9. When a force, ΔP_c, is applied to the piston, it moves by an increment ΔX_c because of the compression of hydraulic oil in the cylinder, in the rigid pipes, and in the hoses and also because of the incremental expansion of their walls, as described by Eq. (6.2.30) in Section 6.2.3. For the considered robot, the cylinder volume in the middle rack position $V_c = 6.43 \times 10^5$ mm^3, $V_p = 5.1 \times 10^5$ mm^3, $V_h = 3.9 \times 10^5$ mm^3, $K_{eff.c} = 0.84K = 1.34 \times 10^3$ N/mm^2, $K_{eff.p} = 0.71K = 1.14 \times 10^3$ N/mm^2, $K_{eff.h} = 0.12 \times 10^3$ N/mm^2 (as calculated from data provided by a hose manufacturer, line 1 in Fig. 6.2.14), and $A_p = 3170$ mm^2. Accordingly, from Eq. (6.2.30)

$$\Delta X_c/\Delta P_c = 4.1 \times 10^{-4} \text{ mm/N.}$$

Since the action line of ΔX_c is at $l_c = 254/2 = 127$ mm distance from the column axis and the end effector is at $l_3 = 1640$ mm, the compliance of the hydraulic drive reduced to the arm end is:

$$(\Delta X/\Delta P)_{ghi} = (\Delta X_c/\Delta P_c)(l^2{}_2/l^2{}_c) = 690 \times 10^{-4} \text{ mm/N}$$

For better visibility of contributions of the cylinder (7), the rigid piping (8) and the flexible hose (9) segments, their compliances reduced to the end of arm can be calculated separately, as

$$(\Delta X/\Delta P)_g = 8 \times 10^{-3} \text{ mm/N}; \quad (\Delta X/\Delta P)_h = 7.5 \times 10^4 \text{ mm/N};$$
$$(\Delta X/\Delta P)_i = 53 \times 10^{-3} \text{ mm/N}$$

Finally, the total compliance for the fully loaded ($m = 50$ kg) manipulator reduced to the arm end is as follows:

$$e = \Delta X/\Delta P = (6.7 + 0.34 + 0 + 1.6 + 2.2 + 1.6 + 80 + 75 + 530)\ 10^{-4} \text{ mm/N}$$
$$= 700 \times 10^{-4} \text{ mm/N} = 70 \times 10^{-6} \text{ m/N}$$

Since full extension of the arm is $l_3 = 1.64$ m, it is equivalent, according to Eq. (6.3.9), to

$$e_\phi = 70 \times 10^{-6} / 1.64^2 = 26 \times 10^{-6} \text{ rad/N-m}$$

Experimental data on static compliance of a Unimate 2000 robot is (in different configurations) $e = 12$ to 105×10^{-6} m/N; thus the calculated data is within the realistic range of compliance.

Since a rough estimation of the structural arm mass, m_s, reduced to the end of arm is $m_s = \sim 40$ kg, thus the total inertia including the payload is $m = 90$ kg and the fundamental natural frequency,

$$f = \frac{1}{2\pi}\sqrt{\frac{1}{70 x 10^{-6} x 90}} \approx 2.0 Hz.$$

This low value of natural frequency is in good correlation with the test results which have shown that it takes 0.5 second for the arm of a Unimate 2000 B robot to respond to a control input [7].

The availability of the breakdown allows a designer to find a simple means to substantially increase natural frequency by:

1. Replacement of the flexible hose with the state-of-the-art hose, line 6 in Fig. 6.2.14 ($K_{eff.h} = 1.4 \times 10^3$ N/mm^2), or with a metal pipe. This would result in $e_\alpha = 21.4 \times 10^{-6}$ m/N and $f_\alpha = 3.6$ Hz.

2. After item 1 is implemented, shortening the total piping length by 50%. The expected result: $e_\beta = 15.3 \times 10^{-6}$ m/N and $f_\beta = 4.3$ Hz.

3. After items 1 and 2 are implemented, introduction of a gear reduction stage with a transmission ratio $i < 1$ between the pinion and the column. Then, compliance of the hydraulic system would enter the breakdown after multiplication by i^2, according to Eq. (6.3.14). For $i = 0.5$, the overall compliance would become $e_\gamma = 4.8 \times 10^{-6}$ m/N and $f_\gamma = 7.6$ Hz.

It can be seen that the very easily attainable modifications (1) and (2) would more than double the natural frequency (and reduce by half the duration of the transient period), and then the design modification (3) would additionally increase natural frequency 25% to 80%. Even after all these modifications are implemented, the mechanical compliances, including the most important contributor — link compliance — do not play a very significant role in the breakdown [less than 1 percent initially, about 15% after modifications (1), (2), and (3) are implemented]. This is typical for hydraulically driven robots.

6.4.2 Electromechanically Driven Robot of Jointed Structure, Fig. 6.4.4

For a comparison, the same mode of motion is considered-rotation around the vertical (column) axis at the maximum arm outreach. The identified contributors to the effective compliance are (Fig. 6.4.4a):

1. Deflection of the forearm under the inertia force;
2. Contact deformations in the joint between the forearm and the upper arm (elbow joint);
3. Deflection of the upper arm;
4. Contact deformations in the joint between the upper arm and the shoulder;
5. Twisting of the vertical column (waist) inside the trunk;
6. Angular deformation of the gear train between the driving motor and the column (waist).

In the previous case, all the components of structural compliances had been reduced to translational compliances at the end effector. In this case, all the components will be reduced to torsional compliances around the vertical axis. Reduction of a linear displacement Δ caused by a force P

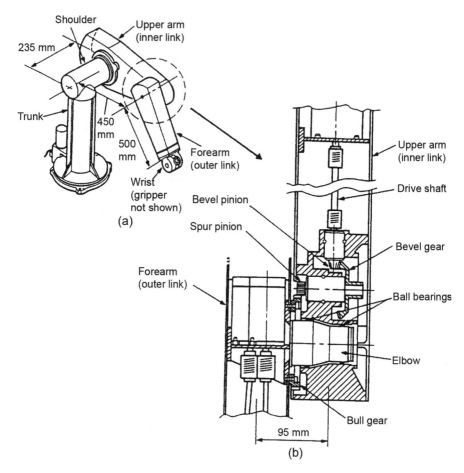

Fig. 6.4.4. Electromechanical jointed robot.

(compliance Δ/P) to torsional compliance e around an axis at a distance, α, from the P and Δ vectors is performed by Eq. (6.3.9).

1. The forearm is an aluminum shell, l_1 = 500 mm long with a 1.5-mm wall thickness, having an average cross-section 90 x 120 mm. Accordingly, I = 8.7 x 10^5 mm^4, EI = 6.1 x 10^{10} Nmm^2, and with the distance between the extended end of the forearm and the waist axis:

$$l_{0,a} = \sqrt{1,016^2 + (235 - 95)^2} = 1,026 \text{ mm.}$$

Thus

$$e_a = (1/l^2)(l^3_1/3EI) = 6.5 \times 10^{-10} \text{ rad/Nmm} = 6.5 \times 10^{-7} \text{ rad/Nm}$$

2. The elbow joint between the forearm and the upper arm (Fig. 6.4.4b) operates with two ultralight ball bearings, which accommodate the moment from the payload transmitted between the end effector and the drive motor. The components of the bearing deformation are calculated by Eqs. (3.4.8) and (3.4.9). The resulting torsional compliance is:

$$e_b = 38.4 \times 10^{-7} \text{ rad/Nm}$$

3. The upper arm is approximated as a hollow aluminum beam 105 mm deep with average width of 227 mm and a 3-mm - thick wall ($I = 2.1 \times 10^6 \text{ mm}^4$ and $EI = 1.46 \times 10^{11} \text{ Nmm}^2$) and with an active length $l_2 = 450$ mm. The effective radius from the waist axis

$$l_0 = \sqrt{450^2 + 235^2} = 508 \text{ mm}.$$

Accordingly, considering loading by the force, P, and by the moment, $M = Pl_1$,

$$e_c = (1/508^2)(1/1.46 \times 10^{11})[450^3/3 + (500 \times 45^2)/2]$$
$$= 21 \times 10^{-10} \text{ rad/Nmm} = 21 \times 10^{-7} \text{rad/Nm}$$

4. The shoulder joint is similar to the elbow joint but with different dimensions. The overall torsional compliance of the shoulder joint is found to be:

$$e_d = 13.9 \times 10^{-7} \text{ rad/Nm}$$

5. The waist is an aluminum tubular part with an outer diameter of 165 mm, a height of $l_3 = 566$ mm, and a wall thickness of $t_3 = 7$ mm. Accordingly, its torsional stiffness is:

$$e_e = l^3/GJ = 9.3 \times 10^{-7} \text{ rad/Nm}$$

6. The torsional compliance of the steel spur gear mesh, reduced to the bull gear attached to the waist, is characterized by $k_m = 6 \times 10^{-5}$

mm^2/N in Eq. (6.7) and also by b = 7.5 mm and the radius of the bull gear, R = 165 mm; α = 20 deg. In this case, the power is transmitted to the bull gear through two preloaded pinions (anti-backlash design), thus the compliance is reduced in half. Since the reduction ratio between the pinions and the bull gear is very large (about 1:12), the compliance of the preceding stages of the motor-waist transmission does not play any significant role. Accordingly,

$$e_f = 1.6 \text{ x } 10^{-7} \text{ rad/Nm}$$

As a result, the breakdown of torsional compliance of the robot in the considered mode is:

$$e = (6.5 + 38.4 + 21 + 12.9 + 9.3 + 1.6)10^{-7} \text{ rad/Nm}$$
$$= 89.7 \text{ x } 10^{-7} \text{ rad/Nm}$$

With the payload, 2.5 kg, and the effective structural mass at the end effector 3 kg (at the distance from the waist axis $R_0 \approx 1$ m), the total moment of inertia, $I = 5.5 \text{ x } 1^2 = 5.5 \text{ kgm}^2$ and natural frequency:

$$f = \frac{1}{2\pi}\sqrt{\frac{1}{5.5 \text{ x } 89.7 \text{ x } 10^{-7}}} = 22.5 \text{ Hz.}$$

The compliance of this electromechanical robot is much lower than that of the hydraulically driven robot considered in Section 6.4.1, which results in a much higher natural frequency (22.5 Hz versus 2.0 Hz) and, accordingly, a much better performance (faster acceleration/deceleration, shorter settling time).

The breakdown of the compliance in this case is much more uniform but still is dominated by one component. This component, 38.4 x 10^{-7} rad/Nm, is associated with the elbow joint equipped with ball bearings; a significant contribution of this joint to the arm deflection was observed also during tests. This defect can be easily alleviated by not very significant modifications of the joint (e.g., a larger spread of the bearings and/or the selection of more rigid bearings). These measures could realistically reduce the elbow joint compliance by about 50%. If this is achieved, the total compliance becomes e = 71.5 x 10^{-7} rad/Nm and the natural frequency increases about 12% up to f = 25.3 Hz. The total (bending) compliance of the links is about 30% of the overall compliance without design modifications and becomes about 38% after the suggested modifications. Thus, in this manipulator, stiffening of the upper arm (inner link) might be warranted after stiffening of the elbow joint is performed.

6.4.3 Electromechanically Driven Parallelogram Robot with Harmonic Drives

The parallelogram structure manipulator is shown in Fig. 6.4.5. A purely rotational movement of the payload in the vertical plane is considered (Fig 6.4.6a). For the rated payload $m = 10$ kg and an assumed deceleration of 1 g, the inertia force of the payload $P = 98$ N. A free-body diagram of the forearm is shown in Fig. 6.4.6b, where the forearm is presented as a beam on two elastic supports. Since the rotational motion of the forearm is driven by the motor-harmonic reducer located in the joint between the crank of the rear upper arm and the base, the torque generated

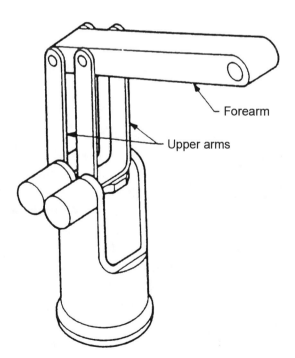

Fig. 6.4.5. Electromechanical robot with parallelogram structure.

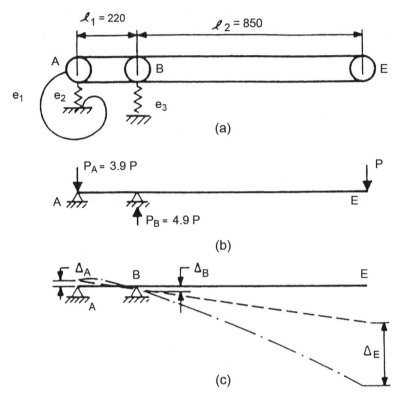

Fig. 6.4.6. Schematics of forearm of parallelogram robot.

by the force P around pivot B is absorbed by the motor-reducer whose torsional compliance is shown as e_1. The rated torsional stiffness of the harmonic drive is $k = 10{,}400$ Nm/rad for transmitted torques $< 0.2\ T_r$, and $k = 58{,}000$ Nm/rad for torques $> 0.2\ T_r$, (manufacturer's data), thus $e_1^a = 9.6 \times 10^{-5}$ rad/Nm at $T < 0.2\ T_r$ and $e_1^b = 1.72 \times 10^{-5}$ rad/Nm at $T > 0.2\ T_r$. Here $T_r = 95$ Nm is the rated torque of the reducer.

Compliance e_2, which is caused by tensile deformations of the rear upper arm, AD, and by deformations of two pivot bearings at A, can be represented as $e_2 = e_2' + e_2'' + e_2''' + e_2''''$, where e_2' is the tensile compliance of the solid part of the rear upper arm, e_2'' is the tensile compliance of two sides of the (upper) section of the rear upper arm, without the middle rib, e_2''' is the tensile compliance of the sections of the arm accommodating the bearings at A, and e_2'''' is the combined compliance of two pivot bearings.

Compliance $e_3 = e_3' + e_3'' + e_3''' + e_3''''$ at support B represents deformations of the front upper arm, BC, and consists of essentially the same components as e_2, with a possibility of e_3'''' having a different magnitude because of its load dependence. Compliance e reduced to end point E is determined by e_1, e_2, and e_3 as well as by the bending compliance of the link (forearm) ABE at E.

Figure 6.4.6b shows the force diagram of the forearm, and Fig. 6.4.6c shows the translational deflection diagram. If oscillations of the payload at E are considered as oscillatory motion around B, the overall angular compliance caused by the translational deflection, reduced to E, is:

$$e' = [(\Delta_A + \Delta_B)/AB + \Delta_E/BE]/Px \ BE, \ \text{rad/Nm} \qquad (6.4.1)$$

where $\Delta_A = 3.86 \ Pe_2$, $\Delta_B = 4.86 \ Pe_3$, and Δ_E is the bending deflection of the forearm caused by P with supports A and B considered as rigid. The total compliance at E is:

$$e = e' + e_1 \qquad (6.4.2)$$

The breakdown of the compliance is as follows:

$e_2' = e_3' = 0.42 \times 10^{-8}$ m/N; $\ \ e_2'' = 0.11 \times 10^{-8}$ m/N; $\ \ e_3''' = 0.61 \times 10^{-8}$ m/N

$e_2^{iv} = 1.03 \times 10^{-8}$ m/N; $\ \ e_3^{iv} = 1.0 \times 10^{-8}$ m/N

$\Delta_A = 8.4 \times 10^{-8} \ P$; $\ \ \Delta_B = 10.2 \times 10^{-8} \ P$; $\ \ \Delta_E = 90 \times 10^{-8} \ P$

$(\Delta_A + \Delta_B)/AB = [(8.4 + 10.2)/0.22] \times 10^{-8} \ P = 84.5 \times 10^{-8} \ P \ \text{rad/N}$

$\Delta_E/BE = (90/0.85) \times 10^{-8} P = 106 \times 10^{-8} \ P \ \text{rad/N}$

Thus, the total structural compliance is $e' = (84.5 + 106)10^{-8}P/0.85P = 22.4 \times 10^{-7}$ rad/Nm.

Since the harmonic drive compliance, $e_1^{a} = 9.6 \times 10^{-5}$ rad/Nm, and $e_1^{b} = 1.72 \times 10^{-5}$ rad/Nm, the total compliance at the end of arm is, per, (6.4.2), for $T < 0.2T_r$ (a), and for $T > 0.2T_r$ (b):

$$e^a = (22.4 + 960) \times 10^{-7} = 982.4 \times 10^{-7} \ \text{rad/Nm};$$

$$e^b = (22.4 + 172) \times 10^{-7} = 194.4 \times 10^{-7} \ \text{rad/Nm}$$

With the rated payload of 10 kg and the effective structural mass reduced to the forearm end also about 10 kg, the total moment of inertia around the pivot B is $I_B = 20 \times 0.85^2 = 14.45$ kgm^2 and the natural frequency is:

$$f^a = \frac{1}{2\pi}\sqrt{\frac{1}{14.45 \times 982.4 \times 10^{-7}}} = 4.2 \text{ Hz for } T < 0.2\, T_r(a), \text{ and}$$

$$f^b = \frac{1}{2\pi}\sqrt{\frac{1}{14.45 \times 194.4 \times 10^{-7}}} = 9.5 \text{ Hz for } T > 0.2\, T_r \text{ (b)}$$

The breakdown of compliance in this case is dominated by the harmonic drive compliance (98% to 89% of the overall compliance, depending on intensity of loading). This situation can be improved by a total redesign or by a mechanical insulation of the harmonic drive from the structure through the introduction of a reducing stage (e.g., gears) or by using an oversized harmonic drive unit. The forearm bending compliance constitutes about 1% to 6% of the overall compliance. This percentage will rise if the drive stiffness is enhanced.

6.4.4 Electromechanically Driven Spherical Frame Robot

The stiffness of the arm in Fig. 6.4.7 in a vertical plane is analyzed for all ranges of its radial positions. The arm is driven by a motor with a harmonic reducer (1 in Fig. 6.4.7). The root segment of the arm consists of tapered tubular member 2 with counterbalance 3 and rigid partitions 4 and 5 into which two guide rods 6 are secured (guide rods 6 are reinforced with gussets 6a). Rods 6 support four open ball bushings 7 attached to intermediate tubular segment 8 with two rigid end walls to which two guide rods 9 are secured. Rods 9 support four open (slotted) ball bushings 10 attached to tubular arm 11 carrying rigid end piece 12 with wrist 13. The radial motion of arm end 12 is accomplished by

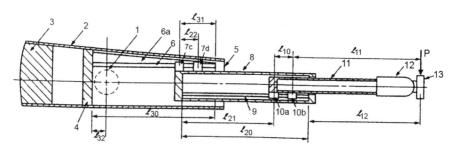

Fig. 6.4.7. Electomechanical spherical robot arm.

a motor-ball-screw-cable system drive (not shown in Fig. 6.4.7), which assures equal relative motion in both link connections.

The main sources of compliance in response to force, P, are as follows:

1. Bending of double-supported cantilever beam (arm) 11;
2. Contact deformations in ball bushings 10 (the open sides of the bushings are directed downward in Fig. 6.4.7);
3. Bending of guide rods 9;
4. Bending of intermediate link 8;
5. Contact deformations in ball bushings 7 (the open sides of the bushings are directed upward.);
6. Bending of reinforced guide rods 6;
7. Bending of root link 2;
8. Torsional compliance of harmonic drive 1.

1. The bending compliance of arm 11 is

$$e_a = l_{11}^2(l_{10} + l_{11})/3(EI)_{11} \qquad (6.4.3)$$

2. To calculate the deflection (compliance) caused by the contact deformations in the bushings, link 11 has to be considered as a rigid beam in compliant supports. For the force, P, direction as shown in Fig. 6.4.7, the left bushing is loaded toward its open side (stiffness $k_a = k_{op}$), and the right bushing is loaded toward its closed side (stiffness $k_b = k_{cl}$). The manufacturer-provided stiffness characteristics for ball bushings in both directions are shown in Fig. 6.4.8c. For the opposite direction of force, P, the stiffness characteristics are reversed. Reaction forces in bushings a and b are, respectively,

$$P_a = P(l_{11}/l_{10}); \qquad P_b = P[(l_{10} + l_{11})/l_{10}] \qquad (6.4.4)$$

Accordingly, the compliance at the arm end caused by the bushings is:

$$e_b = (1/k_a)(l_{11}^2/l_{10}^2) + (1/k_b)[(l_{10} + l_{11})^2/l_{10}^2] \qquad (6.4.5)$$

3. The reaction forces acting on guide rods 9 from the bushings (numerically equal but opposite in direction to forces P_a and P_b) cause deflection of the rods as shown in Fig. 6.4.9. Deflections, y_a and y_b, in points, a and b, are caused by both forces, P_a and P_b, in each. The first subscript in Eqs. (6.4.6) and (6.4.7) below designates the location (point a or b) and the second subscript designates the force (P_a or P_b):

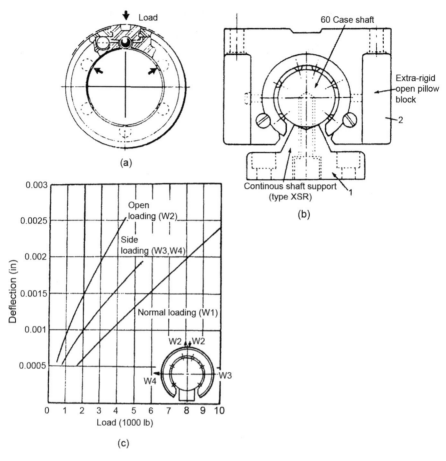

Fig. 6.4.8. "Super ball" bushing (a); open bushing (b); load-deflection characteristic of open bushings (c). (Source: Thomson Industries, Inc.)

$$y_a = y_{aa} + y_{ab} = -\frac{Pl_{11}(l_{20} - l_{21})^2 l_{21}^2}{6l_{10}E_r I_r l_{20}^3}(2l_{21}^2 - 2l_{21}l_{20}) +$$

$$\frac{P(l_{10} + l_{11})(l_{20} - l_{21} - l_{10})^2 l_{21}^2}{6l_{11}E_r I_r l_{20}^3}(2l_{21}^2 + 2l_{21}l_{10} - 2l_{10}l_{20} + 3l_{21}l_{20}) \qquad (6.4.6)$$

$$y_b = y_{ba} + y_{bb} = -\frac{Pl_{11}l_{21}^2(l_{20} - l_{21} - l_{11})^2}{6l_{11}E_r I_r l_{20}^3}(2l_{21}^2 + 2l_{21}l_{10} - 3l_{20}l_{10}) +$$

$$\frac{P(l_{10} + l_{11})(l_{20} - l_{21} - l_{10})^2(l_{21} + l_{10})^2}{6l_{10}E_r I_r l_{20}^3}(2l_{21}^2 + 4l_{21}l_{10} + 2l_{10}^2 - 2l_{20}l_{10} - 2l_{21}l_{10})$$

$$(6.4.7)$$

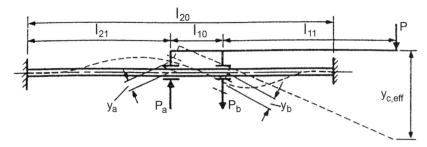

Fig. 6.4.9. Effect of bending of guide rods on the arm-end deflection.

The effective deflection at the arm end because of y_a and y_b is, as it is easy to derive from Fig. 6.4.9,

$$y_{ceff} = \frac{l_{11}}{l_{10}} y_a - \frac{l_{10} + l_{11}}{l_{10}} y_b \tag{6.4.8}$$

and

$$e_c = y_{ceff}/P \tag{6.4.9}$$

4. The bending of intermediate link 8 is caused by the reaction force and the moment in the connection between guide rods 9 and the right end plate of link 8. Link 8 is considered as supported on two bearing blocks (ball bushings) 7c and 7d. Both the force and the moment depend on the relative position of links 11 and 8.

Bending deflection y_s of link 8 end and its end slope θ_8 are causing an inclination of guide rods 9, carrying arm 11, by an angle

$$\alpha_8 = y_8/l_{20} + \theta_8 \tag{6.4.10}$$

and, accordingly, deflection and compliance at the arm end

$$y_{deff} = \alpha_8 (l_{20} + l_{12}); \quad e_d = \alpha_8 (l_{20} + l_{12})/P \tag{6.4.11}$$

5. Compliance, e_e, at the arm end, which is caused by deformations in ball bushings 7c and 7d, is calculated similarly to e_b with the appropriate dimensional changes

$$e_e = \frac{1}{k_c} \frac{(l_{12} + l_{20} - l_{22})^2}{l_{22}^2} + \frac{1}{k_d} \frac{(l_{12} + l_{20})^2}{l_{22}^2} \tag{6.4.12}$$

6. Compliance, e_f, at the arm end, which is caused by the bending of reinforced guide rods 6 and 6a, is calculated similarly to e_c with the appropriate dimensional changes

$$e_f = \frac{y_{feff}}{P} = \left(\frac{l_{12} + l_{20} - l_{22}}{l_{22}} y_c - \frac{l_{12} + l_{20}}{l_{22}} y_d \right) / P \qquad (6.4.13)$$

7. Compliance which is caused by the bending of root link 2, is calculated analogously to e_d with the appropriate dimensional changes, thus

$$\alpha_2 = y_2 / l_{30} + \theta_2 \; y_{geff} = \alpha_2 \, (l_{30} - l_{31} + l_{20} + l_{12});$$
$$e_g = y_{geff} / P \qquad (6.4.14)$$

8. The linear compliance at the arm end, which is caused by the angular compliance e_{hd} of harmonic drive 1, is

$$e_h = e_{hd} \, (l_{30} - l_{31} - l_{32} + l_{20} + l_{21})^2 \qquad (6.4.15)$$

Angular stiffness of the harmonic drive, $k_{hd} = 1/e_{hd}$, is specified by the manufacturer as $k_{hd}' = 5150$ Nm/rad for torques less than 11.5 Nm and $k_{hd}'' = 26{,}750$ Nm/rad for torques above 57 Nm.

Because of the rather cumbersome expressions for force and moment reactions and for deflections and also because of the complex relationships between the compliance components and the arm outreach, the compliance breakdown is plotted in Fig. 6.4.10 as a function of the arm outreach.

Although again the breakdown is dominated by compliance, e_{hd}, of the harmonic drive, especially at low motor torques, compliance component, e_c, which is caused by guide rods 9, is comparable with the component, e_{hd}, and even exceeds it at high torques (where the harmonic drive stiffens) and at short arm configurations. Ball bushings between arm 11 and guide rods 9 also make a noticeable contribution.

In this case, the reduction of the harmonic drive contribution to the overall compliance can be achieved by the same approach as was suggested in Section 6.4.3 (additional reduction stage or an oversized transmission). Components e_c and e_b can be reduced by several design approaches, such as reinforcing of the guide rods, increasing the ball bushing size (i.e., rod diameters), and using other types of linear bearings (e.g., see [7]).

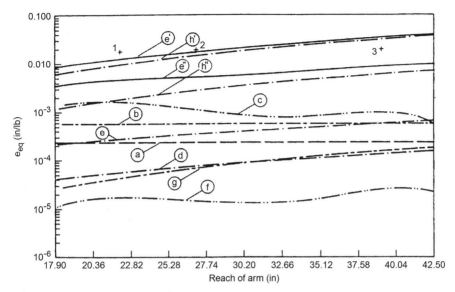

Fig. 6.4.10. Compliance breakdown for robot arm in Fig. 6.4.7 vs. its extension ; letters at the lines designate sources of compliance, as listed in the text ; e′ and e″ are the overall compliance for $T < 100$ lb in. and $T > 100$ lb in. ; experimental points are measured at : (1) $T = 70$ lb in., (2) $T = 90$ lb in., and (3) $T = 127$ lb in.

The overall contribution of links bending $(e_a + e_d + e_g)$ is 3.5% to 8.5% for 17.9 in. outreach, diminishing to 1.4 % to 6.1 % at 42.5 in. outreach.

The natural frequency of the arm with the rated payload of 2.25 kg for a motor torque < 11.5 Nm is 10.7 Hz at 17.9 in. outreach and 4.95 Hz at 42.5 in. outreach. The latter number correlates well with the experimental data. Measured data for static compliance are shown in Fig. 6.4.10 and indicate a reasonably good correlation with the computed values.

6.4.5 Summary

Examples in Sections 6.4.1 through 6.4.4 show that:

1. Computation of the static compliance breakdown allows, with a relatively small effort, to perform evaluation of the design quality and selection of the most effective way to make improvements while in the blueprint stage.

2. Compliance breakdown of the considered robotic structures is usually dominated by one component whose identification allows for a subsequent significant improvement of the system performance.

3. In many instances, a reasonably minor design effort aimed to stiffen the dominating compliant component would lead to a major improvement in the structural stiffness.

4. Accordingly, special efforts directed toward the development of basic mechanical structural units, especially transmissions and joints are warranted.

5. Linkage compliance does not play a critical role in the compliance breakdown of the considered robotic structures; as a result, increased arm outreach is not always accompanied by increased compliance at the arm end.

6.5 MORE ON STIFFNESS AND DAMPING OF ANTIFRICTION BEARINGS AND SPINDLES

Anti-friction bearings, both ball and roller bearings, may play a special role in balance of stiffness and damping of important mechanical systems. For example, an overwhelming majority of machine tool spindles are supported by antifriction bearings, which determine stiffness and damping of the spindle system which, in its turn, frequently determines performance quality of the machine tool. Parameters affecting stiffness and damping of antifriction bearings include rotational speed, magnitudes of external forces and of preload forces, clearances/interferences of fits on the shaft and in the housing, amplitudes and frequencies of vibratory harmonics, viscosity and quantity of lubricants, temperature. Some of these factors are addressed below.

6.5.1 Stiffness of Spindles

Spindles are frequently supported by roller bearings, which are sensitive to angular misalignment between their rollers and races. To maintain a tolerable degree of misalignment, it is recommended that the distance between the bearings does not exceed four to five diameters of the spindle and the length of the spindle overhang in the front (for attachment of tools or part-clamping chucks) is kept to the minimum. To provide for a normal loading regime of the roller bearings, structural stiffness of the spindle itself for light and medium machine tools should be at least 250 N/μm (1.4 x 10^6 lb/in.). This stiffness value is computed/measured if the spindle is considered as a beam simply supported at the bearings' locations and loaded by a concentrated force in the midpoint between the supports. Spindles of high precision machines tools, while usually loaded by cutting forces of small magnitudes, should have at least two times higher stiffness, 500 N/μm (2.8 x 10^6 lbs/in.) to provide adequate working conditions for the roller bearings.

Radial stiffness of a spindle unit can be enhanced by using more rigid bearings and/or by increasing the spindle diameter. The latter approach is less attractive, since it reduces the maximum rpm of the spindle. Effectiveness of using stiffer bearings depends on the role which the bearings play in the overall stiffness of the spindle unit. Smaller spindles (50 mm to 60 mm diameter) contribute so much to the overall compliance at the spindle flange/nose (70% to 80%) that enhancement of the bearings' stiffness would not be very noticeable (e.g., see Fig. 6.5.1).

Stiffness of the spindle at its flange can be significantly enhanced by using short spindles of large diameters. Such an approach resulted in

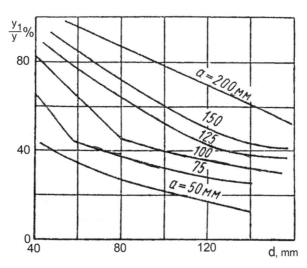

Fig. 6.5.1. Ratio of the spindle proper deformation to total deflection of the spindle unit as function of diameter *d* and overhang *a*. of the spindle.

stiffness 450 N/μm to 2500 N/μm (2.5×10^6 lb/in. to 14×10^6 lb/in.) at the spindle flange for lathes with maximum part diameter 400 mm.

It is important to assure by the design means, that stiffness of the spindle is the same in all directions. Anisotropic stiffness may result in an elliptical shape of the machined part on a lathe, in a distorted surface geometry during milling operations, etc. On the other hand, properly oriented vectors of the principal stiffness may result in enhanced chatter resistance of the machine tool (e.g., [11]).

Optimal dimensioning of the spindle to obtain its maximum stiffness depends significantly on its design schematic. Spindles of high speed/ high power machining centers and of some other machine tools are frequently directly driven by an electric motor whose rotor is the spindle itself. In such cases, the spindle is loaded only by the cutting force, Fig. 6.5.2a, (and also by the driving torque from the motor). However, some machine tools have a driving gear or a pulley located between the spindle bearings. In such cases, two external forces are acting on the spindle: cutting force, P, and the radial force from the pulley or the resultant of the tangential and radial forces from the gear (Q, Fig. 6.5.2b). Radial deflection of the spindle at the application point of the cutting force P considering deformation of the spindle itself and its bearings, and as-

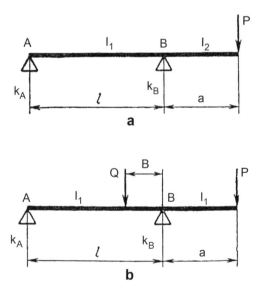

Fig. 6.5.2. Typical loading schematics of spindles.

suming one-step structure of the spindle (constant diameter between the bearing supports and another constant diameter of the overhang section), is:

$$\Delta = P\left[\frac{la^2}{3EI_1} + \frac{a^3}{3EI_2} + \frac{1}{k_B}\frac{(l+a)^2 + a^2(k_B/k_A)}{l^2}\right] \qquad (6.5.1)$$

and angular deflection at the front bearing from the cutting force P is

$$\theta_B = Pal/3EI \qquad (6.5.2)$$

Here I_1, I_2 are cross-sectional moments of inertia of the section between the bearings, of the overhang section, respectively; k_A, k_B are stiffnesses of the rear and front bearing units.

If the driving gear is placed between the bearings like in Fig. 6.5.2b, the total radial force, Q, acting on the spindle from the gear mesh or from the pulley is proportional to the cutting force, P, but has a different direction than P. Accordingly, both P and Q should be resolved along coordinate directions x (vertical) and y (horizontal). The radial and angular components of deformations of the spindle at the application point of the cutting force are:

$$\Delta_{x,y} = P_{x,y}\left[\frac{la^2}{3EI_1} + \frac{a^3}{3EI_2} + \frac{1}{k_B}\frac{(l+a)^2 + (k_B/k_A)a^2}{l^2}\right] +$$

$$+ Q_{x,y}\left[\frac{1}{k_B}\frac{(l+a)(l-b) - (k_B/k_A)ab}{l^2} - \frac{a}{6EI_1l}(b^3 + 2l^2b - 3lb^2)\right],$$

$$(6.5.3)$$

$$\theta_B = \frac{1}{3EI_1}\left[P_{x,y}al - \frac{Q_{x,y}}{2l}(b^3 - 2l^2b - 3lb^2)\right], \text{ rad.} \qquad (6.5.4)$$

It can be seen from Eqs. (6.5.1) to (6.5.4) that distance l between the spindle bearings is critical for its overall stiffness (i.e., for deformation Δ at the given cutting force, P). Fig 6.5.3b illustrates this statement for a typical spindle unit in Fig. 6.5.3a supported by two double-row roller bearings 3182100 series and having a given overhang length, a. In a general case, optimization of ratio $K = l/a$ is important for improving

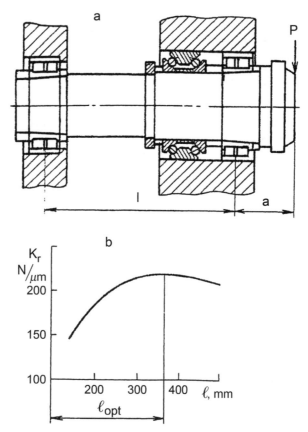

Fig. 6.5.3. Radial stiffness as function of distance between bearing supports (*b*) for spindle unit (*a*).

performance of spindle units. Similar calculations for spindles having non-uniform cross-sections can be easily performed by computer, e.g., [12].

Special attention must be given to reducing the overhang length, *a*. It can be achieved by a careful design of both the spindle and the tool holding/part holding devices which constitute an extension of the spindle and determine the "effective overhang," which can be defined as the distance between the application point of the cutting force and the front bearing. It can be done by making toolholders of a solid design rather than hollow to reduce overhang of the tool. This statement is illustrated by comparing a solid steep (7/24) taper toolholder/spindle interface, Fig. 6.5.4a, and hollow HSK interface, Fig. 8.4.15 [13]. The ultimate stiffening can be

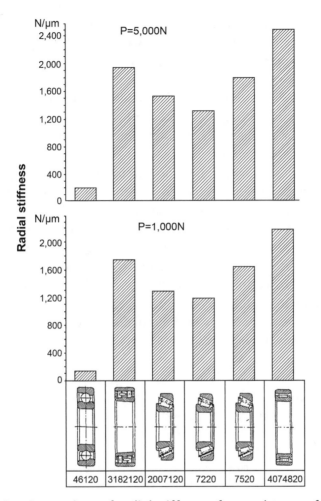

Fig. 6.5.4. Comparison of radial stiffness of several types of bearings
($d = 100$ mm).

achieved by integrating the tool with the spindle or by integrating the part holding chuck with the spindle (there are some commercial designs of lathes using this approach).

6.5.2 Stiffness and Damping of Antifriction Bearings

While stiffness of bearings for critical applications, such as for machine tool spindles, is a critical parameter, it is not the only critical parameter to be considered by the designer. The other very important parameters are the allowable speed (rpm or "dn" where d = diameter of the bore, mm, and n = rotational speed, rpm) and the heat generation. Fig. 6.5.4 [14] compares radial stiffness of different types of anti-friction bearings with d = 100 mm at two radial loads. While the angular contact ball bearing (contact angle, β = 12 deg.) has the lowest stiffness, it is widely used for spindles of high precision and/or high-speed machine tools.

Radial stiffness of both roller and ball bearings depends significantly on their preload (see also Chapter 3). Stiffness of double-row spindle roller bearings of 3182100 type is shown in Fig. 6.5.5 as a function of preload. The radial preloading is achieved by axial shifting of the inner race, having a tapered bore, along the tapered spindle. When the deformation e caused by the preload exceeds $2\delta_r$, where δ_r is the initial radial clearance in the bearing, its stiffness does not change significantly with further increasing of the preload force. The reason for this phenomenon is a rather weak nonlinearity of the radial load-deflection characteristic of the roller bearings (see Chapters 3 and 4).

Determination of radial stiffness of angular contact ball bearing is a more involved process due to their more pronounced nonlinearly and also, due to complexity of their geometry. It is suggested in [14] that the following algorithm provides reasonably accurate data:

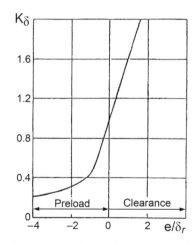

Fig. 6.5.5. Compliance coefficient K_δ for double-row spindle roller bearings 318 2100 and 4162900 as a function of clearance/preload.

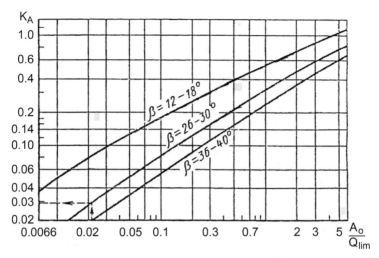

Fig. 6.5.6. Axial compliance factor K_A for determining compliance of angular contact ball bearings.

a. Axial compliance factor, K_A, characterizing the relative axial shift of the races under the axial preload force, A_o, is determined from Fig. 6.5.6 as a function of relative preload, A_o/Q_{lim}, and the contact angle, β; Q_{lim} is the static load capability of the bearing.

b. Radial compliance coefficient, K_r, is determined from Fig. 6.5.7, where P is the radial load on the bearing.

c. Mutual elastic approaching of the races is calculated as:

$$d'_r = 0.03 d_b K_R \qquad (6.5.5)$$

where d_b is diameter of balls in the bearing.

d. Radial stiffness of the bearing itself is:

$$k_R = P/\delta'_r; \qquad (6.5.6)$$

the overall stiffness of the support is:

$$k_s = P/(\delta'_r + \delta''_r), \qquad (6.5.7)$$

where δ''_r is the total sum of contact deformations between the bearing and the housing and between the bearing and the shaft from Eq. (3.4.8). It is important to remember that thermal deformation

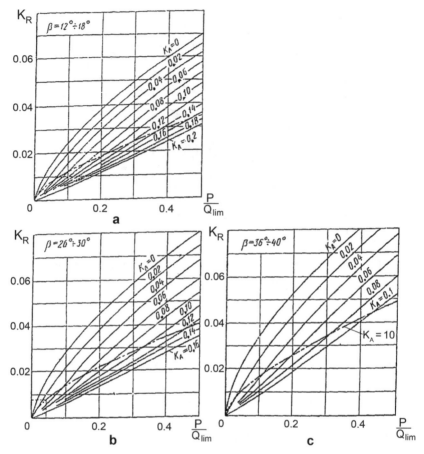

Fig. 6.5.7. Radial compliance factor K_R for determining compliance of angular contact ball bearings. a β = 12 deg. to18 deg.; b β = 26 deg. to 30 deg.; c β = 36 deg. to 40 deg.

of the spindle and centrifugal forces cause changes of the preload forces and of the effective contact angles, β.

The preloading process of angular contact ball bearings is accompanied by increasing of the actual contact angle. Fig. 6.5.8 [15] shows incremental increases $\Delta\beta$ of the contact angle for angular contact bearings with nominal contact angles β_0 = 15 deg. and 26 deg. as a function of $Q_{pr}/z\,d_b^2$, where Q_{pr} is the preload force, z = number of balls, d_b = diameter of the ball. It can be seen that the larger contact angle changes are developing for smaller nominal contact angles. Selection of the preload

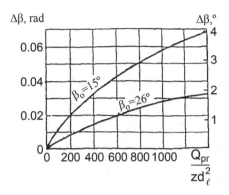

Fig. 6.5.8. Change of actual contact angle versus preload force.

magnitude depends on the load acting on the bearing. Fig. 6.5.9 shows radial stiffness, k_r, of angular contact bearings as a function of preload force, Q_{pr}, and external force, P. The line for $Q_{pr} = 440$ N is typical for the case when all the balls are accommodating force P. Falling pattern of the lines is due to the gradual reduction of numbers of the loaded balls.

An important phenomenon for angular contact ball bearings at high-speed regimes ($dn > \sim 0.5 \times 10^6$ mm rpm) is development of high centrifugal

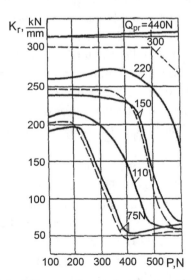

Fig. 6.5.9. Radial stiffness of angular contact ball bearings 36100 (solid lines) and 36906 (broken lines) with contact angle $\beta = 15^0$ as function of external force, P, and preload force Q_{pr}.

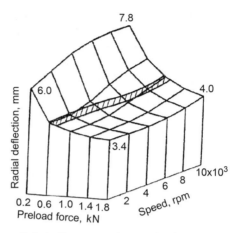

Fig. 6.5.10. Radial deflection of spindle flange (journal diameter, d = 110 mm) under radial load 1000 N as function of rpm and preload force.

forces and gyroscopic moments acting on the balls. These forces are becoming commensurate with the external forces and the preload forces. The centrifugal forces are pressing the balls toward the outer races, thus changing both the effective contact angles and the kinematics of the balls as well as redistributing contact loads in the bearing. These factors lead to reduction of stiffness. Fig. 6.5.10 gives radial deflection of the flange of the spindle with the journal diameter 110 mm supported by two pairs of angular contact ball bearings (in the front and in the rear bearing supports) and loaded by force P = 1000 N. It can be seen that the deflection is increasing (thus, stiffness is decreasing) for the same preload magnitude with increasing rpm. The rate of the deflection increase is steeper when the preload force is lower.

The centrifugal forces can be significantly reduced by using lighter ceramic balls instead of steel balls. This alleviates the stiffness reduction at high rpm and rises the speed limit of the bearings. Fig 6.5.11 [16] shows dependence of radial stiffness of angular contact bearings (d = 50 mm) on rotational speed for steel (a) and ceramic (b) balls. The initial (static) stiffness of the bearings with ceramic (silicon nitride) balls is higher than for the bearing with steel balls due to greater Young's modulus of the ceramic. A more gradual stiffness reduction with increasing rpm is due to much lower (~2.5 times) density and thus, magnitude of centrifugal force

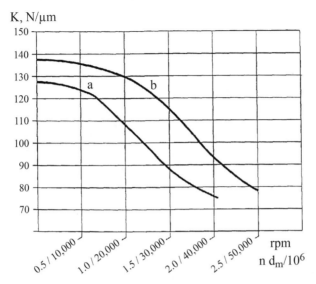

K, N/μm

Fig. 6.5.11. Radial stiffness of angular contact ball bearing ($d = 50$ mm) vs. speed. a Steel balls; b Silicon nitride balls. Abscissa units: rpm (n)/p, where $p = dn/10^6$.

for the ceramic balls. A similar effect can be achieved by using bearings with hollow rollers (Section 8.5) or balls.

In some applications an important factor is *axial stiffness* of spindle bearings which may significantly influence accuracy and dynamic stability of the machining system. While the share of the radial compliance of the spindle bearings in the overall spindle compliance is usually not more than 40% to 60%, the axial stiffness (compliance) of the spindle unit is completely determined by the bearings. The axial stiffness is noticeably influenced by inaccuracies of components of the bearings, especially at low loads.

Since the finishing operations generate very low loads, influence of bearing inaccuracies (such as non-uniformity of race thickness, dimensional variation and non-ideal sphericity of balls, non-perpendicularity of the spindle and headstock faces to their respective axes, etc.) must be considered. Fig. 6.5.12 [14] provides information on axial stiffness of high precision thrust ball bearings (Fig. 6.5.12a), and angular contact ball bearings with contact angle 26 deg. (Fig. 6.5.12b), as functions of preload force, Q_p.

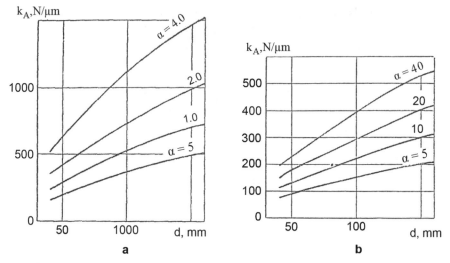

Fig. 6.5.12. Axial stiffness k_a of ball bearings. *a* Thrust bearings of 8100 and 8200 series; *b* Angular contact bearings of 46100 series. Preload force, $Q_p = \alpha d$, N, where *d* = bore diameter in mm.

It is important to understand that stiffness of the spindle unit is determined not only by stiffness of the bearings but also by seemingly minor design issues. Fig. 6.5.13 [14] shows influence of positioning of the bearing relative to the supporting wall on its stiffness. The highest resulting stiffness develops when the outer race of the bearing is fit directly into the housing, without any intermediate bushing, and is symmetrical relative to the median cross-section of the wall (1). Such design techniques as use of an intermediate bushing (2), and especially an asymmetrical installation with a small cantilever of the bearing (3) lead to significant deterioration of stiffness as well as service life of the bearing.

Damping in bearings for spindles and for other critical shafts plays an important role in dynamic behavior of the system. As in many other mechanical systems, various design and performance factors such as fits between the bearing and the shaft/housing, preload, lubrication, rotational speed, etc., influence both stiffness and damping. Frequently, these factors result in very different, sometimes oppositely directed stiffness and damping changes. A survey of three major studies of stiffness and damping of rolling element bearings was presented in [15].

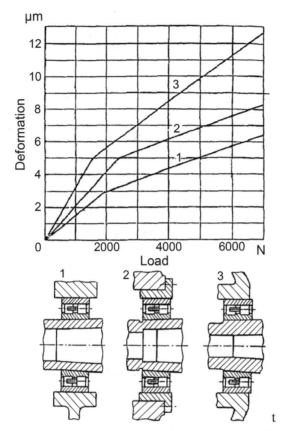

Fig. 6.5.13. Influence of location of the front spindle bearing on compliance of the support.

It was found that stiffness and damping of both roller and ball bearings are significantly dependant on their rotational speed. Fig. 6.5.14 shows stiffness, damping coefficient, and steady state temperature for a grease-lubricated tapered roller bearing for two mounting conditions (0 μm and 5 μm clearance). It can be seen that stiffness, k_L, is increasing with rpm, more for the bearing with 5 μm initial clearance. Although the initial stiffness for the bearing with 5 μm clearance was ~50% lower than for 0 μm clearance, this difference is decreasing with increasing rpm, probably due to thermal expansion causing preload development in both bearings. On the other hand, damping coefficient c_L is decreasing with rpm up to 1600 rpm to 1700 rpm. Actual reduction of damping is even more pronounced than the plot in Fig. 6.5.14b illustrates, since

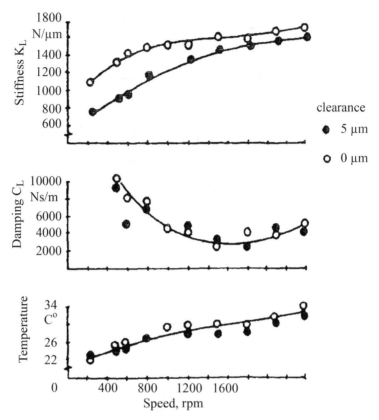

Fig. 6.5.14. Variation of stiffness and damping with speed for tapered roller bearing NN3011 KUP lubricated with grease Arcanol L78.

the damping parameter important for applications is not c_L but log decrement δ related to c_L by Eq. (A1-8). For the same mass m of the vibrating parts and same c_L, δ is decreasing with increasing k_L. This effect results in a smaller increase in δ above 1600 rpm than shown for c_L in Fig 6.5.14b.

Similar results are shown in Fig. 6.5.15 for a heavily preloaded angular contact ball bearing (internal deformation by the preload force 25 μm) with low viscosity (AWS 15) and high viscosity (AWS 32) lubricating oil. Damping is expressed in degrees of phase shift, β, between the exciting force and the resulting deformation under vibratory conditions. Since $\delta = \pi \tan \beta$ and β does not exceed 7 deg., the relative damping changes are proportional to the phase angle readings. Fig. 6.5.15 demonstrates the same correlations as Fig. 6.5.14: stiffness is increasing, while damping is

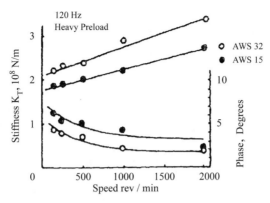

Fig. 6.5.15. Stiffness (a) and damping (b) of angular contact ball bearings.

decreasing with increasing rpm. While the dynamic stiffness is increasing with increasing viscosity of lubricating oil (due to more effort required for squeezing the viscous oil from contact areas under load), damping is increasing with decreasing viscosity (due to more intense motion of oil in the contacts, thus higher friction losses).

Fig. 6.5.16 demonstrates that stiffness of an angular contact ball bearing is decreasing, while damping is increasing with increasing amplitude of vibratory force. These effects are in full compliance with the dependencies derived in Section 4.6.1.

It is interesting to note that angular (tilt) stiffness, k_m, and damping coefficient for angular vibration, c_m, for tapered roller bearing are both

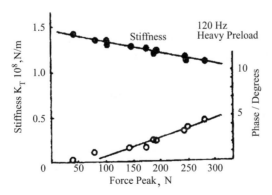

Fig. 6.5.16. Variation of stiffness and phase angle versus amplitude of excitation force.

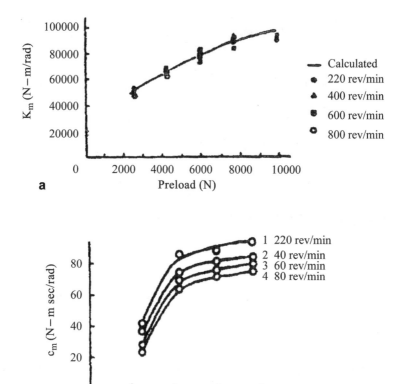

Fig. 6.5.17. Variation of tilt stiffness (a) and of damping coefficient (b) with preload force and speed.

increasing with increasing preload (Fig. 6.5.17). This effect is due to a different mechanism of the angular vibratory motions. These are mostly tangential motions between the rollers and the races, and it is obvious that both resistance to displacement (stiffness) and friction losses (damping) are increasing when the normal forces between the rollers and the races are increasing.

6.6 DAMPING IN POWER TRANSMISSION SYSTEMS

Damping in power transmission systems is determined by damping in the working medium of the driving motor [e.g., by damping of the electromagnetic field between stator and rotor, see Eqs. (6.2.31), (6.2.32)], by energy dissipation in joints (key and spline connections, shaft/bearing units, interference fits, etc.) and in special elasto-damping elements such as couplings, belts, etc. Energy dissipation in materials of the components can be neglected due to its small magnitude. Damping in joints is addressed in Section 4.7.

In many cases, the most important vibratory mode is the mode associated with the lowest natural frequency (the fundamental natural frequency). At this mode, amplitudes of dynamic torque M_{dyn} in the elastic connections can be assumed, in the first approximation, to be constant along the reduced system (see above, Section 6.3). In such cases, the system can be considered, also as a first approximation, as a two-mass system: inertia of the rotor of the driving motor–compliance of the connecting components–inertia of the output member. The latter can be a spindle with a tool in a machine tool, driving wheels of a wheeled vehicle, etc. Compliance of the connecting elements can be presented as $e = e_1 + e_2$. Here e_1 is "elastic compliance," not associated with a significant amount of energy dissipation; usually, it is the sum of the reduced compliances of torsion and bending of shafts. The other component, $e_2 = ae$, is "elasto-damping compliance," which is the sum of the reduced compliances of components whose compliance is due to contact deformations and due to deformations of polymeric materials (couplings, belts, etc.). Energy dissipation for one period of vibratory process (relative energy dissipation) ψ for systems having not a very high damping:

$$\psi = 2\,\delta \qquad (6.6.1)$$

where δ is log decrement of the vibratory process.

In a transmission system composed of shafts, gears, key and spline connections, bearings, etc. but not containing special damping elements such as couplings or belts, the damping is determined only by energy dissipation in joints, δ_j. Relative energy dissipation in joints is $\psi_j = 2\delta_j \approx 0.5 - 0.7$, see Section 4.7, and for the whole transmission system:

$$\psi_j = a\,\psi_j \qquad (6.6.2)$$

In a typical gearbox, compliance of key and spline connections is about 35%; considering also contact compliance of gear meshes and of bearing units, $a \approx 0.45$. Thus, $\psi \approx 0.45\psi_j = 0.22 - 0.31$ and log decrement for a gearbox is $\delta = \psi/2 = 0.11 - 0.15$. These values were confirmed experimentally (e.g., see Fig. 6.3.7b).

In more general cases, the fundamental mode is not as simple as it was assumed above, or there is a need to evaluate damping of higher vibratory modes. In such cases, differences of a and ψ for different segments of the system must be considered, as well as distribution of dynamic elastic moment, M_{dyn}, along the system. Distribution of M_{dyn} along the system is, essentially, a mode shape, which can be computed using a variety of available software packages. Parameters a and ψ for each segment (elastic element) of the system can be easily determined from the original model such as shown in Fig. 6.3.6a, in which the composition of compliance for each component is shown. For example, compliance, e_5, in Fig. 6.3.6a is composed of compliances of two key connections (28 and 11 x 10^{-7} rad/Nm) and one torsional compliance of a shaft, 124 x 10^{-7} rad/Nm. Thus, $a_5 = (28 + 11)/(28 + 11 + 124) = 0.24$. Since the key connections are not tightly fit in this gearbox, from Section 4.7.3, value for δ can be taken as an average between the tight fit key (0.05 to 0.1) and

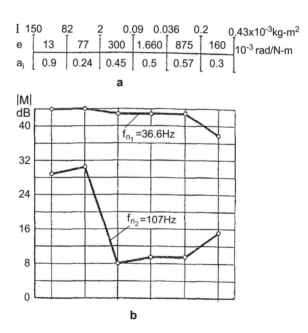

Fig. 6.6.1. Simplified dynamic model (a), and amplitudes of elastic torque (b) in the elastic elements at two lower natural frequencies for the dynamic system in Fig. 6.3.6d.

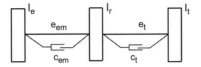

Fig. 6.6.2. Two-degrees-of-freedom model of transmission with driving motor.

the sliding spline (0.3 to 0.4), or $\delta \approx 0.18$ and $\psi \approx 0.36$. Values of a and ψ for other components can be determined in a similar way. After M_{dyn}, a, and ψ are known for each i-th segment of the computational model, the damping parameters for this mode can be calculated as:

$$\psi = 2\delta = \frac{\sum_i M_{dyn_i}^2 \psi_i a_i e_i}{\sum_i M_{dyn_i}^2 e_i} \tag{6.6.3}$$

After these parameters are calculated, resonance amplification factor for this mode can be determined as:

$$Q = \pi/\delta = 2\pi/\psi \tag{6.6.4}$$

Instead of M_{dyn_i}, relative values, M_{dyn_i} / M_{dyn_j}, can be used in Eq. (6.6.3).

Example. *Computing damping characteristics for two lower vibratory modes for gearbox of the vertical milling machine whose mathematical models are given in Fig. 6.3.6d, for n_{sp} = 60 rpm. The computation is performed for a simplified system shown in Fig. 6.6.1a, in which values of a_i are also given for each elastic segment. Simplification is performed using the method described in [1] and briefly in Section 8.6.1, which al-*

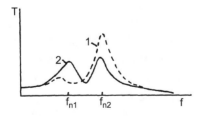

Fig. 6.6.3. Amplitude-frequency characteristic of a motor-driven transmission. 1 – weak coupling between mechanical system and motor field; 2 – strong coupling; f_{n2} — mechanical system–related natural frequency; f_{n1} — motor-related natural frequency.

Table 6.2. Relative dynamic elastic moment for Fig. 6.5.15.

Natural	No. of elastic segment					
Frequency	1	2	3	4	5	6
f_{n1}	0	0	−1.3	−13	−13	−6.0
	1.0	1.0	0.86	0.86	0.86	0.5
f_{n1}	0	+0.5	−21	−19.5	−19.5	−14
	1.0	1.2	0.09	0.106	0.106	0.2

lows to reduce the number of inertias and elastic elements (i.e., degrees of freedom) of a dynamic model without distorting its lower natural frequencies and modes. Vibratory modes of M_{dyn} for two lowest modes corresponding to natural frequencies, f_{n1} and f_{n2}, are given in Fig. 6.6.1b in logarithmic scale. From the plots in Fig. 6.6.1b, values of M_{dyn_i}/M_{dyn_1} can be easily computed. These values are given in Table 6.2 both in dB (upper lines for each f_n) and in absolute numbers (lower lines). Using data from Table 6.2 and values of a from Fig. 6.6.1a, damping can be easily calculated. Two cases are compared below. In the first case (A), ψ_i are assumed constant for each segment of the model in Fig. 6.6.1a, $\psi_i = 0.6$; in the second case (B), damping of coupling ($e_1 = 13 \times 10^{-8}$ rad/Nm in Fig. 6 5.15a) is increased to $\psi_1 = 1.0$. After substituting all the parameters into Eq. (6.6.3):

(A) At natural frequency f_{n1}, $\psi_{f_1} = 0.3$; and at f_{n2}, $\psi_{f_2} = 0.21$;
(B) At natural frequency f_{n1}, $\psi_{f_1} = 0.306$; and at f_{n2}, $\psi_{f_2} = 0.24$.

These are very valuable results. They show that for this system, the overall system damping at the fundamental natural frequency is significantly higher than at the second natural frequency. They also show that a substantial increase of damping in the coupling does not noticeably influence damping at the fundamental mode (2% increase), while increasing damping at the second mode increased by about 15%. These results would be very different for other configurations of the dynamic system. For example, for higher n_{sp} for the system in Fig. 6.3.6, the coupling compliance constitutes a larger fraction of the overall compliance, thus increase of damping in the coupling would increase the system damping at the fundamental mode much more significantly. This statement was validated by experimental results for high damping coupling tested at $n_{sp} = 375$ rpm, which are described in Appendix 5.

REFERENCES

[1] Rivin, E. I., 1980, "Compilation and Compression of Mathematical Model for a Machine Transmission," ASME Paper 80-DET-104, ASME.

[2] Levina, Z.M., 1972, "Ball-Spline Connections," in Components and Mechanisms for Machine Tools, ed. by D.N. Reshetov, Mashinostroenie Publish. House, Moscow, vol. 2, pp. 334–345 [In Russian].

[3] Rivin, E.I., "Key Connection", U.S. Patent 4,358,215.

[4] Rivin, E.I., Tonapi, S., 1989 "A Novel Concept of Key Connection", Proc. of Intern. Power Transmission and Gearing Conf., ASME.

[5] Iwatsubo, T., Arri, S., and Kawai, R., 1994 "Coupled Lateral-Torsional Vibration of Rotor System Trained by Gears," *Bulletin of the JSME*, February, pp. 224–228.

[6] Wang, J., Howard, I., 2004, "The Torsional Stiffness of Involute Spur Gears", Proceed Instn Mech. Engrs, vol. 218, Part C, pp. 131–142.

[7] Rivin E.I., 1988, "Mechanical Design of Robots," McGraw-Hill, N.Y., 325 pp.

[8] Rivin, E.I., 1980, "Role of Induction Driving Motor in Transmission Dynamics," ASME Paper 80-DET-96, ASME.

[9] Asada, H., Kanade, T., and Takeyama, I., 1982, "Control of Direct-Drive Arm," in Robotics Research and Advanced Applications, ASME, New York.

[10] Tooten, K., et al, 1985, "Evaluation of Torsional Stiffness of Planetary Transmissions", Antriebstechnik, vol. 24, No. 5, pp. 41–46 [in German].

[11] Tobias, S.A., 1965, "Machine Tool Vibration", Blackie, London, 351 pp.

[12] Levina, Z.M., Zwerev, I.A., 1986, "Computation of Static and Dynamic Characteristics of Spindle Units Using Finite Elements Method", Stanki i Instrument, No.10, pp. 7–10 [in Russian].

[13] Rivin, E. I., 2000, "Tooling Structure — Interface between Cutting Edge and Machine Tool", *Annals of the CIRP*, vol. 49/2, pp. 591–634.

[14] Figatner, A.M., 1972, "Antifriction Bearing Supports for Spindles" in Components and Mechanisms for Machine Tools, ed. by D.N. Reshetov, Mashinostroenie Publish. House, Moscow, vol. 2, pp. 192–277 [in Russian].

[15] Stone, B.J., 1982, "The State of the Art in the Measurements of the Stiffness and Damping of Rolling Element Bearings", Annals of the CIRP, vol. 31/2, pp. 529–538.

[16] "Bearings for High Speed Operation", Evolution, 1994, No. 2, pp. 22-26

Design Techniques for Reducing Structural Deformations (Stiffness Enhancement Techniques)

$\mathbf{P}$revious chapters have addressed various correlations between design features of structures and their stiffness. While it is rather obvious how to use these correlations for enhancement of the structural stiffness, it is also useful to emphasize some very effective design techniques for stiffness enhancement. This usefulness justifies, in the author's opinion, some inevitable repetitions. In some cases, e.g., Section 7.6 and Appendix 2, issues of stiffness, damping, and mass cannot be separated.

To improve cost-efficiency of stiffness enhancement process, it is important that a breakdown of compliance for the system is performed, computationally (like in Sections 6.3 and 6.4) or by testing (like in Appendix 7). The breakdown allows to select the most compliant (the least stiff) component. Stiffness enhancement of the most compliant components is the most effective and cost-effective way to enhance stiffness of the system.

7.1 STRUCTURAL OPTIMIZATION TECHNIQUES

In many structures, there are critical directions along which deflections must be minimal (i.e., stiffness must be maximized). Frequently, these critical directions relate to angular deformations, usually caused by bending or by angular displacements in joints. Angular deformations are more dangerous, since even small angular deformations may result in large linear deformations if the distance from the center of rotation to the measurement point is significant.

Angular deformations are naturally occurring in non-symmetrical structures, such as so-called "C-frames" Fig. 7.1.1a. C-frame structures are frequently used for stamping presses, drill presses, welding machines, measuring systems, etc. They have an important advantage of easy access to the work zone. However, C-frame machines exhibit large deformations ("opening of the frame") under the work loads. In drill presses, such deformations may cause non-perpendicularity of the drilled holes to the face surfaces; in stamping presses, they may cause non-parallelism of upper and lower dies and other distortions that result in a fast wear of stamping punches and dies as well as in poor quality of parts. To reduce these undesirable effects, the cross-sections of the structural elements (thus, their weights) are "beefed up". Of course, stiffness is increasing, but the rate of performance is often decreasing.

A very effective way of enhancing stiffness of machine structures is to replace C-frames with two-column ("gantry" frames), Fig. 7.1.1b, or with

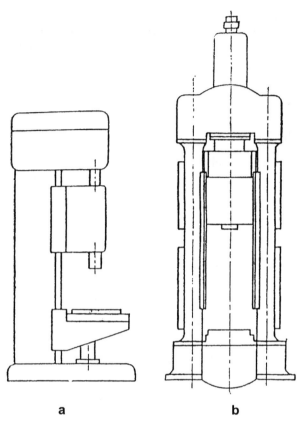

Fig. 7.1.1. Typical schematics of machine frames; (a)-C-frame; (b)-Gantry frame.

three- to four-column systems. Such symmetrical architecture minimizes both the deformations and their influence on part accuracy and on tool life. The difference between maximum deformations δ_{cant} of the overhang (cantilever) cross beam of a C-frame structure (radial drill press, Fig. 7.1.2a and of the double-supported cross beam, $\delta_{d.s.}$, of a gantry machine tool is illustrated in Fig. 7.1.2b.

Stiffness is a very important parameter in machining operations, since relative deformations between the part and the tool caused by the cutting forces are critical components of machining errors. In the case of a cylindrical OD grinder in Fig. 7.1.3a, both the part 1 and the spindle of the grinding wheel 2 can deflect under the cutting forces. On the other hand, in the centerless grinding process in Fig. 7.1.3b, the part 1 is sup-

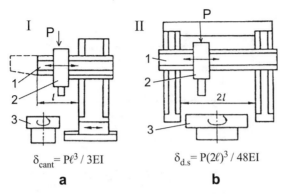

$$\delta_{\text{cant}} = P\ell^3 / 3EI \qquad\qquad \delta_{\text{d.s}} = P(2\ell)^3 / 48EI$$

a   **b**

Fig. 7.1.2. Deformations of C-frame (a) and gantry (b) machine frames.

ported between the grinding wheel 2, supporting wheel 4, and stationary steady rest 3. As a result, the part deformations are reduced, thus greatly improving cylindricity of the part.

In many designs, critical deformations are determined by combinations of several sources. For example, deformation at the end of a machine tool spindle is caused by bending deformations of the spindle as a double-supported beam with an overhang, and by contact deformations of the bearing supports. Changing geometry of the system may influence these sources of deformations in totally different ways. For example, changing spans of the constant cross-section spindle in Fig. 7.1.4 would change its bending deformations, but also changes reaction forces in the supports (thus, their deformations), and also modifies the influence of

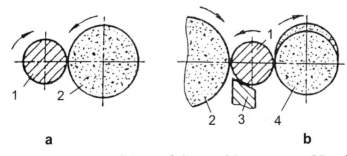

a   **b**

Fig. 7.1.3. Support conditions of the machine parts on OD grinder (a) and centerless grinder (b).

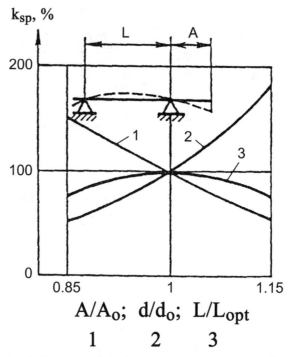

Fig. 7.1.4. Influence of design parameters on spindle stiffness.

the support deformations on deflections at the end. The situation is even more complex for actual spindle designs, which are characterized by varying cross-sections along the axis. These cross-sections and the bearing sizes would be changing if the overall geometry (e.g., distance between the supports) is changing. Thus, optimization of the spindle stiffness becomes a complex multi-parametric interactive problem. While there are many attempts on solving it, e.g. [1], the optimization of the spindle designs still requires a combination of computational results and expertise of the designers (see also Section 6.5.1).

A very powerful design technique resulting in enhancement of the effective structural stiffness is use of rational loading patterns. One approach to the rationalization of the loading pattern (load distribution) is use of supporting/load-bearing devices providing continuous distributed load rather than concentrated forces. For example, hydrostatic guideways provide more uniform load distribution than rolling friction guideways, thus reducing local deformations. Another advantage of hydrostatic guideways is their self-adaptability; positioning accuracy of heavy parts mounted on hydrostatically supported tables can be corrected by monitoring/controlling oil pressures in each pocket.

Another design technique resulting in a desirable load distribution is illustrated in Fig. 7.1.5 [2], showing a set-up for machining a heavy ring-shaped part 1 on a vertical boring mill. If the part were placed directly on table 3, the table would deform as shown by the broken line. The deformation can be significantly reduced by use of an intermediate spacer (supporting ring) 4 with extending arms 2. Ring 4 applies its weight load between guideways 5 and 6, thus eliminating the moment loading of the table 3 on its periphery and reducing its bending deformations.

Fig. 7.1.6 shows a set-up for machining (turning and/or grinding) of a crankshaft having relatively low torsional stiffness. The *effective stiffness* of the crankshaft is enhanced (twisting of the shaft caused by the cutting forces is reduced) by driving the part from two ends using synchronized drivers (bevel gears and worm reducers). This technique was suggested and patented in the U.S. in the beginning of the 20th century and is still widely used on crankshaft grinding machines.

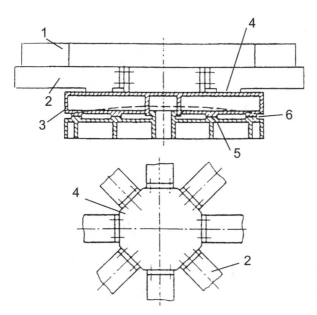

Fig. 7.1.5. Enhancement of effective bending stiffness of rotating round table 3 of a vertical boring mill by intermediate spacer 4 with arms 2, for machining oversized part 1.

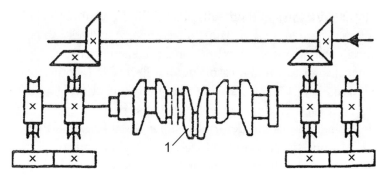

Fig. 7.1.6. Compensation of low torsional stiffness of crankshaft 1 by driving from both ends.

Other systems in which changes in support conditions and load distribution result in reduced deformations and thus, in enhancement of effective stiffness, are shown in Figs. 7.1.7 and 7.1.8 [3]. Both figures show components of power transmission systems. Fig. 7.1.7, similar to Fig. 5.1.5, compares methods of mounting a pulley on the driven shaft. In an embodiment shown in Fig. 7.1.7a, pulley 3 is mounted on the cantilever extension of shaft 4; the deformations are relatively large and stiffness at the pulley attachment point is low. In Fig. 7.1.7b, the loading vector is shifted and passes through the shaft bearing. Thus, although pulley 1 is mounted on the cantilever shaft segment 2, the deformations are significantly reduced as if it were not a cantilever. The highest effective stiffness is in the design of Fig. 7.1.7c, wherein pulley 5 is supported not by the shaft, but by housing 6. The shaft is not subjected to any radial forces but

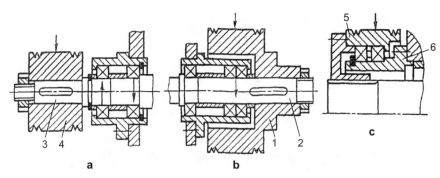

Fig. 7.1.7. Alternative designs of cantilever shafts.

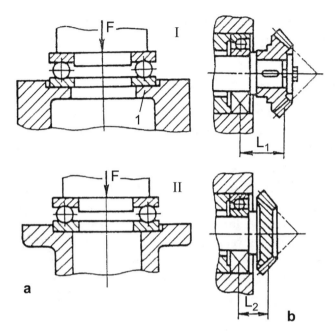

Fig. 7.1.8. Design techniques for avoiding (a) and reducing (b) cantilever loading.

only to the torque, thus the radial stiffness is determined not by the shaft deformations, but by much smaller bearing deformations.

Fig. 7.1.8 emphasizes importance of avoidance or, at least, shortening of loaded overhang segments. In case *aII*, load from the thrust bearing is transmitted through the massive wall whose deformation is negligible, while in *aI*, this load inflicts bending of a relatively thin overhang ring. Fig. 7.1.8b*II* shows how a minor design change — making the bevel gear integral with the shaft — significantly reduces overhang of the gear compared with design in Fig. 7.1.8b*I* in which the gear is fit onto the shaft. Design II is more expensive to manufacture but is much stiffer.

7.2 COMPENSATION OF STRUCTURAL DEFORMATIONS

Structural deformations caused by weight of the components as well as by payloads (e.g., cutting forces) can be reduced by passive and/or active (servo-controlled) systems means. Reduction of the structural deformations is equivalent to enhancement of structural stiffness of the system.

7.2.1 Passive Compensation Techniques

Fig. 7.2.1 shows the work zone of a thread-rolling machine in which two roll dies, 1 and 2, are generating thread on blank 3. To achieve a quality thread, dies 1, 2 must be parallel. However, due to high process loads, initially parallel dies would develop angular misalignment due to structural deformations of the die-holding structure, thus resulting in the tapered thread. To prevent this undesirable effect, the structure is preloaded/predeformed by stud 4 during assembly to create an oppositely tapered (wedged) clearance between the roll dies. The rolling force would make the dies parallel, thus resulting in an accurate cylindrical thread.

In tapered toolholder/spindle interfaces (Fig. 7.2.2), the standard tolerances on the angle of the spindle female taper and on the angle of the toolholder male taper are assigned in such a way that the taper angle of the spindle hole is always smaller than the toolholder taper angle, (International Standard ISO 1947). Pulling the toolholder into the spindle reduces the difference in the angles and guarantees clamping at the front

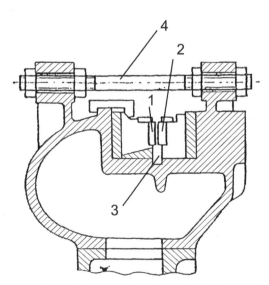

Fig. 7.2.1. Work area of a thread-rolling machine.

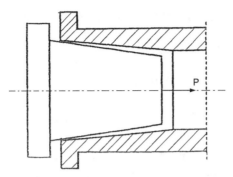

Fig. 7.2.2. Standardized fit of tapered interfaces *(exaggerated).*

part of the connection, thus reducing the effective tool overhang and increasing stiffness (also see Chapter 4).

Many large machine tools and other production machines develop large deformations under weight forces. Since relative positions of the heavy units are not constant (moving tables and carriages, sliding rams, etc.), the weight-induced deformations frequently disrupt the normal operation. It is important to note that, in many cases, the weight-induced deformations cannot be reduced by just "beefing up" the parts. In a radial drill press in Fig. 7.2.3, deformation of the cantilever cross beam 1 caused by weight W of the spindle head 2 is:

$$\delta = Wl^3/3EI. \tag{7.2.1}$$

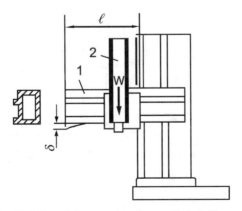

Fig. 7.2.3. Schematic of a radial drill press.

If all dimensions of the machine are scaled up by factor K, then:

$$l_1 = Kl; I_1 = K^4I; W_1 = K^3W, \text{ and } \delta_1 = K^3W\,K^3l/3EK^4I = K^2\delta. \qquad (7.2.2)$$

Thus, a straightforward "beefing up" did not reduce deformations, just the opposite, and more sophisticated design approaches are necessary to achieve the required stiffness. Recently, scaled down machine tools are becoming popular for making high accuracy parts, since in smaller machine tools, the weight-induced deformations are reduced (e.g., [4]).

Deformations in large machines can be very significant. Deformations in a heavy vertical boring mill in Fig. 7.2.4 under weight loads are $\delta_1 = 1.25$ mm (0.05 in.); $\delta_2 = 1.0$ mm (0.04 in.). Deformation of the table under the combined load of cutting force and weight load, $F = 3000$ KN (650,000 lb), is $\delta_3 = 0.05$ mm (0.002 in.).

Many techniques are used to reduce these deformations and, consequently, to enhance the effective stiffness of large machines. A simple and effective technique is "forward compensation" of the potential deformations by intentional pre-distortion of shape of the guideways (Fig. 7.2.5). Here guideways on cross beam 1 for spindle head 2 are made convex

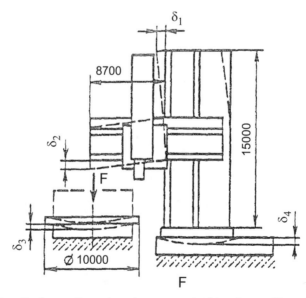

Fig. 7.2.4. Deformations in a large vertical boring mill under weight and cutting forces.

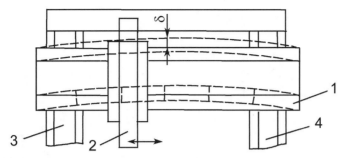

Fig. 7.2.5. Compensation of weight-induced deformations of cross beam by intentional distortion of its original shape.

by a pre-computed increment, δ. Even more effective approach is to use a light material (e.g., hollow structural elements filled with a metallic foam) for critical and/or moving structural parts (e.g., see [5]).

Tolerances on straightness of the long guideways are always assigned to be toward generating a convex shape in the middle of the guideway. Guideways on cross beams have to be machined, while the crossbeam is in place, is attached to columns, and is preloaded. Then, horizontal guideways would stay flat under the gravity force, otherwise they would sag.

There are many fabrication techniques for creating "pre-deformed" frame parts. The most straightforward technique is scraping. A highly skilled operator can relatively easily scrape the required convex or concave profile, only slightly deviating from a flat surface.

While these techniques are "rigid" ones which do not allow for an easy adjustment of the degree of compensation, the techniques illustrated in Figs. 7.2.6 to 7.2.9 [2], [6] are more "flexible," since they provide for adjustments. In Fig. 7.2.6a, the structural member 1 is made hollow. Inside member 1, there are two auxiliary beams 2 of significant rigidity. Each beam 2 can be deformed in the positive and negative z directions by tightening bolts 3 or 4, respectively. Deformation of beams 2 by bolts 3, 4 causes an oppositely directed deformations of structural member 1. A similar system is shown in Fig. 7.2.6b. It has only one auxiliary beam 2, but adjusting bolts 3, 4, 5 can introduce relative deformations between the structural member 1 and the beam 2 in two directions, x and z, and thus, member 1 can be "pre-distorted" in two directions.

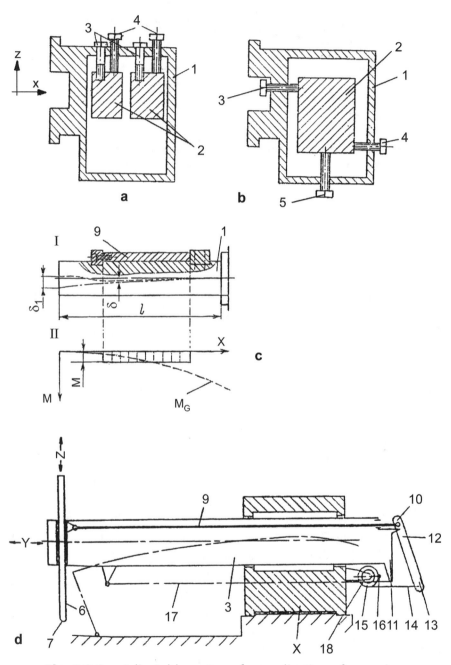

Fig. 7.2.6. Adjustable systems for application of opposing bending moments.

Pre-distortion of cantilever structural member 1 in Fig. 7.2.6c is achieved by tensile loading of auxiliary rod 2, which causes compensation of the weight-induced deflection δ.

A similar system in Fig. 7.2.6d [6] enhances effective stiffness of ram 3 of a coordinate measuring machine. Ram 3 carries quill 6 with measuring probe 7, thus the measurement accuracy is directly dependent on its deformations. Bending of ram 3 caused by its weight is changing with the changing overhang. The changing deformation is compensated by correcting bar 9 placed above the neutral plane of ram 3 and attached to the latter next to quill 6 location. Tensioning of bar 9 causes upward bending of ram 3. To generate the required tensile force, the opposite end of bar 9 is attached to the small arm 10 of lever 12, which is pivoted on the back side 11 of ram 3. Strained wire 14 is attached to the large arm 13 of lever 12 and wrapped on the large diameter 16 of double roller 15 fastened to housing 1. Another strained wire 17 is attached to ram 3 and wrapped on the smaller diameter 18 of roller 15. When overhang of ram 3 is increasing, wire 17 is rotating roller 15, thus forcing wire 14 to pull lever 12, which in turn applies tension to correction bar 9 and straightens ram 3.

The system in Fig. 7.2.7 is using counterbalancing weight 4 connected with moving in and out heavy cantilever ram 1 via cables 2 and 3. Cam

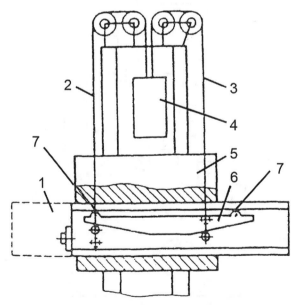

Fig. 7.2.7. Position-dependent weight counterbalancing device.

6 attached to ram 1 is engaged with cables 2 and 3, and transmits load from counterweight 4 to ram 1 via two supporting legs 7. Motion of ram 1 is accompanied by a pre-calculated redistribution of reaction forces between legs 7 thus preventing sagging of ram 1 under weight loads.

The above systems are mechanically operated structures. Also, hydraulic compensating systems applying adjustable forces to the structural members are frequently used. The system in Fig. 7.2.8 has a hydraulic cylinder-piston unit 2 applying counterbalancing force to the heavy spindle head 3. Pressure regulator 1 can be set to compensate the weight load of the spindle head 3 or can vary the counter-balancing force depending on the specified parameters, such as the cutting force. A similar system for a gantry machine tool is shown in Fig. 7.2.9. In this case, hydraulic cylinder 1 "transfers" the weight load of moving heavy carriage 3, which contains the spindle head, from crossbeam 2 to the auxiliary frame 4.

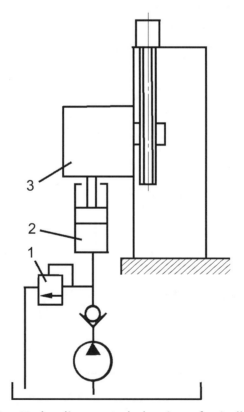

Fig. 7.2.8. Hydraulic counterbalancing of spindle head 3.

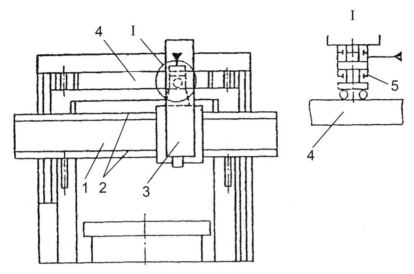

Fig. 7.2.9. Hydraulic system for transference of spindle head 3 weight to auxiliary frame 4.

While the hydraulic systems shown in Figs. 7.2.8 and 7.2.9 are adjustable passive systems, they can be easily transformed into active systems, if sensors and control systems were added (see Section 7.2.2).

A rational design of bearing supports for power transmission shafts can beneficially influence their computational models, thus resulting in significant stiffness increases. Fig. 7.2.10a and Fig. 7.2.10b [7] show two design embodiments of a bearing support for a machine tool spindle. In both cases, two identical bearings are used, the only difference being the

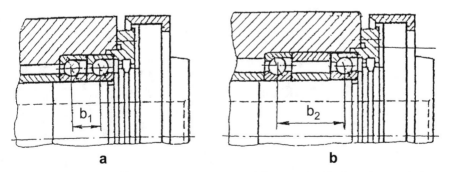

Fig. 7.2.10. Stiffness enhancement of front spindle bearing unit in (a) by a minor design change (b).

distance between the bearings. In the design b, the distance is larger ($b_2 > b_1$), which would create a "built-in" effect for bending of the spindle by the cutting forces, while the design a should be considered as a simple support (low angular stiffness restraint). Thus, design b will exhibit a higher stiffness and a better chatter resistance.

While the design changes reducing both eccentricity of loading conditions and overhang are beneficial for stiffness enhancement, they have to be done cautiously. Fig. 7.2.11 [7] shows three designs of the front spindle bearing. The double row roller bearing is preloaded by tightening bolts 1 acting on preloading cover 2 and through cover 2 — on the front bearing. Stiffness at the spindle end in design a is higher than in design b due to closeness of the preloading bolts to the axis, $C < C_1$. It is also higher than in design c due to the shorter overhang, $a < a_1$. However, due

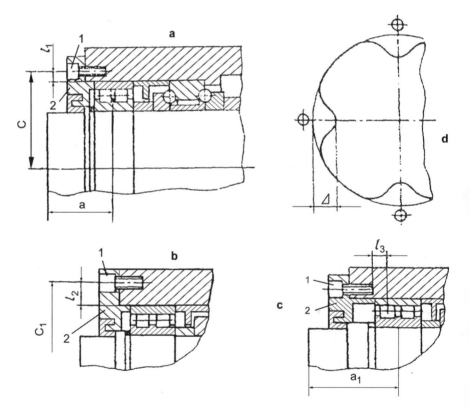

Fig. 7.2.11. Alternative design embodiments of a front spindle bearing unit.

to closeness of preloading bolts to the bore for the bearings, bolt tightening in design a creates large distortions Δ of the housing and of the outer races of the bearings (Fig 7.2.11d), which are detrimental for accuracy. Distortion of the outer race in design c is absent since the bolt 1 is located away from the bearing by a significant distance l_3.

7.2.2 Active (Servo-Controlled) Systems for Stiffness Enhancement

While significant improvements in stiffness characteristics of a mechanical system can be achieved by optimizing its structural design in accordance with the concepts and analyses presented above, there are certain limits for the structural stiffness of mechanical components and systems. For example, the "combination structure" cantilever boring bars described in Section 7.5.2 and in [8] allow for a chatter-free machining of long holes requiring length-to-diameter ratios up to $L/D = 15$. However, while there is no chatter at the extreme L/D, static deflections of such a long cantilever beam by the cutting forces are very significant and may require numerous passes to achieve the required precision geometry of the hole if the original hole had significant variation of the allowance to be removed by the boring operation [see Chapter 1, Eq. (1.3.29)]. The active control is much more effective if applied to well designed mechanical systems. For example, if cutting without chatter at $L/D = 15$ is achieved in a boring bar by passive means, the active system should only compensate static deformations, a relatively simple task. On the other hand, if both static and and dynamic compensation is required, a broadband control system is needed, much more complex and expensive, while less reliable.

In such cases, *active* or *servo-controlled* or *"mechatronic"* systems for compensation of mechanical deflections can be very useful. The term "mechatronics" had been coined in 1969 by Yaskawa Electric Corp. to signify enhancement of performance of mechanical systems by means of electronics. There are many more applications of mechatronic devices for controlling forced and self-excited vibrations than for controlling stiffness. An extensive survey of various applications of mechatronics in mechanical design is given in [9]. In the older active systems for controlling structural deformations, mostly hydraulic actuators had been used, e.g., [10]. It was achieved by adding a controller to a hydraulic compensation system similar to ones shown in Figs. 7.2.8 and 7.2.9. Presently, hydraulic actuators are used in large servo-controlled systems or in cases when the hydraulic system is already used for other purposes, e.g., for hydrostatic bearings and/or guideways. Advancements in piezo-actuator technology led to its wide application in servo-controlled stiffness enhancement systems. In general, hydraulic actuators can develop high forces/high power in a relative small space, while piezoelectric actuators usually have broad frequency range.

Fig. 7.2.12 [2] shows one of the older active systems for reducing deformations of boring bar 1 for machining precision holes. The axis of the machined hole in part 2 should coincide with the beam of laser 5. This beam is reflected from prism 6 rigidly attached to boring bar 1. The reflected beam is reflected again from semi-reflective mirror 7 and its position is compared by photosensor 8 with the reference beam (the primary laser beam refracted by semi-reflective mirror 7). Controller 10 processes the difference between the desired and the actual positions of the beam and generates voltage to activate piezo actuator 4 for a corrective action on boring bar 1. While this system only maintains the axis of the boring bar in a precise position and does not respond to changes in cutting tool 3 due to its wear, it is also possible to use the machined surface as a reference. In such case, both the deflections of the boring bar and the tool wear would be taken into consideration.

There are designs of boring bars with active antivibration devices (e.g., Fig. 7.2.13) [9], which are less complex than alignment-compensated boring bars as in Fig 7.2.12. It is interesting to note that actively controlled chatter-resistant boring bars have maximum values L/D within 10 to 12 range, inferior to $L/D = 15$ for the optimally designed passive boring bars (see Section 7.5.2) [8].

The most complex are laser-based systems for compensation of both boring bar deflections and vibrations (e.g., [11]). The system is using a

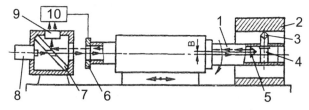

Fig. 7.2.12. Active axis alignment system for a boring bar.

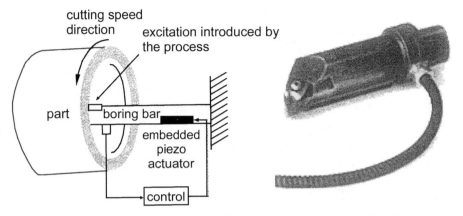

Fig. 7.2.13. Actively controlled chatter-resistant boring bar.

laser simultaneously producing two beams with different wavelengths (blue and green beams). One beam is used to control radial position, and another — to control inclination of the tool. The measured guiding accuracy of piezo actuators is 10 μm. Another system is shown in Fig. 7.2.14 [11]. It is using a combination of laser, strain gage, and capacitance

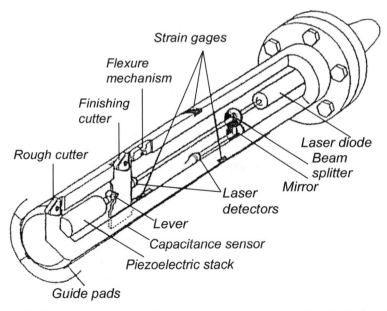

Fig. 7.2.14. Boring bar with active compensation of both deflection and self-excited vibrations.

sensors and a stack of piezoelectric elements for deflection compensation (stiffness enhancement). Complexity of boring bars, especially with active deflection and vibration compensators limits their application to relatively large bars having space for packaging the sensors and the actuators. Since these are not "solid state" devices, reliability of the active compensation systems for boring bars and other tooling structures in the shop-floor environment is limited. Metal chips, abundance of coolant, etc., may lead to fast deterioration of complex electronic devices.

There are many other designs of active boring bars, e.g. [13], as well as active systems for other stiffness and vibration-critical components in which the active systems are used to suppress self-excited vibrations or response of the system to external vibratory excitations (see a survey in [9]). Since the frequency range of the undesirable vibrations can be very broad, up to 300 Hz-to-500 Hz, the servo-controlled vibration suppression system may become very complex and expensive. While in many cases the active systems are certainly superior, it is often definitely more feasible to perform the vibration-suppression tasks by passive means. Such means include rational designs such as the combination structure for the cantilever components (e.g., see Section 7.5 and [8]), use of the "reverse buckling" concept, Section 7.4, use of high damping materials, of dynamic vibration absorbers like in Appendix 6, etc. The active systems are more effective if used only for correction of static or other slow-developing deviations of the critical components from the desired geometrical positions. If an active system is used to control vibrations of a precision low stiffness system, such as a long boring bar, compensation of its static deflections still may be necessary (unless the "reverse buckling" concept is employed). Thus, while the vibration sensor for such system could be very small, a deflection-measuring system is still necessary. It could be laser-based as described above or based on other principles.

If the chatter resistance is enhanced by passive means, like for boring bars described in Section 7.5.2 or [8], a much more rugged system for correcting static deflections can be used, e.g., like in Fig. 7.2.13 [9]. Signal from the strain gage deflection sensor controls an embedded off-axis piezo actuator generating bending moment correcting deflection of the boring bar.

Active systems can be and are successfully used also to correct positions and to enhance effective stiffness of massive frame parts. Fig. 7.2.15 [2] shows a correction system for angular positioning of spindle head 1

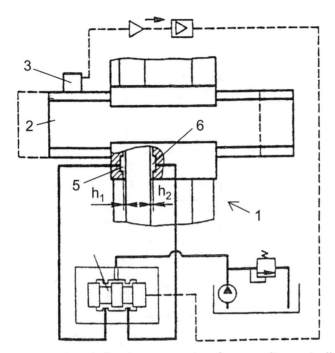

Fig. 7.2.15. Active deflection correction for traveling spindle head.

mounted on horizontal ram 2. Electronic angular sensor (level) 3 measures inclination of ram 2. The signal from level 3 is conditioned and amplified, and used to actuate pilot valve 4 changing the balance of flow between two hydrostatic pockets 5 and 6. This, in turn, changes gaps h_1 and h_2, respectively, in these pockets thus restoring the desired precise orientation of the spindle head.

Hydraulic actuating systems have advantages versus piezoelectric systems in cases when large forces with sizeable strokes/deformations have to be realized. The high pressure hydraulic system similar to one in Fig. 7.2.16 is used for vertical ram 1 carrying milling spindle head 2 in Fig. 7.2.16 [2]. Laser 3 is attached to the main frame of the machine. Deflection δ of milling head 2 caused by the cutting forces results in shifting of cylindrical mirror 4 attached to the head and, consequently, in a changing distance between the primary and reflected beams on array of photo diodes 5. The deviation signal is then conditioned and activates a hydraulic servo-system similar to one in Fig. 7.2.15. While systems in Fig. 7.2.15 and 7.2.16 use hydrostatic guideways for deflection compensation, similar

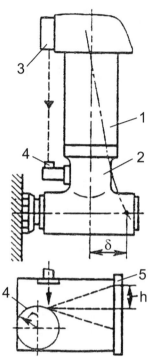

Fig. 7.2.16. Active deflection correction system for a cantilever milling spindle.

schematics can be used for hydrostatic (as well as electromagnetic) bearings for rotating shafts and spindles. Judicious changes of radial forces acting on the shafts at various angular directions and/or bias direction of the axis of rotation can be used, e.g., for machining non-round cross-sections.

Compliance breakdown for a surface grinder in Fig. 7.2.17 [14] indicated two major compliance sources to be the ball screw providing for up-and-down movement of the spindle head along the vertical guideways on the column and deformation of the cantilever spindle head. The latter deflects due to bending of the spindle housing and due to deformations of spindle bearings, caused by cutting forces. These two sources are compensated by two piezo actuators, A and B, respectively. Actuator A is packaged in series with the ball screw and has high stiffness relative to axial stiffness of the ball screw. Accordingly, it does not noticeably reduce stiffness of the ball screw assembly if its control system is "off."

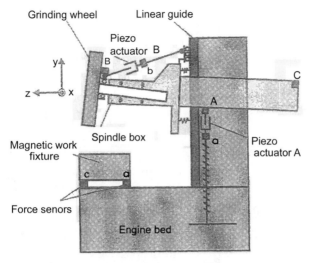

Fig. 7.2.17. Surface grinder with active vibration suppression system.

Actuator B is designed as a brace transforming the support conditions of the spindle head from cantilever attachment to the column to a simple truss. Thus, the brace is also enhancing the spindle head stiffness even when its control system is "off." This prudent design provides stiffening of the system even when the control system is desactivated. Since the system is designed to compensate both static and dynamic deformations, there are two sets of sensors: force sensors a, b, c, d for static compensation and accelerometers A, B, C for dynamic compensation.

Many large machines have inadequate stiffness of their frames and thus require stiffening by massive foundation blocks (see Chapter 5). However, building large foundation blocks is very expensive and time consuming. Use of such blocks impairs flexibility of the production lines, since the machine cannot be easily relocated in response to changing production lines. Sometimes there is a need to install a heavy and not adequately rigid machine on the shop floor directly, e.g., in a location not allowing building a large foundation. In many instances, heavy machines require low–frequency vibration isolation (e.g., see [15]), which calls for low stiffness mounts excluding the stiffening effect of the foundation in case of a direct installation on the isolators, or for very large and expensive spring-suspended foundation blocks. In such cases, active installation systems may economically enhance effective stiffness of the installed machine. There are two basic types of active installation systems [16]:

1. Low-frequency vibration isolators (usually pneumatic) having built-in level- maintaining devices; the latter assure that the distance between the machine bed and the foundation surface remains constant for slow changing forces, such as travel of heavy units inside the machine.
2. Comprehensive installation systems (usually hydraulic) using a separate reference frame and actuators maintaining the required shape of the machine regardless of any changes in foundation deformations and/or changing weight distribution within the machine.

(a) A typical self-leveling vibration isolator is shown in Fig. 7.2.18 [15]. The isolator comprises two basic units: passive damped pneumatic spring *A* and height controller *B*. Both units are mounted on the base 1. Top supporting platform 7 is guided by ring 2 and sealed by O-ring 8. Shaped partition 4 separates the cavity under supporting platform 7 into damping chamber 3 and load-supporting chamber 5, which are connected by a calibrated orifice. Weight load on mounting foot 6 of the installed machine is supported by platform 7, which is resting on pressurized air in chamber 5. Compressed air 17 enters damping chamber 3 and then load-supporting chamber 5 via controller *B*. Control rod 16 carries closing surfaces of both inlet and bleeding valves. Bleeding capillary 13 is attached to elastic membrane 10 separating lower and upper volumes of the controller.

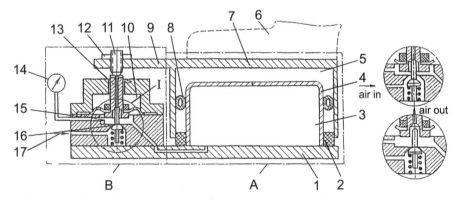

Fig. 7.2.18. Vibration isolation mount Serva-Levl SE (Barry Controls Co.); A Pneumatic spring element; B Height controller.

Feedback is realized by extension 9 of platform 7. When the weight load on platform 7 increases, e.g., due to travel of a heavy unit inside the installed machine, the equilibrium between the weight load on platform 7 and air pressure in chamber 5 is distorted, platform 7 is sagging down, and feedback extension 9 is pushing capillary 13 down. Control rod 16 also moves down and opens the inlet valve. Since air pressure in the compressed air line is higher than in chambers 3 and 5, air pressure inside the isolator is increasing until the higher weight load on platform 7 is counterbalanced. When the weight load on platform 7 is reduced, the excess air pressure in chamber 15 lifts capillary 13 thus opening the bleeding valve until platform 7 returns to equilibrium.

There are numerous commercially available leveling isolator mounts similar to one in Fig. 7.2.18. While they may provide a high leveling accuracy (It is claimed that the height of the isolator platform can be maintained within ~1.0 μm), the settling period is often quite long, up to 10 seconds to 20 seconds. The settling period is increasing with reduction of isolator stiffness (or natural frequency of the isolation system).

Some designs of the leveling isolators use fluidics control systems instead of the valve-based control systems [16, 17]. Both systems have similar overall performance characteristics, but fluidics systems have a smaller "dead zone" (time delay caused by the settling process). It was shown in [18] that dynamic stability of the self-leveling low natural frequency isolators, like the one shown in Fig. 7.2.18 can be significantly enhanced by providing conditions for laminar air flow between chambers 3 and 5.

Since the system does not respond to fast changes in the relative position, it can have a low dynamic stiffness (low natural frequency), while it has a very high static stiffness (level maintenance).

(b) A comprehensive installation system in which positions of the mounting points are precisely maintained in relation to an unloaded reference frame was suggested and tested in [19], (Fig. 7.2.19). Bed 1 of a large machine tool is installed on foundation 4 by means of leveling mounts 8 located at points A_1, A_2, A_3. Distances between the mounting points are made so small that deflections of the frame parts between the mounts are negligible. Reference frame 3 is constructed around the machine and is supported at three points, P_1, P_2, P_3, which are located in relatively undeformed areas of the foundation. Level 2 can be used to adjust and correct, if necessary, horizontality of frame 3. Height

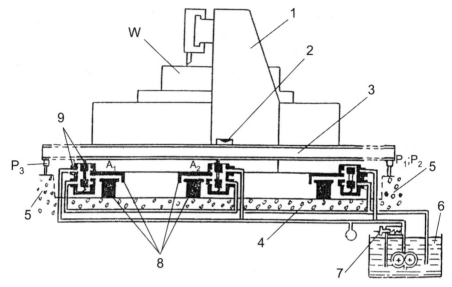

Fig. 7.2.19. Active installation system with a reference metrology frame.

sensors 9 are hydraulic pilot valves regulating flow of hydraulic fluid to mounts 8. Line pressure is generated by pump 6 and adjusted by chock valve 7. When the weight load on a mount increases, bed 1 is bending, the slider in the corresponding pilot valve 9 is shifting, and hydraulic fluid pressure in the affected mount 8 is also increasing until the set distance between the supporting plane of bed 1 and reference frame 3 in that location is restored. Analogously, when the weight load on a mount decreases, the hydraulic pressure in the respective mount also decreases until the distance between the bed and the frame is restored. The tests demonstrated that the level in relation to the reference plane is maintained within ~10 μm, and the settling time is about 0.2 second (for the vertical natural frequency of the machine tool on the mounts 41.5 Hz).

Performance of the system depends on design of the reference frame 3. Large frames may have excessive thermal gradients and ensuing deformations unless special measures (cooling, temperature stabilization) are undertaken. The frame design can be simplified if a "virtual" frame is used which employs system 8 of interconnected fluid chambers (Fig. 7.2.20) [19]. Isolated object 1 is attached to the foundation via mounts 7.

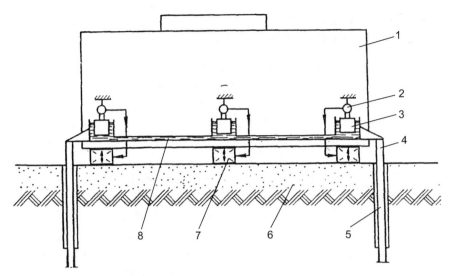

Fig. 7.2.20. Active installation system with reference frame comprising interconnected fluid vessels.

There is a chamber over each mount, and sensors 2 fastened to the object, are following the fluid level in the respective chambers. To reduce influence of the contact force exerted by sensors 2 on the fluid surface, sensors are interacting with the fluid surface via buoys 3. Such system guaranties horizontality of the reference plane, but may be sensitive to air streams, temperature changes, variations in the contact forces of the sensors, etc. It was suggested in [20] to modify the "virtual frame" by filling the system of the interconnected chambers with a low melting point material, which is periodically melted to restore the ideal leveling, but is solid during the operation of the system. Another approach is to use a "pseudo fluid", i.e., to fill the system with a bulk material or with small balls, which are cemented by friction forces in normal condition and thus the surface in the chambers is, effectively, rigid. The system is periodically "liquefied" by applying high frequency vibrations to restore the accurate leveling of all chambers.

Very often, enhancement of stiffness is desired for abatement of forced and self-excited vibrations. In some of these cases, the desired effect can be achieved by using structural elements whose stiffness varies in accordance with a judiciously selected algorithm (e.g., see [21]).

7.3 STIFFNESS ENHANCEMENT BY REDUCTION OF STRESS CONCENTRATIONS

Stress concentrations in structural components may result in large local deformations and thus in increasing overall deformations and reduced structural stiffness. Special attention should be given to alleviation of stress concentrations in stiffness-critical systems. It can be achieved by a judicious selection of design components and/or their shape and dimensions.

Fig. 7.3.1 [3] shows a bearing support for shaft 1. If forces acting on the shaft and the resulting shaft deflections are relatively small, the tapered roller bearing in Fig. 7.3.1a exhibits high stiffness since rollers 2 contact bearing races along their whole length. However, at larger shaft deformations, the length of contact between the rollers and the races is reduced, and the contact stresses are concentrated at the roller edges a, thus creating large local deformations. In this case, it can be advantageous to replace the single tapered roller bearing, with tandem angular contact ball bearings, which share the total reaction force, $R_1 + R_2$, as shown in Fig. 7.3.1b. The shaft deformation is reduced (and its effective stiffness increased) due to increased angular stiffness of the support (close to built-in conditions in Fig 7.3.1b versus single support conditions in Fig. 7.3.1a), and due to elimination of the stress concentrations.

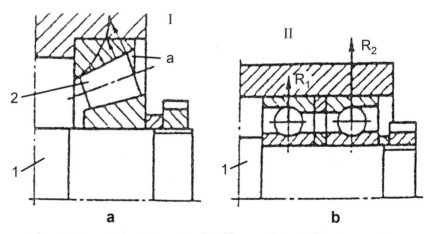

Fig. 7.3.1. Enhancement of stiffness of shaft bearing unit by replacing rigid but misalignment-sensitive tapered roller bearing (a) with two spread-out angular contact ball bearings (b).

Fig. 7.3.2 pictures a roller guideway in which width B of the rollers 1 is less than width of the guideway 2. It results in a non-uniform pressure distribution as illustrated by diagram A. The stress concentrations and the deformations are reduced if widths of the rollers and of the guideway are the same (broken line on the diagram). Since deformations in the guideway are proportional to stresses, reduction of stress concentrations leads to a stiffness enhancement.

Sharp stress concentrations are also characteristic for interference fits (press fit and shrink fit) connections. These stress concentrations and the resulting deformations can be significantly reduced by judicious shape modifications of one or both connected parts. Influence of shape of the bushing 1 on the contact stress distribution in a shrink-fit connection is shown in Fig. 7.3.3.

7.4 STRENGTH-TO-STIFFNESS TRANSFORMATION

While strength of metals is relatively easy to enhance significantly (up to three-to-seven times for steel and aluminum) by alloying, heat treatment, cold working, etc., their stiffness (Young's modulus) is essentially invariant. One exception from this universal rule are aluminum (Al)-lithium (Li) alloys, in which adding 3% to 4% of Li to Al results in 10% to 15% increase of Young's modulus of the alloy (Table 1.1). Fiber-reinforced

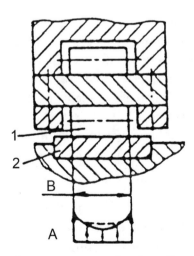

Fig. 7.3.2. Stress concentration in roller guideways.

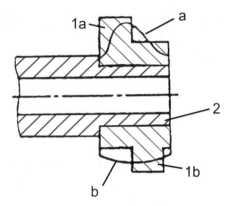

Fig. 7.3.3. Reduction of stress concentration in interference fit connections.

composites can be more readily tailored for higher stiffness, but they cannot be used in many cases. Some of their limitations are discussed in Chapter 1.

The most critical mode of loading of structural components is bending, since bending deformations could be very large even for not very high forces due to transformation of small angular deformations into significant translational deformations which in many cases determines the effective stiffness of the structure. Bending stiffness can be enhanced by reducing spans between supports of the components subjected to bending, by reducing the overhang length of cantilever components, and by increasing cross-sectional moments of inertia ("beefing up"). While two former techniques are frequently unacceptable due to design constraints, "beefing up" of cross-sections can be even counter productive, since it inflates dimensions and increases weights of the components as it was illustrated above in Section 7.1. This Section describes some techniques which can be used for enhancing bending stiffness of components.

7.4.1 Buckling and Stiffness

"Buckling" of an elongated structural member loaded with a compressive axial force, P, is loss of stability (collapse) of the structural member when the compressive force reaches a certain critical magnitude, P_{cr}, which is also called the *Euler force*. Usually, the buckling process is presented as a discrete situation: *stable/unstable*. However, the process of development of instability is a gradual continuous process during which bending stiffness of the structural member is monotonously decreasing with increasing axial compressive force, P. The member collapses at $P = P_{cr}$, when its bending stiffness becomes zero.

This process can be illustrated on the example of a cantilever column in Fig. 7.4.1a. External bending moment, M, causes deflection of the column. If the axial compressive force, $P = 0$, bending stiffness

$$k_o = M/x, \qquad (7.4.1)$$

where $x =$ deflection at the end of the column due to moment, M. If $P \neq 0$, it creates additional bending moment, Px, which further increases the bending deformation and thus, reduces the effective bending stiffness. The overall bending moment becomes [22]

$$M_{ef} = \frac{1}{1 - P/P_{cr}}, \qquad (7.4.2)$$

and the resulting bending stiffness is approximately

$$k_b = \left(1 - P/P_{cr}\right)k_o \qquad (7.4.3)$$

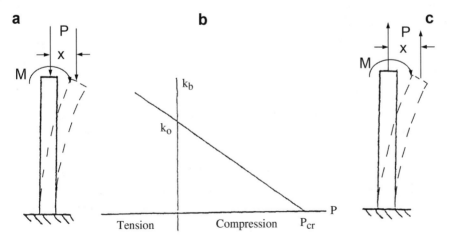

Fig. 7.4.1. Stiffness change of a cantilever beam loaded by a compressive (a) and tensile (b) force.

Fig. 7.4.1b illustrates dynamics of the stiffness change for the column in Fig. 7.4.1a with increasing compressive force, P. Equation (7.4.3) was experimentally validated in [23]. The stiffness-reducing effect presented by this equation must be considered in many practical applications. For example, supporting of a machined part (e.g., for turning or grinding) between the centers involves application of a substantial axial compression force which may result in an additional significant reduction of bending stiffness for slender parts having relatively low P_{cr}.

If the axial force is a tensile force $(-P)$ instead of compressive force P (Fig. 7.4.1c), then the effect is reversed, since the tensile force creates a counteracting bending moment $(-Px)$ on the deflection, x, caused by moment, M. The effective bending moment is thus reduced,

$$M_{ef} = \frac{1}{1 + P/P_{cr}} \qquad (7.4.4)$$

and the effective stiffness is increased,

$$k_b = \left(1 + P/P_{cr}\right) k_o \qquad (7.4.5)$$

Thus, preloading of the structural member loaded in bending by a compressive force results in reduction of its bending stiffness (and in a corresponding reduction of its natural frequencies, see Appendix . . . and [23]). On the other hand, preloading by the tensile force results in enhancement of its bending stiffness (and corresponding increase in natural frequencies, see Appendix 3 and [23]). This is the same effect, which allows to tune *guitar strings* (and strings of other musical instruments) by their stretching. A similar effect for two-dimensional components (plates) is described in [24].

The effect of bending stiffness reduction for slender structural components loaded in bending can be very useful in cases when stiffness of a component has to be adjustable or controllable. An application of this effect for vibration isolators is proposed in [25]. Fig. 7.4.2 shows a device which protects object 14 from horizontal vibrations transmitted from supporting structure 24. The isolation between 14 and 24 is provided by several isolating elements 18 and 16. Each isolator 18, 16 is a thin stiff (metal) post 60, 32, respectively. The horizontal stiffness of the isolation system is determined by bending stiffness of posts 60 and 32. This stiff-

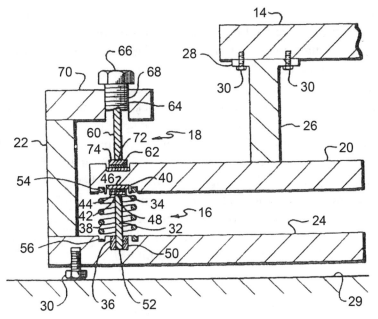

Fig. 7.4.2. Vibration isolator for horizontal motion, based on using buckling effect for stiffness adjustment.

ness can be adjusted by changing compression force applied to posts 60 and 32 (in series) by loading bolt 66. The horizontal stiffness can be made extremely low (even negative if the compressive force applied by bolt 66 exceeds the critical force). In the latter case, the stability can be maintained by springs 38.

7.4.2 "Reverse Buckling" Concept

Very effective stiffness enhancement of structural components subjected to bending can be achieved by axial preloading in tension of the component loaded in bending for reducing its bending deformations. This technique can be called "reverse buckling" or "guitar string" effect.

As it was shown in Chapters 3, 4, preloading of a belt or chain transmission, ball bearings, joints, etc., also results in enhancement of stiffness of the preloaded system. However, belts, chains, balls and races, etc., are subjected to higher forces than the same components in non-preloaded systems. Thus, part of the load-carrying capacity of the system *(strength)* is traded for increase of its effective *stiffness*. The reduction of strength is also the price to be paid for using the "reverse buckling" effect. Thus, all stiffness enhancement techniques based on applying a preload to the system can be named as "strength-to-stiffness transformation."

The tensile preload of a beam not only results in enhancement of its bending stiffness, but also increases its buckling stability, especially for axial compressive forces applied within the span of the beam. Beam in Fig. 7.4.3 is pre-tensioned between its supporting points, 0 and 2, by force P_0. It is also loaded by axial force, P_1, applied at point 1 within its span. The total force acting on the upper section of the beam (l_2) is $P_0 + P_1$, while the total force acting on the lower section (l_1) is $P_0 - P_1$. Thus, application of the compressive force would not generate compressive stresses until the compressive force $P_1 > P_0$. It means that that the critical (Euler) force is, effectively, enhanced by the amount of P_0.

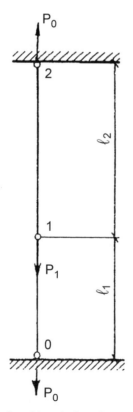

Fig. 7.4.3. Pre-tensioned rod loaded with an axial load within its span.

Both bending stiffness enhancement and buckling stability enhancement are very important for long ball screws which are frequently used in precision positioning systems, e.g., for large machine tools. While bending deformation due to weight and thermal expansion–induced forces can adversely affect the accuracy, high payload forces applied through the nut within the span of the screw could cause collapse of the screw. Both effects can be alleviated by stretching the ball screw.

Fig. 7.4.4b shows design of the bearing support for a ball screw positioning system schematically shown in Fig. 7.4.4a. Two thrust bearings 4, 5 and a long roller radial bearing 3 generate a "built-in"-like supports, which can accommodate a high magnitude of the axial preload. It has to be noted though, that an inevitable temperature increase during intensive operation of the ball screw unit with heavily preloaded bearings may lead to thermal expansion of the screw and, thus, to reduction of the axial tensile force. This effect can be compensated by using springs (for example, Belleville springs) between spacer 7 and tightening nut 8.

Application of this approach (stretching a slender structural member to enhance its bending stiffness) to machining (turning or grinding) of long slender parts without steady rests is described below in Section 7.4.3. It is important to note that stretching of the part during machining results in residual compressive stresses on the part surface after machining is completed and the tensile force is withdrawn. These compressive

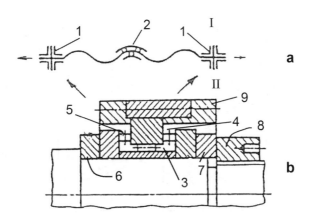

Fig. 7.4.4. Schematic of ball screw translational drive (a) and end support design (b). 1 — end supports; 2 — ball nut; 3 — radial bearing; 4, 5 — thrust bearings; 6, 7 — spacers; 8 — preloading nut; 9 — stationary frame.

stresses are very beneficial for fatigue endurance of the part while in service. The residual compressive stresses reduce or even completely cancel residual tensile stresses, which develop in the part surface during turning operation and are detrimental for the fatigue endurance of the part. Also, there is information that application of tensile force reduces effective hardness of the part surface (surface layer 15 μm to 20 μm deep) and thus, improves machinability. Vickers hardness reduction can be 5% to 10% for steel specimens at 400 MPa tensile stress, and 15% to 20% for titanium specimens at 600 MPa tensile stress.

Application of this effect to two-dimensional components is shown in Fig. 7.4.5 [26]. Grinding wheels having internal cutting edge (Fig. 7.4.5a)

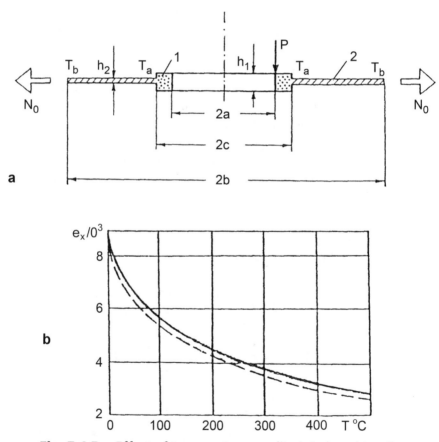

Fig. 7.4.5. Effect of temperature gradient–induced tensile pre-stressing of grinding wheel with internal cutting edge (a) on its compliance coefficient e (b).

are frequently used for slicing semiconductor crystals into wafers. The abrasive ring 1 (the internal part of the wheel) is held by the metal disc 2 clamped in the housing (not shown). Axial (bending) stiffness of the wheel determines its chatter resistance as well as thickness accuracy of the wafers. Since the axial stiffness is proportional to Young's modulus E_2 of the holding disc 2 and to the third power of its thickness h_2, and is inversely proportional to the square of its internal radius, introduction of larger wafer diameters reduces the axial stiffness.

Since thickness and/or its variation lead to increased losses of the expensive crystal material, it was suggested to generate two-dimensional stretching of the holding disc by heating its outside area during clamping (to temperature T_b), thus introducing a temperature gradient along the disc radius. Fig. 7.4.5b shows dependence of *compliance factor e* on the temperature gradient $T^* = T_a - T_b$. Bending stiffness of the wheel is

$$K = E_2 h_2^3 / e \, a^2. \qquad (7.4.6)$$

The above analysis [expressions (7.4.1) to (7.4.5)] assumed that the axial forces have always the same directions regardless of bending deformations of the beam. While this assumption holds for the cases of Figs. 7.4.2 to 7.4.4 and Section 7.4.3, in many cases, such as in the self-contained systems discussed below in Section 7.4.4, directions of the axial forces are changing with the beam deflection, and follow the beam axis inclination at the beam ends (the "following force"). It can be shown (e.g., [27]) that in such case the critical force is significantly higher than for the case of constant directions of the axial forces. For example, for a cantilever column as in Fig. 7.4.6a, the buckling force for compression by a vertical axial force, P, is

$$P_{cr} = \pi^2 E \, I / 4l^2 = 2.47 EI/l^2, \qquad (7.4.7a)$$

while for the case of Fig. 7.4.6b it is:

$$P_{cr}^f = 20.05 \, E \, I/l^2, \qquad (7.4.7b)$$

or about 8.2 times higher.

Expressions for bending stiffness enhancement of beams with the tensile axial preload in Fig. 7.4.7 are given in [28]. For double-supported

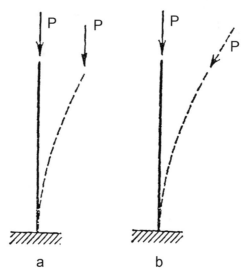

Fig. 7.4.6. Typical cases of compression loading of a column. (a)-Vertical force; (b)-"Follower" force.

beams loaded in the middle by force, P, deflection under the force, while the axial force T is applied is:

$$y_{max} = K \, (y_o)_{max}, \tag{7.4.8}$$

where $(y_o)_{max}$ is maximum deflection without application of the tensile force $(T = 0)$. The *deflection reduction coefficient, K,* for the case in Fig 7.4.7a (axial tensile force) is

$$K' = \frac{\dfrac{\alpha l}{2} - \tanh \dfrac{\alpha l}{2}}{\frac{1}{3}(\alpha l^3)} \tag{7.4.9a}$$

and for the case in Fig. 7.4.7b (following tensile force):

$$K'' = \frac{\dfrac{\alpha l}{2} - \tanh \dfrac{\alpha l}{2}}{\frac{1}{3}(\alpha l^3) \tanh \dfrac{\alpha l}{2}} \tag{7.4.9b}$$

Here parameter α is defined as $\alpha^2 = T/EI$, where EI is bending rigidity of the beam.

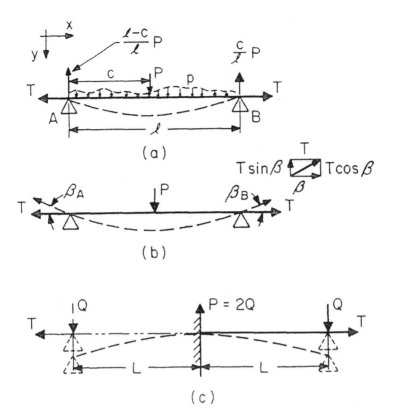

Fig. 7.4.7. Beams under combined bending and axial force loading. (a)-Double-supported beam, axial force; (b)-Double-supported beam, following axial force; (c)-Cantilever beam.

A built-in cantilever beam loaded at the end (solid line in Fig. 7.4.7c) can be considered as one half of a double-supported beam loaded in the middle (dotted line in Fig. 7.4.7c). Instead of force, P, and length, l, as in Figs. 7.4.7a, b, force, $2Q$, and length $2L$ are associated with the simulated double-supported beam in Fig. 7.4.7c. Thus, the deflection reduction coefficient for cantilever beams can be calculated from expressions (7.11) if L is substituted for $l/2$.

The stiffening effect of the tensile force as given by Eqs. (7.4.9) can be assessed from Fig. 7.4.8 in which $\mu = \alpha l/2 = \alpha L$. To better visualize the stiffening effect of the tensile force, a beam of a solid round cross section (diameter, D) can be considered. For such beam, $I = \pi(D^4/64)$, cross-sectional area $A = \pi(D^2/4)$, $I/A = D^2/16$, and

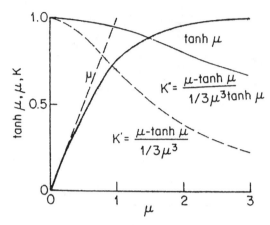

Fig. 7.4.8. Reduction of bending deflection caused by applied tensile force.

$$\alpha l = l\sqrt{\frac{T}{EI}} = l\sqrt{\frac{T}{A}\frac{1}{E}\frac{A}{I}} = l\sqrt{\frac{\sigma_T}{E}\frac{16}{D^2}} = 4\frac{l}{D}\sqrt{\frac{\sigma_T}{E}} = 4\frac{l}{D}\sqrt{\varepsilon_T}, \quad (7.4.10)$$

where σ_T = the tensile stress from the tensile force, T, and ε_T = the relative elongation caused by T.

Some important conclusions can be made from Eqs. (7.4.9) and Eq. (7.4.10) and from, Fig. 7.4.8:

1. The stiffening effect from the tensile force is substantially higher for a cantilever beam than for a double-supported beam.
2. The higher stiffening effect can be achieved if higher tensile stress, σ_T or strain, ε_T, were tolerated. The allowable tensile stress is, in many cases, determined by the yield strength of the beam material.
3. For the same given allowable σ_T, the technique is more effective for lower modulus materials (greater $\sigma_T/E = \varepsilon_T$).
4. The effectiveness quickly increases with increasing l/D (slenderness ratio of the beam). It explains the very high range of stiffness (or pitch) change during tensioning a guitar string.

7.4.3 Stiffening of Slender Parts by Axial Tension during Machining

Low stiffness of a part being machined (e.g., a slender workpiece being turned) leads to geometric inaccuracies, to inferior surface finish, to chatter vibrations, etc. In some cases a judicious "tuning of the tool" may partially compensate the part compliance (see Section 8.1.4). The most widely used are steady rests to provide inter-span support and enhance effective part stiffness. While stationary steady rests interfere with the machining process, traveling steady rests are often used, but they are not suitable if the machined part has abrupt diameter changes, and limit flexibility of CNC lathes. Since installation of a traveling steady rest may require up to three faces of the turret, they also reduce availability of cutting tools.

A significant effort is invested into development of active and semi-active techniques for reducing time of turning long slender shafts. In semi-active techniques the machined parts are measured after the first pass, and the required tool trajectory for obtaining required diameter and cylindricity is computed and entered into the CNC controller. It is important to note that the computation requires a knowledge of magnitudes of cutting forces, not available with high accuracy. In active techniques, the shaft diameter after the tool contact point is continuously measured by a sensor and controlled by the feedback-based servo-controller. The main problem for both active and semi-active systems is chatter, since they are usually based on static measurements and do not consider strong propensity towards chatter development for the low stiffness parts.

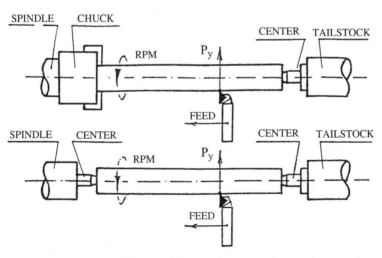

Fig. 7.4.9. Setups for machining elongated parts by turning.

Turning of long parts is usually performed while the part is supported by two centers or is clamped in the chuck and supported by the rear center. In both cases, significant compressive forces are applied to the part which further reduce effective stiffness of the part (Fig. 7.4.1).

Turning of a slender part results in its barrel-like shape, Fig. 1.3.19e. Its diameter is y_x at a distance x from the headstock (chuck), results from superposition of deflections of the headstock and the tailstock and deflection of the part. If the part is supported by two centers, then:

$$y_x = P_y \left[\frac{1}{K_t} + \left(\frac{L-x}{L} \right)^2 \frac{1}{K_H} + \left(\frac{x}{L} \right)^2 \frac{1}{K_T} + \frac{1}{K_w} \right] \qquad (7.4.11)$$

L = workpiece length; K_t = tool stiffness; K_H = headstock stiffness; K_T = tailstock stiffness; K_w (x) = workpiece stiffness; E, I = Young's modulus, cross-sectional moment of inertia of the workpiece. For a slender workpiece, the last term in Eq. (7.4.11) — deflection of the workpiece — is predominant.

For the experimental study in [29], $K_H = 120 \times 10^6$ N/m; $K_T = 11 \times 10^6$ N/m; $K_t = 29 \times 10^6$ N/m; $D = 12.7$ mm (0.5 in.); $L = 387$ mm (15 in.); $L/D = 30$. The maximum deflection of the workpiece is in the midspan (y_{max} without the tensile force, y'_{max} with the tensile force) and their ratio is described by Eq. (7.4.8).

The long workpiece was clamped in the chuck and supported by the tailstock as shown in Fig. 7.4.10. The tensile force was applied to the workpiece through a threaded connection at the tailstock end. Fig. 7.4.11 shows bending stiffness of the workpiece measured by application of a 90 N force perpendicular to the workpiece axis at five points along its

CHUCK BODY SOFT JAWS   SAMPLE   TAILSTOCK CHUCK   TAILSTOCK

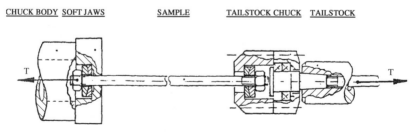

Fig. 7.4.10. Setup for tensile force application to part being machined by turning.

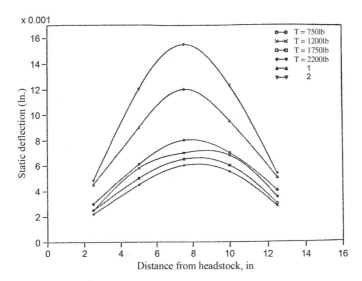

Fig. 7.4.11. Static deflection under 90 N radial force applied at the midpoint of 16 in. long cylindrical part with and without applied tensile force. 1 — chuck and tailstock dead center (no tension); 2 — between two dead centers (no tension).

length. Application of the tensile forces in the range of 3300 N – 9800 N generated tensile stresses 26 MPa to 77 MPa in the workpiece and resulted in a significant increase of stiffness in the midspan, up to 2.5 times.

Dynamic tests were performed on a workpiece having $D = 12.7$ mm (0.5 in.), $L = 432$ mm (17 in.) by impact excitation at 0.25 L from the headstock and measuring the response at 0.125L from the headstock. Typical measured spectra are shown in Fig. 7.4.12 for the workpiece in tension by 9000 N force (line 1, fundamental natural frequency $f_1 = 366$ Hz); workpiece clamped in the chuck and supported by the tailstock dead center, with the compression force 2250 N (line 2, $f_1 = 293$ Hz); and supported by two dead centers with 2250 N compression force (line 3, $f_1 = 171$ Hz).

Conventional machining of such slender workpieces is practically impossible due to bending deformation of the workpiece resulting in its barrel shape, and due to chatter vibrations near midspan, destroying surface finish. Comparison of a conventionally machined shaft (axial force $T = 0$) and the shaft stretched with a relatively small axial tensile force $T \approx 1000$ lbs (4500 N), generating normal stresses in the workpiece $\sigma = 33$

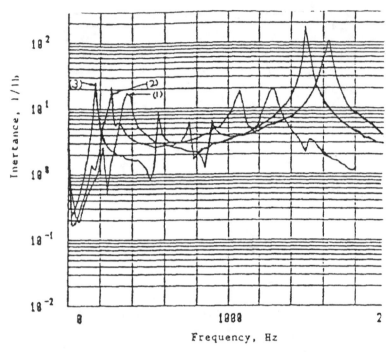

Fig. 7.4.12. Frequency responses of $L = 17$ in., $D = 0.5$ in. part under impact excitation (impact at $0.25L$ from headstock, measuring at $0.125L$ from headstock). 1 — tensile force, $T = 2000$ lb (9000 N), fundamental natural frequency $f_1 = 366$ Hz; 2 — compression force, $T = 500$ lb (2250 N), fundamental natural frequency, $f_1 = 293$ Hz (chuck and dead center); 3-compression force, $T = 500$ lb (2250 N), fundamental natural frequency, $f_1 = 171$ Hz; two dead centers.

MPa (5000 psi) is given in Fig. 7.4.13 for a range of depth-of-cut values. Stiffening of the workpiece by the tensile force allows to obtain good surface finish ($R_a = 1.2$ μm to 1.8 μm or 50 μin. to 75 μin.) and cylindricity (within ~ 5 μm or 0.0002 in.) with depth-of-cut $t = 0.063$ mm (0.005 in.). Greater depths-of-cut are associated with greater cutting forces and require greater T to achieve desirable surface finish and cylindricity.

7.4.4 Self-Contained Stiffness Enhancement Systems

Structural use of the beam-like components preloaded in tension as described above in Section 7.4.3 and in Appendix 4 is limited. There is a

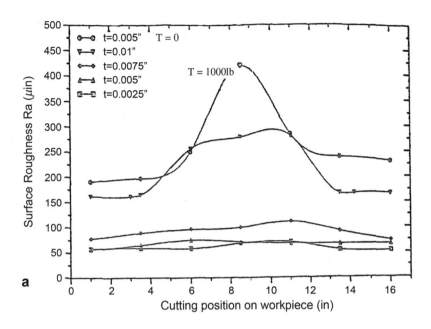

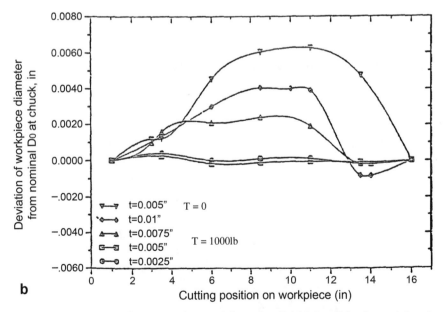

Fig. 7.4.13. Surface roughness (a) and cylindricity (b) of machined long shaft ($L = 17$ in., $D = 0.5$ in., $f = 0.002$ in./rev, $v = 99.5$ sfm) without tensile force and with 1000 lb (4500 N) tensile force with variable depth of cut, t.

need for external force application devices, such as in musical string instruments, or in a ball screw drive in Fig. 7.4.4 or in a machining arrangement for precision turning of slender parts described in Section 7.4.3. A self-contained device in Fig. 7.4.14 is a composite beam having external tubular member 1 and internal core (bolt) 4. Since the cross-sectional moment of inertia of a round beam is proportional to the fourth power of its diameter, both strength and stiffness of the composite beam in bending are determined by the external tubular member whose stiffness is usually 85 % to 90% of the total stiffness. Outside member 1 can be stretched by an axial tensile force P applied by tightening bolt 4. This tensile force is compensated (counterbalanced) by the equal compression force $-P$ applied to internal core 4. These forces would cause stiffening of the external member as indicated by Eq. (7.4.5), as well as reduction of stiffness of the internal core as indicated by Eq. (7.4.3). Since value of the buckling force for the external member, P_{cre}, is much higher than value of P_{cri} for the internal core, the relative stiffness change is more pronounced for the internal core. However, due to the insignificant initial contribution of the internal core to the overall stiffness (10% to 15%), even a relatively small increase in stiffness of the external member would result in an improvement of the overall stiffness even if the internal core stiffness is reduced to a negligible value.

The problem with the design in Fig. 7.4.14 is the danger of collapsing of the long internal core due to the relatively low value of its Euler's force. The collapse could be prevented if the internal core were supported by the inner walls of the external tubular member. However, a precise fitting of a long bolt serving as the internal core is difficult, especially for long slender beams for which this approach is especially effective.

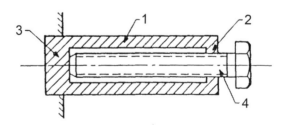

Fig. 7.4.14. Enhancement of bending stiffness by internal axial preload.

It seems that the best way of assuring the required fit between the external and internal components of the composite beam is to form the internal core in place by casting [30], [31]. If the core expands and the expansion is restrained, e.g., by covers or caps on the ends of the external tubular member, this would apply a tensile force to the external member and an equal compressive force to the internal core. While this would generate the overall stiffness enhancement effect as described above, collapse of the internal core would be prevented. Even the stiffness reduction of the internal core per Eq. (7.4.3) would be alleviated, since its Euler's force would be very high due to the supporting effect of the inner walls of the external tubular member.

There are many materials expanding during solidification. The most common material is water; volume of ice is about 9% larger than volume of water before freezing. Another such material is bismuth, which expands 3.3% in volume during solidification. While work with ice requires low temperatures, work with bismuth requires relatively high temperatures (its melting point is 271°C). However, some alloys of bismuth, so-called fusible alloys, have relatively low solidification temperatures and also expand during solidification. Some experiments in [31] were performed with a fusible alloy Asarco LO 158 from Fry Co. (50% Bi; 27% Pb; 13% Sn; 10% Cd). It is used for soldering applications; its melting point is at 70°C (158°F), below the water boiling temperature. This alloy has about 0.6% volumetric growth during solidification, which can be modified by varying the bismuth contents.

If the core is cast inside a tubular rod and is restrained by plugging the internal bore of the rod at its ends, then expansion of the core would be constrained. The process of longitudinal expansion would cause stretching of the tubular rod and compression of the core with the equal forces.

The force magnitude can be determined from the model in Fig. 7.4.15. The initial configuration of the tubular rod (length L_2) and the cast non-restrained internal core after the expansion (length L_1) are shown by solid lines. The equilibrium configuration of the assembly (the case when the ends of internal bore of the rod are plugged after casting) is shown by broken lines. The final length of the assembly is L_3. Magnitude of the resulting force compressing the core and stretching the rod is P_0. In the final condition, the core is compressed by Δ_1, and the rod is stretched by Δ_2.

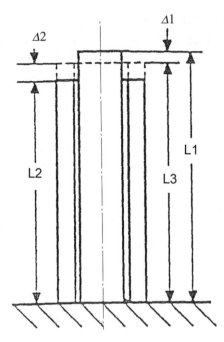

Fig. 7.4.15.   Model of tubular rod with expanding core.

Obviously,

$$\Delta_1 + \Delta_2 = L_1 - L_2 = \Delta; \qquad (7.4.12)$$

$$\Delta_1 = \frac{PL_1}{E_1 A_1}; \qquad \Delta_2 = \frac{PL_2}{E_2 A_2}, \qquad (7.4.13a)$$

where E_1, A_1 are Young's modulus, cross sectional area of the core, respectively, and E_2, A_2 - the corresponding parameters for the rod. Increments Δ_1, Δ_2, Δ are usually very small quantities in comparison with L_1, L_2, and $L_1 \approx L_2 = L$ in Eq. (7.4.13a). Then,

$$\frac{PL}{E_1 A_1} + \frac{PL}{E_2 A_2} = \Delta \qquad (7.4.13b)$$

or

$$P = \frac{\Delta}{L} \bigg/ \left(\frac{1}{E_1 A_1} + \frac{1}{E_2 A_2} \right) = \Delta / \left(L \frac{E_2 A_2}{1 + \frac{E_2 A_2}{E_1 A_1}} \right) \qquad (7.4.14)$$

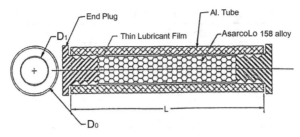

Fig. 7.4.16. General layout of aluminum beam with AsarcoLo-158 alloy.

In this expression, Δ represents expansion of the core in its unrestrained condition. When the expansion is restrained, the material can expand in radial directions and its behavior under pressure might be different from behavior during the free expansion process. Accordingly, expression (7.4.14) must be considered as only an approximate one.

Experimental study was performed in [31] using aluminum tubes L = 0.71 m (28 in.), D_O = 12.7 mm (0.5 in.), D_i = 9.3 mm (0.37 in.) with threaded plugs at both ends as shown in Fig. 7.4.16. The inside volume of the tube was filled with the molten Asarco LO alloy. It was found that frictional conditions on the tube wall have a strong effect on expansion during solidification. Lubrication of the wall with a silicon grease results in a much more consistent behavior.

Changes in both static and dynamic stiffness were monitored during the solidification process. Fig. 7.4.17 shows the setup for monitoring static stiffness change of the tube filled with the constrained fusible alloy

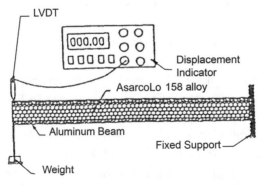

Fig. 7.4.17. Experimental setup for static deflection test.

Table 7.1. Time history of deflection of test tube under 0.25lb weight (lubricated).

Time T, min.	Deflection, in.	Static Stiffness, lb/in.
0	0.05125	4.87
5	0.04915	5.08
10	0.0508	4.92
15	0.04755	5.25
25	0.065	5.37
30	0.0433	5.77

during the solidification process. The beam is clamped at its right end, while the left end is loaded with a weight. Deflection of the left end is monitored by an LVDT probe. Table 7.1 shows results of the static test. Deflection of the tube under 1.1 N (0.25 lb) weight was monitored during the solidification process. After first 10 minutes. during which time the initial shrinkage of the fusible alloy is gradually replaced by its expansion, the deflection is decreasing. Effective stiffness k_{ef} of the beam was calculated by dividing the weight value by the deflection. During the 30 minutes of monitoring, k_{ef} has increased about 20%.

The static stiffness test results reflect not only changes in the beam stiffness due to the internal preload caused by expansion of the core, but also deformations in the clamp. Since deformations in the clamp are not influenced by internal preloading of the clamped beam, the actual stiffness increase may be underestimated. In the dynamic test setup shown in Fig. 7.4.18, the free-free boundary conditions were used for the beam being tested, thus completely eliminating influence of compliance in the clamping devices. The beam (tube) was suspended to the supporting frame by means of a thin string (wire) whose influence on the tube vibrations is negligible.

The tube was excited at one end by an impact hammer. The excitation was applied several times during the cooling /solidification process of the fusible alloy, and the transfer functions were determined by an FFT spectrum analyzer. Plots of the transfer functions recorded at different times after beginning of solidification ($t = 0$) are given in Fig. 7.4.19. At $t = 0$, only four lower natural modes are pronounced in Fig. 7.4.19; higher natural modes are of such low intensity that some of them are lost in the noise, possibly due to increase of damping caused by friction between the liquid or semi-liquid fusible alloy and the walls. The damping is more pronounced at the higher modes since the distance between the nodes is small. With increasing $t \geq 10$ minutes, the higher modes are appearing.

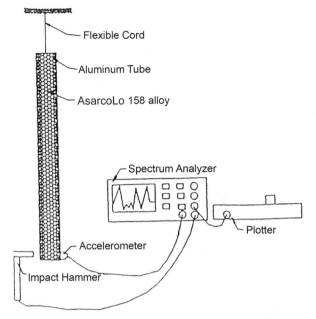

Fig. 7.4.18. Experimental setup for free-free vibration test.

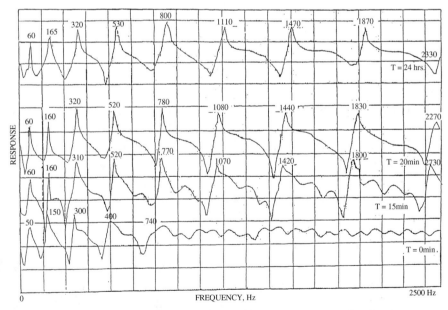

Fig. 7.4.19. Response spectra at different times; 10 dB/div, 0 Hz to 2,500 Hz. range.

Even the lower modes are becoming sharper which indicates damping reduction at these modes. Natural frequencies of all modes are gradually increasing with time due to increasing tensile stresses in the tube caused by increasing volume of the fusible alloy core constrained inside the tube. The final increase (at $t = 24$ hours) was about 20% for the fundamental frequency f_1; 10% for f_2; and about 4% for f_6 – to – f_9. This is in a general agreement with expression (A3.2) derived in Appendix 3. These increases in natural frequencies are corresponding to 44%, 21%, 8% increases in the effective stiffness, respectively.

The expected stiffness increase for the tested beam was calculated using expression (7.4.5). Young's modulus for the Asarco LO 158 alloy is $E_1 = 42$ GPa (6×10^6 psi). The cross-sectional parameters are:

$$I = 9.1 \times 10^{-6}\ m^4;$$
$$A_1 = 5.6 \times 10^{-5}\ m^2\ (0.105\ in^2.);$$
$$A_2 = 5.9 \times 10^{-5}\ m^2\ (0.094\ in^2);$$

critical (Euler's) loads and stiffness values for the fusible alloy core and for the aluminum tube for the dynamically tested free-free condition can be calculated as for the clamped-free columns whose length is $L_c = L/2$ and which are subjected to the "follower force" (see Eqs. (7.4.7)). They are, respectively:

$$P_{cr_1} = 2510\ N\ (530\ lb);$$
$$P_{cr_2} = 10,250\ N\ (2170\ lbs).$$

The relative expansion of the core and the internal preload force were determined in [31] to be $\Delta/L = 0.1\%$ (0.001), 1700 N (377 lb), respectively, and the stiffness increase of the aluminum tube is determined from Eq. (7.4.5) to be

$$1 + 1700/10,250 = 1.17\ \text{times}$$

Actual stiffness increase was measured to be 1.2 to 1.44 times, more than the predicted value. Even larger deviations in the "positive direction" occur for higher modes of vibration as can be seen in Fig. 7.4.19. This phenomenon can be explained by differences in the end conditions which may be not exactly representing the "follower force" model, thus resulting in lower magnitudes of the Euler force. Another reason might be deviations of actual parameters of the tube (E_2, wall thickness) from ones used in the calculations of P_{cr_2}. The amount of stiffness change can be increased by increasing the degree of expansion of the cast core. High strength aluminum can tolerate strains up to 0.003 to 0.005 within its elastic range. At least one half of this range, ~0.002, can be used for transforming into stiffness, thus resulting in a computed value of *1.35* times increase and even higher actual increases, in the range of 1.4 to 1.8 times.

7.5 TEMPORARY STIFFNESS ENHANCEMENT TECHNIQUES

In many cases, high stiffness of a component is required only temporarily, e.g. during machining operations or during dynamic interaction between components. Many mechanical structural components have very low stiffness until/unless assembled with other components. Some examples are thin-walled tubular components, elements of thin-walled shells, etc. Many components of this kind require machining in order to achieve dimensional accuracy and/or specified surface finish. While some types of low-stiffness components can be loaded with high forces in such a way that their stiffness is enhanced, e.g., see Section 7.4.3, in many cases such stiffening procedures can not be used. However, there are very effective techniques not requiring application of high forces but using temporary structural reforming of the low-stiffness components by using controllable temporary reinforcing substances.

These reinforcing substances are securely attached to the components in need for reinforcement for a relatively short time interval, e.g., for machining, and are removed after this interval. Depending on shape and other attributes of the component being reinforced, there are several techniques for attaching and removing the reinforcing substances. The most widely used technique for temporary stiffening/reinforcing low-stiffness structural components is use of easily melting and solidifying materials having melting points preferably within +100°C to 30°C from the room temperature 20°C. Some of such materials are water/ice, urea, fusible alloys (see Section 7.4.4).

An advantage of water/ice material is its obvious availability, high stiffness of the solidified material (ice), and absence of objectionable side effects, such as toxicity. However, a need for a freezing device, as well as significant volumetric expansion of water during its solidification (9% volumetric expansion) creates significant inconveniences in many cases. Use of fusible alloys, especially of Wood alloy (melting temperature ~70°C) does not require bulky freezers, its melting can be easily achieved. However, many fusible alloys contain cadmium (Cd), a highly toxic material; in some cases these alloys can leave patches sticking to the component being reinforced, which require time-consuming cleaning. Urea (carbamide) is a solid material melting at 133°C and widely used as a fertilizer. It was proposed in [32] to use the molten urea for stiffening thin-walled components by filling cavities of the component with the molten urea which quickly solidifies. After the necessary processing of the component is complete, urea can be quickly removed by dissolving it by water without a need for cleaning. Use of urea requires special precautions since it starts disintegrating into toxic gases at ~140°C.

A different approach to stiffening is proposed in [33], (Fig. 7.5.1). Thin-walled component 2 with an arbitrary complex surface 4 made of a metal or ceramic has to be machined. The additional stiffness is provided by a multiplicity of small hard pellets 10 with 2 mm to 4 mm sizes placed between film or fabric 8 and internal surface 6 of component 2. Sand can be used instead of pellets. Pellets 10 fill the cavity between elements 6, 8, and sealing walls 12. A vacuum is created within this cavity resulting in solidification of the pellet massive, thus stiffening the syatem and making machining possible.

An example of a beneficial stiffness change bduring dynamic interaction of components is described in [34]. A major noise source in metal

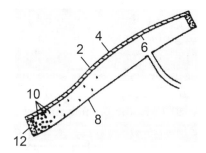

Fig. 7.5.1. Stiffness enhancement of thin-walled parts by vacuuming a bulk-filled cavity.

stamping plants is impacts between parts and scrap pieces discharged from the press and falling on a metal chute. Since the chute stiffness is usually much greater than of the thin metal parts/pieces, ringing of the impacting bparts dominate the environment even if the chute is treated by bapplying damping layers thus preventing the chute ringing.

Fig. 7.5.2 [34] shows a chute design which has solved the problem of parts/scrap ringing. Narrow and thin steel strips are attached to the chute surface via soft foam strips. As`a result, the steel strips can easily deform. When a part or a scrap piece hits the steel strip, the latter deforms and partially conforms to the part/scrap piece shape. This interaction effectively increases the part/scrap piece stiffness during the impact processand prevents its ringing, thus resulting in a dramatic noise reduction. Typical data is given in Fig. 7.5.3, showing that enhancing damping of a conventional chute results in about 3 dB noise reduction, while the stiffness-modifying treatment results in 10 dB to 12 dB noise reductioin.

7.6 PERFORMANCE ENHANCEMENT OF CANTILEVER COMPONENTS

7.6.1 General Comments

Cantilever structures are frequently critical parts of various mechanical systems (boring bars, internal grinding quills, end mills, smokestacks, towers, high-rise buildings, booms, turbine blades, robot arms, etc.). Some of

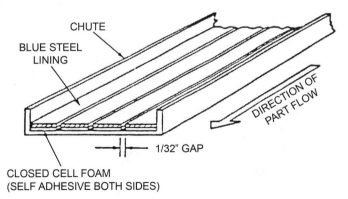

Fig. 7.5.2. Gravity action sliding chute with hard but compliant contact surfaces.

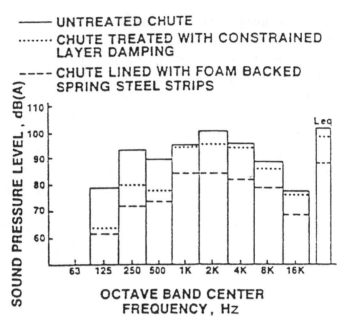

Fig. 7.5.3. Noise spectra for variously treated sliding chutes.

these structures are stationary, like boring bars for lathe use, smoke-stacks, etc.; some rotate around their longitudinal axes, like boring bars and other tools for machining centers and quills; yet other perform a revolute motion around a transverse axis at one end, like turbine blades or robot links. A specific feature of a cantilever structural component is its naturally limited stiffness due to lack of restraint from adjacent structural elements which usually enhances stiffness of non-cantilever components. This feature, together with low structural damping, causes intensive and slow decaying transient vibrations as well as low stability margins for self-excited vibrations. As in many other cases, stiffness, natural frequency(ies), and damping considerations are critical and are interrelated in designing the cantilever components.

Overall dimensions of cantilever components are usually limited. External diameter of a boring bar or cross sectional dimensions of a turbine blade are limited by application constraints. Stiffness enhancement by selecting a material with higher Young's modulus, E, is limited by Young's moduli of available materials, by usually greater density (weight) of materials with high E, and by excessive prices of high-modulus materials.

To enhance natural frequency of a component or to reduce deflection caused by inertia forces of a component rotating around its transverse axis, there is a possibility of shape optimization with the cross-section gradually diminishing along the cantilever, Fig. 7.6.1 [35]. The fundamental natural frequency ω_f can be varied by selecting geometrical parameters of the component per the expression:

$$\omega_f = 8.37 \frac{t}{l^2} \sqrt{\frac{E(3w_2 + w_1)}{\rho(49w_2 + 215w_1)}}, \qquad (7.6.1)$$

where t = thickness of the component. Such approach is, however, limited by other design constraints. For a smokestack, the main constraint is diameter of the internal passage; for a boring bar, some minimum space at the end is necessary to attach a cutting tool and there is often a need for an internal space to accommodate a dynamic vibration absorber (DVA); for a robotic arm, there is a need to provide space for cables, hoses, power transmission shafts, etc.

Damping enhancement, critical for assuring dynamic stability of the structure, is usually achieved by DVAs. Effectiveness of a DVA is determined by the ratio of its inertia mass to the effective mass of the component being treated (the mass ratio, μ). However, the size of the inertia mass in the cantilever systems is limited: in free-standing systems, like towers and high-rise buildings, by economics of huge inertia mass units; in application-constrained systems, like boring bars and grinding quills, by the available space inside the structure. The effective mass of the

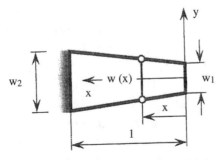

Fig. 7.6.1. Frontal view of constant thickness solid trapezoidal cantilever.

component, on the other hand, can be relatively high if the component is made from a high Young's modulus material, such as steel. Inertia masses made from high specific density materials (e.g., tungsten alloys, $\gamma = \sim 18$) allow for relatively high $\mu = \sim 1.0$ for solid steel boring bars, but even higher ratios are needed to dynamically stabilize boring bars with length, L, to diameter, D, ratio $L/D > 8$ to 9.

According to performance requirements, there are two groups of cantilever design components: (a) stationary and rotating around the longitudinal axis; and (b) rotating around the transverse axis. Components of *group (a)* require high static stiffness (e.g., boring bars for precision machining whose deflections under varying cutting forces result in dimensional errors), as well as high damping to enhance dynamic stability (to prevent self-excited vibrations) and to reduce high-frequency microvibrations causing accelerated wear of the cutting inserts and the resulting deterioration of accuracy (generation of tapered instead of cylindrical bores). The loss of dynamic stability can be caused by cutting forces in cutting tools (chatter), by wind in smokestacks and towers, etc. It can be shown that dynamic stability improves with increasing value of criterion $K\delta$ where K = effective stiffness of the system, and δ = log decrement of the fundamental vibratory mode (Appendix 2). Components of *group (b)* require high stiffness to reduce deflections caused by angular acceleration/deceleration, and high natural frequency in order to reduce time required for transient (usually, start/stop) vibrations to decay below the specified amplitudes.

7.6.2 Stationary and Rotating Around Longitudinal Axis Cantilever Components

To increase stiffness, K, the component of given dimensions must be made of a material with high E. Specific weight (density) of metals is usually increasing with the increasing E (with the exception of beryllium, too expensive for general purpose applications), see Table 1.1. To increase damping by using DVA, its mass ratio, μ must be increased. Since the dimensions and specific gravity of the inertia mass are limited, the effective mass of the cantilever component has to be reduced, thus the "physical contradiction" [36] of this system is the following: the component (e.g., boring bar) must be *in the same time* rigid (thus, *heavy*), and *light*. This contradiction was resolved by *separation of the contradictory*

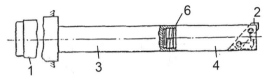

Fig. 7.6.2. Combination boring bar; 1 — clamp; 2 — cutting tool;
3 — rigid root segment; 4 — light overhang segment; 6 — joint.

properties in space [36]. A simple analysis in [37] has shown that stiffness
of a cantilever beam is determined by its root segment (seven-eighths of
the total potential energy in bending is concentrated in the root half of
a uniform built-in/free cantilever beam having a constant cross section
along its length). The effective mass is determined by the overhang segment (three-fourths of the total kinetic energy is concentrated in the end
half of the cantilever beam vibrating in its fundamental mode). Thus,
making the root segment from a high modulus but heavy material would
increase stiffness while not influencing significantly the effective mass
at the end. On the other hand, use of a light material for the overhang
segment assures a low effective mass, while does not affect the stiffness significantly. Such a design, in which the rigid and light segments
are connected by a preloaded joint 6 (Fig. 7.6.2), was suggested in [38].
Fig. 7.6.3b, c, d [37] shows stiffness, effective mass, and natural frequency as functions of segment materials and position of the joint between
the segments as marked on the model in Fig 7.6.3a. It can be seen, that
while the effective stiffness of the combination structure is only insignificantly lower than the stiffness of the structure made of the solid high E
material, the natural frequency (and effectiveness of the DVA mounted
at the overhang end of the structure) are significantly improved. Optimization of the segment lengths (position of the joint) can be performed
using various criteria, such as maximum natural frequency, maximum
effectiveness of DVA, etc., and their combinations with different weight-
ing factors. The in-depth analytical and experimental evaluation of this
concept in application to boring bars (the root segment made of sintered
tungsten carbide, the overhang section made of aluminum), as well as
optimization based on maximization of $K\delta$ criterion are given in [8]. The
optimized boring bars (both stationary and rotating) performed chatter-
free at $L/D = 15$.

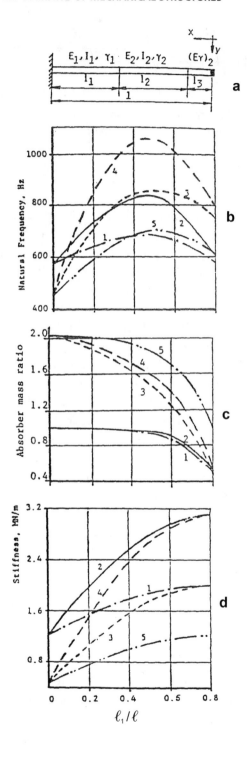

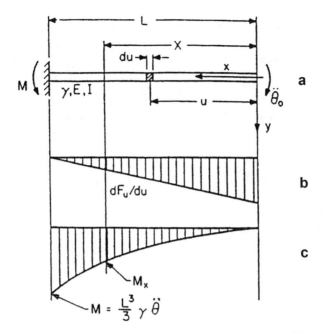

Fig. 7.6.4. Schematic of a revolving massive beam (a) and its loading (b) and (c).

7.6.3 Cantilever Components Rotating Around Transverse Axis

7.6.3a Solid Component

Components of *group b* require high natural frequency (in order to accelerate decay of transient vibrations) and reduction of end point deflections caused by inertia forces associated with acceleration/deceleration of the component. The basic component (robotic arm), made of a material with Young's modulus, E, having a uniform cross section with the cross-sectional moment of inertia, I, and mass-per-unit length, γ, is shown in Fig. 7.6.4a. The arm moves with a constant angular acceleration $\ddot{\theta}_0$. The inertia force acting on an element having an infinitesimal length, du, and located at a distance, u, from the arm end is:

Fig. 7.6.3. Natural frequency (b); absorber mass ratio (c); and effective stiffness (d) of combination boring bar (a). 1 — tungsten/steel; 2 — sintered tungsten carbide/steel; 3 — tungsten/aluminum; 4 — sintered tungsten carbide/aluminum; 5 — steel/aluminum.

$$dF_u = \gamma du[\ddot{\theta}_0(L-u)]. \tag{7.6.2}$$

The intensity of inertia forces described by Eq. (7.6.2) is illustrated by the diagram in Fig. 7.6.4b. The inertia force acting on each element du generates bending moments in all cross sections of the link to the left of the element. At a cross section which is situated at a distance $x > u$ from the end of link, the incremental bending moment generated by dF_u is equal to

$$dM_x = dF_u(x-u) = \gamma\ddot{\theta}_0(L-u)(x-u)du \tag{7.6.3}$$

The total magnitude of moment M_x in cross section x due to inertia forces can be computed by integration of the moment increments generated by all elements du to the right of the cross section x,

$$M_x = \int_0^x \gamma\ddot{\theta}_0(L-u)(x-u)du = \gamma\ddot{\theta}_0\left(\frac{L}{2}x^2 - \frac{x^3}{6}\right). \tag{7.6.4}$$

Moment distribution is illustrated by the diagram in Fig. 7.6.4c. If shear deformations are neglected (which is justified by the slenderness of the component), the bending deflection at the overhanging (free) end is due to the changing curvature of the arm caused by the bending moments. A moment, M_x, acting on a cross section x causes a change in curvature which is equivalent to angular displacement between the end faces of an infinitesimal element dx [39]:

$$d\psi = (M_x/EI)\,dx. \tag{7.6.5}$$

This angular deformation is projected into a deformation of the end of the link in the y-direction:

$$dy = x\,d\psi = (M_x/EI)\,xdx. \tag{7.6.6}$$

The total deflection of the arm end is the sum of increments dy generated by all the cross-sections along the arm or

$$y = \int_0^L \frac{M_x}{EI}xdx = \frac{7}{6}\frac{\gamma\ddot{\theta}_0}{EI}L^5. \tag{7.6.7}$$

Analysis of the derivation steps, Eqs. (7.6.2) through (7.6.7), as well as diagrams in Fig. 7.6.4b and c shows that:

1. The most intense inertia forces are generated in sections near the free end of the link since linear acceleration of a cross-section is proportional to its distance from the center of rotation.
2. The greatest contributors to the bending moment at the built-in end of the arm are inertia forces generated in the sections farthest from the center of rotation since these inertia forces are the most intense, and are multiplied by the greatest arm length.
3. The bending moment magnitude at the center of rotation (at the joint) is the magnitude of the driving torque applied to the link.
4. The deflection at the free end is largely determined by bending moments near the center of rotation since these moment magnitudes are the greatest, and angular deformations near the center of rotation are transformed into linear displacements at the free end through multiplication by the largest arm length.

These observations can be rephrased for design purposes as: a well-designed cantilever component would be characterized by a light free end segment and a rigid segment near the pivot, with not very stringent requirements as to rigidity of the former and/or to the specific weight of the latter.

7.6.3b Combination Link

The above specifications present the same contradiction [36] as was discussed above in Section 7.6.2 for the cantilever components which are stationary or rotate about the longitudinal axis: the component has to be heavy (rigid) and light at the same time. Accordingly, it can similarly be resolved by a combination link design. Such a design is relatively easy to implement by a reliable joining of several segments of the same cross-sectional shape that are made of different materials.

The optimization procedure for such a combination link, however, is very different from the described above and in [8] procedure for a component which is stationary or rotating about its longitudinal axis. To perform the optimization procedure, the equations derived above have to be modified. Firstly, E and γ are no longer constant along the length of the arm, they are constant only within one segment. Secondly, the arm end may carry a lumped mass-payload for the robot arm or the effective mass of the preceding links for the intermediate links, and can be acted upon by the reaction torque from a preceding link. In even further approximation,

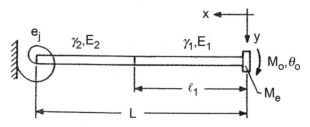

Fig. 7.6.5. A combination arm.

the end-of-link mass may also possess a moment of inertia and be variable (dependent on the system configuration) (Fig.7.6.5). Thirdly, actual links cannot be modeled as built-in beams, since (angular) compliance of the joints can be of comparable magnitude with the bending compliance of the link.

The model in Fig. 7.6.6 incorporates a lumped mass and a moment at the link end, as well as an elastic joint (pivot). For this model, Eq. (7.6.5) will be written as follows [28]:

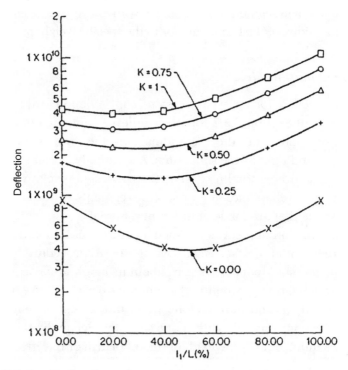

Fig. 7.6.6. Deflection of a combination steel-aluminum arm with various payloads versus length ratio (rigid joint).

$$M_x = M_0 + m_e\ddot{\theta}_0 x + \int_0^{l_1} \gamma_1\ddot{\theta}_0(L-u)du + \int_{l_1}^{x} \gamma_2\ddot{\theta}_0(L-u)(x-u)du \quad (7.6.8)$$

for $L > x > l_1$. If x belongs to another segment $(x < l_1)$, only the first integral would remain, with the integration performed from 0 to x. From Eq. (7.6.8) for $x < l_1$,

$$M_{x1} = M_0 + m_e\ddot{\theta}_0 + \gamma_1\ddot{\theta}_0\left(\frac{L}{2}x^2 - \frac{x^3}{6}\right), \quad (7.6.9)$$

and for $x > l_1$:

$$M_{x2} = M_0 + m_e\ddot{\theta}_0 x + \ddot{\theta}_0\left\{\gamma_1\left[x\left(Ll_1 - \frac{1}{2}l_1^2\right) - \frac{1}{2}Ll_1^2 + \frac{1}{3}l_1^3\right] + \gamma_2\left[-\frac{1}{6}x^3\right.\right.$$
$$\left.\left. + \frac{1}{2}Lx^2 + \left(\frac{1}{2}l_1^2 - Ll_1\right)x + \frac{1}{2}Ll_1^2 - \frac{1}{3}l_1^3\right]\right\} \quad (7.6.10)$$

The moment at the joint $(x = L)$ is:

$$M_L = M_0 + \ddot{\theta}_0\left[m_e L + \gamma_1\left(L^2 l_1 - Ll_1^2 + \frac{1}{3}l_1^3\right)\right.$$
$$\left. + \gamma_2\left(\frac{1}{3}L^3 - L^2 l_1 + l_1^2 L - \frac{1}{3}l_1^3\right)\right] \quad (7.6.11)$$

Analogously, Eq. (7.6.7) must also be integrated in a piecemeal fashion:

$$y = \int_0^{l_1}\frac{M_{x1}}{E_1 I}x\,dx + \int_{l_1}^{L}\frac{M_{x2}}{E_2 I}x\,dx + M_L e_j L = \frac{M_0}{2I}\left(\frac{l_1^2}{E_1} - \frac{l_1^2}{E_2} + \frac{L^2}{E_2}\right)$$
$$+ \frac{\ddot{\theta}_0}{I}\left[\frac{m_e L}{3}\left(\frac{l_1^3}{E_1} - \frac{l_1^3}{E_2} + \frac{L^3}{E_2}\right) + \frac{\gamma_1}{E_1}\left(\frac{Ll_1^4}{8} - \frac{l_1^5}{30}\right)\right.$$
$$+ \frac{\gamma_1}{E_1}\left(\frac{L^4 l_1}{3} - \frac{5}{12}L^3 l_1^2 + \frac{L^2 l_1^3}{6} - \frac{1}{12}Ll_1^4\right)$$
$$\left. + \frac{\gamma_2}{E_2}\left(\frac{11}{120}L^5 - \frac{1}{3}L^4 l_1 + \frac{5}{12}L^3 l_1^2 - \frac{1}{6}L^2 l_1^3 - \frac{1}{24}Ll_1^4 + \frac{1}{30}l_1^5\right)\right]$$
$$+ e_j L\left\{M_0 + \ddot{\theta}_0\left[m_e L^2 + \gamma_1\left(L^2 l_1 - Ll_1^2 + \frac{1}{3}l_1^3\right)\right.\right.$$
$$\left.\left. + \gamma_2\left(\frac{1}{3}L^3 - L^2 l_1 + l_1^2 L - \frac{1}{3}l_1^3\right)\right]\right\} \quad (7.6.12)$$

Equations similar to Eqs. (7.6.11) and (7.6.12) can be derived if the component (link) comprises more than two segments.

If $\gamma_1 < \gamma_2$ and/or $E_2 > E_1$, at a certain l_1 deflection at the free end would be reduced. The reduction will be more pronounced if the mass load at the free end (m_e) is small and/or compliance, e_j, of the joint is small. If the link was initially fabricated from a rigid but relatively heavy material, minimization of its deflection would be accompanied by reduction in its inertia and thus, by reduction in the required joint torque, M_L, required to achieve the prescribed acceleration, $\ddot{\theta}_0$.

Results of computer optimization of the arm structure in Fig. 7.6.5 made of aluminum ($E_1 = 0.7 \times 10^5$ MPa, $\gamma_1 = 2.7$) and steel ($E_2 = 2.1 \times 10^5$ MPa, $\gamma_2 = 7.8$) with different mass loads and joint compliances are given in [28]. Fig. 7.6.6 illustrates these results for the case of rigid joint ($e_j = 0$).

The plots of relative deflection values versus l_1/L in Fig. 7.6.6 are calculated for various mass loads (payloads) characterized by factor $K = m_0/m_{st}$, where m_{st} is the mass of the arm if it were made of steel. For $K = 0$ (no payload) the optimization effect is the most pronounced, with reduction of deflection of 64%, and reduction in driving torque of 66% for $(l_1/L)_{opt} = \sim 0.5$. This case represents circumstances for which reduction of deflections is especially desirable, such as in precision measurements or high-speed laser processing. When a larger payload is used, the effect of the reduced structural mass is less pronounced. However, even at $K = 1$ the end deflection is reduced a noticeable 13%, with the reduction of the required torque of 15%.

An important additional advantage of the optimized combination link is an increase in the fundamental natural frequency of bending vibrations. The increase in natural frequencies (for the case $m_0 = 0$, $e_j = 0$) for the deflection-optimized combination versus a steel link is 59% for $K = 0$, 9% for $K = 0.5$, and 5% for $K = 1.0$. The relationships between the fundamental natural frequency and various parameters of a steel-aluminum combination link are shown in Fig. 7.6.7, in which $Q = \Delta_{st}/\Delta_{ej}$, where $\Delta_{ej} = PL^2 e_j$ is the deflection at the free end loaded by a force P due to the joint compliance alone, and $\Delta_{st} = PL^3/3E_2I$ is the deflection of a solid steel link loaded by the same load due to bending alone.

Although a combination of more than two segments could be beneficial on some occasions, computational analysis of an aluminum-titanium-steel combination versus an aluminum-steel combination has not shown a significant improvement.

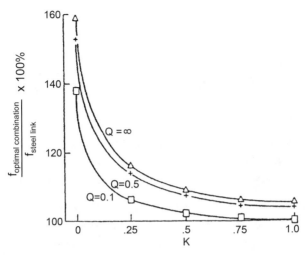

Fig. 7.6.7. Fundamental natural frequency of an optimized combination steel/aluminum link with various payload and joint compliance parameters as a percentage of fundamental natural frequency for a solid steel link.

Use of a light or low-modulus material for the overhang portion of the combination arm reduces its bending stiffness. For example, at $l_1/L = 0.4$ the stiffness of the combination steel/aluminum link is 25% less than the stiffness of the steel link (for a case of loading by a concentrated moment at the link end). In most cases, such a reduction is more than justified by the reduction of the acceleration-induced deflection, reduction in the required driving torque, and by increase in natural frequency, all of which reduce the settling time. In many cases, use of a heavy material, such as steel, is not even considered due to extremely large magnitude of the required driving torque. In such circumstances, the bending stiffness increase in comparison with the solid aluminum link (more than two-fold) would be very significant.

Realization of this concept in actual robots depends on the development of an inexpensive and reliable joining technique for metallurgically different materials comprising the combination component without introducing an extra compliance or extra mass at the joint. One of the approaches is based on using nonlinearity of contact deformations discussed in Chapter 4. Two components pressed together with high contact forces would behave as a solid system. A very light and small tightening device can be (and has been) realized by using a wire made of a shape memory alloy" [38].

7.7 DAMPING ENHANCEMENT TECHNIQUES

7.7.1 Introduction

Damping enhancement in a mechanical system reduces vibration intensity and/or dynamic loads in systems prone to dynamic effects. There are four basic approaches for reducing vibrations/dynamic loads:

1. A major change in the system design: enhancement of structural stiffness and/or modification of natural modes of vibration, see Section 8.6; change parameters of transient processes, such as starting and stopping of driving motors, etc.
2. A change in the working process parameters to avoid a resonance or a self-excitation regime: change of speed of the work organ, e.g. of a cutting tool in a machining center or in a mining machine; change of the work organ design, e.g. change of cutting angles for a tool, of a number or placement of cutting teeth; speed modulation of major motions, such as spindle rpm in machine tools; etc.
3. Use of special anti-vibration devices, usually dampers or dynamic vibration absorbers (DVA), which allow one to enhance dynamic stability, reduce levels of forced vibrations and intensity of dynamic loads, and protect the machine or measuring device from harmful vibrations, all without major changes in the basic design.
4. A simultaneous adjustment of both stiffness and damping in the system in order to increase magnitude of the appropriate stiffness/damping criterion (see Appendix 2).

The anti-vibration devices may be exposed to significant payload forces (e.g., power transmission couplings, see Appendix 5) or weight forces (e.g., vibration isolators [15], or can be installed outside of the load-transmitting path (DVAs, dampers). With the exception of low-damped precisely tuned devices, such as DVAs for systems vibrating with well-defined discrete frequencies, anti-vibration devices are characterized by presence of high-damping elements with or without flexible elements.

Structural damping can be enhanced by a judicious "tuning" of the existing system. The most important high damping "tunable" components of mechanical systems are structural joints. It is shown in Chapter 4 that damping of joints depends on surface finish and lubricated conditions, but also on contact pressures (preload). Usually, damping can be enhanced by reducing the preload force, but this is accompanied by an undesirable stiffness reduction. Thus, "tuning" of a joint involves selecting and realizing such preload force, which results in maximization of a specified criterion combining stiffness and damping parameters, Appendix 2 Frequently, such a criterion is $K\delta$, where K = stiffness and δ = logarithmic decrement associated with the joint. Other examples of "tunable" structural elements include driving motor, whose stronger dynamic coupling with the driven mechanical systems results in damping enhancement, Section 6.6.2 and tuning of high damping tool clamping systems, Section 8.3.1.

Anti-vibration devices have both flexible and damping (energy-dissipating) elements. Although some antivibration devices have only flexible elements (e.g., DVAs without damping) or only damping elements, e.g. Lanchester dampers [40], it is usually more desirable to combine flexibility and damping properties in one device. It can be achieved by using pneumatic, hydraulic, electromagnetic, electrodynamic, and piezoelectric systems, among others, which can relatively easily combine required elastic and damping characteristics, or by using elasto-damping materials.

7.7.2 Dampers

Special damping elements (*dampers*) are used when damping in the elastic element of an anti-vibration device is inadequate. Most frequently used are viscous dampers, Coulomb friction dampers, magnetic and electromagnetic dampers, impact dampers, and passive piezoelectric dampers, among others. An extensive survey of damping systems is given in [41]. Effects of damping on vibration isolation systems are discussed in [15] and [42], and on power transmission systems in Appendix 5. Some practical comments given below may be useful.

1. While *viscous dampers* are not very popular with designers since they usually require a rather expensive hardware (fitted moving parts, seals) and maintenance, they have to be seriously considered in special circumstances. First of all, viscous dampers can be natural in lubricated rotating and translational systems (see Section 5.3).

Secondly, recent progress in development of time-stable electrorheological and magnetorheological fluids, as well as of electronically controlled hydraulic valves, made variation of viscous resistance in the viscous dampers a reasonably easy task. Such damping control frequently represents the easiest implementation of active vibration control systems [43].

2. *Coulomb friction dampers*, e.g., Fig. 7.7.1, use friction between non-lubricated surfaces. Since the friction force is proportional to pressure between the contacting surfaces, there is a need for preloading. Since wear of the non-lubricated contacting surfaces can be significant, it may influence the preload force magnitude if the preloading springs have high stiffness. A low stiffness preloading spring should be deformed for a large amount to create the preload force, thus a small change of deformation due to wear of the frictional surfaces would not noticeably influence the preload force. The Coulomb friction dampers are not effective for small vibration amplitudes. The smaller the vibration amplitude, the stiffer the damper should be. If

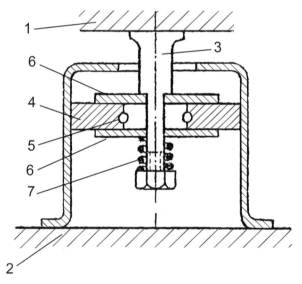

Fig. 7.7.1. 3-D Coulomb friction based damper for translational movements between 1 and 2; 3 — ram; 4 — high friction sectored pads; 5 — ring-shaped spring for horizontal movements; 6 — friction plates for horizontal motions; 7 — preloading spring for horizontal motions.

the stiffness is inadequate, small vibration amplitudes would not result in a relative motion between the frictional surfaces, but would result in elastic deformations in the structure, not associated with a noticeable energy dissipation. No damping action would occur in a Coulomb friction damper if the relative motion between the frictional surfaces is a combination of steady and vibratory motions, and the velocity amplitude of the vibratory motion is less than the velocity of the steady motion. Fig. 7.7.1 shows a Coulomb friction damper providing damping action in three (one vertical and two horizontal) directions.

3. *Elasto-damping materials* usually have relatively low elastic moduli (with some exceptions, see Comment 7 below), allow for large deformations, and have high energy dissipation under vibratory conditions. The most widely used elasto-damping materials are elastomers (rubbers), fiber-based materials (felt), volumetric wire mesh materials, plastics, and some composites. Frequently, both static and dynamic characteristics of such materials are nonlinear. In many cases this is a beneficial factor. Stiffness of elasto-damping materials for dynamic/vibratory loading (dynamic modulus or dynamic stiffness) is usually greater than stiffness at slow (static) loading with frequency below 0.1 Hz. The ratio of dynamic-to-static stiffness is characterized by dynamic stiffness coefficient K_{dyn} (see Chapter 3), which can be up to 5 to 10. Relative energy dissipation in elasto-damping materials in the low frequency range, below ~200 Hz, does not depend strongly on frequency. Accordingly, damping of a material can be characterized by log decrement, δ, which can be as high as 1.5 to 3.0. Usually, both K_{dyn} and δ are amplitude-dependent, see Chapter 3. Thus, such materials can be described as having hysteretic damping with $r = 1$ (see Appendix 1). This fact is very important for applications, especially for designing vibration isolators and vibration isolating systems, since increasing hysteretic damping of isolators, while reducing undesirable resonance amplitudes, does not lead to a significant deterioration of isolation effectiveness (transmissibility) in the after-resonance frequency range (see Appendix 1).

4. *Volumetric wire mesh materials* can be very effective for dampers that have to be used in aggressive environments and under very intense vibrations. They are made from stainless steel cold-drawn

wire, possibly nonmagnetic. In one production process, a net is made from 0.1 mrn to 0.6 mrn diameter wire on a knitting machine. Then the net is wrapped into a "pillow," that is cold pressed in a die under pressure up to 100 MPa (15,000 psi). In another process, a tight spiral is made of 0.15 mrn to 1.0 mrn diameter wire, then the spiral is stretched to 500% to 600% elongation, wrapped into a pillow, and placed into a die where it is compressed to attain the required shape. Wire mesh elements are usually loaded in compression. With increasing load, the number of contacts between the wires is increasing, resulting in stiffness increase. Doubling of the compressive force results in ~1.5 times stiffness increase. The allowable compressive loading Pmax = 3 MPa to 20 MPa (450 psi to 3000 psi) depending on the wire diameter. Allowable dynamic (shock) overloads may reach 8 to 10 P_{max}. Angles between the contacting wires and forces in the contacts are independent random parameters. Any lubrication is squeezed out of the contacts and the friction is, essentially, of dry (Coulomb) type. Due to high friction forces, the volumetric mesh is quite rigid, and its static deformation is qualitatively similar to elasto-frictional connections like one analyzed in Section 4.8.1. For such connections both stiffness and energy dissipation are frequency-independent but strongly amplitude-dependent (energy dissipation increasing, stiffness decreasing with increasing vibration amplitude).

Under dynamic (vibratory) loading the contacting wires are slipping against each other. In the slipping contacts the vibratory energy is dissipated and the lengths of wire segments that can deform in bending and/or tension are increasing. At small relative vibration amplitudes a (amplitude divided by thickness of the element), dynamic loads between the wires are small and slippage occurs only in those contacts where initial contact pressures are low or angles of contacting wires are beneficial for inducing slippage. With increasing a, the number of slipping contacts increases, and amplitudes of the relative motions also increase. This results in increasing energy dissipation (log decrement, δ) and decreasing K_{dyn}. With further increase of a, both compliance and energy dissipation approach their limiting values when all possible slippages are realized. Since the total energy associated with the vibratory motion is proportional to a^2, the relative energy dissipation, $\psi \approx 2\delta$, has a maximum. Figure

3.2 shows typical correlations $\delta(a)$ and K_{dyn} (a) for one type of wire mesh element from recordings of free (decaying) vibrations at $a =$ 0.4×10^{-3}. At low a, δ is small, $\delta = 0.15$ to 0.2, and $K_{dyn} = 8$ to 10. With increasing a, δ quickly rises, reaching $\delta = 1.5$ to 2.0 and peaking at $a = 7 \times 10^{-3}$, and then slowly decreasing. K_{dyn} is monotonously decreasing with increasing a, asymptotically approaching $K_{dyn} = 1$.

It can be concluded that the wire mesh materials are very effective at large vibration amplitudes since they have very high δ at the large amplitudes, can withstand high temperatures and aggressive environments, have low creep rate, and can be loaded with significant forces. However, at low amplitudes they are very stiff and have low damping. A quantitative comparison of wire mesh and elastomeric materials for vibration isolation applications at various vibration amplitudes is given in [15].

5. *Felt* is a fabric produced by combining fibers by application of mechanical motions, chemicals, moisture, and heat, but without waving or knitting. Felt is usually composed of one or several grades of wool with addition of synthetic or plant fibers. The best grades of felt are resistant to mineral oils, greases, organic solvents, cold/dry environment, ozone, and UV light. Felt structure is similar to that of the wire mesh (chaotic interaction between the fibers) but the felt fibers are much more compliant than the steel wires and have their own material damping, while material damping of the steel wires is negligible. As a result, the amplitude dependencies for both K_{dyn} and δ are less steep (Fig. 3.2). The allowable compression loads on felt pads are much lower than for wire mesh, $P_{max} = 0.05–0.35$ MPa (7.5 psi to 53 psi), and up to 0.8 MPa to 2.0 MPa (12 psi to 30 psi) for thin pads.

6. *Rubber* (Appendix 4) is composed of a polymeric base (gum) and inert and active fillers. Active fillers (mostly, carbon black) are chemically bonded to the polymeric base and develop complex interwoven structures that may break during the deformation process and immediately re-emerge in another configuration. The character of these breakage-reconstruction events is discrete, similar to Coulomb friction, whereas a body does not move until the driving force reaches the static friction (stiction) force magnitude. When the body stops, the static friction force is quickly reconstructed. Thus, the deformation mechanism of the active carbon black structure in

a filled rubber is somewhat analogous to the deformation mechanism of the 3-D wire mesh structures. Accordingly, dynamic characteristics of rubber are similar to those of felt (discussed above). They are composed of dynamic characteristics of the active carbon black structure (K_{dyn} and δ are strongly amplitude-dependent and frequency-independent), and of dynamic characteristics of the polymer base (K_{dyn} and δ are amplitude-independent; frequency-dependency in 0.01 Hz to 150 Hz range is different for different types of rubber). The relative importance of these components depends on contents of active fillers; for lightly filled rubbers the amplitude dependencies are not noticeably pronounced. Some rubber blends have ingredients preventing building of the carbon black structures [30c]; these blends do not exhibit the amplitude dependencies even with heavy carbon black content.

Fig. 7.7.2 [44] shows amplitude dependencies for shear modulus, G, and for log decrement, δ, for butyl rubber blends which differ only in percent contents of carbon black. These dependencies are very steep in their areas of change (up to 15:1 change in G; up to 8:1 change in δ) and demonstrate "peaking" of δ at a certain amplitude. A lack of consideration for the amplitude dependencies of the dynamic characteristics of elasto-damping materials while designing damping-enhancement means may result in a very poor correlation between the expected and the realized system characteristics, as well as in an inadequate performance.

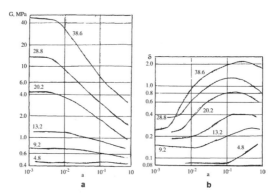

Fig. 7.7.2. Amplitude dependencies of (a) shear modulus G and (b) log decrement δ for butyl rubber for different percentage content of carbon black.

7. *High modulus elasto-damping materials* are also available. They are represented by metals having high internal energy dissipation. The high damping metals group includes lead, some magnesium alloys, nickel-titanium (NiTi) alloys ("shape memory" or "superelastic" alloys, e.g., see Section 2.7.4), etc. The highest modulus and strength are associated with NiTi alloys, which demonstrate high damping (log decrement $\delta = 0.5$) when prestressed to -70 MPa (10,000 psi) (see Section 8.3.1). Due to high modulus/strength, the high damping metals can be used as structural materials for critical parts, such as clamping devices.

8. Optimization of dynamic behavior of dynamically loaded or vibrating mechanical systems can be helped by using, wherever possible, *criterial expressions connecting stiffness* and *damping* parameters, Appendix 2. In some cases, an indiscriminate increase in damping is desirable. Such cases are usually associated with resonating systems. The resonance can occur in a structure experiencing translational vibrations or in a torsional (e.g., power transmission) system. Such cases in which δ can be considered as a criterion are represented by power transmission couplings as described in Appendix . . . .

In many cases, which are addressed in several chapters here, the criterion connecting stiffness and damping is $K\delta$. It is very important for optimization of dynamic stability of a system (its resistance to development of a self-excitation process). Although direct applications of this criterion have been addressed in several sections, a more involved practical case of using this criterion for structural optimization of cantilever boring bars is described in [8]. In this case, shifting the joint in Fig. 7.6.2 between the "stiff" and the "light" segments of the bar is changing both effective stiffness and effective mass of the bar. Increase in the length of the "light" segment reduces the stiffness but can be used to enhance effectiveness of the dynamic vibration absorber, thus enhancing the damping capacity. However, the problem is complicated since the dynamic system has more than one degree of freedom.

It is shown in Appendix 2 that quality of vibration isolation of precision objects can be characterized in some typical cases by the criterion K/δ. Use of this criterion allows one to optimize selection of materials for vibration isolators depending on amplitude and frequency, which are prevalent in the specific cases.

The influence of mount parameters on chatter resistance of the mounted machine tool is shown to be characterized by criterion $K^{3/2}\delta$. Knowledge of this criterion allows one to optimize material selection for mounts (vibration isolators), which are used for installation of chatter-prone machines and equipment.

Stiffness/damping criteria are given in Appendix 2 also for some other cases of machinery installation.

7.7.3 Dynamic Vibration Absorbers

A *dynamic vibration absorber* (DVA) is a dynamic system (frequently a single- degree-of-freedom mass/damped spring oscillator) attached to a body whose vibration level has to be reduced. An appropriate tuning of a DVA results in reduction of vibration amplitudes of the vibrating body, while the mass of the absorber ("inertia mass") may exhibit high vibration amplitudes. The basics of operation of a DVA are provided in textbooks and handbooks on vibration, e.g. [40]. The state of the art in design and application of DVAs is extensively presented in [45]. The latest developments are surveyed in [46]. Several practical comments on design of effective DV As are given below.

1. *Tuning parameters* needed for the most effective performance of a DVA are derived in vibration handbooks for the case of a sinusoidal (harmonic) vibration of the vibrating body. However, it is usually not explained that these *tuning parameters* are not universal and *depend on characteristics of the vibration* that needs to be suppressed. It is important to understand that if vibrations of the body whose vibrations are to be suppressed by attaching a DVA are not sinusoidal (e.g., random vibrations), or if the absorber has to enhance dynamic stability of the system rather than to suppress vibration amplitudes, the tuning parameters may significantly change, [8]. Another special case that may require a special dynamic analysis to develop optimal tuning conditions is reduction (acceleration of settling process) of transient vibrations.

2. Effectiveness of a properly tuned DVA is determined mostly by its *mass ratio* μ, which is the ratio between the inertia mass of the absorber and the effective mass of the body whose vibration characteristics have to be modified. Although the inertia mass size is

limited by packaging constraints, as in the case of boring bar (see Section 7.5.2), or by economics, it is often forgotten that the mass ratio has two components, one being the inertia mass and the other, the effective mass of the vibrating body. The weight/mass reduction of the vibrating body is frequently feasible and, if realized, it allows to significantly enhance effectiveness of the absorber. An example of such an approach is given in Sections 7.6.2 and 7.6.3 and in [8].

3. In many applications, it is desirable to suppress vibrations of the vibrating body in a wide frequency range. This can be achieved by using a high damping connection between the inertia mass and the vibrating body. However, elastomeric or polymeric materials possessing high damping capacity have their damping and stiffness parameters influenced by many factors: amplitude and frequency of vibrations; temperature; process variation in making the material; etc. As a result, computational optimization of the absorber can hardly be realized at the "first try." It is beneficial to use tunable connections for a DVA that allow one to correct the tuning imperfections. The "tunability" can be easily attained by using nonlinear elastomeric elements.

REFERENCES

[1] Levina, Z.M., Zwerev, I.A., 1986, "Finite Element Analysis of Static and Dynamic Characteristics of Spindle Units", Stanki i instrument [Machines and Tooling], No. 10, pp. 7–10 [in Russian].

[2] Bushuev, V.V., 1991, "Compensation of Elastic Deformations in Machine Tools", Stanki i instrument, No. 3, pp. 42–46 [in Russian].

[3] Bushuev, V.V., 1991, "Design and Loading Patterns", Stanki i instrument, No. 1, pp. 36–41 [in Russian].

[4] Microlution Co., Catalog, 2007.

[5] Neugebauer, R., Hipke, T., 2006, "Machine Tool with Metal Foam", Advanced Engineering Materials, vol. 8, No. 9, pp. 858–863.

[6] Wieck, J., "Compensator for Coordinate-Measuring Machine", Patent of German Democratic Republic (East Germany) No. 133,585.

[7] Bushuev, V.V., 1989, "Paradoxes of Design Solutions", Stanki i instrument, No. 1, pp. 25–27 [in Russian].

[8] Rivin, E.I., Kang, H., 1992, "Enhancement of Dynamic Stability of cantilever Tooling Structures", Intern. J. of Machine Tools and Manufacture, vol. 32, No. 4, pp. 539–562.

[9] Neugebauer, R., Denkena, B., Wegener, K., 2007, "Mechatronic Systems for Machine Tools", Annals of the CIRP, Vol. 56/2, pp. 657–686.

[10] acticut.com.

[11] Catskill, A., et al, 1997, "Development of a High-Performance Deep-Hole Laser-Guided Boring Tool: Guiding Characteristics", Annals of the CIRP, vol. 46.

[12] Koren, Y., Pasek, Z. J., Szuba, P., 1999, "Design of a Precision Agile Line Boring Station", Annals of the CIRP, Vol. 48/1, pp. 313–316.

[13] Dornhöfer, R., Kemmerling, K., 1986, "Boring with Long Bars", VDI Z., vol. 128, pp. 259–264 [in German].

[14] Simnofske, M., Hasselbach, J., 2006, "The Increase of the Dynamic and Static Stiffness of a Grinding Machine", ASME IDETC/CIE Conference, DETC2006-99740, pp. 23–27.

[15] Rivin, E.I., 2005, "Passive Vibration Isolation", ASME Press.

[16] Rivin, E.I., 1971, "Active Vibration Isolators and Installation Systems", NIIMASH, Moscow, 80 pp. [in Russian].

[17] Push, V.E., Rivin, E.I., Shmakov, V.T., 1970, "Vibration Isolator", USSR Certificate of Authorship 261,831.

[18] DeBra, D.B., 1992, Vibration Isolation of Precision Machine Tools and Instruments", Annals of the CIRP, vol. 41/2, pp. 711–718.

[19] Hailer, J., 1966, "A Self-Contained Leveling System for Machine Tools – an Approach to Solving Installation Problems", Maschinenmarkt, vol. 72, No. 70 [in German].

[20] Rivin, E.I., 1970, "A Device for Automatic Leveling", USSR Certificate of Authorship 335,448.

[21] Onoda, J., Endo, T., Tamaoki, H., Watanabe, N., 1991, "Vibration Suppression by Variable Stiffness Members", AIAA Journal, Vol. 29, No. 6, pp. 977–995.

[22] Timoshenko, S.P., Gere, J.M., 1961, "Theory of Elastic Stability", McGraw-Hill, N.Y., 541 pp.

[23] Jubb, J.E.M., Phillips, I.G., 1975, "Interrelation of Structural Stability, Stiffness, Residual Stress and Natural Frequency", J. of Sound and Vibration, vol. 39 (1), pp. 121–134.

[24] Ivanko, S., Tillman, S.C., 1985, The Natural Frequencies of In-Plane Stressed Rectangular Plates", J. of Sound and Vibration, vol. 98 (1), pp. 25–34.

[25] Platus, D.L., "Vibration Isolation System", U.S. Patent 5,178,357.

[26] Petasiuk, G.A., Zaporozhskii, V.P., 1991, "Enhancement of Axial Stiffness of Grinding Wheels", Sverkhtverdie materiali [Superhard Materials], No. 4, pp. 48–50 [in Russian].

[27] Feodosiev, V.I., 2005, "Advanced Stress and Stability Analysis: Worked Examples", Springer.

[28] Rivin, E.I., 1988, "Mechanical Design of Robots", McGraw-Hill, N.Y., 328 pp.

[29] Rivin, E.I., Karlic, P., Kim, Y., 1990, "Improvement of Machining Conditions for Turning of Slender Parts by Application of Tensile Force", Fundamental Issues in Machining, ASME PED, vol. 43, pp. 283–297.

[30] Rivin, E.I., "Method and Means for Enhancement of Beam Stiffness", U.S. Patent 5,533,309.

[31] Rivin, E.I., Panchal, P., 1996, "Stiffness Enhancement of Beam-Like Components", ASME.

[32] Kuklev, L.S., Bobreshov, S.A., Sokolov, A.V., "Self-Hardening Liquid", USSR Certificate of Inventorship 500,007.

[33] Philippe, F., Roland, D., Didier, L., "Method of Machining of Thin-Walled Parts", French Patent Application 2,664,522.

[34] Rivin, E.I., 1983, "Cost-Effective Noise Abatement in Manufacturing Plants", Noise Control Engineering Journal, No. 6, pp. 103–117.

[35] Lobontiu, N., 2008, "Dynamics of Microelectromechanical Systems", Springer, 402 pp.

[36] Fey, V.R., Rivin, E.I., 2005, "Innovation on Demand", Cambridge University Press.

[37] Rivin, E.I., 1986, "Structural Optimization of Cantilever Mechanical Elements", ASME J. of Vibration, Acoustics, Stress and Reliability in Design", vol. 108, pp. 427–433.

[38] Rivin, E.I., Lapin, Yu.E., "Cantilever Tool Mandrel", U.S. Patent 3,820,422.

[39] Rivin, E.I., et al, 1987, "A High Stiffness/Low Inertia Revolute Link for Robotic Manipulators", in Modeling and Control of Robotic Manipulators and Manufacturing Processes, ASME, pp. 253–260.

[40] DenHartog, J.P., 1985, "Mechanical Vibrations", Dover Publications, 436 pp.

[41] Johnson, C.D., 1995, "Design of passive damping systems", Trans. of ASME, Special 50th Anniversary Design Issue, Vol. 117, pp. 171–176.

[42] Rivin, E.I., 2007, "Vibration isolation of precision objects", S)V Sound and Vibration, #7.

[43] Karnopp, D., 1995, "Active and semi-active vibration isolation", Trans. of ASME, Special 50th Anniversary Design Issue, Vol. 117, pp. 177–185.

[44] Davey, A.B., and Payne, A.R., 1964, "Rubber in Engineering Practice", McLaren & Sons, London.

[45] Korenev, B.G., Reznikov, L.M., 1993, "Dynamic Vibration Absorbers", John Wiley & Sons, 312 pp.

[46] Sun, J.Q., Jolly, M.R., Norris, M.A., 1995, "Passive, adaptive, and active tuned vibration absorbers—a survey," Trans. of ASME, Special 50th Anniversary Design Issue, Vol. 117, pp. 234–242.

CHAPTER 8

Use of "Managed Stiffness" in Design

In many cases mechanical systems benefit from increasing stiffness of their critical components. However, stiffness increases are often achieved by "beefing up" the structural components, increasing contact areas and preloading forces (to increase contact stiffness), thus requiring larger preloading devices. Such changes usually lead to greater costs and to undesirable increases of size and weight of the unit being designed. While in many cases the required stiffness can be reduced if the damping is enhanced, see Appendix 2, the desirable results, usually achieved by stiffness enhancement, can be also achieved by much less costly techniques if the role of the stiffness parameter were clearly understood. Often, stiffness is enhanced to just reduce deformations in the system. Obviously, the deformations can be reduced either by stiffness enhancement, or by reduction of forces causing the deformations.

There are many important cases where stiffness reduction is beneficial or where there is an optimal range of stiffness values. Some examples of such cases are as follows: generation of specified, constant in time, forces, e.g., for preloading bearings, cam followers, etc.; vibration isolators, force measuring devices (load cells) in which a compromise must be found since the reduction of stiffness improves sensitivity of the device but may distort the system and/or the process being measured; compensating resilient elements for precision overconstrained systems; reduction of stress concentration; improvement of geometry and surface finish in metal cutting operations by intentional reduction of stiffness of the machining system; use of elastic elements for limited travel bearings and guideways; use of anisotropic components having significantly different stiffness in different directions; trading off stiffness for addition of damping into the system; etc. This chapter addresses some of these issues many of which deal with machining systems, but also with general mechanical systems and design components.

8.1 CUTTING EDGE/MACHINE TOOL STRUCTURE INTERFACE

8.1.1 Introduction

Modern machine tools are characterized by high stiffness, high installed power, and high spindle rpm [1]. These expensive features are implemented in order to better realize beneficial properties of state-of-the-art cutting materials, such as coated carbides and high speed steels (HSS), cubic boron nitride (CBN), and polycrystalline diamond (PCD). While a significant progress had been achieved in both cutting materials and machine tool designs, in many cases the weakest link in the machining system is the tooling structure which serves as an interface between the cutting insert and the machine tool. Inadequacy of the tooling structure results in excessive static deflections limiting the achievable accuracy, and in forced and self-excited vibrations limiting the cutting regimes as well as surface finish of the machined surface. Stiffness and damping parameters as well as performance characteristics of machine tools are continuously advancing, but the cutter-machine tool interface (the tooling structure) characteristics are lagging, thus the importance of tooling structures for advanced machining systems is increasing.

The tooling structure comprises attachment devices for the cutting inserts; the tooling proper, which can be a solid structure or a modular system comprising several joined components; toolholder with a clamping device for the tool; and the toolholder-machine (or toolholder-spindle) interface (tapered, cylindrical, toothed, etc.). In automated machine tools (Computer Numerical Controlled or CNC machine tools), the tools are changed in accordance with the programmed sequence, and the majority of tools are of cantilever design, with their external dimensions determined by the machined part design and by process limitations. This, together with inevitable compliances in the numerous joints and the very heavy cutting regimes typical for the state-of-the-art cutting materials, leads to inaccuracies, to microvibrations resulting in poor surface finish and shorter life of the inserts, and also to chatter vibrations whose onset during an automated machining process can lead to serious damage both to the tool and to the machined part.

The latest trend in machine tool design is to increase spindle rpm in order to utilize more fully the enhanced cutting capabilities of advanced cutting materials and to reduce the cutting forces.

These and other issues indicate the need for development of tooling designs which do not degrade the performance of advanced machine tools and cutting materials. These are difficult problems, which call for novel design concepts. Also, investment in tooling structures and cutting tools amounts to at least 10% of the total cost of a CNC machine tool, thus it is important to utilize tools with a maximum efficiency, since they represent a large portion of the overall investment.

It is universally accepted that high stiffness of a machining system is a necessary condition for successful performance of a cutting process. However, this statement is one-sided and in some cases incorrect.

Broadening and correcting this statement may have serious implications for development of tooling and of machining systems in general. Larger deformations due to cutting forces degrade accuracy of the machined surface and require additional machining passes to achieve required dimensions. Potential chatter vibrations call for reduction of machining regimes, and high frequency micro-vibrations cause reduction of life of the cutting edge (wear, cracking).

While increase of stiffness would improve machining performance (with all other parameters being the same), it may cause undesirable changes in other parameters. If stiffness is enhanced by "beefing up" the tool and the spindle and its bearings, then masses, natural frequencies, damping, properties and overall costs deteriorate. If stiffness is changed by tightening of structural joints, damping is usually declining (Chapter 4 and Appendix 2). For majority of machining operations the product of stiffness and damping parameters $K\delta$, is determining chatter resistance of the system, see Appendix 2. Since both accuracy and dynamic stability are determined by deformations caused by cutting forces, reduction of cutting forces must be considered, along with increasing the $K\delta$ criterion. The final decision is determined by economics. Some techniques for reduction of cutting forces are described below. Some of these effective techniques were known for a long time but rarely, if at all, used.

8.1.2 Techniques for Reduction of Cutting Forces

Inaccuracies due to static or qusi-static deformations in the machining system are especially important for finishing passes with relatively small cutting forces. During roughing regimes with large forces and deformations, accuracy is often only a second concern as compared with the rate of metal removal (productivity) limited by onset of chatter. It is worthwhile to survey some techniques for reduction of cutting and/or friction forces in the cutting zone for both roughing and finishing operations, in addition to selecting cutting angles, cutting insert materials and coatings, and dividing cutting forces by using multi-edge cutting heads. It should be noted, however, that multi-edged cutting heads can be associated with stiffness problems in edge-adjustment mechanisms. Their deformations can be as great as 25 μm. However, with a proper design, critical components of these mechanisms caused by contact and bending deformations can be adequately corrected and their effect on machining accuracy reduced.

PCD and, to a lesser degree, HSS and carbide cutting inserts with hard coatings have low friction coefficients and generate reduced cutting forces. A similar effect and increased tool life were achieved by implanting atoms of elements Cl, Br, I, S, In, Ga, Sn to the depth of about 0.4 μm [2] into the cutting edge. A similar treatment for HSS inserts described in [3] resulted in 20-30% reduction of the friction coefficient and in two-to-three times longer life of the cutting edge. The implantation is performed by gas ion generators with or without magnetic field.

Introduction of ultra-high pressure (70MPa to 280 MPa) lubricoolant into the cutting zone reduces the cutting force by ~50% while improving other performance parameters.

Significant reduction of the cutting force in machining titanium and its alloys, notorious for their toughness and friction in the cutting zone can be achieved by saturating the surface layer of the workpiece with hydrogen at 700°C to 800°C (0.2%-0.8% by weight). After machining, the remaining hydrogen is removed by heating the workpiece in vacuum. Besides a drastic reduction of the friction forces, a four-fold increase in tool life and a sixteen-fold reduction in machining time have been recorded. This technology was extensively used in aerospace industry in the former Soviet Union.

The cutting force can be reduced by heat-induced softening of the workpiece material in the cutting zone. Use of CO_2 laser for the cutting zone heating of hardened steel, high strength alloys, and ceramics resulted in 55% to 90% reduction of cutting forces, three-fold reduction of vibration amplitudes, reduced tool wear and improved geometry of the machined part. Similar results can be obtained with plasma heating. Significant reduction of the cutting forces allowed turning of slender shaft $L = 2700$ mm, $D = 65$ mm without a steady rest.

Conventional lubricoolants significantly reduce both cutting and friction forces. Similar results can be achieved by using chilled ionized air (a technique extensively used in military industry in the former Soviet Union).

Significant reduction of both cutting and friction forces in the cutting zone can be realized by introduction of additional motion(s) into the cutting zone. A 35% to 40% reduction was achieved by using self-propelled rotary tools for turning titanium alloys and AlSi composites.

There were many studies surveyed in [1] and [4], showing large reductions of cutting and friction forces when low frequency vibrations (20 Hz to 1,000 Hz) and /or ultrasonic vibrations (20 KHz to 40 KHz) had been applied, simultaneously or separately), to the cutting zone.

8.1.3 Influence of Stiffness and Damping in the Cutting Zone on Cutting Forces and Tool Life

Influence of stiffness on life of cutting inserts is not always straightforward. In addition to direct influence of stiffness on accuracy due to reduction of deformations caused by cutting forces, also chatter vibrations, and frequently micro-vibrations, are critical for understanding tool wear and cutting forces. The latter factors are determined by a combination of stiffness and damping usually expressed as the $K\delta$ criterion. This understanding helps to explain some inconsistencies in studies of influence of stiffness at the cutting zone on cutting forces and toot life. The most comprehensive studies of correlation between stiffness and tool life are reported in [5] and [6]. Wear of superhard tool inserts was studied in [5] for two machining conditions: turning using CBN inserts and milling using PCD inserts. In the first (*turning*) set-up three cases of machining alloyed steel were studied: stiff workpiece/stiff toolholder (workpiece stiffness $k_w = 3$ N/μm, toolholder stiffness $k_t = 65$ N/μm); compliant workpiece/stiff toolholder ($k_w = 0.35$ N/μm to 2 N/μm); compliant workpiece/compliant toolholder ($k_t = 3.5$ N/μm). Although crater wear was increasing with reducing stiffness, flank wear was minimal for the second case (about one half of flank wear for the first and third cases). The static component of cutting force (for the same cutting regimes) was 870 N for the first case; 370 N for the second case, and only 270 N for the third case. In the second set-up (milling with PCD) cutter stiffness was $k = 9$ N/μm to 24 N/μm for the first case, 7 N/μm to 10 N/μm for the second case, and $k = 3.5$ N/μm to 4 N/μm for the third case. The lowest flank wear was observed for the first case (about one half of the flank wear for the second and third cases). The static cutting force component was the lowest in the second case. A significant, and also somewhat contradictory, case of influence of the tool compliance on cutting forces is described in Section 8.1.4 below.

   To determine influence of stiffness of the clamping device for carbide cutting inserts on wear, milling with a two-teeth milling cutter was performed in [6] with different overhangs of the inserts, thus changing their stiffness within the range $k_t = 18$ N/μm to 54 N/μm. Fig. 8.1.1 shows correlation between k_t and flank wear h_f, number of loading cycles N before a microcrack occurs, and length l_c of the microcrack. It can be seen that there is an optimal stiffness (in this case about 36 N/μm) which is associated with the minimum flank wear. This effect is not the same for all grades of carbide inserts. For some grades, increase of stiffness was always correlated with reduction of flank wear. It was observed that fracture mechanism is different at the maximum stiffness (ductile intra-grain fracture) and at the minimum stiffness (brittle intra-grain fracture). Development of microcracks may be caused by high frequency vibrations since they increase concentration of vacancies in the crystallic structure, increase energy dissipation on its defects, and impede heat transfer from the cutting edge. These effects selectively depend on frequency of micro vibrations, thus explaining the non-monotonous dependance of wear on stiffness. While similar effects in [5] can be to a certain extent explained by damping variation for the studied cases, set-ups in [6] seem to be characterized by the same damping. Damping is positively influencing tool life. However, details of its effects on high frequency micro-vibrations of cutting inserts are not yet totally clear.

Stiffness and damping characteristics of the attachment system between the cutting insert and the toolholder are quite important. Using a

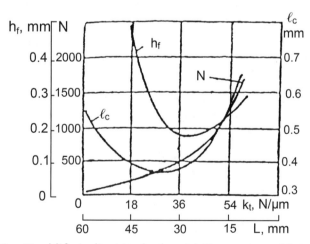

Fig. 8.1.1. Tool life indicators h_f, l_c, and N versus machining system stiffness k_t (cutting insert overhang L of a face milling cutter).

steel shim between the insert and the holder reduces the effective stiffness at the edge by half, while a carbide shim results in a much lesser stiffness reduction. Bending stiffness of the tool is greatly influenced by contact deformations in the joint between the tool and toolholder causing deviation of clamping condition from the ideal "built-in" condition. Damping characteristics of the tool at large vibration amplitudes are due to normal contact deformations, and at small amplitudes are due to tangential contact deformations. Carbide shims do not significantly influence stiffness but enhance damping. They also improve consistency of the insert clamping conditions. This may explain frequent application of carbide shims in commercial milling cutters. Life of cutting inserts also can be extended by a judicious selection of shims. Similar results can be achieved by attaching inserts to the toolholder by gluing. The adhesive layer has elevated damping and its relatively low stiffness alleviates or eliminates altogether stress concentrations due to clamping.

8.1.4 Machining Systems with Intentionally Reduced Tool Stiffness

8.1.4a Cutting Tools with Reduced Normal Stiffness

While the common approach is to increase stiffness of the machining system as much as possible, there are many cases in which a judicious reduction of stiffness resulted in improvements of the machining process. The most common example is a "goose neck" or "swan neck" tool (Fig. 8.1.2). A possible explanation of chatter-abatement properties of such tool is in

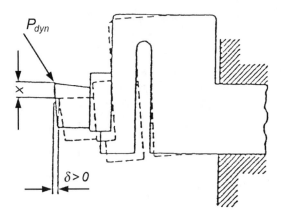

Fig. 8.1.2. Parting "goose neck" tool holder deflected (x) by dynamic cutting force P_{dyn}, so that cutting edge backs away from workpiece by δ.

"retraction" of the cutting edge out of the workpiece when the dynamic force is increasing, thus stabilizing the chip thickness (Fig. 8.1.2 [7]).

A successful application of this concept to milling is illustrated by Fig. 8.1.3 [8]. The face milling cutter has spring-shaped inserts (Fig. 8.1.3a). Kinematics of the cutter is shown in Fig. 8.1.3b. When an insert enters the front segment A to B in Fig. 8.1.3c, the nominal depth of cut is large and the spring is deformed by cutting forces. The initially machined surface is traveled by the cutter again when it is at the back (C to D in Fig. 8.1.3c),

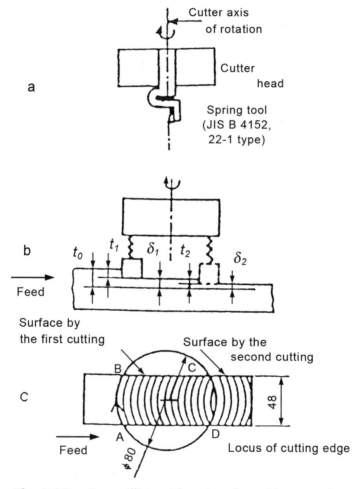

Fig. 8.1.3. Face milling with spring-shaped inserts. a-Insert; b-Schematic of the cutting process; c-Tooth marks and cutting directions.

with much smaller nominal depth of cut (finishing pass). Two passes, the second one in the opposite direction to the first (Fig. 8.1.3c), result in a much better surface finish as compared with conventional milling (R_a is reduced three-fold), and also in greatly reduced residual stresses.

Compliant mounting/clamping devices for grinding wheels using super hard (diamond and CBN) abrasives were shown to be free from chatter vibrations as well as from wheel waviness generated by chatter vibrations. One example of resilient mounting is shown in Fig. 8.1.4. Wheel 5 is mounted on tapered bushings 2, 6. When bushing 6 is shifted in axial direction by activating nut 7, damping ring 4 is deformed, thus changing its radial stiffness. Axial stiffness is provided by Belleville springs 3, adjustable by nut 8. Damping in the axial direction is generated by slipping of springs 3 against faces of wheel 5. The design was used successfully for electrochemical-abrasive grinding of hard to machine materials.

8.1.4b Cutting Tools with Reduced Tangential Stiffness

Enhancement of chatter resistance of the machining system by reduction of stiffness in the direction tangential to the machined surface had been first proposed in [9]. Fig. 8.1.5 shows turning (a) and boring (b) tools tested in [9]. In the test set-up for turning (grooving) in Fig. 8.1.5a, tool 9

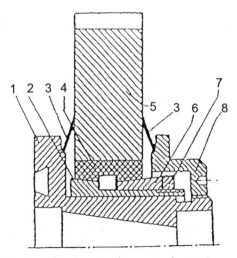

Fig. 8.1.4. Elastic attachment of grinding wheel.

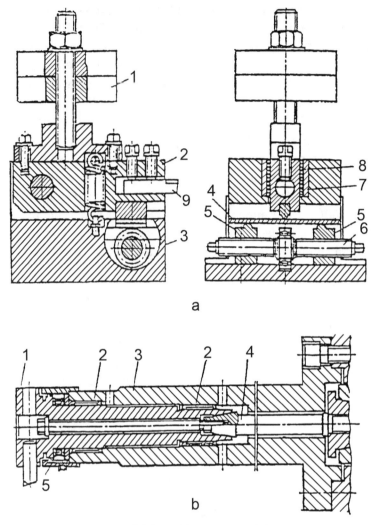

Fig. 8.1.5. Turning tool (a) and boring bar (b) with high tangential compliance to reduce chatter [9].

is clamped in holder 2 having angular mobility provided by flat spring 4. Its stiffness can be adjusted (tuned) by screw 6 changing positions of its supports 5. Variation of size and location of weight 1 results in changing moment of inertia of tool 9 assembly, thus allowing to tune the tool natural frequency. Friction between toolholder 2 and housing 3 is reduced by roller guides 7, 8. Use of this system resulted in increase of the stable depth of cut by up to 2.5 times.

Cutting insert in Fig. 8.1.5b is attached to boring head 1 connected to toolholder/bar 3 ($L/D = 9$) via torsional spring 4. Head 1 can move with low friction relative to bar 3 on hydrostatic connections 2 and thrust bearing 5. Test results demonstrated that stable cutting regimes for aluminum and steel workpieces were limited only by strength of the cutting insert, no chatter was observed.

The tool designs used in [9] are quite complicated and bulky, using hydrostatic connections and precision springs. Analytical treatment is very involved. A more practical design of boring bars using the tangential stiffness reduction concept is proposed and tested in [10]. Since tangential motion of the tool tip is considered, it adds a degree of freedom to the boring bar model (Fig. 8.1.6), and equations of motion become as follows:

$$M\ddot{X} + C_x\dot{X} + K_xX = F\cos\beta$$
$$I\ddot{\Phi} + C_t\dot{\Phi} + K_t\Phi = FR\sin\beta. \tag{8.1.1}$$

Here X = linear vibratory displacement of the bar at the tool tip; Φ = torsional vibratory displacement of the tool head; M = equivalent mass of the boring bar (including the tool head) reduced to the tool tip position; I = moment of inertia of the tool head; C_x = translational damping coefficient of the bar; C_t = torsional damping coefficient of the tool head subsystem; F = dynamic cutting force; R = distance from the tool tip to the axis of the boring bar; β = angle between the resulting cutting force direction and X-direction, approximately 60 deg [11].

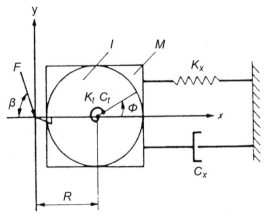

Fig. 8.1.6. Model of boring bar with torsionally compliant head.

The radial and tangential motions in Eq. (8.1.1) are dynamically coupled only through the cutting force F and can be transformed into a set of coupled homogeneous equations [12]:

$$\ddot{X} + 2(\xi_x + \xi_{cx}\cos\beta)\omega_{nx}\dot{X} + (\omega_{nx}^2 + \omega_c^2\cos\beta)X + \frac{2\xi_{c\varphi}\omega_{n\varphi}\cos\beta}{\mu}\dot{\Phi} = 0$$

$$\ddot{\Phi} + 2(\xi_\varphi + \xi_{c\varphi}\sin\beta)\omega_{n\varphi}\dot{\Phi} + \omega_{n\varphi}^2\Phi + 2\xi_{cx}\omega_{nx}\mu\sin\beta\dot{X} + \omega_c^2\sin\beta\dot{X} = 0$$

$$(8.1.2)$$

where:

$$\mu = \frac{M}{I}R^2; \frac{C_x}{M} = 2\xi_x\omega_{nx}; \frac{K_x}{M} = \omega_{nx}^2; \frac{C_t}{I} = 2\xi_\varphi\omega_{n\varphi};$$

$$\frac{K_t}{I} = \omega_{n\varphi}^2; \frac{C_c}{M} = 2\xi_{cx}\omega_{nx}; \frac{K_c}{M} = \omega_c^2; \frac{C}{I} = 2\xi_{c\varphi}\omega_{n\varphi}$$

μ = mass ratio of the torsional head; ω_{nx}, $\omega_{n\phi}$ are partial angular natural frequencies of the boring bar and torsional head subsystems in X and torsional directions, respectively; ζ_x, ζ_ϕ are the respective damping ratios of these subsystems; ω_c = the angular cutting frequency; and ζ_{cx}, $\zeta_{c\phi}$ are cutting damping ratios in X and the torsional directions.

Stability of equations (8.1.2) was analyzed using Routh-Hurvitz criteria [12] with the assumption that the tangential loop of the machining system is always stable, $\zeta_{c\phi} > 0$. Effectiveness of introducing compliance into the tangential loop ("torsionally compliant head", TCH) can be characterized by the critical value of ζ_{cx}, which corresponds to the stability limit of the system. The TCH is the more effective, the larger the magnitude of the negative value of ζ_{cx} for which the TCH can compensate. Thus, smaller (more negative) values of ζ_{cx} correspond to better stability margins.

An analysis of equations (8.1.2) shows that the critical value of ζ_{cx} does not depend on mass ratio, μ. Its magnitude and effectiveness of the TCH is increasing with increasing R. The optimum TCH tuning corresponds to the frequency ratio $\omega_{n\phi}/\omega_{nx} = 0.9$ to 1.0. Effectiveness of the TCH *increases with reduction of damping in the torsional system* (Fig. 8.1.7).

Design of a TCH should provide the required torsional stiffness and natural frequency $\omega_{c\phi}$ and retain high stiffness K_x in radial (x) direction. These features were attained by varying preload of proprietary nonlinear rubber elements, also providing torsional guidance and high radial stiffness.

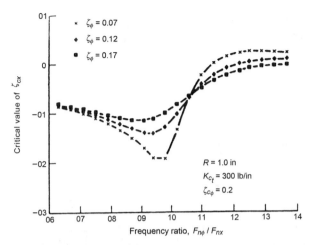

Fig. 8.1.7. Stability boundaries of boring bar with TCH at various frequency and damping tuning.

Vibration and cutting tests to validate tuning conditions for the boring bar with the TCH were performed on a tungsten carbide/aluminum combination boring bar (see Section 7.6.2) having diameter $D = 1.25$ in, $L = 9$ in (total overhang), and aluminum segment 2.5 in long. The natural frequencies had been determined by impact testing, cutting tests were performed by boring 3.5 in. to 4.0 in. diameter holes in St. 1045 specimens. Test results in Fig. 8.1.8 show that minimum relative vibrations between the tool and the workpiece are at $\omega_{n\phi}/\omega_{nx} \approx 0.9$, in agreement with the analytical prediction.

Other cutting tests were performed with the boring bar having $D = 1.25$ in, $L = 14$ in, $L/D = 11.2$, boring 3.5 in. diameter holes in St.1045. Elastic connections for the TCH were made, in one case, from high damping rubber ($\zeta_\phi = 0.4$), and in the other case, from a low damping rubber ($\zeta_\phi = 0.07$). Surface finish, R_a, for both cases is shown in Fig. 8.1.9 versus rpm of the bar (cutting speed in sfm is about 10% less than the numerical value of rpm). The results validate the analytical prediction of a detrimental effect of high damping in the TCH, as shown in Fig. 8.1.7. However, when TCH is used on a boring bar equipped with a dynamic vibration absorber (DVA), this conclusion is changing. The study in [12] has shown that the combination has a synergistic effect, it is not just a summation of the partial effects. The tuning of TCH and DVA when both are present should be different than in cases where only

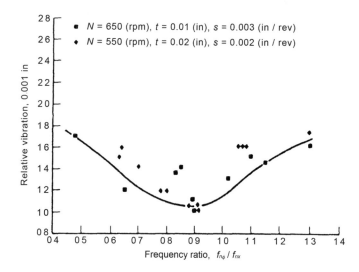

Fig. 8.1.8. Relative vibrations between tool and workpiece for various TCH tuning.

one device is used. The optimal frequency ratio for the DVA is slightly lower than for the case when only a DVA is used, while the optimal frequency ratio for TCH is 50% to 100% higher than for the case when only the TCH is used. While for the cases when only TCH is used its damping has to be reduced, it can be seen in Fig. 8.1.9 that a combination of DVA with high damping TCH resulted in the best performance of the bor-

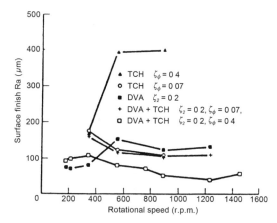

Fig. 8.1.9. Surface finish versus cutting speed for boring bar with various vibration control devices; $s = 0.002$ in/rev; $t = 0.01$ in; $L/D = 11.2$.

ing bar. It can be seen that effectiveness of TCH is improving at higher cutting speeds. This fact is very important since conventional "chatter resistant" boring bars with various DVA designs can function only at low cutting speeds (25 m/min to 30 m/min), as also illustrated by plot ■ in Fig. 8.1.9.

This bar (L/D =11.2) was also used for machining 6061 aluminum (2.45 in hole diameter; depth of cut, $t = 0.01$ in.; feed, $s = 0.002$ in./rev). The surface finish, $R_a = 20$ μin. (0.5 μm) at 890 rpm (700 sfm), 19 μin. (0.48 μm) at 1230 rpm (965 sfm), and 23 μin. (0.54 μm) at 1430 rpm (1120 sfm). Such surface finish is the limit for an old lathe used for the tests.

During the cutting process, the TCH vibrates in torsion, which results in a cutting speed variation in the range of 5% to 50%. This variation is also contributing to enhancement of chatter resistance of the tool.

8.1.4c Trading-Off Stiffness for Damping to Improve the Overall Machining Performance

Performance of machining systems (as well as of many other mechanical dynamic systems) is determined by many parameters, the most important being stiffness and damping, see Appendix 2. Understanding of their roles allows to achieve significant overall performance improvements if some degrading of one parameter is accompanied by substantial gains in the other.

It is universally accepted that performance of a machining system (chatter resistance, accuracy, tool wear) is improving with increasing stiffness of its components. Increasing stiffness of machine tool frames, tooling and fixturing structures, structural connections, etc. is always very expensive but not always very useful.

Since the chatter resistance of a machining system improves with increasing value of the criterion $K\delta$, where K is effective stiffness of the systems and log decrement δ is a measure of its effective damping, it can be concluded that some *stiffness reduction can be tolerated* if it is *accompanied by a more significant increase in damping*. This approach was, indirectly, used in designing the composite boring bar shown in Fig. 7.6.1a and discussed in detail in [10]. Making the overhang part of the boring bar from a light material resulted in some reduction of its stiffness.

This reduction is about 15% for a tungsten carbide-aluminum bar as compared with a solid tungsten carbide bar. However, since the mass ratio μ of DVA was disproportionally increased, and subsequently the effective damping was at least doubled, the value $K\delta$ and, accordingly, chatter resistance, were also significantly increased.

The simplest application of this approach to machining systems can be achieved by modifying tool clamping devices. Three generic cases were reported in [13, 14], and [15].

Turning of Low Stiffness Parts. In [13], turning of a long slender part clamped in the chuck and supported by the tailstock is considered. Dynamic behavior of the machine frame does not significantly influence stability of the cutting process when a slender bar is machined. The equivalent stiffness of the workpiece and its end supports (chuck, spindle, tailstock) are considerably lower than the structural stiffness of the machine, thus the effective (equivalent) stiffness, which is determined by the weakest element in the force transmission path, is also relatively low. Under the chatter conditions the system "spindle-workpiece-tailstock" is vibrating. The conventional approach in machining such parts is to provide additional support means, such as steady rests, which are bulky, expensive and do not perform well for stepped or axisymmetric shafts. Another technique is described in Section 7.4.3, wherein stiffness of the part was enhanced by application of a tensile force. While effective, this technique requires special means for applying the tensile force. In some cases, it is desirable to achieve the stable no-chatter cutting as well as improved accuracy and surface finish, with minimum changes in the machining system. Since the workpiece has low stiffness and damping, and since these parameters are difficult to modify (unless external devices like steady rests or the tensioning means are used), a natural way to improve the stability is from the cutting tool side. An effective approach to doing it is to add damping to the cutting tool. However, stiffness of the tool is very high as compared with the workpiece stiffness, and its vibratory displacements are very small. As a result, damping enhancement of the tool would not have a noticeable effect on the overall damping of the machining system since the energy dissipation is proportional to vibratory velocity and/or displacement of the damping element. Thus, to achieve enhancement of damping in the machining system, enhancement of the cutting tool damping must be accompanied by reduction of its stiffness. With a proper tuning of the dynamic system, an additional damping then could be pumped into the workpiece subsystem and stability of the cutting process would be increased.

The following factors are influencing stability of the cutting process:

1. workpiece material and geometry;
2. tool geometry and stiffness;
3. cutting regimes such as cutting speed, feed and depth.

All these factors can be studied using the model in Fig. 8.1.10 where k_w, c_w are stiffness and damping coefficients of the workpiece, k_t, c_t — same for the tool, and k_c, c_c — effective stiffness and damping of the cutting process (see Section 1.4). These are dependent on tool and workpiece materials, tool geometry, and cutting regimes. The most important parameter is the damping associated with the cutting process, which is related to the cutting conditions. Under certain conditions the damping of the cutting process becomes negative, and if damping of the workpiece and tool subsystems is not high enough, this negative damping overcomes the positive damping in the system and instability occurs. The system stability can be analyzed for any given set of parameters in the model in Fig. 8.1.10 [13]. Optimum values of the combination of tool parameters, k_t and c_t (or δ_t, log decrement of the tool system) can be obtained by such analysis.

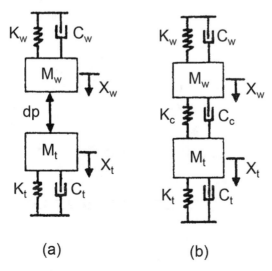

Fig. 8.1.10. Mathematical model of the cutting system.

While the tool stiffness in the direction radial to the workpiece has to be reduced, it must stay high in all other directions. It can be achieved by mounting the tool in a sleeve made of the thin-layered rubber-metal laminate material described in Section 3.3.3 and in [14] (Fig. 8.1.11). The radial component, P_x, of the cutting force causes shear deformation of the laminate (low stiffness direction), while the components P_y and P_z, orthogonal to P_x, load this laminated sleeve in compression which is characterized by a very high stiffness. The radial dynamic stiffness of the "reduced stiffness" cutting tool tested in [13] was K_t = 10,350 lb./in., while the original tool stiffness was k_t = 162,000 lb./in., (about 15 times reduction). The rubber in the laminates had high damping corresponding to δ_t = 1.8, which is about 10 to 15 times higher than damping of the conventional tool. It is important to note that the overall stiffness of the machining system was reduced to a much lesser degree as shown in Fig. 8.1.12, since the low workpiece stiffness was the determining factor in the original system. The workpiece was a bar 0.7 in. (18 mm) diameter and 15 in. (376 mm) long. An important feature of plots in Fig. 8.1.12 is a dramatic improvement in uniformity of stiffness along the work piece which resulted in a good cylindricity of the machined bar, within 0.07 mm (0.0028 in.) versus 0.125 mm (0.005 in.) with the conventional tool. Relative vibration amplitudes between the tool and the workpiece have diminished about two-to-four times for cutting with the "reduced stiffness" tool as compared with cutting with the conventional tool. Surface finish with the "reduced stiffness" tool was acceptable R_a = 2.2 μm versus

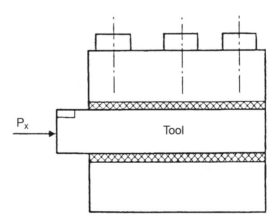

Fig. 8.1.11. "Reduced Stiffness" tool in the laminate sleeve.

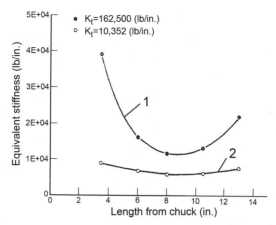

Fig. 8.1.12. Equivalent static stiffness between workpiece and tool along the work piece length; 1-with conventional tool; 2-with reduced stiffness tool.

unacceptable $R_a > 10$ μm for the conventional tool, and chatter resistance of the process was significantly improved.

Modification of Tool Clamping Systems for Cantilever Tools. High stiffness of tool-clamping systems for cantilever tools (boring bars, end mills, etc.) is specified in accordance with two requirements. High *static stiffness* is required in order to reduce deformations of the machining system under cutting forces (which result in dimensional inaccuracies of the machined surfaces), and high *dynamic stiffness* is needed in order to reduce self-excited (chatter) and forced vibrations resulting in a poor surface finish and reduced tool life. Static stiffness of the tool clamping devices is, in most cases, higher than is needed for machining with the required tolerances since at the roughing regimes, characterized by high cutting forces, the requirements to geometric accuracy are not very stringent, while at the finishing regimes the cutting forces are very low. It is not always understood that high stiffness is frequently combined with very low damping as illustrated in Chapter 4. This combination has a negative impact on the surface finish and the tool life. Since high damping is usually characteristic for polymeric materials whose very low Young's moduli make them unacceptable for the tool clamping devices, the absolute majority of the clamping devices are made of steel having very high Young's modulus but very low loss factor $\eta = \tan \beta = 0.001 - 0.003$ ($\delta = \pi\eta = 0.003 - 0.009$).

It was discovered in [15] that an alloy of ~50% Ni and ~50% Ti (NiTi or Nitinol), which is known for its "shape memory effect", has very high damping when pre-stressed in tension or compression. Fig. 8.1.13 presents the loss factor of the NiTi specimen as a function of the pre-stress magnitude and the cyclic stress amplitude. One can see from the plot that the test specimen has extremely high damping even at small cyclic stress amplitudes when subjected to the optimal pre-stress of 10,500 psi (70 MPa). The loss factor of $\eta = \tan \beta = 0.06 - 0.1$ ($\delta = 0.19 - 0.3$) is 20 to 100 times higher than the loss factor of steel. It is important that such a high loss factor develops even at low cyclic stress amplitudes (1000 psi or 6.5 MPa), since vibrations of tools, both self-excited and forced, are associated with relatively small stresses.

Cross-sections of 3-D plot (loss factor versus pre-stress and versus cyclic stress, respectively) are shown in Fig. 8.1.14 and 8.1.15. Figs. 8.1.13 to 8.1.15 show that at certain combinations of pre-stress and cyclic amplitudes the internal damping of this material is becoming very high, (log decrement δ up to 0.33).

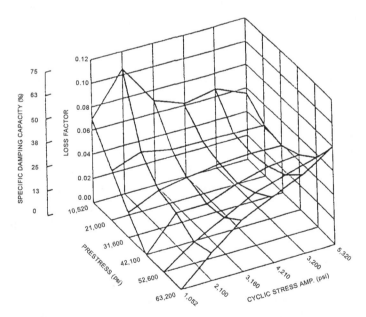

Fig. 8.1.13. Loss factor versus pre-stress and cyclic stress for NiTi specimen 0.5 in. diameter, 2.5 in. long under compression.

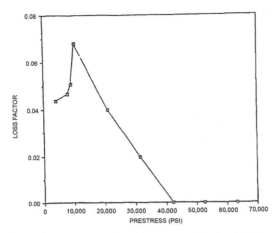

Fig. 8.1.14. Loss factor at cyclic stress amplitude 1,050 psi as function of pre-stress.

Fig. 8.1.16 shows that the dynamic modulus of elasticity (stiffness) is decreasing with the increasing stress amplitude, similarly to dependence of stiffness on vibration amplitude for other high-damping materials, see Chapter 3 and Section 4.6.2. The dynamic modulus is in the range of 5.5 – 10.5 x 10^6 psi (0.37 x 10^5 MPa – 0.7 x 10^5 MPa), or three to six times lower

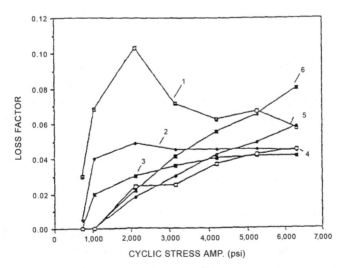

Fig. 8.1.15. Loss factor at various pre-stress magnitudes as function of cyclic stress amplitude; 1 — prestress 1050 psi, 2 — 2100 psi, 3 — 3160 psi, 4 — 4210 psi, 5 — 5260 psi, 6 — 6320 psi.

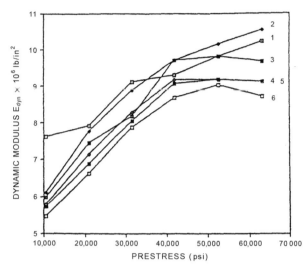

Fig. 8.1.16. Dynamic modulus versus prestress for various values of cyclic stress amplitude; 1 — cyclic stress amplitude 1050 psi, 2 — 2100 psi, 3 — 3160 psi, 4 — 4210 psi, 5 –5260 psi, 6 — 6320 psi.

than Young's modulus of steel. Since damping of this material is 20 to 100 times higher than steel, the magnitude of criterion $K\delta$ can be higher than for steel. However, in a structural use of NiTi both stiffness and damping of a device using this material would certainly be modified by other components of the device, by joints, etc.

The described effect was used in [15] for making a clamping device for a small cantilever boring bar from the high damping pre-stressed NiTi alloy. The approach described in Section 7.5.2 and in [10] allows to enhance effectiveness of dynamic vibration absorbers which are used to increase damping of cantilever structures, such as boring bars. However, the proposed techniques require design modifications which are feasible only for tools having external diameter not less than 15 mm to 20 mm (0.625 in. to 0.75 in.). There is no available technique to improve chatter resistance of small cantilever tools rather than making them from a high Young's modulus material (e.g., sintered tungsten carbide).

Since small boring bars cannot be equipped with built-in vibration absorbers, the damping enhancement should be provided from outside, e.g., by making a clamping device from the high-damping material. The tool, a boring bar with nominal diameter 0.25 in. (6.35 mm) was clamped in a NiTi bushing (Fig. 8.1.17).

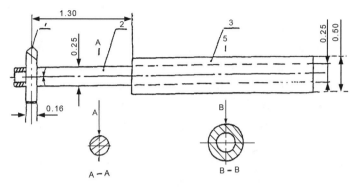

Fig. 8.1.17. Boring bar with high-damping NiTi clamping bushing; 1 — tool bit; 2 — boring bar; 3 — NiTi bushing.

The pre-stress was adjusted by changing amount of interference in the "bar – NiTi bushing" connection. The reference (conventional) boring bar was mounted in the steel bushing. These NiTi and steel bushings were clamped in the standard steel slotted square sleeve with a 0.25 in. diameter bore which was, in turn, clamped in the tool holder of a 12 in. lathe. Dynamic characteristics (natural frequency and log decrement) were determined by a free vibration test.

The test results are shown in Table 8.1. It can be seen that at high pre-stress levels, damping of the boring bar clamped in the SMA bushing is 5.5 times higher than for the similar bar clamped in the steel bushing. However, clamping in a steel bushing results in ~15% higher natural frequency (about 30% higher effective stiffness).

Dynamic stability of cantilever tools is determined by the criterion

$$K_{ef}\delta = (2\pi)^2 m_{ef} f_n^2 \delta = A f_n^2 \delta \qquad (8.1.3)$$

where m_{ef} = effective mass of the system, f_n = natural frequency, $\delta = \pi\eta$ = log decrement, η = loss factor. Since m_{ef} is constant for all the tests

Table 8.1. Free vibration test results for boring bar in SMA and steel clamping bushings.

Prestrain		NiTi bushing			Steel bushing	
ε, 10^{-6}	δ	f_n, Hz	$f_n^2\delta$	δ	f_n, Hz	$f_n^2\delta$
850	0.058	515	15,400	0.016	602	5,800
1,050	0.088	531	24,800	0.016	638	6,500

summarized in Table 8.1, A is also constant and values of $f_n^2\delta$ listed in Table 8.1 are proportional to $K_{ef}\delta$. It can be seen that at the high prestrain the dynamic stability criterion with the NiTi holder bushing is about four times higher than with the steel holder. While in the machine tool setting this difference may be reduced due to influence of stiffness (compliance) and damping in the attachment of the tool holder to the machine tool, a significant improvement of cutting conditions can still be expected.

Cutting tests were performed on a lathe with a non-rotating boring bar (diameter $D = 0.25$ in., overhang length, $L = 1.9$ in., $L/D = 7.5$). Natural frequencies measured on the lathe were $f_{st} = 1315$ Hz, $f_{NiTi} = 855$ Hz (average between two directions) and log decrement, respectively, $\delta_{st} = 0.025$, $\delta_{NiTi} = 0.113$. Accordingly, $(f^2\delta)_{st} = 43,230$ and $(f^2\delta)_{NiTi} = 80,500$, about two times difference.

Cutting tests were performed on workpieces made of medium carbon St.1045 and stainless steel 316. Surface finish (R_a) was compared for the cutting tests when the boring bar was clamped in the NiTi (stressed to the strain value $\varepsilon = 1.3 \times 10^{-3}$) and in the steel bushings. All tests were performed with feed 0.0015 in/rev (0.038 mm/rev) by boring 1.8 in. (46 mm) diameter hole. The tests have demonstrated that the boring bar clamped in the pre-stressed NiTi bushing is more dynamically stable, resulting in significant improvements in surface finish, up to two times reduction in R_a [15]. A similar approach can be used for clamping end mills of both

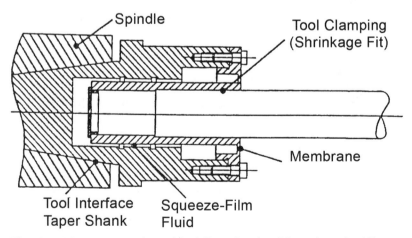

Fig. 8.1.18. Damped workholding chuck with reduced stiffness.

small and large size (since it is very difficult to insert dynamic vibration absorbers into end mills), as well as for other cantilever systems.

A similar concept, also for boring, was utilized in [16], Fig. 8.1.18. The tool is connected to the toolholder via membrane assuring co-axiality between the tool and the toolholder while allowing angular mobility of the tool and creating "squeeze-film" damping effect in the oil film at the end of the tool. While static stiffness of the tool is reduced by about 40%, it is more than compensated by very high damping.

8.2 STIFFNESS OF CLAMPING DEVICES

8.2.1 Introduction

While the above section mostly dealt with stiffness and damping of tools proper, in many cases devices for attaching the tool to toolholder and for attaching toolholder to spindle or tool carriage have critical importance for dynamics of the machining system [1]. However, such interfaces are also important for general mechanical systems, the reason why this section is separated from the above section. Still, since the R&D efforts as well as research publications on clamping devices mostly deal with machining systems/tooling applications, the corresponding terminology is used below.

A tool clamping device provides an interface between the tool and the toolholder, which in its turn is connected to the spindle or to the tool carriage. Until recently, tool clamping devices used on machining centers and other automated machine tools were collets, Weldon clamps, Fig. 8.2.1, and chucks for not-critical applications, such as drilling. These devices

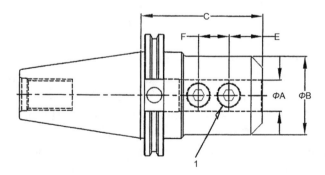

Fig. 8.2.1. Weldon chuck.

are challenged by high speed/high power/high accuracy requirements to modern machining systems. The Weldon clamps are asymmetrical and thus prone to runout and unbalance. Their stiffness is compromised by presence of several joints and by relatively small clamping bolts; their stiffness anisotropy may reduce chatter resistance. Conventional collets and chucks also have several clamping components, thus lacking a solid structure. They are prone to shifts and deformations induced by assembly uncertainties, as well as by cutting and centrifugal forces. Due to presence of lubricants and lubricoolants, friction coefficients on the clamping surfaces are low and very high external forces are required to produce the required clamping forces. New and advanced generations of tool clamping devices in which these shortcomings are alleviated or eliminated, are continuously appearing.

The important performance characteristics of tool clamping devices are accuracy, stiffness, grip (torque capacity), insensitivity to centrifugal forces, repeatability (consistency) of maintaining the listed characteristics, convenience of use, costs. It can be stated that Weldon connections cannot be used above 10,000 rpm, general purpose collets cannot be used above 15,000 rpm, specially designed collets or shrink fit should be used between 15,000 and 25,000 rpm, and shrink fit connections above 25,000 rpm.

8.2.2 General Purpose Clamping Devices

This group is mainly represented by three-jaw chucks, Weldon clamps in which a cylindrical tool shank with machined longitudinal flat is inserted with a small clearance, usually 10 μm to 30 μm per diameter, into cylindrical receptacle A (Fig. 8.2.1) of the holder and clamped by one or two set screws 1 acting on the flat made on the tool shank and by collet chucks. *Three-jaw chucks* are mostly used for relatively low accuracy non-critical applications on lathes and for clamping drills.

Weldon clamps are popular for clamping drills and end mills for not very high speed/high power/high accuracy applications. They have inherent substantial radial runouts, approximately equal to the diametral clearance, since the shank is clamped by pressing it to one wall of the holder receptacle. Stiffness of Weldon chucks depends on the number of the clamping bolts. Stiffness of a two-bolt 12.7 mm Weldon chuck is about 15% lower than stiffness of a monolithic shrink fit connection, but about 60% lower for a one bolt Weldon chuck. The stiffness also depends

Table 8.2. Angular Compliance of Weldon Chucks.

Clamping Diam.	Clearance	Compliance, rad/KN m	
mm	μm	‖ to bolts	⊥ to bolts
36	14	0.0032	0.0060
37	37	0.0041	0.0079
48	14	0.0010	0.0014
	26	0.0012	0.0032
	48	0.0016	0.0062

on the tool shank fit in the chuck receptacle. The stiffness is anisotropic, higher in the plane of the clamping bolts. This may cause parametric vibrations in the machining system, see Section 3.4.4, and also may reduce chatter resistance. Table 8.2 [17] gives angular compliance of Weldon chucks $e_\phi = \phi/M$, where ϕ = angular deformation caused by bending moment, M.

In a modified Weldon chuck for small (6 mm to 12 mm diameter) tools (Fig. 8.2.2) the screws are replaced by a clamping wedge, "Holding Pin" in Fig. 8.2.2, engaging with the flat on the tool shank. The wedge is actuated by Pressure Screw via Pressure Bar. Due to the use of two wedge mechanisms (Pressure Screw/Pressure Bar and Holding Pin), the mechanical force amplification is increased, resulting in a greater clamping force than in conventional Weldon chuck, a greater stiffness, and better balance. This chuck is more complex and more expensive.

The most versatile general purpose clamping device is **collet chuck**. Majority of collet chucks use solid thin slotted clamping sleeves made of

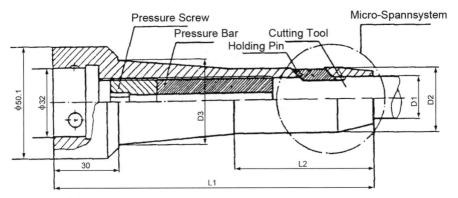

Fig. 8.2.2. Modified Weldon chuck for small tools (B+W Micro Vario).

hardened steel and precisely ground both on internal cylindrical surface and on external tapered surface(s). It is very expensive to mass produce tool shanks and internal cylindrical surfaces of collet sleeves with diametral accuracy better than 5 μm to 10 μm, thus actual contact areas between the tool shank and and each segment of the collet sleeve are relatively narrow, thus making the system sensitive to cutting and centrifugal forces. These forces reconfigure the contact areas, resulting in radial and angular displacements of the clamped tool in relation to the spindle. It is usually assumed that collet chucks grip the tool at the front edge of the collet sleeve, thus creating a "built-in" boundary condition and not adding to the effective length of the clamped tool or part. However, it was experimentally demonstrated that usually the clamping effect starts at about 25% of the collet length from the edge, about 10 mm for 42 mm diameter tool. Deformations of collet chucks are mostly due to contact deformations between the collet and its housing and between the clamped object and the collet. Static stiffness of a collet chuck can be high, only about 5% less than monolithic or shrink-fit connection, but relatively minor design and fabrication deviations may result in significant reduction of stiffness. Table 8.3 [17] gives experimentally determined angular compliances for joints "collet-housing" and "tool-collet" for two chuck sizes. Each chuck can accommodate a range of collet sizes.

The angular displacements caused by moments of cutting forces and angular compliance of collets result in additional runout of the clamped tool. High precision collets may have a low runout, about 5 μm. However, it frequently deteriorate as much as to 30 μm after a machining process using the clamped tool causes its angular misalignment which becomes frozen by friction forces.

Table 8.3.   Angular Compliance of Collet Chucks.

Model	Clamping Diam. mm	Compliance, rad/KNm	
		Collet/Housing	Tool/Collet
1	10	0.0019	0.0084
			0.0027
	12		0.0028
	16		0.0021
	20		0.0019
2	20	0.0010	0.0019
	40		0.0016

Many general purpose collet chucks use tightening nuts for clamping. At high clamping forces they can cause misalignment and/or twisting of the collet segments, thus negatively affecting concentricity and distribution of clamping forces. An exaggerated sketch of the deformations is shown in Fig. 8.2.3. Collet chucks for high speed/high precision operation may have a wedge-operated axial tightening system preventing the distortions. The conventional collets with thin slotted sleeves have poor repeatability of clamping force distribution and, consequently, poor stiffness repeatability. Better performing are collets with solid (not segmented) thick-walled sleeves having long longitudinal holes, Fig. 8.2.4a or grooves, Fig. 8.2.4b. The sleeves are externally tapered and the tightening force is applied via needle rollers effectively reducing friction in the tapered wedge connection. The inner sleeve diameter can be reduced as much as 0.1 mm by the tightening torque and can generate a very high clamping pressure, three to four times greater than in thin segmented collet sleeves.

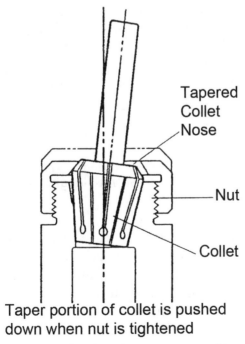

Tapered
Collet
Nose

Nut

Collet

Taper portion of collet is pushed
down when nut is tightened

Fig. 8.2.3. Distortion of collet alignment caused by tightening nut (Heartech Co.).

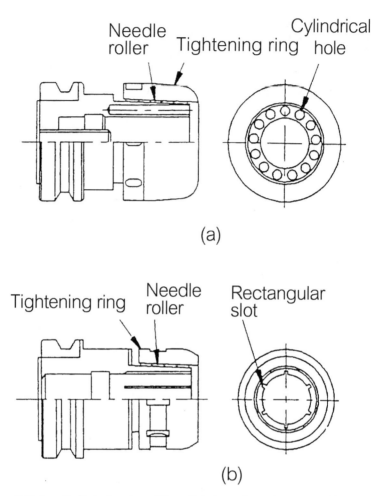

Fig. 8.2.4. Collet chucks with solid clamping sleeves; a-Sleeve with longitudinal holes; b-Sleeve with longitudinal grooves.

Both stiffness and damping of the collet-clamping system are determined by contact interactions between the clamped shank and the collet receptacle. Accordingly, stiffness is increasing and damping decreasing with increasing clamping force, with damping (and criterion $K\delta$) decreasing much faster than stiffness is increasing. Thus, from chatter resistance point of view it is better to use lower clamping (gripping) forces, unless damping is enhanced by other means. However, the lower clamping forces are associated with reduced repeatability of both stiffness and damping values. In tests, clamping of a 6.25 mm diameter tool with 40 Nm clamping torque resulted in 3% variation of stiffness and ~50% variation of damping for repeated clamping/unclamping; with 60 Nm — 2% and 15%, respectively, and at 90 N-m no variation had been detected.

If a collet is constructed from rigid metal components transmitting clamping forces, even minute microslips in the multiple joints, e.g., caused by shank diameter variations, may cause a major force redistribution. It is especially pronounced when an oscillating cutting force is applied to the collet. This effect can be reduced by introducing compliant elements into the system. Fig. 8.2.5 shows a collet with rubber base 1 holding steel clamping blades 2. Mobility of blades 2 provides for their better contact with the shank, even with a relatively large shank diameter variation from its nominal value, thus potentially having a greater stiffness and a better grip.

Comparison of several tool-clamping devices for a 20 mm diameter shank and 80 mm overhang has shown that stiffness at the end is 6.1 N/μm for collet clamping; 6.8 N/μm for Weldon chuck clamping; and 9.1 N/μm for a solid structure directly attached to the spindle.

8.2.3 "Solid State" Tool Clamping Devices

Presence of multiple contacting components in the above described tool-clamping devices increases the role of contact deformations, creates a

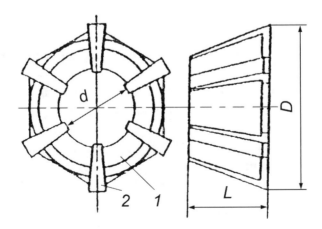

Fig. 8.2.5. "Rubber Flex" collet (Jacobs Chuck Co.).

potential for uneven force and displacements distribution in addition to built-in asymmetry (Weldon), etc. These devices are not well suited for high speed applications. These shortcomings led to development of "solid state" tool clamping systems, such as "power shrinking" toolholders and shrink-fit toolholders.

Thermal Shrink-Fit Toolholders. In this method, the tool shank diameter is 12 μm to 75 μm larger than diameter of a smooth coaxial cylindrical hole in the toolholder. For assembly, the toolholder is heated, until the hole is expanded so the shank can be easily inserted. After cooling down, the tool is structurally connected to the toolholder. For disassembly, the toolholder is heated again. Since the tool is located farther from the heat source than the toolholder housing and due to thermal resistance of the contact surface between the shank and the hole, the hole is expanding faster than the tool shank and the latter can be easily withdrawn. This connection has an excellent co-axiality, is not affected by centrifugal forces, has stiffness close to that of a solid tool (made as one piece with the toolholder). Tests were performed in [18] of a tool 20 mm diameter, 50 mm overhang in a hydraulic toolholder (see below); Weldon chuck; collet chuck; and shrink-fit holder. The tests have shown that natural frequencies and dynamic compliances H at resonance are, respectively, 3700 Hz, $H = 1.6$ μm/N; 772 Hz, $H = 1.45$ μm/N; two resonances: 799 Hz, $H = 1$ μm/N and 5328 Hz, $H = 0.9$ μm/N; no significant resonance, $H < 0.3$ μm/N. In other tests, a tungsten carbide bar was shrink-fit into a steel sleeve with high interference, 0.1 mm for 12.7 mm diameter. While stiffness of this connection was the same as of a single piece of the same materials, its damping was about one half of damping of other three connections.

"Power Shrinking" Toolholders. These toolholders achieve similar or even higher clamping forces (and stiffness) than shrink-fit toolholders without application of high temperatures.

The hydraulic toolholder, Fig. 8.2.6, has thin cylindrical membrane 2 (expansion sleeve) located within housing 1and surrounded by a co-axial hydraulic chamber, which is a part of the closed hydraulic system 7 filled with oil. When pressure is applied by turning actuating screw 5 connected wth an actuating piston, oil pressure `is increasing and membrane 2 is deforming inside the hole, thus clamping the inserted tool shank. Tuning of the device is performed by adjustment screw 4, also connected to piston 3. Seals 6 contain the pressure which is reaching 200 MPa to 300 MPa. Such chucks have high accuracy and concentricity,

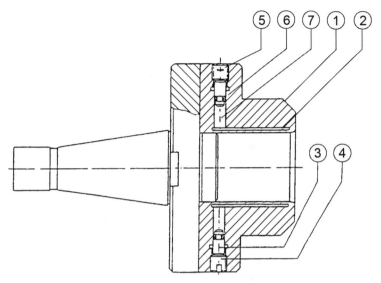

Fig. 8.2.6. Hydraulic toolholder (HDT Co.).

they are easy to manipulate, but they have serious shortcomings. Their stiffness, especially angular stiffness, is not high due to presence of fluid (oil) which is not adequately resisting the external forces due to its high mobility. The expanding sleeve (membrane) starts clamping 10 mm to 15 mm from the hole mouth, thus increasing the effective overhang. The test results listed above show the greatest dynamic compliance (lowest stiffness) of hydraulic chucks as compared with Weldon, collet chucks and shrink-fit chucks. Accordingly, hydraulic chucks are not suitable for heavy cutting regimes and are used mostly for finishing operations. Use of high viscosity silicon gel-like or rubber-like compounds instead of oil, together with making the hydraulic chamber hermetic eliminates the seal leakage. Hydraulic toolholders are very expensive and very sensitive to proper handling. The reduced stiffness of hydraulic chucks may result in greater deformations of the tool and their displacements. Fig. 8.2.7 [19] compares location accuracy of drilled holes while 12.7 mm diameter and 90 mm overhang tungsten carbide drills had been clamped in hydraulic chucks and collet chucks at various rpm and feed rates.

There are also several purely mechanical systems for power shrinking. "Tribos" system, Fig. 8.2.8, is using a solid sleeve (collet), which is, in its un-clamped condition, cylindrical outside and having a polygonal hole (Fig. 8.2.8a). To condition it for clamping, the collet is inserted into a

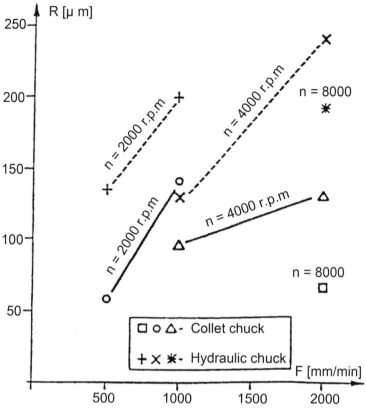

Fig. 8.2.7. Average radial hole location error with hydraulic and collet chucks used.

stationary power (hydraulic) device which applies three high radial forces along vertices of the polygon (Fig. 8.2.8b). The forces deform the collet until the hole aquires a cylindrical shape. At this stage the tool is inserted (Fig. 8.2.8c), and the external radial forces are removed, thus resulting in a secure three-line clamping of the shank along its length. Both run-out and repeatability of the system is very good, within 3 μm. Since the clamping starts at the edge of the collet, the effective and the actual over-hangs are the same, thus no stiffness is lost.

The "CoroGrip" system, Fig. 8.2.9, employs wedge mechanism principle for clamping. Due to high force amplification by mechanical advantage of the wedge mechanism ($\sim 2°$ wedge angle), the clamping force is claimed to be a double of the shrink-fit clamping force. Fig. 8.2.9a shows the chuck in the process of clamping and Fig. 8.2.9b -in the process of unclamping. The tool shank is fit inside of stationary thin-walled sleeve

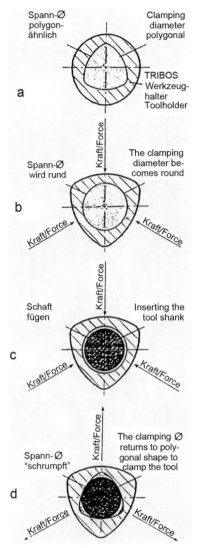

Fig. 8.2.8. Clamping concept of Tribos" chuck (Schunk Co.).

A whose external tapered surface is mated with movable sleeve B. For clamping, oil at 70 MPa from an external oil pump is supplied through inlet D between end faces of sleeve B and stationary locking ring C, thus pushing movable sleeve B up along the tapered surface. This results in radial contracting of stationary sleeve A and in the clamping action. For unlocking, the same pressure is supplied through inlet E to the joint

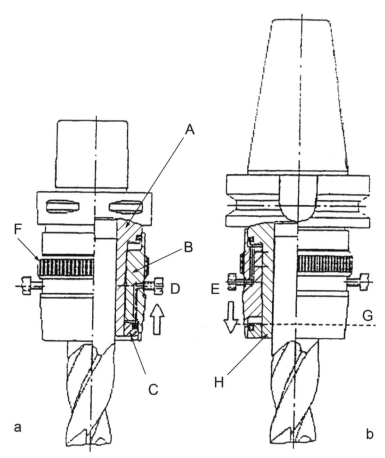

Fig. 8.2.9. CoroGrip clamping chuck; a Clamping, b Unclamping (Sandvik Corp).

between stationary A and movable B sleeves, causing downward motion of movable sleeve B, and thus unlocking the clamp.

8.3 MODULAR TOOLING

Modular tooling systems allow for quick constructing of toolholders for a wide variety of operations, thus justifying higher initial prices for the collection of modules and sometimes lower machining regimes while using modular systems due to its lower stiffness than the solid tooling because of many joints. Also, the modules often cannot be used in an optimal configuration, since there is a good chance that a modular cantilever tool will be longer and/or will have a smaller diameter than one required by the specifications. On the other hand, damping of the modular toolholders is usually higher due to the presence of multiple joints.

Prices of modular systems are high, largely due to complexity of joints which have to be a part of each module. Major requirements for the joint design are accuracy (co-axiality and straightness) of two connected modules; easiness of assembly; high rigidity and damping of the connection to achieve acceptable chatter resistance of the tool; the assembled tool may have up to two to four joints. The joint usually incorporates a cylindrical fit between a protrusion (male pilot) on one module and a cylindrical bore (receptacle) on its counterpart. The connection is tightened to achieve a rigid face contact between the two connected modules, e.g., see Fig. 8.3.1. In this design the tightening is achieved by using two set screws with tapered ends. The screws are positioned at 90 deg. to each other in the cross-section perpendicular to the longitudinal axis of the connection. The taper angle is 30 deg, resulting in 3 to 3.5 times force amplification for creating the face contact pressure. Some systems have only one set screw, in other systems the tapered connection(s) are replaced by mechanical amplification due to contact between the spherical end of the set screw and the spherical socket on the pilot. The wedge- and

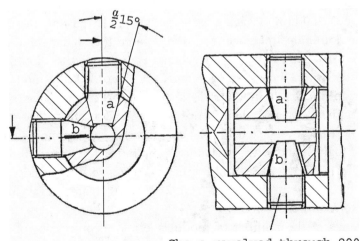

Shown revolved through 90°

Fig. 8.3.1. Joint between modules in Multibore modular tool system (Wohlhaupter GmbH).

ball-based force amplification devices are surveyed and analyzed in [20]. The radially positioned set screws are pushing the pilot to a line contact with the receptacle. To simplify the assembly, there is a sliding fit between the pilot and the receptacle with radial clearance 2.5 μm to 5.0 μm; this clearance adds to runout of the tool in addition to other factors. If more than two modules (more than one joint) are used, the total runout at the cutting edge is correspondingly larger. This effect is typical for the so-called "radial clamping".

Some modular tool systems are using axial clamping, e.g., see Varilock sytem in Section 4.7, Fig. 4.7.1. The modules are connected by face clamping (surface 3) using axially located bolt 2, while the radial location is attained by double cylindrical fit 5. Torque is transmitted by face key connection 4 between the connected modules. While the runout is potentially less in such a design, it is shown in Section 4.7 that it may have even more undesirable angular distortions.

Since stiffness and damping of modular tooling systems depend in a complex way on their design, especially on their joint designs, their performance can be evaluated only by testing. The most extensive comparative testing of twelve modular tooling designs was performed in [21]. All tested systems had diameter 50 mm, were assembled from three modules (toolholder/spindle interface ISO taper 40, one spacer, and toolhead), and had 200 mm and 300 mm overhangs. The overhang was measured from the gage line of the taper. Dynamic compliance measured at the tool edge at 200 mm overhang varied widely, from 1.1 μm/N to 5 μm/N, with the dominant influence from the tapered toolholder/spindle interface. Its influence at 300 mm overhang was expected to diminish, but still scatter from 3.6 μm/N to 12.0 μm/N, was recorded. The modular system design noticeably influenced the cutting force for cutting (with the same regimes) by end mills clamped in a Weldon clamp, from 470 N to 850 N, with lowest and highest magnitudes corresponding to cases with the highest and lowest dynamic compliance; the surface finish (R_a) varying from 0.7 μm to 3.6 μm. For finish boring with single edge cutter, cutting force magnitudes varied from 28 N to 70 N, and R_a varying from 0.4 μm to 7.0 μm.

Influence of the number of modules/joints on properties of the assembled toolholder in comparison with a solid toolholder of the same dimensions was studied in [17] (Fig. 8.3.2). The modules were assembled by a axial bolt with maximum tightening torque 94 Nm. The modular

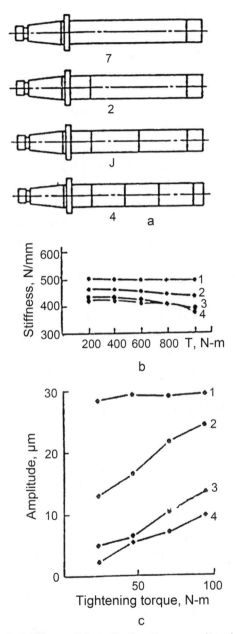

Fig. 8.3.2. Static stiffness (b) and vibration amplitudes (c) for various modular assemblies of the same geometry (a).

toolholders have lower bending stiffness than the solid toolholder (1 in Fig. 8.3.2a), about 10% reduction for the three-module structure (2 in Fig. 8.3.2), another 10% reduction for 3 (four-module structure), and no additional reduction for 5 (five-module structure). Static bending stiffness of multi-module assemblies is decreasing with the increasing bending moment (a softening nonlinearity), probably due to partial opening of face contacts in the joints.

Vibration amplitudes measured during dynamic testing of toolholder assemblies in Fig. 8.3.2a are decreasing with increasing number of joints, since the joints introduce damping in the system (see Chapter 4). It was reported in [17] that dynamic stability during cutting was greater for toolholders 2 to 4 in Fig. 8.3.2, than for the stiffer solid toolholder 1. Both stiffness and damping of the joints as well as chatter resistance of the modular tooling can be adjusted by changing the tightening action of the bolts (or other tightening means); the stiffer assemblies would result in higher accuracy while smaller tightening pressures, thus lower stiffness but higher damping would result in better chatter resistance.

Performance characteristics of modular tooling would be improved if the module (spacer) closest to the clamp (the spindle in machining centers) were made from a high modulus material, such as tungsten carbide. However, making high precision intricate surfaces characteristic for complex joints of the commercially available modular tool systems from tungsten carbide, would be prohibitively expensive.

8.4 TOOL/MACHINE INTERFACES. TAPERED CONNECTIONS

Static indeterminacy and/or imperfect dimensional accuracy can adversely affect performance characteristics of mechanical systems as illustrated in Chapters 4 and 5. In many cases, effects of both static indeterminacy and of inaccuracies can be alleviated or completely eliminated by introducing low stiffness compensating elements, e.g., such as shown in Fig. 5.3.6. While the system in Fig. 5.3.6 is over-constrained by an excessive number of positive (hardware) restraints, there are numerous cases when the excessive restraints can be resolved by using compliant elements in the system. Use of the compensating elements and other application-related issues are illustrated below for tapered (conical) connections in-depth analyzed above in Chapter 4.

8.4.1 Managed Stiffness Connections to Reduce Friction-Induced Position Uncertainties

Friction-induced position uncertainties in mechanical connections develop due to inevitable variation of the friction coefficients and of the normal forces in the frictional contacts. These uncertainties are especially important (and objectionable) in precision devices where in many cases position uncertainties even within fractions of one μm cannot be tolerated. The important case of such a device is the so-called "kinematic coupling," which is used as a repeatable connection between the tool and the tool carriage in a precision lathe [22].

The "kinematic coupling" concept is used for providing statically determined connection between two mechanical components. The statically determined connection provides six restraints for six degrees of freedom of the component. Frequently, it is realized by using three-grooves/three-balls connection (Fig. 8.4.1). Each ball has two contacts (one with each side of the respective groove). While, theoretically, these contacts are points, actually they are contact areas, which become relatively large if the connection is preloaded for enhancing its stiffness. As a result of

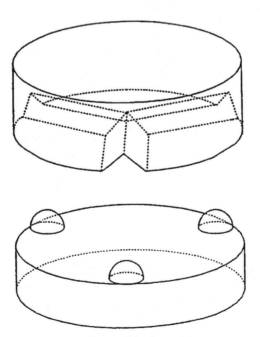

Fig. 8.4.1. Kinematic coupling with three V-grooves and three balls.

finite sizes of the contact areas, friction forces are developing along these areas which cause uncertainty of relative positioning between the two connected components. In fact, the system becomes a statically indeterminate one on the micro-level. It was found in [22] that this positional uncertainty (hysteresis) could be 0.4 μm to 0.8 μm. The solution for this problem proposed in [22] is to introduce compliance in the design by making the contact areas "floating," (Fig. 8.4.2). The compliance was introduced on each side of each groove by making two longitudinal holes along each groove and by machining slots, which partially separate the contact areas from the grooved plate. In such a design, tangential forces generated by the preload force, F_p, due to inclination of the contact surface do not generate dynamic friction forces and micro-displacements, but slightly elastically deform the contact segments of the groove wall under the static friction forces.

This approach resulted in reduction of hysteresis down to 0 μm to 0.1 μm, albeit with reduction of the overall system stiffness.

8.4.2 7/24 Steep Taper Connections

8.4.2.1 Definition of the Problem

Tapered connections are frequently used for fast and repetitive joining of precision parts without pronounced radial clearances. The alternative to the tapered connection is connection by fitting precisely machined

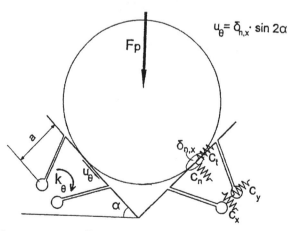

Fig. 8.4.2. Ball in V-groove with "elastic hinges".

cylindrical surfaces. However, for easy-to-assemble cylindrical connections without flexible compensating elements, the diameter of the male component must be somewhat smaller than the diameter of the female component. This results in an inevitable radial clearance. Even with the clearance, especially when its magnitude is kept small, assembly of a cylindrical connection requires precision positioning of the components before the assembly. A tapered connection fabricated with the same degree of precision is easy to assemble and, when preloaded with an axial force, it is guaranteed to have a contact at the large or small diameter of the connection depending on the tolerance assignment. As the result, there would be no radial clearance along the whole length of the connection but there would always be some angular play between the connected components. This angular play can be eliminated if a very expensive individual matching of the components is performed. The angular play results in an indeterminate angular position within the tolerance of the male component in relation to the female component and thus, in an undesirable runout increase and stiffness reduction. Since frequently the components have to be interchangeable within a large batch of spindles and tool holders for machine tools, the matching is only rarely feasible.

While it is not very difficult to maintain very stringent tolerances on angles of the taper and the hole in the tapered connection, it is difficult (unless expensive individual matching is used) to maintain in the same time a high dimensional accuracy on diameters and axial dimensions of the connected parts. These difficulties are due to a statically indeterminate character of the connection since the simultaneous taper and face contact are associated with an excessive number of constraints. As a result, a simultaneous contact of the tapered and the face (flange) surfaces is economically impossible to achieve for interchangeable parts. Because of this, standards on tapered connections (such as U.S. National Standard ASME B5.50-2009 for "V" Flange Tool Shanks for Machining Centers With Automatic Tool Changers") specify a guaranteed clearance between the face surfaces of the connected parts. Fig. 8.4.3 shows the standard arrangement for the most frequently used 7/24 taper connectors (taper angle 16° 36' 39") The axial clearance between the face of the spindle and the flange of the male taper is specified to be about 3 mm (0.125 in.). Axial position of the male taper in relation to the tapered hole for the tool holder/spindle connections can vary for the same components within 15 μm to 20 μm due to variations of the dimensions, of the

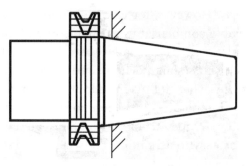

Fig. 8.4.3. Positioning of standard 7/24 tapered connection.

friction conditions of the contacting surfaces, and of the magnitude of the axial force [23, 24].

8.4.2.2 Tapered Toolholder/Spindle Interfaces for Machine Tools. Practical Sample Cases

Until recently, the steep (7/24) taper interfaces satisfied basic requirements of machining operations. However, fast development of high accuracy/high speed/high power machine tools caused by increased use of hard-to-machine structural materials, by proliferation of high performance cutting inserts, and by required tight tolerances of the machined parts, posed much more stringent requirements to machining systems. Modern machine tools became stiffer while the interfaces did not change. Now, the toolholder and the toolholder/spindle interface are the weakest links in the machining system.

Standard 7/24 Connection. The standard steep taper connections are discussed in detail in Section 4.4. They have many positive features for machine tool applications. They are not self-locking thus allowing fast connections and disconnections with a simple drawbar design not requiring a "kick out" device. The taper is secured by tightening the toolholder taper in the tapered hole of the spindle. The solid taper allows clamping of the tool inside the tapered part, thus reducing its overhang from the spindle face. Only one dimension, the taper angle, has to be machined with a high degree of precision. As a result, the connection is inexpensive and reliable.

Many shortcomings of the 7/24 taper interface are caused by the fact that the taper surface plays two important roles simultaneously:

precision location of the toolholder relative to the spindle, and clamping inside the spindle in order to provide adequate rigidity to the connection. Since it is practically impossible to make interchangeable tapers which have both face and taper contact with the spindle, the standard 7/24 toolholder/spindle interface has a guaranteed face-to-flange clearance (Fig. 8.4.3). Radial location accuracy is not adequate since standard tolerances specify a "minus" deviation of the hole angle and a "plus" deviation of the toolholder angle, resulting in a clearance in the back of the connection between the male taper and the spindle hole (see Section 4.4). A typical AT4 qualitet (ISO IS 1947) has 13 ang. sec tolerance for the taper angle of both male and female tapers, which may result in radial clearance as great as 13 μm (0.0005 in.) at the back end of the connection. This radial clearance leads to a significant runout and to worsening balancing condition of the tool. The absence of face contact between toolholder and spindle leads also to micro-motions between the male and female tapers within the radial clearance under heavy cutting forces, resulting in accelerated wear of the front part of the spindle hole ("bell mouthing") and in extreme cases - to fretting corrosion of the spindle which results in even faster wear. The axial positioning of the tool is uncertain within 15 μm to 20 μm.

It is important to discriminate between two major shortcomings of conventional 7/24 taper interfaces:

1. Radial clearance in the back part of the tapered connection due to the taper tolerancing, which reduces stiffness, increases runout, introduces unbalance, and creates the potential for micro-motions causing fretting corrosion; and

2. Mandated axial clearance between the flange of the toolholder and the face of the spindle creates uncertainty in the axial positioning of the toolholder. This is undesirable when the machined part is measured on the machining center using touch probes carried by the tapered itoolholder; in such cases, axial calibration of the probe is required. This shortcoming can be eliminated by assuring the face contact between the toolholder and spindle flanges. However, due to the steepness of the 7/24 taper, providing simultaneous contact at both taper and face requires fractional μm tolerances. This is difficult but possible to achieve for a given toolholder/spindle combination, and some toolholder manufacturer and user companies

provide such precision fitting for critical machining operations. However, it is impractical for the huge inventory of toolholders and spindles, since variations in the spindle gage diameter are in the range of tens of μm.

Requirements to Toolholder/Spindle Interfaces. The three categories of machining conditions - high accuracy, high power of the driving motor, and high spindle rpm are not always used in the same time and pose different requirements to the toolholder/spindle interface.

High accuracy machining operations require high accuracy of tool positioning both in radial and in axial directions. However, both radial and axial accuracy of standard 7/24 interfaces are adequate for many operations, such as drilling, many cases of end milling, etc.

High power cutting is frequently performed at not very high spindle rpm. The most important parameters of the interface for such regimes are high stiffness and/or damping assuring good chatter resistance, and high stability of the tool position. For standard 7/24 interfaces, the latter requirement is not satisfied due to the clearance in the back of the tapered connection. This allows for small motions under heavy cutting forces, and leads to fretting corrosion and fast bell mouthing of the spindle.

High speed operations require assurance that the interface conditions are not changing with the increasing spindle rpm. The front end of high speed spindles expands due to centrifugal forces, as much as 4 μm to 5 μm at 30,000 rpm for the taper # 30 [25]. Since the toolholder doesn't expand as much, the spindle expansion increases the effective length of the cantilever tool, reduces its stiffness, and changes the axial position of the toolholder. Maintaining a reliable contact in the connection under high rpm requires a very significant initial interference, as much as 15 μm to 20 μm for the taper #40. The clearance in the connection caused by the taper tolerances can also be eliminated by introduction of preloading (interference) along the whole length of the tapered connection. However, for the standard solid steel taper the interference has to be at least 13 μm (for AT4 grade) to eliminate the clearance. Such magnitudes of interference are impractical for conventional connections since they require extremely high drawbar forces, and disassembly of the connection with the high interference for the tool changing operation would be close to impossible. In addition, high magnitudes of interference for solid toolholders would result in bulging of the spindle, similar but larger than bulging shown in Fig. 4.5.18 and resulting from the interference fit of a *hollow* toolholder. This may damage the spindle bearings. A better alternative technique for holding toolholder and spindle together at high rpm is to generate axial face/flange contact between them with high friction forces, exceeding centrifugal forces. In such case, the spindle expansion would be prevented.

High speed machining requires very precise balancing of the toolholder. Precision balancing of standard 7/24 taper interfaces can be disrupted by the inherent unbalance of the large keys and keyslots, and also by the discussed above clearance in the back of the connection causing radial runout and thus, unbalance. The keys are required to transmit torque (in addition to the tapered connection) and to orientate the angular position of the toolholder relative to the spindle which may be desirable for some operations (such as fine boring). The need for the torque-transmitting keys can be eliminated if enhanced friction forces were generated in the connection. The desired tool orientation, if needed, can be maintained by other means.

Alternative Designs of 7/24 Toolholder/Spindle Interfaces. The Theory of Inventive Problem Solving (TRIZ) [26] teaches that if a technological system becomes inadequate, attempts should be made to resolve its contradictions by using its internal resources, without major changes in the system (i.e., to address a so-called *"mini-problem"*). Only if this approach has proven to be unsuccessful, major changes in the technological system should be undertaken, i.e., a *"maxi-problem"* should be addressed. However, in solving the toolholder/spindle interface problem, the mini-problem until recently was not addressed. Due, largely, to competitive considerations, several manufacturers undertook drastic changes, see Section 4.4.2d and Section 8.4.3, thus making existing spindles and the present huge inventory of 7/24 toolholders obsolete.

The described below advanced toolholder designs with 7/24 taper resulted from addressing the "mini-problem" related to the toolholder/spindle interface. This goal can be formulated as development of an interface system exhibiting higher stiffness, better accuracy, and better high-speed performance than conventional interfaces while being fully compatible with existing toolholders and spindles, and not prohibitively expensive. This effort took two directions in accordance with two sets of requirements formulated above in Section 8.4.2b2. The first direction was development of a toolholder providing for taper/face interface with the spindle. This eliminates indeterminacy of axial position of the conventional 7/24 taper interface, improves its stiffness and radial runout, and creates better balancing conditions for high rpm use due to elimination of the need for torque-transmitting keys. The second direction was modification (retrofit) of existing toolholders in order to improve their stiffness and runout characteristics. The second direction is important, since it has the potential for upgrading performance characteristics of millions of existing toolholders with absolutely minimal changes in the system. Since the problems of the standard connections are mostly due to the static indeterminacy of the system, there is a definite need for an elastic link in the system. However, the elastic elements must have an acceptable ("managed") stiffness in order to satisfy the performance requirements, and must be very accurate in order to satisfy the accuracy requirements for the connection.

After an extensive survey of the state of the art [24] and analysis based on TRIZ [26], seven designs were developed and tested [27]. These designs are based on a "virtual taper" concept, in which there are discrete points or lines defining a tapered surface and contacting the tapered spindle hole. All the design elements providing the contacts are elastic; thus axial force from the drawbar brings the toolholder into face contact with the spindle by deforming only the elastic elements and not the whole shank of the toolholder as in other designs, e.g., described in Section 4.4.2d. Such an approach allows to achieve the desired interference but without absorbing a significant portion of the drawbar force and also without creating a high sensitivity to contamination. The latter is characteristic for the designs having hollow deforming tapers contacting with the spindle hole along the whole surface. Most of the designs were rejected because they introduced new critical design dimensions in addition to the taper angle, thus requiring expensive precision machining.

The final version (WSU-1), shown in Fig. 8.4.4 [27], requires the same accuracy of machining as what is needed to manufacture the standard 7/24 toolholders. It has a tapered (7/24) shank 60d whose diameter is smaller than the tapered shank of a standard toolholder with the same diameter of flange 60b. A metal or plastic cage 62c containing a number of precision balls 68 of the same diameter is snapped on (or attached to) shank 60d. In the free condition, the gage diameter of the virtual taper defined by the balls is 5 μm to 10 μm larger than the gage diameter of the spindle. Under the axial force applied by the drawbar 22, the balls

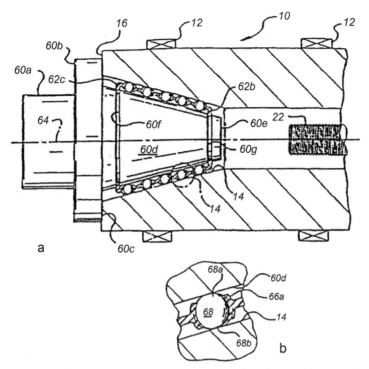

Fig. 8.4.4. Taper/face interface WSU-1 with spherical elastic elements; a General layout; b Ball/tapered Surfaces interaction.

undergo elastic deformation and the toolholder moves inside the spindle, stopping after flange face 60c of the toolholder touches face 16 of the spindle. The cage 66 has lips 62b and 62c interacting with grooves 60f, 60g, thus protecting its inside area from dirt.

This system uses precision balls as precision elastic elements. Balls made of various metals (e.g., steel, titanium, aluminum), glass, plastics, etc. are available with sphericity and diameter accuracy within fractions of 1 μm, at very reasonable prices. The high dimensional accuracy of the balls guarantees the high accuracy of their elastic characteristics. The deformation of the ball-taper connection in Fig. 8.4.4 is composed of Hertzian deformations between the balls and two tapered (male and female) surfaces, and the solid body deformation of the ball, see Table 4.2. It was found that plastic balls do not have an adequate stiffness to support a long and heavy tool during its insertion into the spindle, thus steel and titanium balls are being used.

If steel balls are used (e.g., 6 mm diameter balls for a #50 spindle taper), then with a safety factor of 2.0 for contact stresses, axial motion δ_{ax} of the toolholder for up to 35 μm (~0.0015 in.) is allowable, see Table 8.4. If precision titanium or glass balls are used, then axial motion up to 70 μm (~0.003 in.) is allowable. After the face contact is achieved, it is tightened by the drawbar force. Since a relatively low force is required to deform the balls, the axial force required for the axial motion to achieve the face contact is significantly lower than in the "shallow hollow taper" designs. Thus, a larger part of the drawbar force can be used for the face clamping. There were expressed concerns about denting of the spindle and toolholder tapered surfaces in the contact areas with the balls. It was shown in [28] that a small permanent deformation (1/4 wavelength of green light) of a flat steel plate in contact with a steel ball develops at the load $P = SD^2$, where D = diameter of the ball. Coefficient S is 2.4 N/mm^2 for the plate made from hardened steel with 0.9%C, 7.2 N/mm^2 for the plate made from superhardened steel with 0.9%C, and 5 N/mm^2 for the chromium-alloyed ball-bearing steel plate tempered at 315°C (spring temper). For a 6 mm ball and $S = 2.4$ N/mm^2, $P = 86$ N. Actual testing (6 mm steel ball contacting a hardened steel plate) did not reveal any indentation marks at loads up to 90 N (20 lb). This force corresponds to a deformation about 10 μm and axial displacement of the toolholder of about 70 μm (~0.003 in.). It assures a normal operation of the system even with steel balls. No danger of denting exists when the balls are made from a lower Young's modulus material, such as titanium.

Since only a small fraction of the drawbar force is spent on deformation of balls, friction forces in the flange/face contact are very high;

Table 8.4. Deformation and stress parameters of WSU-1 toolholder with 6 mm balls.

| Ball material | Load per ball, N | | | | | | | | | | | |
| | 4.5 | | | | 36 | | | | 121 | | | |
	δ	δ_{ax}	k	σ	δ	δ_{ax}	k	σ	δ	δ_{ax}	k	σ
Steel $\sigma_{al} = 5.3$GPa	1.3	9	6.8	1.1	5.0	32	13.6	2.2	12	82	20.4	3.3
Glass $\sigma_{al} = 4$GPa	2.5	16	1.5	0.6	9.5	66	3.0	1.2	21	150	4.6	1.9

δ — radial deformation of ball, μm; δ_{ax} — axial shift of toolholder, μm; k — stiffness per ball, N/mμ; σ — contact stress, GPa; σ_{al} — allowable contact stress, GPa.

the friction coefficient is about $f = 0.15$ to 0.2. Thus, in many cases the keys are needed only as a safety measure since an instantaneous sliding can occur at dynamic overloads during milling. However, it would be beneficial to further enhance friction at the face contact by coating the toolholder face by tungsten carbide (using electro-discharge apparatus) to eliminate the need for keys altogether. Measured static friction coefficients, f, for one type of coating are given in Fig. 8.4.5 versus normal pressure in the contact. It can be seen that f does not depend significantly on the presence of oil in the contact area. Due to the high friction, the connection can transmit very high torques without relying on keys. For a #50 taper, axial force 25 KN (5600 lb), and $f = 0.35$, such a connection can transmit 360 Nm torque which translates into 180 KW (250 HP) at $n = 5000$ rpm. Even higher f, up to $f = 0.9$, are realizable by a judicious selection of the coating [29].

The measured runout of the toolholder in Fig. 8.4.4 was less than that of any standard toolholder (about one half), since there is no clearance between the toolholder and the spindle hole. Balls with sphericity within 0.25 μm (medium accuracy grade) had been used. A significant stiffness enhancement due to face contact was measured, in line with the data on stiffness of the shallow hollow taper connections with the face contact.

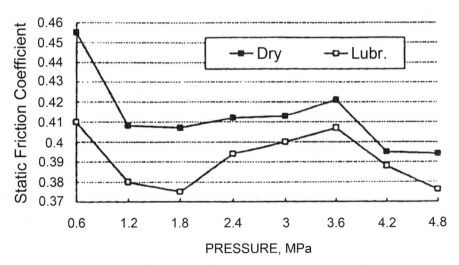

Fig. 8.4.5. Static friction coefficient between coated flat surfaces.

The design in Fig. 8.4.4 solves the same problems as the other alternative designs listed in Section 4.4.2d and Section 8.4.3 (axial registration, high stiffness, insensitivity to high rpm), but without their shortcomings. No changes in the spindle design are required; manufacturing of the new shank is no more complex than manufacturing of the conventional standard shanks; the cost differential is minimal; shrink-fit tools or other tool clamping devices can be located deep inside the tapered shank of the holder, thus reducing the necessary tool overhang and enhancing the effective stiffness; and no bulging of the spindle occurs.

Performance testing of the WSU-1 design was performed on a milling machine equipped with a manually operated drawbar, instrumented with strain gages to measure the axial force. Face and slot milling operations were performed at the regimes creating maximum allowable load on the cutters. Significant improvements in flatness and surface finish of the machined surfaces over conventional interfaces have been observed [30].

As it was noted in Section 8.4.2b, not all applications require the axial indexing of the toolholder. If the axial indexing is not required, the main problems of the standard 7/24 taper toolholders are: reduced stiffness, large runout, and fretting. All these shortcomings are due to the clearance at the back of the taper connection. Successful testing of the design in Fig. 8.4.4 led to the design of interface WSU-2 (Fig. 8.4.6) [31], which solves these problems.

The toolholder in Fig. 8.4.6 has taper 1 of the standard dimensions. At the back side of the taper a coaxial groove 2 is machined. Inside this groove one or more rows of precision balls 3 are packed. The balls protrude out of the groove by an amount slightly exceeding the maximum possible clearance between the male and female tapers in the connection, and are held in place by rubber or plastic filling 4. The balls deform during the process of inserting the toolholder thus "bridging" the clearance. This deformation assures the precise location of the toolholder in the tapered hole, and also provides additional stiffness at the end of the tool since it prevents "pivoting" of the toolholder about its contact area at the front of the connection. This eliminates micro-motions of the shank inside the spindle hole and the resulting wear.

This modification can be applied both to existing and to newly manufactured toolholders. Although the modification is very simple and inexpensive, it results in a very significant increase in effective stiffness and reduction of runout of the interface. The stiffness increase is espe-

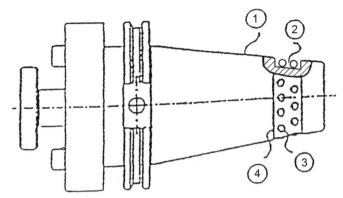

Fig. 8.4.6. Modified 7/24 taper interface WSU-2 (with an elastic bridge for backside clearance).

cially pronounced at low drawbar forces. Fig. 8.4.7 presents typical plots of stiffness vs. axial force for conventional and modified #50 toolholders. Stiffness was measured under vertical (Y-direction) load applied 40 mm (~1.5 in.) in front of the spindle face in the direction of the keys, and in the perpendicular direction. While at low axial forces the stiffness increases as much as three-fold, even at high axial forces the increase is still significant. The stiffness of WSU-2 is comparable to the stiffness of the taper/face interfaces. As a result, this simple modification may be sufficient for majority of machining operations without the need for radical changes in the toolholder design.

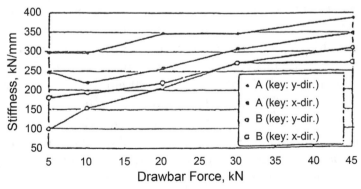

Fig. 8.4.7. Stiffness of 7/24 #50 toolholders; A with two rows of 6 mm balls (WSU 2, Fig. 8.4.6); B without balls.

Runout reduction was measured by comparing runout of the toolholder with the machined groove with and without balls. Reductions in the range of 10% to 50% have been observed, depending on the fabrication quality of the toolholder. Flatness and surface finish for face milling with the WSU-2 interface are significantly better than with the conventional 7/24 interface, although not as good as with the taper/face WSU-1 interface [30].

Toolholder/Spindle Interfaces Using the Giant Superelasticity Effect. While precision steel balls were successfully used as precision springs in the above described WSU-1 and WSU-2 systems, their shortcomings are a limited displacement range (up to 10 μm for a 6-mm diameter ball) and high hardness creating a potential danger of denting the mating surface of the tapered hole in the spindle. These shortcomings can be eliminated by using a recently discovered so-called "Giant Superelasticity Effect" (GSE), [32] and Section 2.7.4. This effect, due to combination of material properties of superelastic materials (SEM) and geometry of design components, states that recoverable (elastic) deformations of cylindrical or tubular specimens made from a NiTi SEM and radially compressed between loading surfaces having initial linear contact with the specimen are ε_{max} = 10% to 22%. Such values of ε_{max} are more typical for elastomers than for metals. These values of ε_{max} are two-to-three times greater than the elastic strain limits of the SEM measured in tension. Thus, radially loaded wires and tubes can be used as compact springs. While a 6 mm diameter ball used as a spring has allowable deformation ~10 μm, a 0.2-mm diameter radially loaded tube made from NiTi SEM has allowable deformation of 20 μm.

A WSU-1 – like taper using such a tubular spring is shown in Fig. 8.4.8 [33]. Toolholder 82 has two rings 86 and 87 made from a SEM tubing with outer diameter D = 0.2 mm to 0.5 mm. The external surfaces generate a "virtual taper" surface conforming with tapered surface 84 inside spindle 81. If only 10% (20 μm to 50 μm) radial compression of the rings is allowed, it allows for 20(24/3.5) $\approx$ 140 μm to 50(24/3.5) $\approx$ 350 μm axial compensation, respectfully, thus allowing for the taper/face contact even with very loose axial tolerances for the connection. Use of such "virtual taper" concept also allows to replace the hardened and precisely machined taper in conventional toolholders (and in other applications) by a simple mandrel with two narrow machined surfaces for SEM rings.

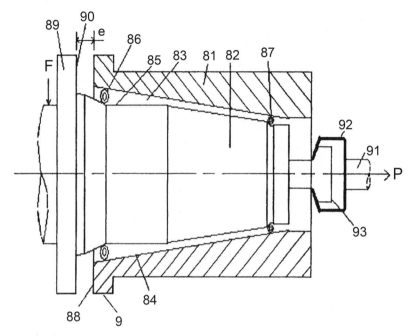

Fig. 8.4.8. Taper/face assembly with GSE compensators.

Bridging the backside clearance (a WSU-2 like design) can be achieved by a simple retrofitting of the standard tapered connection (Fig. 8.4.9). One tubular ring 3 made from an SEM is inserted in groove 2 machined in the body 1 of the male taper. The size of the ring protrusion h from the male taper body is selected so that the radial compression of ring 3 is 0.05 to 0.1D.

An important issue is attachment of the SEM tubular rings to the taper body. Their direct welding would distort SEM properties in the welding area. The distortions are prevented if a strong steel or tungsten wire is threaded through the SEM tube and welded to the taper body [34]. Also, this technology allows to maintain tension in the wire and a uniform pressure between the ring and the groove.

Fig. 8.4.10 shows typical static load-deflection plots for steep 7/24 taper toolholders #40 at two drawbar forces 1.6 kN and 8.3 kN, loaded in bending by the force perpendicular to the toolholder axis at 100 mm from the gage line. Three toolholder designs had been tested: standard; with the bridged back clearance per Fig. 8.4.9 ("one ring"); and with taper-face contact per Fig. 8.4.8 ("two rings"). For both axial preloads, stiff-

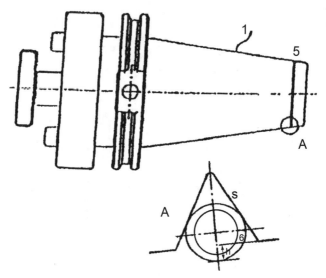

Fig. 8.4.9. "Steep taper toolholder with bridged backside clearance using GSE compensators.

ness of the bridged (one ring) connection is 24% to 29% higher than the standard one, while for the taper/face connection the stiffness increase is 58% to 65%. For all three toolholders change of the axial preload from 1.6 kN to 8.3 kN results in 15% to 25% stiffness increase.

Dynamic testing of the bridged toolholder in comparison with the standard one at the 22 kN axial preload indicated increase in the fundamental natural frequency from 1857 Hz to 2362 Hz and increase of dynamic stiffness of 36%. The damping ratio has changed from $\zeta = 0.53\%$ to $\zeta = 1.49\%$. During these tests the runout of the standard toolholder was measured to be 10 μm, and of the bridged toolholder only 4 μm.

8.4.3 Other Tapered and Geared Toolholder/Spindle Interfaces

The technical challenges for toolholder/machine interfaces include a need for radical improvements in stiffness and damping of the connection; low radial runout; high axial repeatability; high torque-transmitting capability both for static and dynamic loading; easy balancing; reduction of the required axial travel of the tool changing device; reasonable cost. The most widely used interface systems developed during the last 15 years to 20 years are shallow 1/10 taper KM and HSK interfaces, polygonal/shallow taper Capto interfaces, and flat gear (Curvic) interface.

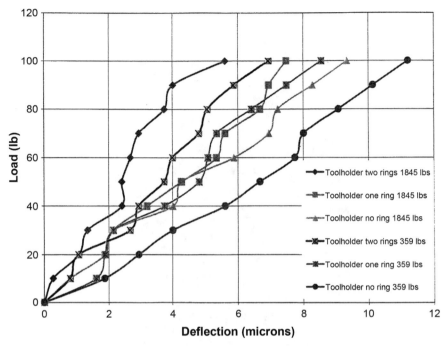

Fig. 8.4.10. Load-deflection plots of standard steep taper toolholder (no rings), design of Fig. 8.4.9 (one ring), and design of Fig. 8.4.8 (two rings) at two drawbar forces, 359 lb and 1845 lb.

8.4.3.1 Curvic Coupling Connection

The Curvic coupling [35] comprises two identical flat spiral teeth gears, one attached to the machine (to spindle in machining centers), and another attached to the flange of the tool (Fig. 8.4.11). This connection satisfies practically all the abowe listed challenges. Stiffness and damping of the Curvic coupling connection is certainly the best in comparison with other interfaces, see Figs. A2.1 and A2.2. Dynamic properties of the Curvic coupling connection are far superior to flat joints and to standard taper connection due to large contact area (many contacting teeth), so the high stiffness develops at low drawbar force, which is correlated with high damping, see Chapter 4. Very high chatter resistance of systems with this connection was proven by cutting tests. Radial repeatability of this connection is within ±0.4 μm, and axial repeatability within ±1.5 μm. Torque transmitting capability is superior to the tapered connections due to a large number of load transmitting teeth. The design is highly

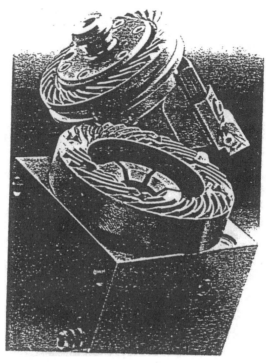

Fig. 8.4.11. Curvic coupling connection (ITW Woodworth Co.).

symmetrical; its drawbar is short and rigid, thus together with the light weight assuring good balancing properties. The multiple teeth contact results in a very low wear rate; the connection accuracy, stiffness, and torque capacity is even improving with time due to a "work in" process. The connection requires a very short travel for the tool change since the axial dimension of the teeth is only 5 mm to 10 mm. The connection does not require a big hole in the spindle, thus in some cases allowing to use smaller diameter spindle bearings. On the negative side, the Curvic coupling is about two times more expensive than the taper connection.

8.4.3.2 KM System

The KM system, Fig. 8.4.12, was developed initially for clamping lathe tools and later was adopted for machining centers applications [36]. It develops a simultaneous taper/face contact with the tapered hole in the receptacle (turret or spindle) due to an interference fit between the radi-

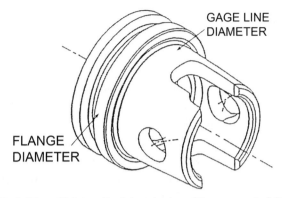

Fig. 8.4.12. KM toolholder design (Kennametal Corp.).

ally contracting hollow tapered shank and expanding spindle under high axial force. The shallow 1/10 taper and ball-wedge axial force enhancer (6.5:1 mechanical advantage, Fig. 8.4.13) result in significant radial force amplification. This leads to expansion of the back part of the hollow tapered shank, notwithstanding its relatively thick walls. The expansion is assisted by deep slots in the back of the taper. Tolerances for both the male taper and the receptacle provide for their initial guaranteed contact at the front part of the connection and for a small clearance in the back;

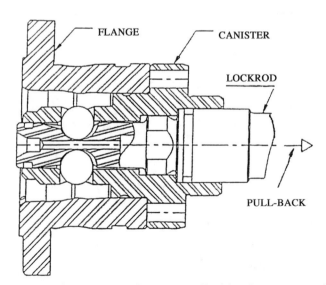

Fig. 8.4.13. Clamping unit for KM toolholder (Kennametal Corp.).

the midsection of the connection is slightly recessed. Shrinkage of the front part under the axial force and expansion of the back part leave a significant portion of the axial force (about two thirds [20]) for the face contact thus providing for its high stiffness (see Chapter 4). The relatively thick walls of the male taper and large interference (15 μm to 36 μm for KM40; 20 μm to 45 μm for KM50; 25 μm to 50 μm for KM60) cause a significant bulging of the spindle, so the spindle bearings must be located beyond the taper area. Drawbacks of the KM design are its hollow structure resulting in increased overhang of the clamped tools, similar to HSK toolholders, Fig. 8.4.15, below and small (self-locking) taper angle requiring a "kick-off" mechanism in the drawbar.

8.4.3.3 HSK System

The HSK system, Fig. 8.4.14, is similar to KM system [20]. Deformation characteristics of the HSK system are discussed in-depth in Section 4.4.2, here some applications-related issues in comparison with other interface systems are addressed. The simultaneous taper/face contact in the HSK system is achieved by deforming the hollow taper and the receptacle, serving as compensation elements, by drawbar forces amplified by wedge action of the drawbar's end effector. The actual face-clamping forces are much lower than the effective axial clamping force in KM toolholders. Because of this, tolerances on critical dimensions (ta-

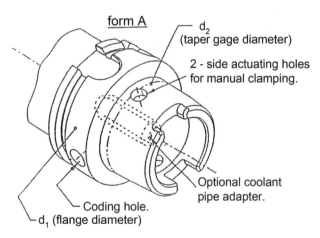

Fig. 8.4.14. HSK toolholder design.

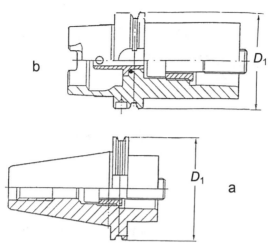

Fig. 8.4.15. Positioning of tool-clamping device of Coromant Capto modular tool system with steep taper (a) and with HSK (b).

pered surfaces) of both the toolholder and the receptacle are extremely tight, within 2 μm to 6 μm. Even with such tight tolerances, variation of interference between the male taper and the receptacle is very significant, up to 12 μm. The variation can be even wider due to difficulties in measuring the toleranced surfaces. The taper/face contact is partly due to shrinking of the male taper, and partly due to bulging of spindle (Fig. 4.5.18).

While the obvious advantage of the HSK system (and of the similar KM system) is realization of the simultaneous taper/face contact, it has serious shortcomings. The intricate design together with very tight tolerances make these toolholders very expensive, at least 1.5 to 2 times more expensive than 7/24 taper toolholders. Since the HSK taper is hollow, the tool clamping part, such as a collet or a shrink-fit supporting surface has to be located totally in front of the spindle (Fig. 8.4.15). This increases the tool overhang, thus reducing the end-of-tool stiffness and increasing bending moment at the face, causing separation at the face contact at lower cutting forces. The drawbar for HSK toolholders is more complicated and more expensive since it needs a powerful mechanism for "kicking out" the shallow taper (and thus self-locking) toolholder from the spindle. Spindle expansion, Fig. 4.5.18, may adversely affect spindle bearings. The transitional area between the hollow taper and the massive flange may develop stress concentrations and eventually cause breakage of the toolholder. Thin walls of the hollow tapered shell are not very

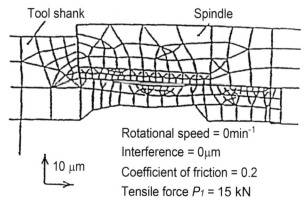

Rotational speed = 0min⁻¹

Interference = 0μm

Coefficient of friction = 0.2

Tensile force P_1 = 15 kN

Fig. 8.4.16. Deformation of HSK interface due to clamping force (Tsutsumi, et al).

stable. They are bulging inwardly both under the axially applied drawbar force, Fig. 8.4.16, and due to centrifugal forces at high rpm (Fig. 8.4.17). A combination of these effects may cause a situation when the toolholder is not contacting the receptacle surfaces at all or has two narrow bands of contact. Such pattern of contact may become detrimental to stiffness of the connection, especially if face separation is developing under high cutting surfaces.

Fig. 8.4.18 compares load-deflection characteristics of various tapered interfaces (tested by Prof. S. Smith). CAT is a standard 7/24 taper toolholder, Big Plus is a steep taper interface with simultaneous taper/face

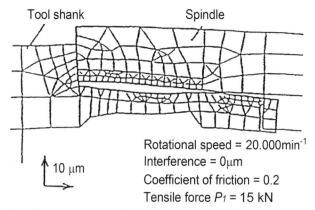

Rotational speed = 20.000min⁻¹

Interference = 0μm

Coefficient of friction = 0.2

Tensile force P_1 = 15 kN

Fig. 8.4.17. Deformation of HSK interface due to centrifugal force (Tsutsumi, et al).

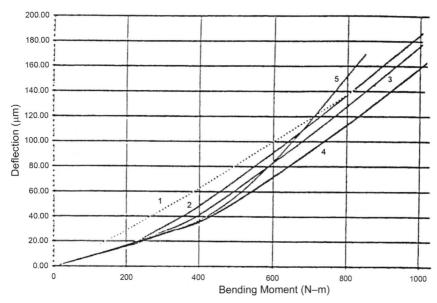

Fig. 8.4.18. Bending deflection at end-of-tool; 1 — CAT #40, axial force $P_a = 17$kN; 2 — BigPlus #40, $P_a = 7$ kN; 3 — same, $P_a = 12$ kN; 4 — same, $P_a = 17$ kN; 5 — HSK 63A, $P_a = 10$ kN.

contact. It can be seen that the standard CAT #40 interface has the same inclination of the load-deflection plot in the whole range 0 Nm to 1,000 Nm of bending moment, while all taper/face interfaces "open up" at 220 Nm to 400 Nm, depending on the axial drawbar force.

8.5 BENEFITS OF INTENTIONAL STIFFNESS REDUCTION IN DESIGN COMPONENTS

Our "common sense" tells us that solid components can tolerate higher loads than the same components but "weakened" by holes or cuts. However, this belief does not consider very complex interactions between the components which involve contact deformations on very small surface areas, stress concentrations, nonuniform load sharing between several contact points (e.g., in power transmission gears with contact ratios exceeding 1.0). Quite frequently, stiffness reduction for some components may result in increasing of the overall stiffness and/or in improved performance due to changing load and deformation distribution patterns.

8.5.1 Hollow Roller Bearings

A very important mechanical component is roller in contact with a flat surface (guideways) or with a round surface of much smaller curvature (bearings). It was shown in [37, 38] that a hollow roller (roller with an axial bore) develops significantly lower contact (Hertzian) stresses due to the larger contact area resulting from deformation of the roller's body. With increasing parameter $a = r/R$, where R = radius of the roller, and r = radius of the bore, the wall is getting so thin that its bending stresses may become a critical parameter. Fig. 8.5.1 [38] displays dimensionless plots of σ_{max}/E_1 and P/RLE_1 as functions of angular width $2\phi_0$ of the contact rectangle. Here σ_{max} = maximum contact stress, P/RLE_1 = dimensionless load on the roller, E_1 = Young's modulus of the roller; P = force acting on the roller; L = length of the roller; ϕ_0 = measured in angular minutes. The same width of the contact area $2\phi_0$ for a roller with larger a would develop at a lower force, P. These plots can be used to determine σ_{max} using

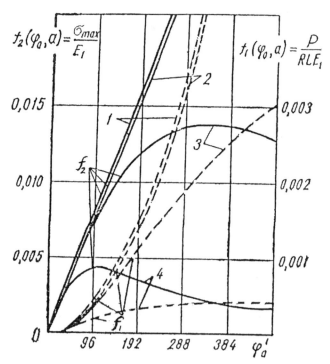

Fig. 8.5.1. Dimensionless stresses in contact of hollow rollers with a flat surface. 1 — $a = 0$; 2 — $a = 0.5$; 3 — $a = 0.75$; 4 — $a = 0.875$.

the following procedure: calculate dimensionless load on a roller P/RLE_1; draw the line parallel to abscissa from this value of dimensionless load on the right ordinate to the line representing $f_1(\phi_o,a) = P/RLE_1$ for the given a; this determines ϕ_o, thus intersection of vertical line from this ϕ_o with a plot for $f_2(\phi_o,a) = \sigma_{max}/E_1$ for the given a solves the problem.

Table 8.5 [37] compares magnitudes of P, which cause a certain maximum contact stress ($\sigma_{max} = 590$ MPa) for a steel roller $R = 50$ mm, $L = 200$ mm at various a. A thin-walled roller ($a = 0.875$) can absorb 14% higher load while being about four times lighter. Of course, the thin-walled roller is more compliant overall due to its bending deformation. But even relatively small holes, $a \approx 0.5$, which do not influence significantly the overall deformation, could be very beneficial for roller bearing and guideway applications.

In a roller bearing for an aircraft gas turbine rotor [37], rollers with $R = 10$ mm and $L = 20$ mm are placed around the circumference with diameter $D_o = 200$ mm. The typical cage design allows to place 24 rollers with angular pitch $\alpha = 15$ deg. The bearing is loaded with radial force, $P = 5000$ N at $n = 10,000$ rpm; it is not preloaded, and the load on the most loaded roller can be calculated, using Eq. (A6.5) as:

$$P_o = \frac{P}{\left[1 + 2\sum_{i=1}^{12} \cos^{5/2} i\alpha\right]} = P/6.01 = 832 \text{ N} \tag{8.5.1}$$

Centrifugal forces press the roller to the outer race. Mass of one roller is $M = \gamma\pi R^2 L = 0.05$ kg where $\gamma = 7.8 \times 10^3$ kg/m^3 is density of steel. Linear velocity at the center of the roller is $V = 0.5(10,000/60)2\pi(D_o - R) = 47.1$ m/s and centrifugal force on one roller is $P_{c.f.} = mv^2/(D_o/2) = 1110$ N,

Table 8.5. Compressive load P_{max} causing stress 590 MPa in contact of roller with flat surface depending on degree of hollowness, a, of steel roller $R = 50$ mm, $L = 200$ mm.

a	P_{max}, N	P_{max}, %	Weight, %
0	95,100	100	100
0.8	98,000	102.9	36
0.875	108,800	119.2	23.4

which is more than the payload per one roller. Thus, the maximum radial force on the most loaded roller is:

$$P_{max} = P_o + P_{c.f.} = 1942 \text{ N.}$$

If $a = 0.5$, then the contact stresses are, practically, not affected (the maximum contact stress, σ_{max}, would decrease, but rather insignificantly, see Fig. 8.5.1). However, the presence of the hole reduces mass of the roller by 25%, and $P'_{c.f.} = 0.75 P_{c.f.} = 833 \text{ N.}$

The payload on the most loaded roller in this case is:

$$P_o' = P_{max} - P_{c.f.}' = 1109 \text{ N}$$

and the corresponding allowable external force

$$P = P_o' \times 6.01 = 6665 \text{ N.}$$

Thus, the rated load on the bearing is 33% higher while its weight is reduced by ~5%. In many cases, increase in the rated load is not as important as increase of the life span of the component. The correlation between the load and the length of life for roller bearings is

$$Ph^{0.3} = \text{const,} \tag{8.5.2}$$

where h = number of hours of service. Thus, the length of service h' of the bearing with hollow rollers can be found from

$$(h'/h)^{0.3} = P'/P = 1.34$$

or $h' = h \times 1.34^3 = $ ~2.6 h, and the life resource of the bearing is 2.6 times longer.

There is an important secondary effect of this design change. Use of hollow rollers allows to change design of the cage by using the holes to accommodate pins of the modified cage. This allows to increase the number of rollers to 30 ($\alpha = 12°$), and also reduces friction between the rollers and the cage. Using Eq. 8.5.1, it is easy to find that the allowable external force can be increased to $P'' = 8580 \text{ N}$ for the same maximum load on the roller (1942 N). This would increase the rated load by 67% as compared with the original solid roller bearing, or prolong its life by a factor of 4.5.

Similar roller bearings, Fig. 8.5.2 [39, 40] were tested (and are mar-keted) for high-speed/high-precision machine tool spindles. In this case, values of $a = 0.6$ to 0.7 are considered as optimal, since it was found that at such values of a the balance between contact and bending deforma-tions of the rollers is the best. The bearings are uniformly preloaded by using slightly oversized rollers; they are used without cages. It was estab-lished, that if the rollers are initially tightly packed, they become sepa-rated by small clearances after a few revolutions. Such bearings have only about 50% of the maximum load capacity of the solid roller bearings (partly due to use of a part of strength to enhance stiffness by preload). However, it is compensated by the combination of high stiffness, high rotational accuracy (runout less than 1 μm is reported), and high-speed performance up to $dn = 3.5 \times 10^6$ mm-rpm (due to reduced weight and centrifugal forces from the rollers). It is interesting to note that the simi-lar results both in effective mass and in stiffness can be achieved by using high Young's modulus, low density ceramic balls or hollow steel rollers with intentionally reduced stiffness.

8.5.2 Stiffness Reduction in Power Transmission Gears

Two principal challenges in designing power transmission gears are: in-creasing payload for given size/weight and reliability, and reduction of vi-bration and noise generation for high-speed transmissions. The payload

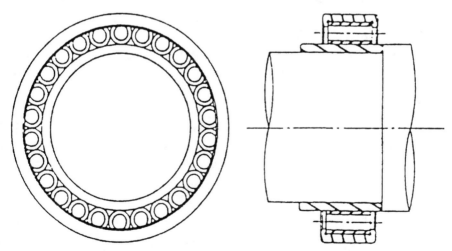

Fig. 8.5.2. Spindle bearing with hollow rollers.

capacity is determined by contact stresses in the mesh; by bending stresses in the teeth (especially, stress concentrations in the fillets connecting the teeth with the rim); and by dynamic loads generated due to deviations from the ideal mesh kinematics, especially at high speeds. These deviations are caused by: imperfect uniformity of the pitch; deviations of involute tooth surfaces from their ideal shapes; deformations of shafts, bearings, and connections; changing mesh stiffness during each mesh cycle; non-whole number contact ratio causing abrupt changes in number of engaging teeth in each mesh cycle. Vibration and noise generation is also closely correlated with the dynamic loads in the mesh.

Conventional techniques for handling the above-listed factors are modification of the tooth geometry (intentional deviation from the ideal involute profiles by flanking and/or crowning the tooth surfaces, optimization of fillet shapes, etc.); improvements in gear material (use of highly alloyed steels with sophisticated selective heat treatments, use of high purity steel, hard coatings, etc.); and tightening manufacturing tolerances for critical high speed gears. While significant improvements of the state-of-the-art gears were achieved by these approaches, new developments along these lines are bringing diminishing returns for ever increasing investments. For example, while higher accuracies in pitch and profile generation lead to reduction of dynamic loads, thus to increasing payloads and to noise reduction, costs of further tightening of tolerances for already high precision gears are extremely high. However, even ideal gears would deviate from the ideally smooth mesh due to deformations of teeth which vary during the mesh cycle and due to distortions caused by deformations of shafts, bearings, and connections.

Because of these complications, another approach is becoming popular – intentional reduction of stiffness of the meshing gears (this development is in compliance with the universal *Laws of Evolution of Technological Systems*, e.g. see [26]).

Introduction of elastic elements into power transmission gears is a subject of many patents, starting from the last century [41]. Some of the typical approaches are shown in Fig. 8.5.3 [42]. Five design groups shown in Fig. 8.5.3 as *a – e* achieve different effects. In designs in Fig. 8.5.3a, compliant teeth experience smaller contact (Hertzian) stresses, similarly to the hollow roller described in Section 8.5.1. Since the greatest bending stresses develop in the fillets between the teeth and the rim, they are not significantly affected by a slot in the "upper body" of the

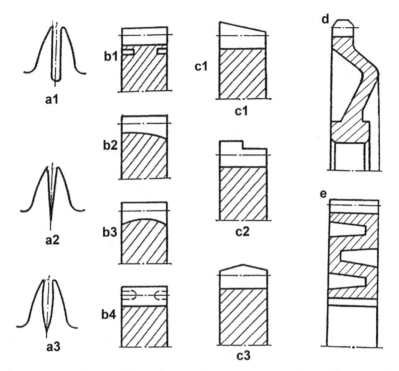

Fig. 8.5.3. Gears with enhanced compliance. a-Compliant teeth; b-Compliant interface between teeth and gear hub; c-Shaped teeth to compensate stiffness non-uniformity of mesh; d-Compliant gear hub; e-Rim elastically connected to hub.

tooth. However, by enhancing the tooth compliance by means of introduction of groove within the tooth ("artificial stress concentrator"), Fig. 8.5.3, a1–a3, peak stresses (in the fillet) can be reduced by 20 % to 24% with the optimal dimensioning of the groove. Use of compliant teeth may also result in a dramatic reduction of dynamic load amplitude P_d which is expressed as:

$$P_d = \psi v_0 \sqrt{km} \qquad (8.5.3)$$

Here v_0 = tangential velocity of the meshing gears which determines impact velocity, ψ = coefficient reflecting gear parameters, accuracy, etc., k = stiffness of impacting pair of teeth; m = effective mass of the teeth.

Designs in Fig. 8.5.3b are characterized by the same contact stresses, same bending stresses in designs b1 and b4, and somewhat increased bending stresses (due to increased effective height of the teeth) in cases b2, b3. The effect of these designs is reduction of dynamic loads due to the reduced stiffness in accordance with Eq. 8.5.3. The similar effects are realized by designs in Fig. 8.5.3d, e in which the enhanced radial compliance of hubs is equivalent to enhanced tangential compliance of teeth in designs of Fig. 8.5.3b, c.

Periodic variation of stiffness of the meshing teeth, both due to the constantly changing radial position of the contact point on each tooth and due to the non-whole value of the contact ratio, is a powerful exciter of parametric vibrations, especially in high-speed gears. The stiffness variation can be, theoretically, compensated by modification of the tooth shape in the axial direction, e.g. as in Fig. 8.5.3c.

While designs in Fig. 8.5.3a to e are rather difficult for manufacturing, designs in Fig. 8.5.4 are the most versatile. These designs use conventionally manufactured gear rims connected with the gear hub (which in this case can be made from a low-alloyed inexpensive steel or even from cast iron) by a special flexible connection. Both contact and bending strength of the teeth under a static loading are the same as for a conventional (solid) gear. However, self-aligning of the rim in relation to the hub results in a more uniform load distribution along the teeth. Another important effect is reduction of dynamic stresses due to reduction of both k and m in Eq. 8.5.3. Such designs were extensively tested in [42]. These tests demonstrated that ratio of maximum-to-minimum stresses along the tooth were 1.1 to 1.4 (depending on the total load magnitude) for the composite gear with the flexible connection versus 1.65 to 1.85 for

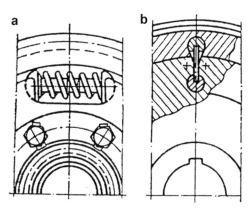

Fig. 8.5.4. Composite gears with elastic rim-hub connections; a Coil spring connection; b Flat spring connection.

the solid gear. This difference was maintained both in static tests and in the working transmissions. It results in much slower wear rates of teeth profiles in the composite gears. Fig. 8.5.5 shows a typical comparison of *rms* of vibration acceleration for the solid and composite (per Fig. 8.5.4) gears. A very significant difference (three times or 10 dB) is characteristic for dynamic loading in the mesh.

The flexible rim-hub connection presents a very simple and economical way to reduce dynamic loads in the mesh as well as noise levels of power transmission gears. However, conventional flexible connectors have flexibility not only in the circumferential (desirable) direction, but also in the radial direction. An excessive radial rim-hub compliance may distort the meshing process and is undesirable. Frequently, the hub and the rim have a sliding circumferential connection, like in Fig. 8.5.4. The inevitable friction and clearances may introduce performance uncertainty. This situation can be corrected by *introducing elements with anisotropic stiffness* which have a low stiffness value in one direction and a high stiffness value in the orthogonal direction. Such stiffness characteristics are

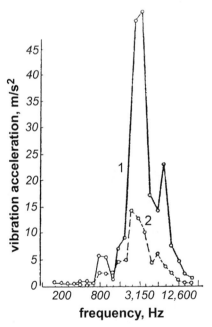

Fig. 8.5.5. Spectra of rms vibration acceleration of the gear housing for solid gears (1) and composite gears (2) tested on a four-square test rig.

typical, for example, for thin-layered rubber-metal laminates described in Section 3.3.2.

While the anisotropic elements may be designed to fully satisfy requirements to the stiffness values in two orthogonal directions, it is often useful to separate these function between two distinctly different design components. It was suggested in [43, 44] to use high compression stiffness of the thin-layered rubber-metal laminates for the radial restraint, while designing them with very low shear (circumferential) stiffness. The required circumferential stiffness is provided by a flexible torsional connector (coupling) (Fig. 8.5.6). Superior performance characteristics of the torsional connection (coupling) using multiple rubber cylinders compressed in the radial direction as described in [45], allows to package the connection into the available space between the rim and the hub of even heavy duty power transmission gears.

A similar effect can be achieved by a totally different approach – modification of the meshing system by using elastic elements to separate sliding and rolling in the mesh. Two such systems were proposed in [46] and [47].

The [46] system is most applicable to gear teeth characterized by a constant curvature of the tooth profile. This feature is typical for so-called *conformal* or *Wildhaber/Novikov* (W/N) gears, e.g., [48]. Metal layer 1 in Fig. 8.5.7 has the same shape as tooth profile 2 and is attached to it by

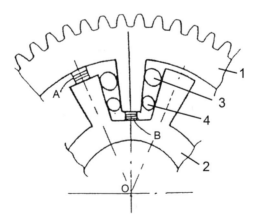

Fig. 8.5.6. Torsionally flexible rim-hub connection for a power transmission gear with radial restraint by rubber-metal laminates. 1 — rim, 2 — hub; 3, 4 — flexible coupling elements, A, B — rubber-metal laminates.

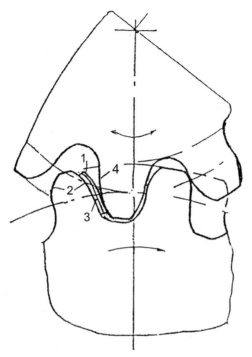

Fig. 8.5.7. Gears in which sliding in the mesh is accommodated by internal shear in rubber-metal laminate coating.

rubber-metal laminate or rubber layer 3. During the mesh process, counterpart tooth profile 4 engages without sliding with 1. Sliding between the meshing profiles which is necessary for the meshing process is accommodated by shear deformation of laminate 3 while the tangential load is transmitted by compression of laminate (layer) 3. The contact pressures between the meshing profiles, which are relatively low for the W/N gears (about six to ten times lower than for the contact between conventional involute profiles) is further reduced *due to a higher compliance of element 1 and laminate 3 in compression than the compliance of the steel-to-steel contact*. This, combined with the high load-carrying capacity of thin-layered rubber-metal laminates in compression, Section 3.3.2, allows to improve load capacity of the gears while, in the same time, to compensate for center distance inaccuracies and to reduce noise generation (15 dB to 20 dB, see [49]).

The system in [47] resolves the combined rolling/sliding motion between the two meshing involute profiles into separate rolling and sliding motions.

A pure rolling motion takes place between involute profile 13 of one gear and the specially designed profile 18 of slider 20 on the counterpart gear (Fig. 8.5.8). Slider 18 is attached to tooth core 14 of the counterpart gear with a possibility of sliding relative to core 14. The sliding is realized by connecting slider 18 with cylindrical or flat surface of tooth core 14 via rubber-metal layer or laminate 20. The sliding motion of slider 18 relative to core 14 is accommodated by shear deformation of laminate 20. Again, the incremental compression compliance of rubber-metal laminate 20 results in substantial reduction of dynamic loads and noise generation as well as in very low sensitivity of the mesh to manufacturing inaccuracies [50].

8.5.3 Stiffness Reduction of Chain Transmissions

While reduction of dynamic loads is important for power transmission gears in order to enhance their load-carrying capacity and reduce noise generation, it is even more important for power transmission chains. Operation of the chain drives is naturally associated with generation of relatively high dynamic loads when an approaching chain link is engaging with the sprocket, e.g., see [51]. The dynamic loads are responsible

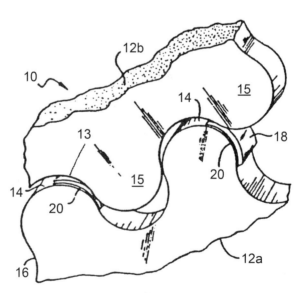

Fig. 8.5.8. Composite gears with separation of sliding and rolling motions.

for high noise levels of the chain drives; they limit the load carrying capability and, especially, maximum speeds of chains. It was suggested in [52, 53] to reduce dynamic loads inherent to chain drives by compliant attachment of sprocket teeth to the sprocket body.

A normal engagement between the chain and the sprocket would take place if deformation of the sprocket tooth is not excessive and does not prevent engagement of the next tooth with the next link of the chain. An experimental study in [53] was performed with the driving sprocket design shown in Fig. 8.5.9. Each tooth 1 of the sprocket is connected with hub 2 by pivots 5 and is supported by rim 3. The sprocket with rigid teeth (reference) had a metal rim, while the sprocket with compliant teeth had the rim made of rubber, as shown in Fig. 8.9 by 3. Teeth 1 are held in contact with rim 3 by rubber bands 4.

The test results demonstrated a three to six-fold reduction of dynamic loads associated with entering new links into engagement with the sprocket teeth for the compliant sprocket. It was also concluded that the payload is more uniformly distributed between the teeth of the compliant sprocket.

8.5.4 Compliant Bearings for High-Speed Rotors

Rotational speeds of machines such as turbines, machine tool spindles, etc. are continuously increasing. While balancing, both static and

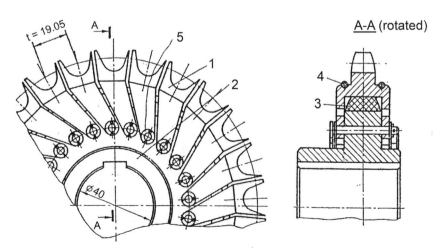

Fig. 8.5.9. Chain sprocket with compliant teeth.

dynamic, has become a routine procedure, there are situations where balancing of rotors before their assembly with bearings is not adequate for assuring low vibration levels of the machine and low dynamic loads on the bearings. In turbines, the balancing conditions of high-speed rotors may change due to thermal distortions, especially for horizontal rotors, which sag when stopped while their temperature is still elevated [54]. Machine tool spindles carry tools whose balance is changing due to wear of cutting inserts or grinding wheels, variations in clamping conditions resulting in slight eccentricities (runouts), change of mass distribution during dimensional adjustments, etc. The resulting unbalance exhibits itself in high levels of vibration and high dynamic loads transmitted through the bearings to housings and other frame parts. These dynamic loads reduce life span of the bearings and also result in undesirable temperature increments which, in turn, increase highly undesirable thermal deformations of spindles.

While the conventional approach to bearing designs is to increase their stiffness, significant benefits can be often obtained by an *intentional reduction of the bearing stiffness*. It is well known (e.g., [54]) that after a rotor passed through its first critical speed, its center of mass tends to shift in the direction of its rotational axis. If some masses attached to the rotor have mobility, a self-balancing effect can be realized. Since the first critical speed of a rigid rotor is usually very high (e.g., for machine tool spindles), it can be artificially reduced by using compliant bearings. The existing auto-balancing devices use special bearing systems which sustain high stiffness of bearings at working conditions and reduce stiffness of bearings when the balancing is required.

In many cases, reduction of dynamic loads on high-speed bearings is the most important goal. It was demonstrated in [55] that if a rotor is supported by compliant bearings, Fig. 8.5.10, the dynamic forces between the rotor and the bearings disappear if the following conditions are satisfied:

$$m_1 = k_1/\omega^2; \qquad m_2 = k_2/\omega^2. \qquad (8.5.4)$$

Here m_1, m_2 are masses (nonrotating) of bearings A, B, respectively; k_1, k_2 are stiffness coefficients of the bearings, and ω, rad/second is rotational speed of the rotor. Fig. 8.5.11 shows the load per unit length of the bearing for a high-speed rigid rotor in rigid bearings (line 1) and

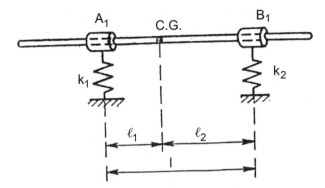

Fig. 8.5.10. Rigid rotor rotating in two compliant bearings.

rotor in compliant bearings for which conditions Eq. 8.5.4 are satisfied at $n = \omega/2\pi = 60{,}000$ rpm (line 2). It can be seen that the force acting on the bearing is greatly reduced at rotational speeds around 60,000 rpm. Use of this approach to turbine rotors and to machine tool spindles (e.g., [56]) demonstrated significant reductions in vibration levels as well as temperature reduction of the bearings.

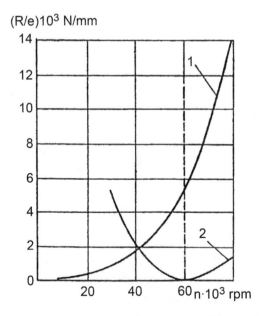

Fig. 8.5.11. Dynamic pressure between rigid rotor and bearing versus rpm; 1 — rigid bearings; 2 — compliant bearings tuned for 60,000 rpm.

8.6 CONSTANT FORCE (ZERO STIFFNESS) VIBRATION ISOLATION SYSTEMS

Effectiveness of vibration isolation systems in filtering out unwanted vibrations can be improved by reducing natural frequencies of the isolated objects supported by vibration isolators [57]. If the isolators support weight of the isolated object, reduced natural frequency in vertical directions and the corresponding reduced stiffness of the isolator lead to an unacceptably large static deflection of the isolator(s) and to their static instability and packaging problems. Static deflection Δ in cm of a vertical spring caused by weight of the supported object can be expressed as:

$$\Delta = \frac{25}{f_n^2}; \quad f_n = \frac{5}{\sqrt{\Delta}} \tag{8.6.1}$$

where f_n = the vertical natural frequency. Very low natural frequencies in the range of 1 Hz to 4 Hz, which are often desirable for vibration protection of humans as well as for vibration protection of ultra-precision equipment, such as photolithography tools, are associated with static deflections 1.5 cm to 25 cm (0.6 in. to 10.0 in.). Such deflections are hardly attainable since they require very large dimensions of the isolators. For example, a rubber isolator of a conventional design and loaded in compression has to be at least 6.5 to 10 times taller than the required static deformation, which results in highly unstable systems. The stability can be enhanced by introduction of a bulky and expensive "inertia mass" (foundation block) or an active leveling, Fig. 7.2.19, and/or an active isolation system.

In some cases, the problems associated with isolating systems characterized by such and even lower natural frequencies can be alleviated by using so-called constant force (CF) elastic systems [58]. The "Constant Force" F as a function of deflection is equivalent to "zero stiffness,"

$$K = dF/dz = 0. \tag{8.6.2}$$

CF systems can provide zero stiffness at one point of their load-deflection characteristic or on a finite interval of the load-deflection characteristic. Since the stiffness values vary, the CF systems are always nonlinear. Besides vibration isolation systems, CF systems are very effective for shock absorption devices where the CF characteristic is the optimal one (e.g., see [59]).

Fig. 8.6.1a illustrates a CF system having a stiffness compensator. Main spring 1 having constant stiffness, k_1, (linear load-deflection characteristic in Fig. 8.6.1b) cooperates with compensating springs 2 having total stiffness k_2. Due to geometry of the device, load-deflection characteristic of springs 2 is nonlinear as shown in Fig. 8.6.1c. The effective load-deflection characteristic of the system is a summation of load-deflection characteristics for springs 1 and 2 (Fig. 8.6.1d). The basic effective load-deflection characteristic shown as the solid line in Fig. 8.6.1d has zero stiffness at deflection $x = 0$ and very low stiffness on the interval *m-n* (working interval). Weight of the supported object (not shown in Fig. 8.6.1a) is compensated by an initial preload F_{10} of the main spring 1. Change of weight can be accommodated by a corresponding change of the pre-load, as shown by chain line in Fig. 8.6.1d. Effective stiffness of the device can be adjusted by changing preload F_{20} of compensating springs 2 (two preload magnitudes are shown by broken lines). All initial values are indicated by subscript "0".

Fig. 8.6.2a shows application of the concept illustrated by Fig. 8.6.1a for a vibration-protecting handle for hand-held impact machines (jack hammer, concrete breaker, etc.). The handle consists of handle housing 1, and two links 2, 3 which are connected via pivots 0', 0" with the machine 4 and via rollers 5, 6 with the handle. Links 2 and 3 are engaged by gear sectors 6, 7 which assure their proper relative positioning. The elastic connection is designed as two pairs of springs 8, 9 and 10, 11, with springs 10, 11 (stiffness, k_s) being shorter than springs 8, 9 (stiffness, k) and the latter having progressively decreasing pitch (like in Fig. 3.2.2c).

Fig. 8.6.2b shows the measured load-deflection characteristic of the handle. An increase of stiffness of the elastic connection (due to "switch-out" of some coils in the progressively coiled springs 8, 9 and also due to "switching in" of springs 10, 11 after some initial deformation of springs 8, 9), results in a two-step characteristic in Fig. 8.6.2b having two CF (low stiffness) sections $k - l$ and $k'- l'$. The second CF section serves as a safety device in case the specified operator pressure F_s on the handle is exceeded. The test results confirmed effectiveness of this system for vibration protection.

Fig. 8.6.3a shows another embodiment of a similar system which is self-adjusting for changing static (weight) loads. The main difference between the systems in Figs. 8.6.1 and 8.6.3 is a frictional connection between load-carrying bar 1 attached to main spring 2 and sleeve 3

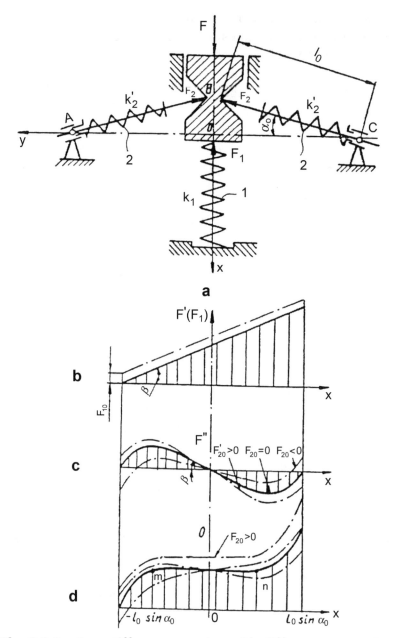

Fig. 8.6.1. Zero stiffness suspension with stiffness compensator.

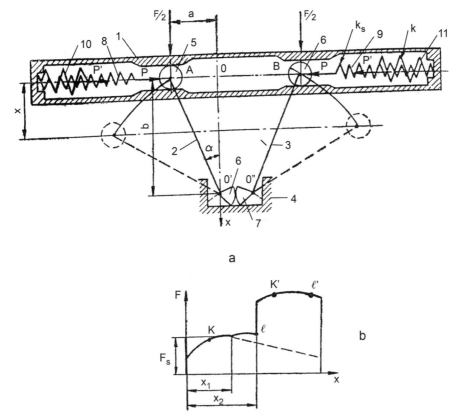

Fig. 8.6.2. Handle for zero-stiffness vibration protection system for jack hammer.

contacting with poles 4 transmitting forces from compensating flat springs 5. Sleeve 3 is preloaded on bar 1 but when the load, F, is increasing beyond a pre-assigned increment, sleeve 3 is slipping along bar 1 and stops in a new position. Fig. 8.6.3b illustrates the load-deflection characteristic of the system.

Another type of CF devices is based on using a specially shaped elastic element. The elastic element in Fig. 8.6.4 is a complex shape spring in which the side parts, AM and CH, act like the main spring 1 in Fig 8.6.1a, and top part ABC acts as compensating spring 2 in Fig. 8.6.1a. Interaction between the top and the side parts generates reaction forces, F'', whose resulting force, F', has a characteristic similar to Fig. 8.6.1c.

The third group of CF devices involves linkage-based compensating devices like in Fig. 8.6.5. Increase of restoring force F' of the spring with

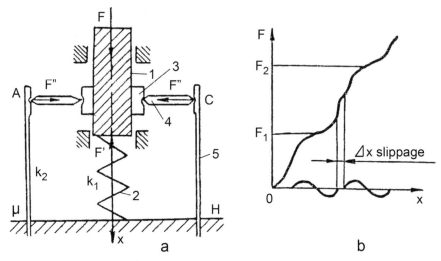

Fig. 8.6.3. Stiffness-compensated zero-stiffness system with automatic height adjustment.

increasing displacements of handles A, B is compensated by decreasing (shortening) of arm b associated with the force, F. Approximately, in the working interval

$$W = 2F(b/a) \approx \text{const.} \qquad (8.6.3)$$

Use of the linkage allows to reduce friction in the system.

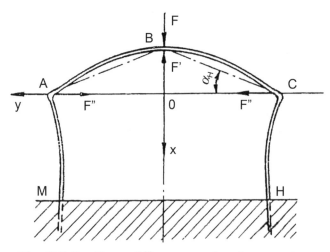

Fig. 8.6.4. Zero-stiffness complex shape spring.

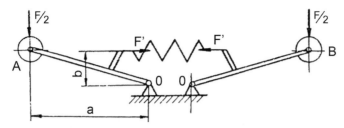

Fig. 8.6.5. Zero-stiffness combination spring/linkage system.

The fourth group is characterized by using cams for the compensating devices (Fig. 8.6.6). By selecting an appropriate profile, any shape of the load-deflection characteristic can be realized. However, the characteristic is very sensitive to relatively minor errors of the cam profile.

The fifth group of CF devices employs the buckling phenomenon. When a mechanical system buckles, its resistance to external forces (stiffness) ceases to exist, see Chapter 7. Special shapes of elastomeric devices, such as an "inverted flower pot" shock absorber shown in Fig. 8.6.7a [59] can be used for a large travel while exhibiting the constant resistance force (Fig. 8.6.7b). While the total height of the rubber element is 130 mm, deflection as great as 100 mm can be tolerated. A similar effect, within a somewhat smaller range, can be achieved by axial [59] or radial (Section 3.3.2) compression of hollow rubber cylinders.

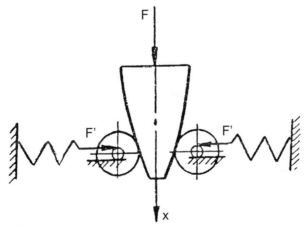

Fig. 8.6.6. Cam-spring variable stiffness system.

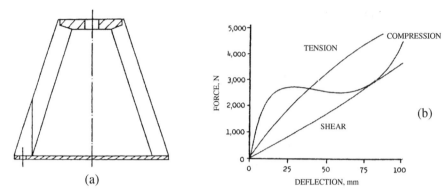

Fig. 8.6.7. Buckling-type shock mounting (a) and its force-deflection characteristic (b).

A low-stiffness system with widely adjustable load-deflection characteristic is shown in Fig. 8.6.8a [60]. It comprises housing 1 and T-shaped drawbar 2. The "shelf" of drawbar 2 applies force to leaf springs 3 having an initial curvature which results in their reduced resistance to buckling. Between the convex surfaces of springs 3 and housing 1, auxiliary

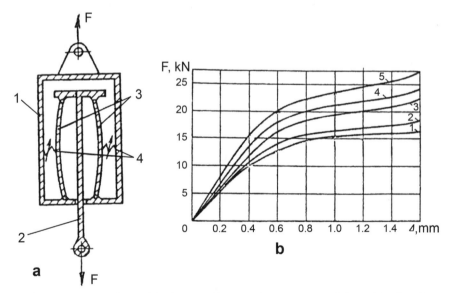

Fig. 8.6.8. Adjustable low stiffness elastic system (a) and its load-deflection characteristics (b). 1 — without pneumatic sleeve; 2 — with pneumatic sleeve, air pressure $p = 0$; 3 — $p = 0.2$ MPa (30 psi); 4 — $p = 0.4$ MPa (60 psi); 5 — $p = 0.6$ MPa (90 psi).

elastic elements 4 are placed. Presence of elements 4 with stiffness k leads to increase in critical buckling force P_{cr} for leaf springs 3 as [60]

$$P_{\mathrm{cr}} = P_{\mathrm{cr_0}} + 2lk/\pi^2, \qquad (8.6.4)$$

where l = length of leaf springs 3 and $P_{\mathrm{cr_0}}$ = critical (buckling) force without auxiliary elements 4. It is convenient to use an annular pressurized pneumatic sleeve between springs 3 and housing 1 as auxiliary element 4. In this case, change of pressure in the sleeve would change its stiffness and consequently, load-deflection characteristic of the system. Fig 8.6.8b shows a family of load-deflection characteristics for l = 238 mm and cross-section of each spring 3 (width x thickness) 30 x 4.5 mm^2 at various magnitudes of air pressure in the annular sleeve.

CF or low stiffness characteristic can be obtained also from specially designed pneumatic and hydraulic devices. Such systems can be equipped with servo-control systems accommodating the static (weight) loads, thus the low stiffness for small incremental dynamic motions (low dynamic stiffness) is not accompanied by large deformations, e.g., see Fig. 7.2.18.

8.7 ANISOTROPIC ELASTIC ELEMENTS AS LIMITED TRAVEL BEARINGS (FLEXURES)

Usually, high stiffness of a structural component in some direction is associated with a relatively high load-carrying capacity in the same direction. For example, both stiffness and load–carrying capacity in compression for the thin-layered rubber-metal laminates are increasing with decreasing thickness of the rubber layers, see Section 3.3.3. The rubber-metal laminates as well as some types of metal springs (e.g., flat springs) may have very different stiffness values in different directions. For thin-layered rubber-metal laminates the stiffness ratios between the compression and shear directions can be in the range of 3000 to 5000 and even higher [14]. These ratios have to be considered together with the fact that the stiffness values in different directions are reasonably independent on loads in the perpendicular directions. For thin-layered rubber-metal laminates the shear stiffness may increase ~15% for a change in the compression force 100:1. This allows to use anisotropic elastic elements to accommodate limited displacements between the structural components.

Such elements are in many ways superior to bearings and guideways of conventional designs since they do not have external friction and thus are responsive to even infinitesmal forces and displacements. On the contrary, the friction-based bearings do not respond to the motive forces below the static friction (stiction) force. Another advantage of the elastic guideways is absence of clearances (backlashes). The elastic connections can even be preloaded without increasing the resistance force for the useful motion.

A desirable performance for some mechanical systems depends not only on stiffness values of the stiffness-critical components, but also on a properly selected stiffness ratio(s) in different linear and/or angular directions. While the chatter resistance of a metal cutting machining system can be improved by increasing its stiffness (as referred to the cutting tool), it also depends significantly on orientation of the principal stiffness axes and on the ratio of the maximum and minimum principal stiffness, e.g., see [7]. A proper selection of stiffness ratios in the principal directions is important for assuring a satisfactory performance of vibration isolation systems. Such systems are often subjected to contradictory requirements: on one hand, they have to provide low stiffness for the high quality isolation, and on the other hand they should have a relatively high stiffness in order to assure stability of the isolated object from the rocking motion caused by internal dynamic loads, spurious external excitations, etc. A typical example is a surface grinder which requires isolation from the floor vibration to produce high accuracy and high surface finish parts, but which also generates intense transient loads due to acceleration/deceleration of the heavy table. An effective way of achieving both contradictory goals (good isolation and high stability) is to use isolators with judiciously selected stiffness ratios [57].

Optimal stiffness ratios for the situations described in the above paragraph usually do not exceed 0.3 to 3.0. A much higher stiffness ratio may be called for in cases when a mechanical connection is designed to accommodate some relative motion between the connected components. Significant displacements (travel) between the components are usually accommodated by continuous motion guideways with sliding or rolling friction. However, limited travel motions are better accommodated by elastic connections having distinctly anisotropic stiffness values in the principal directions. These connections have such positive features as: "solid state" design not sensitive to contamination and not requiring lubrication; absolute sensitivity to even minute forces and displacements which some continuous motion guideways lack due to effects of static friction (on a macro-scale for sliding friction guideways, on a micro-scale for rolling friction guideways); absence of backlash and a possibility of preloading in the direction of desirable high stiffness without impairing the motion-accommodating capabilities; usually more compact and light-weight designs.

Two basic types of the anisotropic guideways/bearings are elastic "kinematic" suspensions for precision (e.g., measuring) devices in which the forces are very small and the main concern is accuracy, and rubber-metal laminated bearings/guideways, which can be used for highly loaded connections.

8.7.1 Elastic Kinematic Connections (Flexures)

Elastic kinematic connections using metal, usually spring-like, constitutive elements provide low stiffness (low resistance to motion) in one direction while restraining the connected components in the perpendicular direction. Their main advantage is extremely high sensitivity to small magnitudes of forces and displacements and very low hysteresis. Such devices are most suitable to be used in precision instruments where accommodation of high forces is not required. Depending on the application, the designer may select out a huge variety of designs for accommodating rotational (revolute) motion (pivots or hinges), for accommodating translational motion, or for transforming from translational to rotational motion. Some typical devices are described below [61].

8.7.1a Elastic Connections for Rotational Motion

The simplest revolute connection is shown in Fig. 8.6.1a. It consists of frame 1, elastic (spring) strip 2, and connected (guided) link 3. Usually, strip 2 is initially flat (leaf spring), although it can be initially bent. Moving link 3 may rotate by angle, $\theta \leq 15$ deg. about axis parallel to the long side of cross-section of strip 2. However, position of this axis may shift depending on the displacement of link 3 and on the active forces.

Double-band pivot in Fig. 8.7.1b comprises frame 4, elastic bands 1 and 2, and connected (guided) link 3. Bands 1 and 2 can be flat or bent (shaped). Position of the intersection axis I-I and the angle between bands

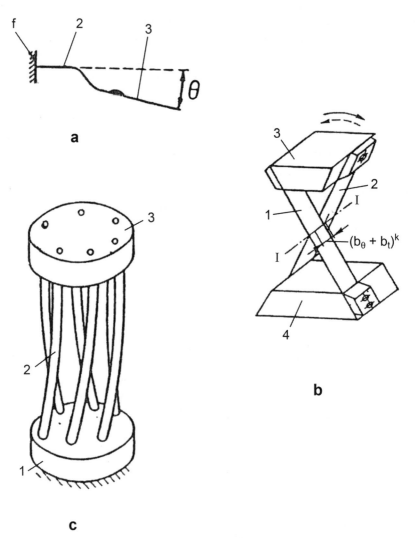

Fig. 8.7.1. Elastic revolute motion guideways; a-Single strip pivot; b-Cross-strip pivot; c-Torsional connection with several elastic rods.

1 and 2 can vary. This pivot can also accommodate rotational angle $\theta \leq$ 15 deg.

Torsional guideways in Fig. 8.7.1c comprise frame 1 and moving element 3 connected by several slender torsionally elastic rods 2. Rotation of element 3 relative to frame 1 is accommodated by torsional deformations of rods 2. This device is very rugged and may accommodate large forces/torques.

8.7.1b Elastic Connections for Translational Motion

Spring parallelogram in Fig. 8.7.2a, consists of frame 1 connected with moving element 3 by two flat or shaped (as shown) springs 2, 4. In the neutral (original) condition springs 2 and 4 are parallel and lengths $AB = CD$. The main displacement s_1 of element 3 is always accompanied by a smaller undesirable displacement s_2 in the perpendicular direction. The maximum allowable displacement is usually $s_1 < 0.1L$. The mechanism is very simple but s_2 may be excessive, and its performance is sensitive to variations in forces applied to moving elements 3.

The more stable design is the reinforced spring parallelogram shown in Fig. 8.7.2b. Its springs 2 and 4 are reinforced by rigid pads 5. Pads 5 substantially increase buckling stability of the device under compressive forces applied to element 3, and also improve consistency of the motion parameters under moments/torques applied to element 3. The devices both in Fig. 8.7.2a and in Fig. 8.7.2b have a disadvantage of not providing for a straight path of the moving element 3 due to a noticeable magnitude of the "parasitic" motion s_2. A "double parallelogram" in Fig. 8.7.2c allows for a relatively long travel of the moving element (slider) 7, $s_1 \leq 0.25\ L$, with a greatly reduced s_2. The elastic connection comprises frame 1, elastic bands 2, 3, 5, 6, moving element 7, and intermediate moving element 4. The resulting motion of

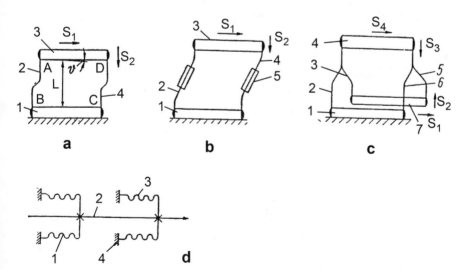

Fig. 8.7.2. Elastic translational motion guideways; a-Spring parallelogram; b-Spring parallelogram with reinforcements; c-Double (series) spring parallelogram; d-Guideways with bellows.

slider 7 is very close to the straight motion since its "parasitic" motion s_2 is a sum of transfer motion s_3 of intermediate element 4 and oppositely directed motion s_{74} of slider 7 relative to element 4,

$$|s_2| = |s_3 + s_{74}| < |s_3|$$

Usually, magnitudes of s_3 and s_{74} cannot be made exactly equal due to different longitudinal forces acting on elastic bands 2, 3, 5, 6. However, their differences are not very substantial and s_2 is greatly reduced.

A very accurate motion direction can be realized by using axisymmetrical systems, such as the double-bellows system in Fig. 8.7.2d. This device has moving element 2 attached to the centers of two bellows 1 and 3, which are fastened to frame 4.

8.7.1c Elastic Motion Transformers

Elastic motion transformers are used in precision devices in which backlashes and "dead zones" (hysteresis) are not tolerated. The motion transformer in Fig 8.7.3a can be called a "double reed" mechanism. It transforms a small rectilinear motion, s_1, of driving element 5 into a significant rotation angle, θ, of driven link 3. The device comprises two leaf springs (reeds) 2 and 4, frame 1, and driving and driven elements 5 and 3. Reed 2 is fastened to frame 1; the corresponding end of reed 4 is connected to driving element 5. Initially, reeds 2 and 4 are parallel with a small distance (which determines the transmission ratio) between them. This simple device may have a transmission ratio up to 0.1 deg/μm with the range of rotation of element 3 up to 5 deg.

The device in Fig. 8.7.3b is utilizing post-buckling deformation of leaf spring 2. Driving element 1 pushes leaf spring 2 in the longitudinal

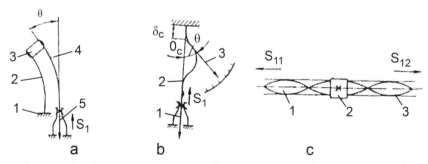

Fig. 8.7.3. Elastic motion transformers; a-Double reed device; b-Buckled strip device; c-Twisted strip device.

direction, and driven element 3 is attached to the opposite end of spring 2. While the transmission ratio is not constant (a nonlinear mechanism), position of the center of rotation is reasonably stable. The range of rotation angle is $\theta < 20°$.

The greatest transmission ratio together with large rotation angles can be realized by a "twisted strip" motion transformer in Fig. 8.7.3c. A small linear displacement of the ends of a pre-twisted elastic strip composed of two identical segments 1 and 3 with opposing (left and right) twist direction is transformed into large rotation of driven element 2. The relative displacement $s_1 = s_{11} - s_{12}$ of the ends of elastic strip 1, 3 is transformed into rotation angle, θ, of driven element 2 with transmission ratio $i_\theta = d\ \theta / ds_1 = 0.8$ deg/μm to 10 deg/μm.

8.7.2 Elastic Kinematic Connections Using Thin-Layered Rubber-Metal Laminates

While the elastic kinematics connections are mostly used for light load applications, mostly in precision instruments, thin-layered rubber-metal laminates can be and are used in heavy-duty devices subjected to very high preloads and/or payloads.

The most heavy-duty applications of rubber-metal laminates are in helicopters for connecting rotor blades with rotor hub, with three degrees of mobility while subjected to very intense centrifugal forces, and in civil engineering — for supporting bridges [59] and for protecting buildings from earthquakes. Bridge bearings have to accommodate expansion and contraction of the structure caused by temperature and humidity variations, and also allow for small rotations caused by bending of the bridge span under heavy vehicles. Although relative changes of the longitudinal dimensions due to variations in temperature and humidity are small, the absolute displacements can be significant in bridges with large spans. The bridge bearings must have high vertical stiffness to prevent excessive changes of pavement level caused by traffic-induced loads, and low horizontal stiffness in order to minimize forces applied to the bridge supports by expansion and contraction of the span. Previously, rolling or sliding bearings were used for this purpose. However, the rollers are not performing well for very small displacements due to a gradual development of small dents ("brinelling") causing increases in static friction and wear of the contact surfaces. While sliding bearings (usually,

Teflon-stainless steel combinations) are competing with rubber-metal laminated bearings, they require more maintenance since the sliding zone must be protected from contamination. They may also require special devices for accommodating small angular motions, which can be naturally accommodated by properly dimensioned laminates.

The laminates used for supporting bridges (as well as for earthquake protection of buildings) have rubber layers 5 mm to 15 mm thick with the aspect ratio (ratio of the smaller dimension in the plane view to thickness) usually not less than 25. While the specific load-carrying capacity of the rubber-metal laminates with such relatively thick rubber layers is smaller than that of the thin-layered rubber-metal laminates described in Section 3.3.3 and in [14], specific compressive loads as high as 15 MPa to 30 MPa (2250 psi to 4500 psi) can be easily accommodated. Shear deformations of the bridge bearings exceeding 100 mm do not present a problem for the rubber-metal laminates, while the compression deformation does not exceed 3 mm.

Use of rubber-metal laminates for mechanical devices is based on the same properties which made them desirable for the bridge and building supporting bearings. These properties include combination of high stiffness in one (compression) direction and of low stiffness in the orthogonal (shear) directions. Usually, high stiffness is associated with high allowable loading in the direction of high stiffness. Thin-layered rubber-metal laminates can be used in four basic applications: (a) Anisotropic elastic elements; (b) Bearings or guideways for limited travel; (c) Compensation elements; (d) Wedge mechanisms.

8.7.2a Rubber-Metal Laminates as Anisotropic Elastic Elements

This application is illustrated in Fig. 8.7.4 on the example of vibration-stimulated gravity chute for conveying parts and scrap from the work zone of stamping presses. If inclination of a chute is less than 15 deg. to 20 deg., part and scrap pieces may stick to the sliding surface of a gravity chute. In such cases, vibration stimulation of the chute is used to assure easy movement of the stampings and of scrap pieces. A pneumatic vibration exciter (vibrator) having a ball forced around its raceway by compressed air or having an unbalanced impeller is attached to the side of the chute (Fig. 8.7.4). The useful stimulating effect of vibration is accompanied by excessive noise levels due to resonance amplification

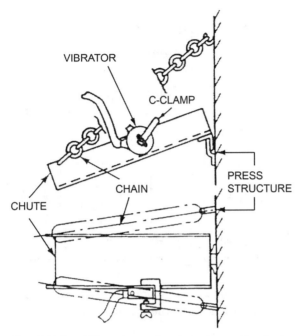

Fig. 8.7.4. Vibration-stimulated gravity chute.

by the chute structure of high frequency harmonics of the intense vibratory force from the vibrator. Reduction of the noise levels was attempted to achieve by enhancing effectiveness of the vibration-stimulation effect and, consequently, by reducing the magnitude of the required vibratory force.

It is known that the most effective vibration-assistant conveyance of particles along a flat surface develops when the rotating force vector describes an elliptical trajectory. However, ball and turbine vibrators generate circular trajectories of the vibratory force vector. To transform the circular trajectory into an elliptical trajectory of the vibratory force, a force vector transformer in Fig. 8.7.5a was proposed [62]. Vibrator 1 is attached to mounting bracket 3 via two rubber pads 2 with a large aspect ratio. Due to a significant anisotropy of pads 2 (high compression stiffness in the direction of the holding bolt and low shear stiffness in the orthogonal directions), the respective natural frequencies, f_c and f_s, of vibrator 1 mounted on pads 2 in these two directions are very different (Fig. 8.7.5b). Stiffness values in the compression and shear directions of the mounting device are designed in such a way that the fundamental

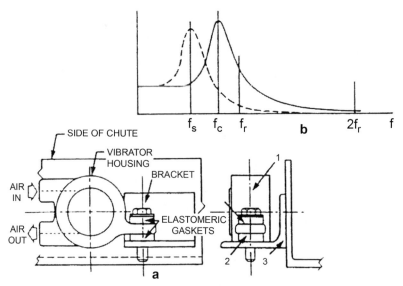

Fig. 8.7.5. Vibration force vector transformer (a) and its transmissibility curves (b).

frequency of vibratory force f_r (rpm of the ball or of the turbine) is correlated with f_c and f_s as shown in Fig. 8.7.5b. Then the compression component of the vibratory force is amplified, and the shear component is attenuated, thus creating the required elliptical trajectory of the vibratory force vector. Orientation of this trajectory relative to the chute surface can be adjusted by positioning bracket 3. Fig. 8.7.6 shows the time of part travel along the chute as a function of orientation of the force vector transformer. The minimum time is 1.8 s versus 20 s for the same vibrator without the force vector transformer (12 times improvement). Such improvement allows to reduce pressure of the compressed air and reduce noise by 5 dBA to 6 dBA, additional noise reduction (also about 5 dBA) is achieved due to isolation of higher harmonics of the vibratory force from the chute structure by rubber pads 2. This sizeable noise reduction is accompanied by a significant reduction in energy consumption (savings of compressed air).

8.7.2b Use of Rubber-Metal Laminates as Limited Travel Bearings

The most typical application of thin-layered rubber-metal laminates is for accommodation of small displacements ("limited travel bearings").

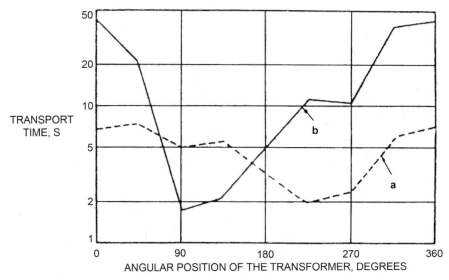

Fig. 8.7.6. Part transport time with vibration force vector transformer; a-Counterclockwise rotation; b-Clockwise rotation.

Advantages of such bearings are their "solid state" design not sensitive to contamination; no need for lubrication; absolute sensitivity to small forces and displacements; possibility of preloading due to virtual independence of their shear resistance from compressive forces, low energy losses [14]; generation of restoring force due to elastic character of the connection. Some such applications are described in Section 8.5.2, Figs. 8.5.7, 8.5.8 in which generation of the restoring force is important for returning the sliders to their initial positions.

Important examples of mechanical components whose performance is based on accommodation of small displacements are U-joints and misalignment compensating couplings (such as Oldham coupling), see Appendix 5.

U-joint (or Cardan joint) allows transmitting rotation between two shafts whose axes are intersecting but not coaxial. Fig. 8.7.7 shows a U-joint with rubber metal laminated bushings serving as the trunnion bearings [63]. Efficiency of such U-joint can be analyzed as follows. For angle α between the connected shafts, each elastic bushing is twisted $\pm\alpha$ per a revolution of the joint. With angular stiffness k_α of each bearing, maximum potential energy stored in one bushing during the twisting cycle is:

$$V = k_\alpha \frac{\alpha^2}{2} \qquad (8.7.1)$$

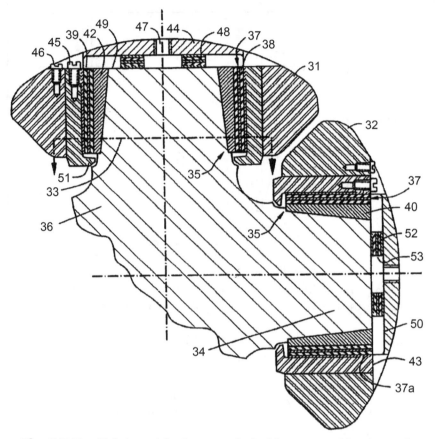

Fig. 8.7.7. U–joint with elastomeric (rubber – metal laminated) trunnion bearings.

and the energy dissipation during one twisting cycle of one bushing is:

$$\Delta V_1 = \psi k_\alpha \frac{\alpha^2}{2} \tag{8.7.2}$$

where $\psi = \delta/2$ is relative energy dissipation of the rubber blend used for the laminate, δ is its log decrement. The total energy dissipation in four bushings during one revolution is:

$$\Delta V = 4\Delta V_1 = 2\psi k_\alpha \alpha^2 \tag{8.73}$$

The total energy transmitted by the joint in one revolution is:

$$V = 2\pi T, \tag{8.7.4}$$

where T = transmitted torque. The efficiency of the joint is:

$$\eta = 1 - \frac{\Delta V}{V} = \frac{\psi k_\alpha \alpha^2}{\pi T} \qquad (8.7.5)$$

It can be compared with efficiency of a conventional U-joint [16]:

$$\eta = 1 - f \frac{d}{R} \frac{1}{\pi} \left(2 \tan \frac{\alpha}{2} + \tan \alpha \right), \qquad (8.7.6)$$

where d = effective diameter of the trunnion bearing, $2R$ is the distance between the centers of the opposite trunnion bearings, and f is friction coefficient in the bearings.

It can be seen from Eqs. (8.7.5), (8.7.6) that while efficiency of a conventional U-joint is a constant, the losses in the U-joint with elastic bushings are constant while its efficiency increases with increasing load (when the energy losses are of the greatest importance). The losses in the elastic U-joint at the rated torque can be one to two decimal orders of magnitude lower than the losses for conventional U-joints. Due to high allowable compression loads on the laminate (in this case, high radial loads), the elastic U-joints can be made smaller than the conventional U-joint with sliding or rolling friction bearings for a given rated torque.

Fig. 8.7.8a shows a compensating (Oldham) coupling which allows to connect shafts with a parallel misalignment between their axes without inducing non-uniformity of rotation of the driven shaft and without exerting high loads on the shaft bearings. The coupling comprises two hubs 1 and 2 connected to the respective shafts and an intermediate disc 3. The torque is transmitted between driving member 1 and intermediate member 3, and between intermediate member 3 and driven member 2, by means of two orthogonal sliding connections a-b and c-d. Because of the decomposition of a misalignment vector into two orthogonal components, this coupling theoretically assures ideal compensation while being torsionally rigid. The latter feature may also lead to high torque/weight ratios. However, this ingenious design finds only an infrequent use, usually for non-critical low speed applications. Some reasons for this are as follows:

1. Since a clearance is needed for the normal functioning of the sliding connections, the contact stresses are non-uniform with high peak values (Fig. 8.7.8b). This leads to a rapid rate of wear.

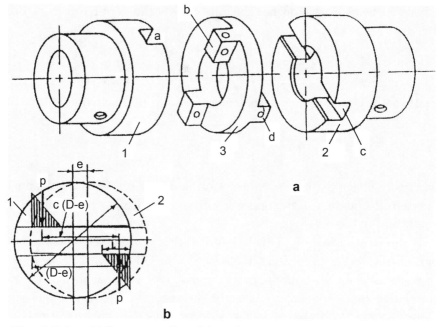

Fig. 8.7.8. Oldham coupling (a) and contact stress distribution in its sliding connections (b).

2. The lubrication layer in the highly loaded contact areas is squeezed out, thus the effective friction coefficient is high, at least $f = 0.1$ to 0.2. As a result, the coupling exerts high forces,

$$F = Pf = 2fT/D \qquad (8.7.7)$$

on the connected shafts. Here P = tangential force acting in each sliding connection, and D = external diameter of the coupling.

3. The coupling does not compensate misalignments less than ~0.5 x $10^{-3}D$-1.0 x $10^{-3}D$. At smaller misalignments hubs 1, 2 and intermediate disc 3 stay cemented by the static friction forces and sliding/compensation does not develop.

4. The coupling component must be made from a wear-resistant material (usually, heat-treated steel) since the same material is used both for the hub and disc structures and for the sliding connections.

Since displacement in the sliding connections a-b and c-d in Fig. 8.7.8a are small (equal to the magnitude of the shaft misalignment), the Oldham coupling is a good candidate for application of the thin-layered rubber metal laminates. Fig. 8.7.9 [64] shows such application, which was extensively studied in [65].

In Fig. 8.7.9a, hubs 101 and 102 have slots 106 and 107, respectively; whose axes are orthogonal. Intermediate disc can be assembled from two identical halves 103a and 103b. Slots 108a and 108b in the respective halves are also orthogonally oriented. Holders 105 are fastened to slots 108 in the intermediate disc and are connected to slots 106 and 107 via thin-layered rubber-metal laminated elements 111 and 112 as detailed in Fig. 8.6.9b. These elements are preloaded by sides 125 of holders 105, which spread out by moving preloading roller 118 radially toward the center.

This design provides for the kinematics advantages of the Oldham coupling without creating the listed above problems associated with the conventional Oldham couplings. The coupling is much smaller for the same rated torque than the conventional one due to the high load-carrying capacity of the laminates and the absence of the stress concentrations shown in Fig. 8.7.8b. The intermediate disc (the heaviest part of the coupling) can be made from a light strong material, such as aluminum. This makes the coupling suitable for high-speed applications.

The misalignment compensation stiffness and the rated torque can be varied by proportioning the laminated elements (their overall dimensions, thickness and number of rubber layers, etc.). The loads on the connected shafts are greatly reduced and are not dependent on the transmitted torque since the shear stiffness of the laminates does not depend significantly on the compression load. The efficiency of the coupling is similar to efficiency of the elastic U-joint described above.

8.7.2c Wedge Mechanisms

A special case of limited travel mechanical connections is presented by wedge mechanisms. These are widely used for various clamping devices transforming small input forces, P_{in}, with relatively large input displacements into high output (clamping) forces, P_{ou}, associated with very small displacements. The mechanical advantage is increasing with the decreasing wedge angle α. However, with an inevitable friction in the wedge connection the mechanical advantage, M, is decreasing.

$$M = \frac{P_{out}}{P_{in}} = \frac{1}{2\tan(\alpha + \varphi)} \qquad (8.7.8)$$

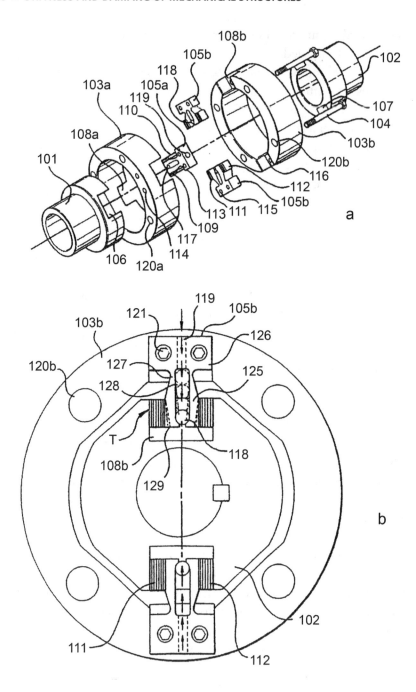

a

b

Fig. 8.7.9. Oldham coupling with elastomeric (rubber-metal laminated) connections.

where $\tan\phi = f$, and f is the friction coefficient. One typical embodiment of the wedge mechanism (a "double wedge") is represented by the model in Fig. 8.7.10, where symmetrical moving wedge 1 is pushed down by force P_{in} between stationary wedge 2 and output wedge 3, the latter being free to move on rollers 4 but overcoming the resistance (output) force P_{ou}. The mechanism is attached to base 5. It can be seen from Eq. (8.6.8) that friction is especially influential at small wedge angles α thus, effectively limiting enhancement of M by reducing α. For example, for $\phi = 8.5$ deg. ($f = 0.15$), if $\alpha = 5$ deg., then $M = P_{ou}/P_{in} = 5.7$ for $f = 0$ and $M = P_{ou}/P_{in} = 2.1$ (2.7 times less) for $f = 0.15$. Changing α from 3 deg. to 2 deg., or by 50%, results in increasing of M by 50% for the mechanism without friction, and by $\tan(8.5$ deg. $+ 3$ deg.$)/\tan(8.5$ deg. $+ 2$ deg.$) = 0.203/0.185 = 1.09$, or only by 9%, for the mechanism with $f = 0.15$. It can be also noted that reduction of f by using lubricants (e.g., lubricating oils) can be counterproductive. It increases M but reduces gripping forces between the clamping surfaces and the object being clamped. This is a typical "contradiction," using terminology of the Theory of Inventive Problem Solving (TRIZ) [26]: the system has to have low friction (for a better performance of the wedge mechanism) but it has to have high friction (for a better performance of the clamping mechanism).

Friction-based wedge mechanisms are not sensitive to very low input forces when the force generated in the wedge connection is less than the static friction force in this connection.

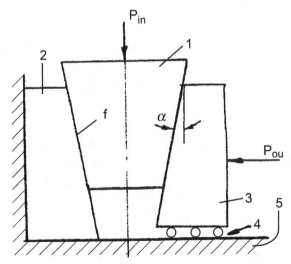

Fig. 8.7.10. Double wedge mechanism with frictional connection.

Yet another shortcoming of the friction-based wedge mechanisms is their "self locking" if $\alpha < \phi$, thus making unclamping a very time-consuming procedure. To prevent self locking, wedge mechanisms for clamping devices are usually made with $\alpha \geq 15$ deg., thus significantly limiting the mechanical advantage M.

Significant improvement in performance of wedge mechanisms both in clamping devices and in other applications can be achieved by replacement of the frictional wedge connection by connection through a thin elastomeric layer or a thin-layered rubber-metal laminate [66], a in Fig. 8.7.11. The expression for mechanical advantage of such system can be derived for a model in Fig. 8.7.11, which is a modification of the system in Fig. 8.7.10 wherein the wedge surfaces are separated by rubber layers or rubber-metal laminates a, each having compression stiffness, k_c and shear stiffness, k_s. When wedge 1 is moving downwards (displacement z, caused by compression deformations, Δ_c, and shear deformations, Δ_s, of laminates a, the equilibrium of the right half of the system can be expressed as:

$$P_{in} /2 = P_c \sin \alpha + P_s \cos \alpha$$
$$\Delta_c = z \sin \alpha \qquad \Delta_s = z \cos \alpha \qquad (8.7.9)$$
$$P_c = k_c \Delta_c \qquad P_s = k_s \Delta_s$$

Thus,

$$P_{in} = 2z(k_c \sin^2 \alpha + k_s \cos^2 \alpha) \qquad (8.7.10)$$

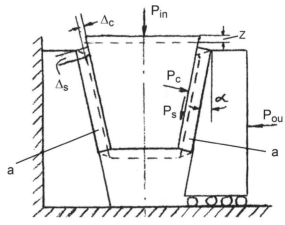

Fig. 8.7.11. Double wedge mechanism with elastomeric connection.

Similarly,

$$P_{ou} = P_c \cos \alpha + P_s \sin \alpha = z \cos \alpha \sin \alpha \, (k_c + k_s) \qquad (8.7.11)$$

Accordingly,

$$M = \frac{P_{ou}}{P_{in}} = \frac{1 + k_s/k_c}{2[\tan \alpha + (k_s/k_c)\cot \alpha]}. \qquad (8.7.12)$$

It can be seen that instead of friction, critical for the mechanism in Fig. 8.7.10, the critical factor influencing the mechanical advantage of the mechanism in Fig. 8.7.11 is the ratio between the shear and compression stiffness constants of the rubber layers/laminates. For small wedge angles $\alpha \leq 5$ deg., typical for real-life wedge mechanisms per [66], relatively large magnitudes (and rising fast with decreasing of α) of $\cot\alpha$ ($\cot\alpha = 11.4$ for $\alpha = 5$ deg.) ruin the magnitudes of M, unless counteracted by small magnitudes of k_s/k_c indicative for thin rubber layers or thin-layered rubber-metal laminates. For example, for a square-shaped (13 x 13 mm) soft rubber layer (durometer $H = 35$), 0.63 mm thick, $k_s/k_c \approx 0.005$, M of the mechanism in Fig. 8.7.11 is $M = 3.54; 3.52; 2.96$ for $\alpha = 5$ deg., 3 deg., 2 deg., respectively, decreasing with decreasing angle α (!). However, for a square 26 x 26 mm (same thickness) $k_s/k_c \approx 0.0015$, and $M = 4.80; 6.25; 6.41$ for the same wedge angles. For a square 52 x 52 mm $k_s/k_c \approx 0.0007$, and $M = 5.26; 7.58; 8.93$ for the same wedge angles. Thus, selection of the optimal angle α for given dimensions and other specifications of the mechanism is a critical issue. It also should be noted that Eq. (8.7.12) does not consider increasing k_c and thus, decreasing k_s/k_c, with increasing load as shown in Fig. 3.3.12. This effect beneficially increases effectiveness (the mechanical advantage) of this mechanism at the final stage of the clamping action. Obviously, M can be significantly increased by preloading of the laminates.

The effect of this design concept was studied on the setups similar to mechanisms shown in Figs. 8.7.10 and 8.7.11 with $\alpha = 5$ deg. The input (tangential) force, $P_{in} = 0$ N to 9000 N, was applied, while the output force, P_{ou}, was measured by another load cell. $M = 2.94$ was measured for the oil-lubricated wedge, while $M = 3.85$ (~31% improvement) was measured with two-layered laminate of $H = 35$ rubber layers (35 x 50 x 0.63 mm). Two-layered laminate has the same k_s/k_c ratio as a single layer of the same proportions, but allows greater displacements of the moving wedge.

Since the wedge mechanism containing the laminates does not need lubrication, the friction in the clamping device will be greater thus further improving performance of wedge-actuated clamping devices. Another important feature of mechanisms like one in Fig. 8.7.11 is the fact that the wedge mechanism, even with very small wedge angles, is absence of self-locking and thus, self-unclamping after the force, P_{in}, is removed.

8.7.2d Use of Rubber-Metal Laminates as Compensators

Use of the laminates as compensators is exemplified by the above example of bridge bearings. Similar applications are also important for precision mechanical devices, such as long frames for machine tools and measuring instruments. It is required that the frame is always parallel to the supporting structure and is connected with the supporting structure by very rigid (in compression) elements. However, these rigid elements must exhibit very low (in fact, as low as specified) resistance for in-plane compensatory movements caused, for example, by temperature changes and/or gradients.

8.8 MODIFICATION OF PARAMETERS IN DYNAMIC MODELS

Dynamic performance of a mechanical system having more than a single degree of freedom depends on stiffness values of its components in a nontrivial way. The same is true for masses and damping parameters of the components. Thus, even a substantial change (e.g., increase) of stiffness (as well as of mass or damping) of some components and/or of their connections in order to achieve a desirable shift in values of natural frequencies and/or vibration amplitudes might be ineffective if the "wrong" stiffness were modified. However, the expenses associated with stiffness and other parameter modifications are similar for either "right" or "wrong" components being modified. To maximize effectiveness of the modifications, the role of the stiffness/inertia/damping component slated for modification must be clearly understood. This is similar to modifications of static compliance breakdowns (see Chapter 6 and Appendix 7), whereas the effectiveness of the design modification is determined by importance of the stiffness component to be modified in the compliance breakdown. Two techniques, briefly described here, allow to perform modifications of dynamic performance of the system more effectively by "managing" the parameters being modified.

8.8.1 Evaluation of Stiffness and Inertia Components in Multi-Degrees-of-Freedom Systems

Dynamic models of real-life mechanical systems usually have many degrees of freedom, sometimes up to 100 and more, and a corresponding number of natural frequencies and vibratory modes. However, the practically important ones are in most case only two to three lowest natural frequencies and modes. When these lowest natural frequencies/modes have to be modified, it is important to know which stiffness and/or inertia components of the model significantly influence the frequencies/modes of interest and which ones do not. These conclusions are not obvious and usually can not be arrived at by analyzing the full model.

However the stiffness and inertia components of the dynamic model which do not influence the selected lowest natural frequencies and modes can be quickly and easily identified by using Rivin's Compression Method developed initially for chainlike dynamic models [67, 68]. The method is based on the fact that any chain-like model (e.g., see Figs. 6.3.6b, c, and d) can be broken into single-degree-of-freedom (SDOF) partial subsystems of two types as shown in Fig. 8.8.1: two-mass systems (Fig. 8.8.1a) and single-mass systems (Fig. 8.8.1b). If a chain-like dynamic model is broken into such partial dynamic systems, the latter become *free body diagrams,* with the remainder of the model on both sides replaced by torques, *T,* and position angles, *φ,* (for a torsional transmission system), forces and linear coordinates for a translational systems, etc.

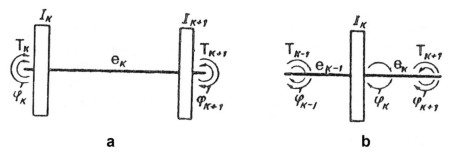

Fig. 8.8.1. Chain-like dynamic model broken into two single-degree-of-freedom partial susbsystems: *a*-Two-mass, *b*-Single-mass systems.

It was shown in [67, 68] that within a specified frequency range 0 to f_{lim} the complexity of the dynamic model can be alleviated (its number of degrees of freedom reduced) without introducing significant errors in the natural modes and natural frequencies of vibration. To achieve such "compression," some partial subsystems of type a in Fig. 8.8.1 have to be replaced with subsystems of type b, and vice versa. There are several conditions that must be observed for such transformation. First of all, the natural frequency of the partial subsystem to be replaced must be much higher than the higher limit, f_{lim}, of the frequency range of interest. Natural frequencies of the systems in Fig. 8.8.1 are, respectively,

$$n_{ak} = \sqrt{\frac{I_k + I_{k+1}}{e_k I_k I_{k+1}}} \qquad n_{bk} = \sqrt{\frac{e_{k-1} + e_k}{I_k e_{k-1} e_k}} \qquad (8.8.1)$$

If the vibratory modes below f_{lim} have to be maintained within ±2 dB in the compressed system, then subsystems being transformed should have natural angular frequencies $n \geq 3.5$ to $4.0\ \omega_{lim}$ where $\omega_{lim} = 2\pi f_{lim}$. In such a case, the natural frequencies of the compressed system will be within 2% to 3% of the corresponding natural frequencies of the original system. If only values of the natural frequencies in the frequency range 0 to f_{lim} are of interest, they will remain within 5% to 10% of the corresponding natural frequencies of the original system if the partial subsystems having $n \geq 2$ to $2.5\ \omega_{lim}$ are transformed.

Fig. 8.8.2 illustrates this compression algorithm. The initial dynamic model in Fig. 8.8.2a, as a first step, is broken into partial subsystems type A (Fig. 8.8.2b), and partial subsystems type B (Fig. 8.8.2c). Then, for each partial subsystem in Figs. 8.7.2b and c the value $n^2 = 1/I_k^* e_k^*$ is calculated, where $I_k^* = I_k I_{k+1}/(I_k + I_{k+1})$, $e_k^* = e_k$ for a kth subsystem type a, and $I_k^* = I_k$, $e_k^* = e_{k-1} e_k/(e_{k-1} + e_k)$ for a subsystem type e. As a next step, subsystems having $n^2 >> \omega_{lim}^2$ (i.e., $I^* e^* << \omega_{lim}^2$ are replaced with *equivalent subsystems of the opposite type*. The equivalent subsystem of type B for the subsystem in Fig. 8.8.1a would have its parameters as:

$$I_k' = I_k + I_{k+1}; \quad e_{k-1}' = \frac{I_{k+1}}{I_k + I_{k+1}} e_k; \quad e_k' = \frac{I_k}{I_k + I_{k+1}} e_k \qquad (8.8.2)$$

The equivalent subsystem of type a for the subsystem in Fig. 8.8.1b would have its parameters as

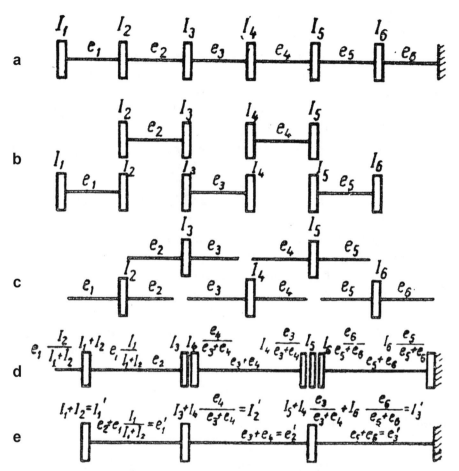

Fig. 8.8.2. Compression algorithm: a-Initial six-degrees-of-freedom dynamic model; b-Partial susbsystem a; c-Partial subsystem b; d-Intermediate stage of the transformation; e-Final "compressed" three-degrees-of-freedom dynamic model.

$$I_k'' = \frac{e_k}{e_{k-1} + e_k} I_k \qquad I_{k+1}'' = \frac{e_{k-1}}{e_{k-1} + e_k} I_k \qquad e_k'' = e_{k-1} + e_k; \quad (8.8.3)$$

After all allowable substitutions are performed; the dynamic model is rearranged. In the Fig. 8.8.2 example, the first type *a* subsystem and the fourth and sixth type *b* subsystems have been replaced with the equivalent subsystems of the opposite types. The intermediate stage of the compression process is shown in Fig. 8.8.2d and the final version of the "compressed" dynamic model is shown in Fig. 8.8.2e. Instead of six degrees of freedom as in the initial model, the final model has only three degrees of freedom. However, natural modes and/or natural frequencies of both systems in the 0 to f_{lim} range are very close if the stated above conditions for the subsystems transformation had been complied with.

If the transformed subsystem is at the end of the chain, e.g., $I_1 - e_1 - I_2$ or $e_5 - I_6 - e_6$ in Fig. 8.8.2a, then there appears a "residue" after the transformation. Such "residues" are represented by a "free" compliance $e_1 \dfrac{I_2}{I_1 + I_2}$ or a "free" inertia $I_6 \dfrac{e_5}{e_5 + e_6}$ in Fig. 8.8.2d. These components do not participate in the vibration process and have to be abandoned. This means that the importance of the corresponding parameters in the original system (compliance e_1, moment of inertia I_6) for the dynamics of the model in Fig. 8.8.2a in the specified frequency range 0 to f_{lim} is limited. It is especially so when the abandoned segment represents a substantial part of the original component.

The transformation process can be performed very quickly, even with a pocket calculator. If it demonstrates that a certain elastic (compliance) or inertia component does not "survive" the transformation, then its modification in the original system would not be an effective one for modification of dynamic characteristics in the specified frequency range. This compression algorithm was extended to generic (not chain-like) dynamic systems in [69].

8.8.2 Modification of Structure to Control Vibration Responses

Vibratory behavior of structures and other mechanical systems can be modified by the so-called modal synthesis based on experimental modal analysis. This approach can provide desirable resonance shifts, reduction of resonance peaks, shifting and optimal placement of nodal points, etc. However it can not identify values of such structural parameters as mass or damping, which have to be added to or subtracted from values of these parameters of the current structural components in order to achieve desired vibratory responses. This task can be realized by using the Sestieri / D'Ambrogio modification algorithm [70].

This algorithm is based on experimentally determined frequency response functions (FRF) between selected points in the structure. After the matrix $\boldsymbol{H}_o(\omega)$ of all FRF is measured, it can be modified by structural modifications at the selected points (ω is angular frequency). These modifications include adding and/or subtracting stiffness, mass (inertia) and/or damping at the selected points. The matrix of such modifications can be written as $\Delta B(x \cdot \omega)$, where x is a vector of values of these modifications. The matrix of FRF of the modified system is $H(x, \omega)$ and it can be expressed as

$$H(x. \omega) = [I - H_o(\omega) \Delta B(x \cdot \omega)]^{-1} H_o(\omega) \qquad (8.8.4)$$

This technique allows one to determine the required $\Delta B(x \cdot \omega)$, i.e., to identify the required changes in the structural parameters, in order to realize the required modified FRF matrix $H(x. \omega)$. The latter may represent the system modification in which magnitudes of selected FRF are limited (thus, vibration responses are constrained) or natural frequencies are changed in a specified manner (some increased, some reduced), etc. The importance of this technique is the fact that it results in well-defined requirements for changes in stiffness, inertia, and damping parameters in order to obtain the desirable changes in the FRF.

This can be demonstrated on an example of two-degrees-of-freedom system in Fig. 8.8.3. It was assumed that the structural parameters of this system (stiffness values k_1, k_2, k_3 and mass values m_1, m_2) are not known, but only its FRF matrix (inertance matrix) is known, which also identifies its angular natural frequencies to be $\omega_1 = 364$ rad/sec and $\omega_2 = 931$ rad/sec. It was required to modify this system to shift the natural frequencies to be $\omega_1 = 300$ rad/second and $\omega_2 = 1000$ rad/second. Application of the frequency-response modification technique resulted in recommending the following modifications: $\Delta m_1 = 0$; $\Delta m_2 = 1.46$ kg; $\Delta k_1 = 0$; $\Delta k_2 = 2 \times 10^5$ N/m; $\Delta k_3 = 0$. Thus, the solution demonstrated very different influence of various structural parameters on the structural

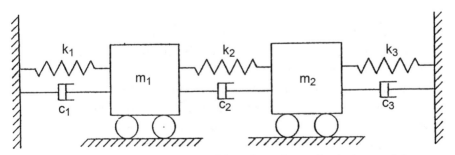

Fig. 8.8.3. Two-degrees-of-freedom dynamic chain model.

dynamic characteristics of interest. Both stiffness and mass values must be managed to achieve the desired dynamic effects.

This technique was applied to modification of a high-speed machining center to reduce magnitude of amplitude-frequency characteristic of the spindles [71]. *It was found that adding stiffness, mass, or damping at the selected points would be useless.* The only recommendation from the evaluations using the Sestieri/D'Ambrogio algorithm [70] was to install a dynamic vibration absorber on the spindle sleeve. The effectiveness of this approach was proven experimentally.

REFERENCES

[1] Rivin, E.I., "Tooling Structure: Interface between Cutting Edge and Machine Tool", *Annals of the CIRP*, 2000, Vol. 49/2, pp. 591–634.

[2] Walter, J.L., Skelly, D.W., "Wear Resistant Cutting Tools and Shaping Method". U.S.Patent 5,038,645.

[3] Pleshivtsev, N.V., Bondarev, D.V., Stanki i Instrument, 1994, No. 6, pp. 21–23 [in Russian].

[4] Astashev, V.K., Babitsky, V.I., "Ultrasonic Processes and Machines", 2007, Springer Verlag, 330 pp.

[5] Chryssoluris, G., "Effects of Machine-Tool-Workpiece Stiffness on the Wear Behavior of Superhard Cutting Materials", 1982, Annals of the CIRP, Vol. 31/1, pp. 65–69.

[6] Fadeev, V.S., Petridis, A.V., "Influence of Stiffness of the System Machine-Fixture-Tool-Workpiece on Strength of Carbide Tools", 1985, Stanki i Instrument, No. 6, pp. 30–31 [in Russian].

[7] Tobias, S.A., "Machine Tool Vibration", 1965, Blackie, L.

[8] Minato, J., Ammi, S., Okamoto, S., "Application of Spring Tool to Face Milling", 1984, Bull. Japan Soc. of Precis Engng, vol. 18, No. 1, pp. 35–36.

[9] Elyasberg, M.I., Demchenko, V.A., Savinov, I.A., "A Method for Structural Improvement of Machine Tool Vibration Stability during Cutting", 1983, Soviet Engng Research, vol. 3, No. 4, pp. 59–63.

[10] Rivin, E.I., Kang, H., "Enhancement of Dynamic Stability of Cantilever Cutting Structures", 1992, Intern. J. of Machine Tools and Manufacture", vol. 32, No. 4, pp. 539–562.

[11] Kapoor, S.G., Zhang, G.M., Bahney, L.L., "Stability Analysis of Boring Process System", 1986, Proc. 14th North American Manufacturing Research Conference, pp. 454–459.

[12] Kang, H.-L., "Enhancement of Dynamic Stability and Productivity for Machining Systems with Low Stiffness Components", 1990, PhD Thesis, Wayne State University.

[13] Rivin, E.I., Kang, H. "Improving Machining Conditions for Slender Parts by Tuned Dynamic Stiffness of Tool", Intern. J. of Machine Tools and Manufacture, 1989, vol. 29, No. 3, pp. 361–376.

[14] Rivin, E.I., "Properties and Prospective Applications of Ultra-Thin-Layered Rubber-Metal Laminates for Limited Travel Bearings", 1983, Tribology International, vol. 16, No. 1, pp. 17–26.

[15] Rivin, E.I., Xu, L., "Damping of NiTi Shape Memory Alloys and Its Application for Cutting Tools", in *Materials for Noise and Vibration Control*, 1994, ASME NCA-Vol. 18/DE – vol. 80, pp. 35–41.

[16] Weck, M., Hennes, N., Krell, M., "Spindle and Tool Systems with High Damping", 1999, Annals of the CIRP, vol. 48/1, pp. 297–302.

[17] Fadyushin, I.L., Musykant, Ya.A., Messheryakov,A.I., "Tooling for CNC Machine Tools", 1990, Mashinostroenie Publ. House, [in Russian].

[18] Tlusty, J., "High Speed Milling", 1994, Proceed. of the 6th Machine Tool Engng Conf., Osaka, Japan, pp. 35–60.

[19] Rotberg, J., Lenz, E., Levin, M., "Drill and Clamping Interface in High Performance Drilling", 1998, Inter. J. of Advanced Manufacturing Technology, vol. 14, pp. 229–238.

[20] Lembke, D., Weck, M., "Study on Design Possibilities for the ConnectionMachine/Tool", 1991Final Research on Research Project, WZL [Laboratory for Machine Tools and Applied Economics], TU Aachen.

[21] Hijink, J.A.W., Van der Wolf, A.C.H., "Measurements on the Dynamic Behavior sof Modular Milling Tools", 1992, Annals of the CIRP, vol. 41/1, pp 113–116.

[22] Schouten, C.H., Rosielle, P.C.J.N., Schellekens, P.H.J., "Design of a Kinematic Coupling for Precision Applications", Precision Engineering, 1997, vol. 20, No. 1, pp. 46–52.

[23] Tsutsumi, M., et al, "Study of Stiffness of Tapered Spindle Connections", Nihon Kikai gakkai rombunsu [Trans. Of the Japan. Society of Mechanical Engineers], 1985, C51(467), pp. 1629–1637 [in Japanese].

[24] Rivin, E.I., "Trends in Tooling for CNC Machine Tools: Tool-Spindle Interfaces", ASME Manufact. Review, 1991, vol. 4, No. 4, pp. 264–274.

[25] Meyer, A.,"Werkzeugspannung in Hauptspindeln für hohe Drehfrequenzen [Holding Tools in Spindles Rotating with High Speeds]", Industrie-Anzeiger, 1987, vol. 109, No. 54, pp. 32–33 [in German].

[26] Fey, V.R., Rivin, E.I., "Innovation on Demand," 2005, Cambridge Univ. Press.

[27] Rivin, E.I., U.S. Patent 5,322,304 "Tool Holder-Spindle Connection".

[28] Braddick, H.J.J., "Mechanical Design of Laboratory Apparatus", Chapman & Hall, London, 1960.

[29] Gangopadhyay, A., "Friction and Wear of Hard Thin Coatings", *in Tribology Data Handbook*, ed. by E.R. Booser, CRC Press, Boca Raton, 1997.

[30] Agapiou, J., Rivin, E., Xie, C., "Toolholder/Spindle Interfaces for CNC Machine Tools", 1995, Annals of the CIRP, vol. 44/1, pp. 383–387.

[31] Rivin, E.I., U.S. Patent 5,595,304 "Improvements Relating to Tapered Connections".

[32] Rivin, E.I., Sayal, G., Johal, P.R.S., ""Giant Superelasticity Effect" in NiTi Superelastic Materials and Its Applications", 2006, J. of Materials in Civil Engineering, vol. 18, No. 7, pp. 1–7.

[33] Rivin, E.I., "Precision Compensators Using Giant Superelasticity Effect", 2007, Annals of the CIRP, vol. 56/1/2007, pp. 391–394.

[34] Rivin, E.I., "Mechanical Contact Connection", U.S. Patent 6.779,955.

[35] St. Henry, C., Huber, W.G., "Quick Change Tooling and Workholding — the QCS Coupling", 1987, SME Technical Paper ", MR87-434, 1987 SME Intern. Tool and Mfg. Engng Conference.

[36] Erickson, R., "Toolholder and Method for Releasably Mounting", U.S. Patent 4,747,735.

[37] Grigoriev, A.M., Putvinskaya, E.I., "Rational Geometric Parameters of Hollow Supporting Rollers", in *Detali mashin*, Tekhnika Publish. House, Kiev, 1974, No. 19, pp. 72–78 [in Russian].

[38] Grigoriev, A.M., Putvinskaya, E.I., "Contact between Hollow Cylinder and Flat Surface", Ibid., pp. 79–83.

[39] Bhateja, C.P., Pine, R.D., "The Rotational Accuracy Characteristics of the Preloaded Hollow Roller Bearings", ASME J. of Lubrication Technology, 1981, vol. 103, No. 1, pp. 6 – 12.

[40] Holo-Rol Bearings, Catalog of ZRB Bearings, Inc., 1997.

[41] Sawer, J.W., "Review of Interesting Patents on Quieting Reduction Gears", Journal of ASNI, Inc., 1953, vol. 65, No. 4, pp. 791–815.

[42] Berestnev, O.V., "Self-Aligning Gears", Nauka i tekhnika Publ. House, Minsk, 1983, 312 pp [in Russian].

[43] Rivin, E.I., "Torsional Connection with Radially Spaced Multiple Flexible Elements", U.S. Patent 5,630,758.

[44] Rivin, E.I., "Conceptual Developments in Design Components and Machine Elements", ASME Transactions, Special 50th Anniversary Design Issue, 1995, vol. 117, pp. 33–41.

[45] Rivin, E.I., "Shaped Elastomeric Components for Vibration Control Devices", 1999, Sound and Vibration, vol. 33, No. 7, pp. 18–23.

[46] Rivin, E.I., "Gears Having Resilient Coatings", U.S. Patent 4,184,380.

[47] Rivin, E.I., "Conjugate Gear System", U.S. Patent 4,944,196.

[48] Chironis, N., "Design of Novikov Gears", in Gear Design and Application, McGraw-Hill, N.Y., 1967.

[49] Rivin, E.I., Wu, R.-N., "A Novel Concept of Power Transmission Gear Design", SAE Tech. Paper 871646, 1987.

[50] Rivin, E.I., Dong, B., "A Composite Gear System with Separation of Sliding and Rolling", Proc. of the 3rd World Congress on Gearing and Power Transmission, Paris, 1992, pp. 215–222.

[51] Shigley, J.E., "Mechanical Engineering Design", 3rd Edition, McGraw-Hill, N.Y., 1977, 695 pp.

[52] Zvorikin, K.O., "Engagement of Chain Links with Compliant Sprocket Teeth", in Detali Mashin, No. 40, Tekhnika Publ. House, Kiev, 1985, pp. 3–8 [in Russian].

[53] Bondarev, V.S., et al, "Study of Chain Drive Sprockets with Compliant Teeth", Ibid., pp. 8–13 [in Russian].

[54] Den Hartog, J.P., "Mechanical Vibrations", McGraw-Hill, NY 1956, 436 pp.

[55] Kelzon, A.S., Zhuravlev, Yu.N., Yanvarev, N.V., "Design of Rotational Machinery", Mashinostroenie, Publ. House, Leningrad, 1977, 288 pp. [in Russian].

[56] Kelzon, A.S., et al, "Vibration of a Milling Machine Spindle Housing with Bearings of Reduced Static Rigidity", Vibration Engineering, 1989, vol. 3, pp. 369–372.

[57] Rivin, E.I., "Passive Vibration Isolation", 2003, ASME Press, N.Y., 426 pp.

[58] Alabuzhev, P.M., Gritchin, A., Kim, L., Migirenko, G., Chon,V., Stepanov, P., "Vibration Protecting and Measuring Systems with Quasi Zero Stiffness", Hemisphere Publishing, N.Y.

[59] Freakley, P.K., Payne, A.R., "Theory and Practice of Engineering with Rubber", Applied Science Publishers, London, 1978.

[60] Rogachev, V.M., Baklanov, V.S., "Low Frequency Suspension with Stabilization of Static Position of the Object", Vestnik Mashinostroeniya, 1992, No. 5, pp. 10–11 [in Russian].

[61] Tseitlin, Ya., "Elastic Kinematic Devices", 1972, Mashinostroenie Publ. House, Leningrad, 296 pp. [in Russian].

[62] Rivin, E.I., "Noise Abatement of Vibration Stimulated Material-Handling Equipment", Noise Control Engineering, 1980, No. 3, pp. 132–142.

[63] Rivin, E.I.,"Universal Cardan Joint with Elastomeric Bearings", U.S. Patent 6,926,611.

[64] Rivin, E.I., "Torsionally Rigid Misalignment Compensating Coupling", U.S. Patent 5,595,540.

[65] DeSousa, V., "Prototype Development and Performance Characteristics of a Novel Misalignment- Compensating Coupling", 18994, MS Thesis, Wayne State University.

[66] Rivin, E.I., "Wedge Mechanism", U.S. Patent 7,465,120.

[67] Rivin, E.I., "Dynamics of Machine Tool Drives", Mashinostroenie Pub. House, Moscow, 1966 [in Russian].

[68] Rivin, E.I., "Compilation and Compression of Mathematical Model for a Machine Transmission", ASME Paper 80-DET-104, ASME, N.Y., 1980.

[69] Banakh, L., "Reduction of Degrees-of-Freedom in Dynamic Models", Mashinovedenie, 1976, No. 3, pp. 77–83 [in Russian].

[70] Sestieri, A., and D'Ambrogio, W., "A Modification Method for Vibration Control of Structures", Mechanical Systems and Signal Processing, 1989, Vol. 3, No. 3, pp. 229–253.

[71] Rivin, E.I., and D'Ambrodgio, W., "Enhancement of Dynamic Quality of a Machine Tool Using a Frequency Response Optimization Method", Mechanical Systems and Signal Processing, 1990, Vol. 4, No. 3, pp. 495–514.

APPENDIX

1

Single-Degree-of-Freedom Dynamic Systems with Damping

Viscous Damping. This brief description attempts to illustrate differences and special features of some of the various damping mechanisms typical for mechanical systems. A classic single-degree-of-freedom (SDOF) mechanical system in Fig. A1.1 comprises mass m, spring k, and viscous damper c. The equation of motion of this system at free vibration condition (no external force) can be written as

$$m\ddot{y} + c\dot{y} + ky = 0, \tag{A1.1}$$

where y = displacement of mass, m. When viscous friction in the damper is not very intense, $c < 2\sqrt{km}$, the solution of (A1.1) is

$$y = e^{-nt}(C_1\sin\omega^*t + C_2\cos \omega^*t), \tag{A1.2}$$

where

$$n = \frac{c}{2m}, \omega^* = \sqrt{\omega_0^2 - n^2}, \tag{A1.3}$$

ω_0 = natural frequency of the system without damping ($c = 0$), and constants C_1 and C_2 are determined from initial conditions $y(0) = y_0$, $\dot{y}_0 = \dot{y}_0$ as:

$$C_1 = (\dot{y}_0 + ny_0)/\omega^*, \qquad C_2 = y_0. \tag{A1.4}$$

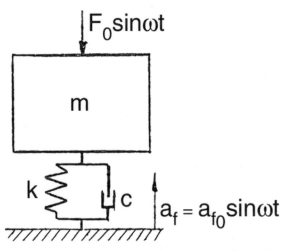

Fig. A1.1. Basic single-degree-of-freedom (SDOF) mechanical system.

Another format of the solution is:

$$y = Ae^{-nt}\sin(\omega^*t + \beta),\qquad\text{(A1.5)}$$

where

$$A = \sqrt{\frac{(\dot{y}_0 + ny_0)^2}{\omega_0^2 - n^2} + y_0^2}, \tan\beta = \frac{y_0\sqrt{\omega_0^2 - n^2}}{\dot{y}_0 + ny_0}\qquad\text{(A1.6)}$$

It can be seen from Eqs. (A1.2) and (A1.5), that variation of y in time (motion) is a decaying oscillation with a constant frequency ω^* and gradually declining amplitude, Fig. A1.2. Parameter β is called "loss angle", and $\tan\beta$ is "loss factor".

Envelopes of the decaying time history of y are described by functions

$$A = \pm\, A_o e^{-nt},\qquad\text{(A1.7)}$$

where A_o = the ordinate of the envelope curve at $t = 0$.

The ratio of two consecutive peaks $A(t):A(t + T^*)$ which are separated by time interval $T^* = 2\pi/\omega^*$ (*period* of the vibratory process) is e^{nT^*} = constant. The natural logarithm of this ratio is called "*logarithmic (or log) decrement*". It is equal to:

$$\delta = nT^* = \frac{2\pi c}{\sqrt{4mk - c^2}} \approx \frac{\pi c}{\sqrt{mk}}\qquad\text{(A1.8)}$$

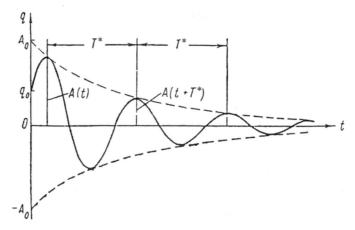

Fig. A1.2. Decaying oscillations with constant frequency, w^*, and gradually decreasing amplitude.

Thus for the same damper (*damping coefficient c*), *log decrement* of a system with viscous damper *depends on stiffness k and mass m*, of the system.

For forced vibrations (excitation by a harmonic force generated within the object $F = F_o \sin \omega t$ applied to mass, m), the equation of motion becomes

$$m\ddot{y} + c\dot{y} + ky = F_o \sin \omega t \tag{A1.9}$$

and the response amplitude is:

$$A = \frac{F_o}{k\sqrt{\left(1 - \dfrac{\omega^2}{\omega_o^2}\right)^2 + \left(\dfrac{c}{\sqrt{mk}}\dfrac{\omega}{\omega_o}\right)^2}} = \frac{F_o/k}{\sqrt{\left(1 - \dfrac{\omega^2}{\omega_o^2}\right)^2 + \left(\dfrac{\delta}{\pi}\right)^2 \left(\dfrac{\omega}{\omega_o}\right)^2}}. \tag{A1.10}$$

Expression (A1.10) is plotted in Fig A1.3 for various δ.

Frequently, there is a need for vibration isolation. Two basic cases of vibration isolation are: (1) protection of foundation from force $F = F_o \sin \omega t$ generated within the object (machine) represented by mass m in Fig. A1.1; and (2) protection of vibration sensitive object (machine) represented by mass m from vibratory displacement of the foundation $a = a_f \sin \omega t$.

In the first case, the force transmitted to the foundation is $F_{fo} \sin \omega t$ and the quality of vibration isolation is characterized by force transmissibility $T_F = F_{fo}/F_o$. In the second case, the displacement $a_{mo} \sin \omega t$ transmitted to mass m is characterized by displacement transmissibility $T_a = a_{mo}/a_{fo}$. For the SDOF isolation system, which can be modeled by Fig. A1.1

$$T_F = T_a = \frac{\sqrt{1 + \left(\dfrac{\delta}{\pi}\right)^2 \left(\dfrac{\omega}{\omega_o}\right)^2}}{\sqrt{\left[1 + \left(\dfrac{\omega}{\omega_o}\right)^2\right]^2 + \left(\dfrac{\delta}{\pi}\right)\left(\dfrac{\omega}{\omega_o}\right)^2}} \tag{A1.11}$$

Expression (A1.11) is plotted by solid lines in Fig. A1.4 for several values of δ. It is important to note that while higher damping results in decreasing transmissibility T around the resonance ($\omega = \omega_o$), it also leads to a significant deterioration of isolation (increased T) above $\omega = 1.41\,\omega_o$ (isolation range).

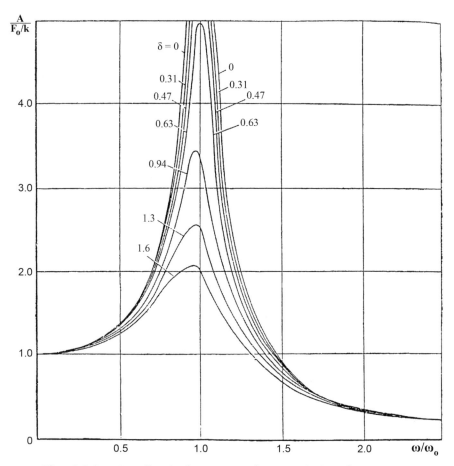

Fig. A1.3. Amplitude-frequency characteristics of mass m in Fig. A1.1 acted upon by a sinusoidal force, as function of frequency ω and damping δ.

Hysteresis-Induced Damping. Deformation of mechanical components and joints between them (contact deformation) is not perfectly elastic. It means that deformation values during the process of increasing the external force (loading) and decreasing the external force (unloading) are not the same for the same magnitudes of the external force. This effect results in developing of a "hysteresis loop," Fig. A1.5, which illustrates change of deformation y for the processes of increasing (loading) and decreasing (unloading) force P. The area of the hysteresis loop represents energy lost during one loading/unloading cycle. It is established by numerous tests that for majority of structural materials, as well

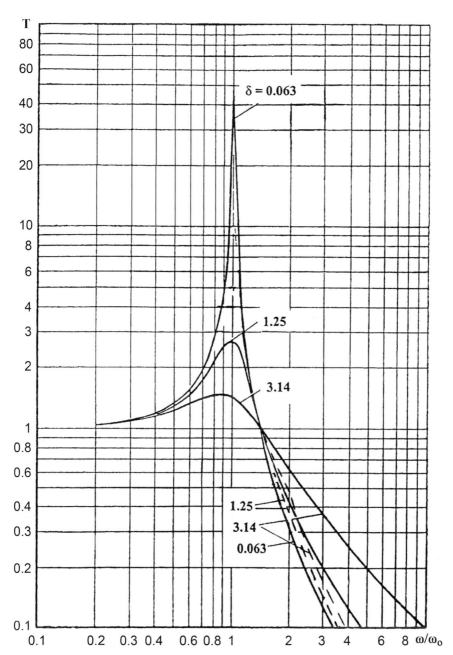

Fig. A1.4. Transmissibility of a SDOF isolation system in Fig. A1.1 as function of frequency, ω, and damping, δ; Solid lines — viscous damping; broken lines — hysteretic damping.

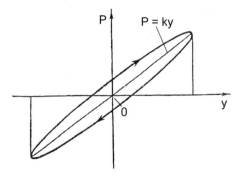

Fig. A1.5. Hysteresis loop associated with deformation process.

as for joints between components, the area of the hysteresis loop *does not strongly depend on the rate of force change* (i.e., on frequency of the loading process), but *may depend on amplitude of load/deformation*. This statement can be formalized as an expression for the energy Ψ lost in one cycle of deformation as function of amplitude of deformation, A,

$$\Psi = \alpha A^{r+1},\qquad(\text{A1.12})$$

where α and r are constants.

To derive the law describing the decaying oscillatory process when the damping is due to hysteresis, the loss of energy during one cycle can be equated to change of system energy during one cycle of vibration. Let's consider one period of the process in Fig. A1.2 and start the period when the displacement peaks at amplitude $A(t)$. At this moment the *kinetic energy* of the mass is $K = 0$, and all energy is stored in the spring k, Fig. A1.1, as the *potential energy V*. In the beginning of the period:

$$V_o = \tfrac{1}{2} kA^2 (t)\qquad(\text{A1.13})$$

At the end of the period

$$V_{T*} = \tfrac{1}{2} kA^2 (t + T^*)\qquad(\text{A1.14})$$

The increment of the potential energy is

$$\Delta V = V_{T*} - V_o = \tfrac{1}{2} k[A^2 (t + T^*) - A^2 (t)] = \tfrac{1}{2} k[A (t + T^*) \\ + A (t)][A(t + T^*) - A (t)].\qquad(\text{A1.15})$$

The sum inside the first square brackets in right hand side is $\sim 2A(t) = 2A$ if the energy loss during one cycle is not very large, the difference inside the second square bracket is ΔA, and

$$- \Delta V = kA\Delta A. \tag{A1.16}$$

This increment or the potential energy is equal to the energy loss (A1.12), or

$$- \alpha A^{r+1} = kA\Delta A \tag{A1.17}$$

or

$$\Delta A = -(\alpha/k)A^r \tag{A1.18}$$

This expression defines the shape of the upper envelope of the oscillatory process. Considering this envelope as a continuous curve $A = A(t)$, approximately

$$\Delta A = T^*(dA/dt) = (2\pi/\omega^*)(dA/dt). \tag{A1.19}$$

From (A1.19) and (A1.20),

$$dA/dt = -(\alpha\omega^*/2\pi k)A^r \tag{A1.20}$$

If $r = 1$, solution of (A1.20) is an exponential function

$$A(t) = A_0 e^{-(\alpha\omega^*/2\pi k)t} \tag{A1.21}$$

The ratio of two peak displacements $A(t)/A(t + T^*) = e^{\alpha/k}$ is again constant, as in the case of viscous damping, but in this case the *log decrement does not depend on mass m,*

$$\delta = \alpha/k \tag{A1.22}$$

While for $r = 1$ the log decrement does not depend on amplitude of vibration, for some real life materials and structures value of r may deviate from 1, and then the log decrement would be changing in time with the changing vibration amplitude. The character of the change is illustrated in Fig. A1.6 by envelopes for the vibratory process at $r = 0$ and $r = 2$, in comparison with the exponential curve (amplitude-independent log decrement) for $r = 1$. Dependence of log decrement on amplitude for fibrous

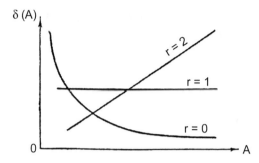

Fig. A1.6. Amplitude dependencies of log decrement, δ, for different damping mechanisms.

and elastomeric materials is illustrated in Fig. 3.1.2. For many elastomeric materials (rubber blends) $r = {\sim}1$.

For forced vibrations in the system like in Fig. A1.1 but with the hysteretic damping, the response amplitude is:

$$A = \frac{F_o}{\sqrt{\left(1 - \dfrac{\omega^2}{\omega_o^2}\right)^2 + \left(\dfrac{\alpha A^{r-1}}{\pi k}\right)^2}} \qquad (A1.23)$$

For $r = 0$ (Coulomb friction – induced damping), for $r = 2$, and for other r not equal 1, the amplitude of the forced vibrations can be found after solving Eq. (A1.23). However for $r = 1$, which is typical for many rubber blends,

$$A = \frac{F_o}{k\sqrt{\left(1 - \dfrac{\omega^2}{\omega_o^2}\right)^2 + \left(\dfrac{\alpha}{\pi k}\right)^2}} = \frac{F_o/k}{\sqrt{\left(1 - \dfrac{\omega^2}{\omega_o^2}\right)^2 + \left(\dfrac{\delta}{\pi}\right)^2}} \qquad (A1.24)$$

Equation (A1.24) is similar to expression for the response to external harmonic excitation of a system with viscous friction damping described by (A1.10) with the only difference being the second (damping) term under the radical sign. For a system with viscous friction this term is frequency dependent, while for a system with hysteresis damping it does not depend on frequency.

Transmissibility T between mass m and foundation for the force excitation of the mass by $F = F_{fo} \sin \omega t$ or between foundation and mass m for the "*kinematic*" excitation of the foundation $a_f = a_{fo} \sin \omega t$ for the system with hysteretic damping when $r = 2$, is

$$T_F = \frac{\sqrt{1 + \left(\dfrac{\delta}{\pi}\right)^2}}{\sqrt{\left(1 - \dfrac{\omega^2}{\omega_0^2}\right)^2 + \left(\dfrac{\delta}{\pi}\right)^2}}.$$ (A1.25)

The differences between (A1.10) and (A1.25) are due to presence in the latter and absence in the former of the frequency ratio multipliers in the second terms under the radical signs both in the numerator and in the denominator. Effect of these seemingly subtle differences is practically not noticeable in the frequency range $\omega = 0 - 1.41\omega_0$, as well as for small damping (δ) values. However, for larger damping values the difference in isolation effectiveness (at $\omega > 1.41\,\omega_0$) *is very pronounced, as shown by broken lines in Fig. A1.4.* Although vibration isolators having significant viscous damping demonstrate deteriorating performance in the "isolation range" $\omega > 1.41\,\omega_0$, vibration isolators with the hysteretic damping characterized by $r = 1$ provide effective performance in the "isolating range" while limiting transmissibility around the resonance frequency.

A system with hysteretic damping can be described by Eqs. (A1.1) and (A1.9) if the damping coefficient c is replaced by a frequency- and amplitude-dependent coefficient:

$$c_h = \alpha A^{r-1}/\pi\omega,$$ (A1.26)

which for $r = 1$ becomes only frequency- dependent,

$$c_h = \alpha/\pi\omega,$$ (A1.27)

The loss factor for a system with hysteresis damping is

$$\tan\beta = \frac{\alpha A^{r-1}}{\pi k\left(1 - \dfrac{\omega^2}{\omega_0^2}\right)}$$ (A3-28)

Impact Damping. Impact interactions between mechanical components cause energy losses [1]. Such impact interactions can occur in clearances between components in cylindrical joints, guideways, or in specially designed impact dampers. During a typical vibratory process in a system with clearances, impacts occur every time the system passes its equilibrium configuration. The amount of energy loss during impact can be presented as a function of relative velocity v of the co-impacting components before the impact occurence,

$$\Psi = bv^2, \tag{A1.29}$$

where b is a constant having dimension of mass.

Let's consider a half-period of oscillation which commences at a maximum displacement $A(0)$. In the first quarter of the period the mass is moving with a constant energy $\frac{1}{2} kA^2(0)$ and the square of velocity at the end of this quarter of the period is

$$v^2 = (k/m) A^2(0) \tag{A1.30}$$

At this moment, an impact occurs and, consequently, a loss of energy by the amount of (A1.29). After the impact, the system moves with energy:

$$kA^2(0)/2 - (bk/m)A^2(0) = \frac{1}{2}kA^2(0)(1 - 2b/m). \tag{A1.31}$$

This energy remains constant during the second quarter of the period. Accordingly, at the end of the second quarter the potential energy is equal to

$$\frac{1}{2} k A^2(T/2) = \frac{1}{2} kA^2(0)(1-2b/m) \tag{A1.32}$$

Thus, the ratio of maximum displacements is:

$$\frac{A(0)}{A\left(\dfrac{T}{2}\right)} = \frac{1}{\sqrt{1 - \dfrac{2b}{m}}} \tag{A1.33}$$

The same ratio would materialize for the next half-period. Thus, for the whole period,

$$A(0)/A(T) = 1/(1 - 2b/m) = \text{const.} \tag{A1.34}$$

Since the ratio of sequential maximum displacements is constant, the envelope of the time plot of the decaying vibratory process is an exponential curve,

$$A = A_0 e^{-nt} \tag{A1.35}$$

This process is associated with the log decrement

$$\delta = nt = \log_e [1/(1 - 2b/m)] = \sim \log_e (1 + 2b/m) = \sim 2b/m, \tag{A1.36}$$

if $2b/m$ is small.

REFERENCES

[1] Panovko, Ya.G., 1971, "Introduction to Theory of Mechanical Vibration", Nauka Publ. House, 240 pp [in Russian].

[2] Rivin, E.I., 2003, "Passive Vibration Isolation", ASME Press, N.Y.

APPENDIX

2

Stiffness/Damping/ Natural Frequency Criteria

A2.1 INTRODUCTION

Vibration control in mechanical devices usually involves in-depth study of mechanisms of vibration generation and transmission as well as attempts to change dynamic characteristics (stiffness, damping, natural frequencies) of the controlled object. Often, improvement of one characteristic, such as stiffness or damping, leads to deterioration of other characteristics. In many cases of vibration control requirements to stiffness, natural frequencies, and damping are interrelated and can be expressed as criteria. Derivation and importance of such criteria is illustrated here for two large groups of systems: vibration isolation of vibration sensitive and vibration-producing objects, and machine tool and tooling systems. Use of such criteria may solve a vibration control problem or provide guidance for more elaborate efforts.

Generally, vibration control is required to reduce vibration levels of certain components of the system or to prevent development of self-excited vibrations. Two generic techniques used for solving these vibration control problems are increasing stiffness or the fundamental natural frequency of the specified components/subsystems, and increasing their damping. These directions for vibration control are often contradictory to each other. Increase of stiffness of a component or of a subsystem results in reduction of deformations of the component or in reduction of relative motions between the components in a subsystem, or both. Because of this, usually the stiffness increase is accompanied by reduction of damping in the stiffened component/subsystem, since energy dissipation constituting the damping effect results from deformations of and relative displacements (motions) between the components. Thus, expensive modifications aimed to enhance stiffness, may be very successful in achieving this aim but in the same time fail in controlling vibrations since these modifications may result in reduction of damping. There are known observations that an older machine tool with worn guideways, having lower stiffness due to loosened connections between the guided and the supporting parts, may have a more stable cutting process (is less prone to developing self-excited chatter vibrations) than a similar new machine tool. It is important to note here that the old machine may be, in the same time, much less accurate than the new one because of its wear; in this case, the accuracy is a "static" effect, while the chatter resistance is a "dynamic" effect.

Since stiffening a component or a subsystem is a time-consuming and expensive undertaking, it is important to have an understanding as to what degree the stiffness should be enhanced without losing too much of damping capacity of the system. In cases when the fundamental natural frequency needs to be increased, it is usually achieved also by increasing stiffness, although in some cases it also can be achieved by reducing effective masses of the system, e.g., see [1].

Increase of damping can be achieved in two modalities. One is to use damping devices not influencing stiffness, such as add-on dynamic vibration absorbers (DVA). Although DVAs can be very effective, they are rather expensive additions to the system and require some additional packaging space. Highly effective low-damping/low mass ratio DVAs need precision tuning in order to enhance their effectiveness. It limits use of passive DVAs for systems whose dynamic characteristics may change due to adding/removal of massive components, and due to movements within the system (e.g., movements of a massive carriage in a lathe). More robust high mass ratio and/or high damping DVA are not very sensitive to tuning but are usually more bulky and expensive. Another modality of damping enhancement is by introducing micro motions or deformations into the system (e.g., by reducing preload forces in joints between structural components). Such an approach is, of course, contradictory since it usually results in stiffness reduction.

It is shown here, that in many situations wherein a vibration control is required, the requirements to stiffness/natural frequency and to damping are interrelated and can be formulated as criteria, different for different cases. These criteria combine stiffness/natural frequency and damping qualifiers and give a clear indication how strong is influence of stiffness and/or damping on achieving the vibration control goals in each case. When such criteria can be formulated, the vibration control practitioner gets a convenient tool for estimating a degree to which stiffness/natural frequency magnitude can be sacrificed for enhancing damping, and vice versa. The criteria can be formulated both for self-excited vibrations and for forced vibrations.

A2.2 SELF-EXCITED VIBRATIONS-DYNAMIC STABILITY CRITERION

Two frequently encountered cases of self-excited vibrations in mechanical systems are self-excited (chatter) vibrations during cutting processes in metal cutting machine tools, in mining and rock cutting/drilling machines, etc., and self-excited vibrations associated with friction (stick-slip vibrations, brake squeal, etc.).

The greatest advances in understanding self-excited cutting-related vibrations have been achieved in studies of chatter using the stability analysis apparatus of automatic control theory. It was originated in [2], where the expression was formulated for the maximum unconditionally stable depth of cut $t_{\lim}$ as:

$$t_{\lim} = CRe(G), \tag{A2.1}$$

where C = "cutting force coefficient," and $Re(G)$ is the real part of transfer function G of the machining system at the natural frequency of the potential chatter mode. In the further development in [3] G was expressed via the system parameters: effective stiffness K and damping ratio $\eta = 2\pi\delta$ of the chatter mode, where δ is log decrement, so that

$$t_{\lim} = 2\pi CK\delta. \tag{A2.2}$$

If the goal is "vibration control," i.e., enhancement of chatter resistance of the machining system, all three parameters in Eq. (A2.2) should be increased. These parameters represent two independent systems: parameter C represents cutting conditions (workpiece material, cutting tool material and design, cutting regimes), and K and δ represent the mechanical system of the machine. Leaving C to cutting specialists, $K\delta$ can be considered as a *criterion* whose increase would benefit *dynamic stability* of the machining system. Obviously, an increase in damping, δ, can be beneficial even if the stiffness, K, is reduced (by a smaller percentage), and vice versa. A similar statement about interrelation of K and δ can be also made for self-excited friction-induced vibrations.

Although the above interrelation between stiffness K and damping δ has been known for quite a while, no publication could be found until late nineteen eighties on considering it as a criterion and on its subsequent use for enhancement of dynamic stability in structures prone to development of self-excited vibrations. The main techniques were (and still are) increase of stiffness and a judicial use of "stability lobes" to find relatively stable operating regimes (e.g., [4]).

Enhancement of dynamic stability by an *intentional reduction of stiffness* and a greater simultaneous increase of damping was first proposed in [5]. This concept was successfully tested for a case of turning a highly flexible bar while the radial (x-direction) stiffness of the cutting tool was reduced by clamping it via a thin layer of a highly damped ($\delta = 1.8$) elastomeric material (see Section 8.1, Fig. 8.1.11).

Another application of the concept of interrelation between stiffness and damping is presented in [6] and Section 8.1, Fig. 8.17 for a small boring bar. Two clamping conditions were compared: rigid clamping in a steel sleeve with low damping, and low stiffness clamping in a Nitinol clamping sleeve with high damping and greater $K\delta$.

A similar concept for clamping boring bars was tested in [7], also see Section 8.1, Fig. 8.1.18. In this setup, the boring tool is clamped in a sleeve attached to a membrane having a relatively low angular stiffness but having "squeeze-film" damping effect. Static stiffness of the tool is reduced by 40%, but it is more than compensated by high damping thus resulting in doubling of t_{lim}.

Clearly, use of this concept is not limited to boring bars; it can also be used for end mills and other applications.

These cases involve introduction of special devices (guiding sleeve for the tool in Fig. 8.1.18, special holding devices in Figs. 8.1.17. and 8.1.18). In structural designs, especially but not only for machine tools, significant enhancements/reductions in stiffness and damping values can be achieved by design and manufacturing approaches. Since metal cutting machine tools usually have very stringent accuracy requirements, reduction of stiffness can be tolerated only if static deformations of the system by cutting forces do not exceed the allowable values. In real life, however, the main attention by the designers (at least, the machine tool designers) is still paid to increasing stiffness.

Stiffness at the spindle end exceeds 200 N/μm for modern high speed machining centers. Since the cutting force at finishing high speed regimes is usually much below 200 N, these high stiffness values are often excessive. On the other hand, these high stiffness values may lead to negative dynamic effects, such as reduction of damping and decreasing the $K\delta$ criterion. Stiffness increase of spindle bearings by preloading above certain levels may result not only in reduction of damping, but also in high dynamic loads on the rolling bodies and the races, especially at very high speeds. These dynamic loads may lead, in turn, to significant temperature increases in the bearings. High preload forces lead to high frictional losses, to temperature increase, and to shortened bearing life.

Similar effects (damping reduction by stiffness-enhancing preload) may develop in rolling friction guideways and ball screws.

Until recently, very little attention was paid to damping enhancement of the mechanical structures proper. When low damping values were becoming a critical issue, DVAs are used for machine tool and tooling structures. They are not subjected to the process loads and do not cause reduction of stiffness.

Damping in mechanical structures is due, mostly, to material damping in the components and to energy dissipation in the joints between the components. The latter is, usually, the main source of damping. Stiffness of mechanical structures is also determined by deformations of the structural components and by deformations in the joints. Joints are the most important stiffness and damping determining factors for many precision structures. Data on joint damping and stiffness in mechanical structures had been available for many years, but used mostly for analytical purposes. Preloading of a joint results in increase of its stiffness, e.g., see Eq. (4.1.1), but also in reduction of its energy dissipation (damping), e.g., see (4.8.1). Dependences of both stiffness and damping of joints on the preload force are nonlinear. At certain preload forces (dependent on the joint design and on macro- and micro-geometry of the contacting surfaces), the rate of stiffness increase with increasing preload force is slowing down. At about the same preload, the rate of damping decrease with increasing preload is also slowing down.

The above facts are seldom considered in assembling the frames of machine tools and other machines. Usually, the permanent joints between the constitutive components of the frame structure are preloaded to very high contact pressures in the joints, thus resulting in the highest stiffness. However, a very significant role in generating the overall stiffness and damping characteristics of machines, such as machine tools, is played by non-permanent joints, such as toolholder-spindle interfaces. In some cases, this interface is responsible for more than 50% of deformations at the tool-workpiece contact, e.g., see Section 4.5.2. Since the preload forces in the interfaces are usually adjustable, it is important to define the optimal preload conditions for the interfaces.

The most extensive information on stiffness and damping of toolholder-spindle interfaces is provided in [8], Figs. A2.1 and A2.2. The measurements had been performed for 7/24 taper interfaces (#30, #40, #45), flat flange-to-face interface, and so-called "Curvic coupling" interface (see

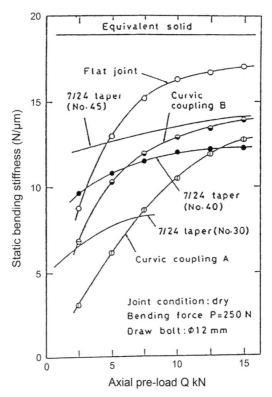

Fig. A2.1. Static bending stiffness of various interface systems versus axial prerload.

Section 8.4) which comprises all-teeth engagement between two flat gears with spiral teeth. The solid (integral) toolholder-spindle is given for comparison. The preloading of the toolholder-spindle connection is provided by a "drawbar" applying axial force to the toolholder. The character of both stiffness and damping plots in Fig. A2.1 and A2.2 confirms the above statements — stiffness is increasing and damping is decreasing with the increasing preload force, with the rates of change for both parameters starting to slow down at about the same axial preload forces.

Use of the **Kδ** criterion allows to easily perform optimization of dynamic performance of this important interface. Table A2.1 gives K, δ, and **Kδ** for high (15 KN), low (5 KN), and medium (10 KN) preload forces. The greatest values of **Kδ** for each interface are shown in bold

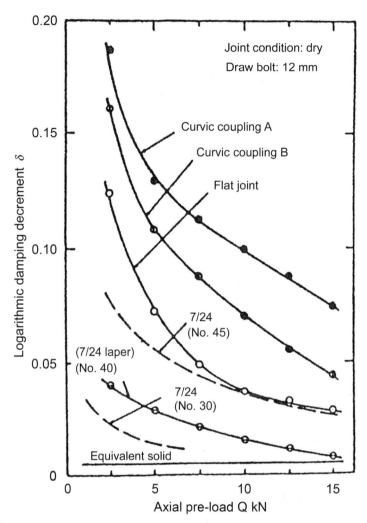

Fig. A2.2. Damping of various interface systems versus axial prerload.

print $K\delta$. K and δ vary for various interface designs and various preloads up to 3:1, values of $K\delta$ vary up to ~2.5:1. For each interface, the optimal preload force is easy to select; in most cases it is low (for taper and flat interfaces) or medium (for Curvic coupling) axial force magnitude. This fact is explained by a much stronger dependence of damping than stiffness on the preload force. Data in Table A2.1 also guides towards selection of the best interface designs (flat joint and Curvic coupling).

Table A2.1.   Chatter-Resistance Criterion K𝛿 for Toolholder/Spindle Interfaces.

	K, N/𝜇m			𝛿			K𝛿		
Q, KN	5	10	15	5	10	15	5	10	15
Integral		18.5			0.006			0.11	
Flat joint	13	16.5	17.3	0.075	0.04	0.03	**0.98**	0.66	0.52
#45	12.5	13	14	0.06	0.04	0.03	**0.75**	0.52	0.42
Curvic coupling B	7.6	12.5	14	0.11	0.072	0.045	0.84	**0.9**	0.63
#40	10.5	12	12.5	0.03	0.02	0.01	**0.32**	0.24	0.125
Curvic coupling A	5.5	10.2	13	0.13	0.1	0.075	0.72	**1.0**	0.98
#30	7.5	7.9	----	0.02	0.01	----	**0.15**	0.08	----

A2.3 VIBRATION ISOLATION OF MECHANICAL OBJECTS

Installation of mechanical objects, such as machine tools on the floor/foundation, is very important for obtaining their optimal static and dynamic performance. While influence of installation on static performance (effective static stiffness) is addressed in Chapter 5, dynamic performance depends on dynamic stiffness and damping of the mounting elements. Criteria describing influence of dynamic stiffness and damping of optimal mounting elements on various performance requirements of the object can be very useful since they make selection of the optimal elements a much more flexible process. Some of the cases are considered below. They relate to chatter resistance of the installed object, its behavior under forced vibrations, and vibration isolation.

A2.3.1 Influence of Isolation on Chatter Resistance

Since modern machine tools often combine high performance characteristics with high accuracy and surface finish, vibration isolation of a machine tool is often required [9]. If the machine tool is not very large (weight ≤ ~10 ton), it often can be installed directly on vibration isolating (flexible) mounts without reduction of its effective structural stiffness. However, sometimes installation of a machine tool on low stiffness mounts may lead to deterioration of its chatter resistance. On the other hand, installation on high stiffness mounts does not provide the desirable isolation from external excitations. The mounting system may influence effective damping in the working zone (between the tool and the workpiece) of a machine tool or a photolithography apparatus. The effective structural stiffness in the working zone is not noticeably influenced by the installation technique, this type of mounting changes only δ and, accordingly, the $K\delta$ criterion.

Analysis of the influence of mount parameters on the chatter resistance can be performed on a simplified two-mass/two-degrees-of-freedom model in Fig. A2.3a. Here m_1 is generalized mass of the machine frame (bed) on which the workpiece is mounted, m_2 is generalized mass of the toolholding unit of the machine, $k_1 = k_v$ is stiffness of the mount system, $k_2 = k_m$ is generalized structural stiffness of the machine. Damping of each spring (log decrement) is designated as δ_m, δ_v, respectively. Dynamic influence of the installation system m_1-k_1 on the machining system m_2 – k_2 – m_1 can be evaluated by analyzing dynamic coupling between these systems with a reasonable assumption that the partial natural frequency f_v of the installation (vibration isolation) system is much lower than the partial natural frequency of the machine structure f_m, or $f_v << f_m$. In most cases, the mounts do not influence K. However, it is shown in [9] that the mounts may "pump" some damping into the working zone. The incremental increase of the working/machining system damping due to mounts can be expressed as [9]

$$\Delta\delta_m = \frac{m_2}{2\pi(m_1 + m_2)f_m^3} f_v^3\delta_v \qquad (A2.3)$$

Thus, *influence of the installation system on chatter resistance* can be characterized by the *criterion $f_v^3\delta_v$*. Since f_v is proportional to the root square of stiffness, k_v, of isolators, this criterion can be rewritten as $k_v^{3/2}\delta_v$. This criterion states that if softer mounts (lower f_v or k_v) are used instead of stiffer mounts for installing a machine tool (to achieve its vibration isolation), then chatter resistance will not deteriorate if damping of the isolation system (of vibration isolators) is increased by at least the third power of the ratio of natural frequencies between the stiffer and softer installations.

While in the chatter resistance criterion from Section A2.2.1 above stiffness K and damping δ are of the same importance, stiffness is more influential in the criterion for influence of installation on chatter resistance (power 3/2 for k versus power 1 for δ).

This analysis was validated by comparing performance of a lathe installed on six different mounting systems A to F, Table A2.2. The test results correlate reasonably well with this criterion as can be seen in Fig. A2.3b. Although the natural frequency and damping are significantly different between installations on mounts A and B, C and D, the criterion

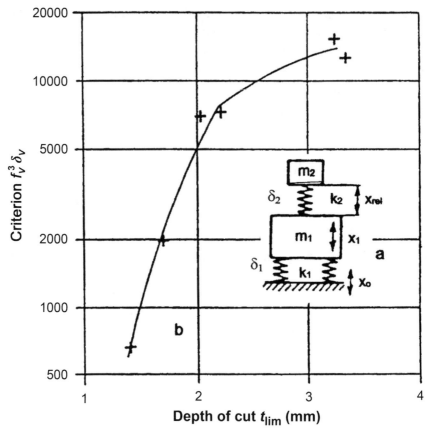

Fig. A2.3. Dynamic model of machine on isolators, (a); (b)-Maximum stable depth of cut on a lathe versus criterion $f_v^3 \delta_v$.

magnitudes are very close for each of these pairs, and the chatter resistance t_{lim} is about the same.

The existence of such criterion does not mean that all machine tools are sensitive to the installation conditions. If the original structural stiffness K and/or structural damping δ_m are very large and/or structural natural frequency f_m is very high, then the installation influence for such an object can be neglected.

A2.3.2 Forced Vibrations

While forced vibrations caused by internal excitation sources can be controlled by design modifications, these modifications can be made only

Table A2.2.   Mounting Systems for Lathe Chatter
Tests.

Mounts	f_v, Hz	δ_v	$f_v^3 \delta_v$
A	32	0.39	12,800
B	30	0.57	15,400
C	24	0.5	6,900
D	20	0.9	7,200
E	17	0.42	2,060
F	12	0.38	660

by the designers of the units, not by the users. However, a significant degree of control may be available to the user by means of a proper installation of the machine. Two situations can be considered: reduction of relative vibration amplitudes in the working zone (between the tool and the workpiece in a machine tool, between the stylus and the part in a coordinate-measuring machine, etc.), and reduction of intensity of the machine frame vibration causing a discomfort for the operator.

Obviously, for a detailed analysis in both cases the frequency content and the amplitudes of the (internal) excitation should be known. Usually, there are two frequency ranges of the internally generated vibrations: a low frequency range, about 10–100 Hz, in which vibrations are caused mostly by centrifugal forces due to unbalance and/or runout of the rotating components, and a high frequency range, about 100 Hz to1000 Hz, in which vibrations are mostly caused by impacts in connections, gear errors, electromagnetic fluctuations in motors, etc. Only the low frequency range can be noticeably influenced by installation techniques.

A2.3.2a Vibration Level Criteria

Vibration level on the frames of mechanical objects is often determined by excitations from rotating unbalanced components in their mechanisms. The amplitude of the centrifugal force generated by a rotating component is proportional to its unbalance and to the second power of its rpm. Thus, the fastest rotating component should make the largest contribution to the vibration intensity. However, the fast rotating components (such as pulleys, gears, grinding wheels, etc.) are usually the better balanced ones. Analysis of the available data in [9] has shown that the rated amplitudes F_o of centrifugal forces of the rotating components can be assumed to be not dependent on rpm (or frequency, f) within the range 10 Hz to 100 Hz. If $F_o(f)$ = const, and m is the machine mass, the resonance amplitude of the machine on its mounts (the most dangerous case) is:

$$x_{res} = \frac{F_o}{k_v} \frac{\pi}{\delta_v} = \frac{F_o}{4\pi f_v^2 m \delta_v} \qquad (A2.4)$$

where k_v, δ_v are stiffness and damping of the mounting system, and f_v = natural frequency of the machine on its mounts. The amplitude of relative vibrations in the working zone for the Fig. A2.3a model is then [9]:

$$x_{rel} = \frac{F_o}{4\pi f_v^2 m_1 \delta_v} \frac{1}{\dfrac{m_1}{m_1 + m_2} \dfrac{f_m^2}{f_v^2} - 1} \qquad (A2.5)$$

If $f_v \ll f_m$, which is usually true for installation on vibration isolators, then:

$$x_{rel} = \frac{m_1 + m_2}{m_1^2} \frac{F_o}{4\pi f_m^2 \delta_v} \frac{1}{} \qquad (A2.6)$$

Thus the maximum amplitudes of relative displacement in the working area with the above assumptions do not depend on the natural frequency of the isolation system, but only on its damping, δ_v, which can be considered as *a vibration level criterion.*

In case of stiff mounts, when f_m and f_v are commensurate, the relative displacements would substantially increase. Fastening the machine to a massive foundation block is equivalent to increasing m_1 in (A2.5) with a corresponding reduction of relative vibrations.

Often the *vibratory velocity level* of a machine frame is considered as an indicator of the machine vibration status. An equally hazardous or annoying action of vibration on personnel corresponds to an equal vibratory velocity level at frequencies higher than 2 Hz to 8 Hz (International Standard ISO 2631). From Eq. (A2.4), the maximum velocity amplitude:

$$(v_o)f_v = 2\pi f_v \, x_{res} = \frac{F_o}{2m} \frac{1}{f_v \delta_v} \qquad (A2.7)$$

Thus, a comparison of alternative mounting systems by *vibratory velocity level* can be performed by the *criterion $f_v \delta_v$.* Thus, to reduce the vibratory velocity level of an isolated object, increase of the natural frequency (use of stiffer mounts) and of damping in vibration isolators have similar effects which can be added together.

A2.3.2b Vibration Isolation Criteria for Vibration-Sensitive Objects

Vibration isolation is especially critical for precision machine tools and measuring equipment, such as coordinate measuring machines [9]. Precision machine tools and equipment should be protected from vibratory excitations generated by vibration-producing equipment, such as stamping presses, cranes, fans, etc. and modified by dynamic characteristics of the soil, floor, etc. Vibration spectrum of the floor in both vertical and horizontal directions usually contains components in a broad frequency range, starting from very low frequencies (2 Hz to 4 Hz). Analyses of basic models of vibration isolation as described in all vibration textbooks imply that the natural frequency of the vibration isolation system in a given direction should be significantly less than the lowest excitation frequency, which leads to very low stiffness of vibration isolators and the ensuing rocking sensitivity of the installation. To enhance stability, heavy and expensive inertial foundation blocks supported by low stiffness isolators (springs) are often used. The textbooks analyses also conclude that damping in the vibration isolation system has a detrimental effect on its performance (higher transmissibility of the unwanted floor vibrations to the isolated object in the after-resonance frequency range, see Appendix 1). As a result, extensive surveys of floor vibrations are performed at the sites of intended installation of precision equipment and the foundations are designed in accordance with the survey results. While such an approach is very time consuming and expensive, it often does not bring desirable results. One reason for this — vibration levels at the site change when some surrounding equipment units (including the units to be protected) are added, replaced, or relocated.

To simplify design of the vibration isolation system and to make it more reliable, results of numerous measurements of the floor vibration at various representative sites were combined and analyzed. It was observed that the upper boundaries of the combined spectra for both vertical and horizontal vibration amplitudes can be approximated by "constant amplitude" lines in rather narrow frequency ranges. These spectra can be used as a conservative representation for any industrial site.

It was also suggested to control amplitude of vibration in the working zone of the equipment, e.g., between the tool and the machined surface [9]. This can be looked at as vibration of the frame filtered by the dynamic system of the equipment unit; its amplitude should not exceed a specified fraction of the tolerance. Based on such generalized excitation spectra and on consideration of the "working zone" vibration, it was shown that *a resonance of the vibration isolation system can be a working regime* and, consequently, that quality of a vibration isolating installation is determined not by the natural frequency, f_v, of the vibration isolation system in the considered direction, but by the *vibration isolation criterion* $\frac{f_v}{\sqrt{\delta}}$. Since f_v is proportional to $\sqrt{k_v}$, criterion $\frac{f_v}{\sqrt{\delta}}$ can be rewritten as:

$$\frac{f_v}{\sqrt{\delta}} = \sqrt{\frac{K_{dyn}k_{st}}{M\delta}} = \sqrt{\frac{k_{st}}{M}}\sqrt{\frac{K_{dyn}}{\delta}} \tag{A2.8}$$

where k_{st} is static stiffness of the isolators. Eq. (A2.8) shows that for a given specified value of criterion $\frac{f_v}{\sqrt{\delta}}$ for isolation of a certain object, minimization of ratio K_{dyn}/δ_v results in increasing static stiffness of the isolators without a detrimental effect on the isolator effectiveness..

The meaning of this criterion: if the isolators have higher damping, their *stiffness can be proportionally increased* without detrimental effects on isolation effectiveness. The main advantages of higher stiffness isolators are higher installation stability; lesser reduction of the effective frame stiffness of the object; possibility of reduction (or, even, elimination) of the inertia block; lesser amplitudes and faster decay of transients caused by internal causes, such as stoppage or reversal of heavy moving parts. This criterion, as well as the concept of vibration isolation on which it is based, were validated by successful installations of many thousands of various precision objects (machine tools, measuring instruments, etc.). Extensive tests demonstrated that stiffness values K of numerous commercially available vibration isolators recommended for the same types of precision equipment differ by ±150%, while values of K/δ calculated for the same isolators differ only by ±25%. Obviously, these recommendations were based on practical application experiences.

A somewhat different *criterion*, $f_v/\delta_v^{0.25}$, or $K_v/\delta_v^{0.5}$, is derived in [9] for isolation of the ground from dynamic excitation by forging hammers or other pulse-generating machinery. Although the effect of damping increase in isolators is not as strong in this case as in the case of vibration isolation of precision devices, increase in damping allows to increase stiffness of isolators and/or to reduce the size of the inertia block without negative effects on the isolation effectiveness. Usually, forging hammers are rigidly mounted on huge concrete blocks suspended by large steel coil springs (f_v = 2 Hz to 4 Hz, $\delta_v \approx$ 0.1 to 0.2). It was demonstrated on several installations [9], that the same quality of isolation is achieved by suspending smaller blocks on rubber mats with f_v = 5 Hz to 6 Hz, δ_v = 0.5 – 0.7, in a general compliance with the criterion $K_v / \delta_v^{0.5}$.

A2.4 USE OF STIFFNESS–DAMPING CRITERIA

It is interesting to note that while increase in structural damping allows *to reduce structural stiffness* according to the "chatter resistance" criterion $K\delta$; increase in isolator damping per criterion $K_v^{3/2}\delta_v$. allows *to reduce the isolator stiffness* for machines whose chatter resistance is sensitive to installation, thus allowing effective vibration isolation of such machines; and increase in isolator damping per criterion $\dfrac{f_v}{\sqrt{\delta}}$ allows to *increase the isolator stiffness* without losing effectiveness of vibration isolation.

Another important use of the criterion k/δ is for selection of an optimized material for vibration isolators [9]. Dynamic stiffness (effective stiffness under vibratory conditions) of materials used in isolators (elastomers, wire mesh elements, etc.) is always higher than their static stiffness measured at slow loading, at frequency $\leq$ ~0.1 Hz (see Chapter 3). The dynamic stiffness is frequently dependent on vibration amplitude, a (usually, the dynamic-to-static stiffness ratio K_{dyn} is decreasing with increasing a). Damping, e.g., log decrement δ is also dependent on a. Damping is usually increasing with increasing a.

Use of the criterial ratio K/δ allows to optimize isolators for a specific application depending on the range of a for this application. For example, high precision production equipment is exposed to floor vibration in 1 μm to 3 μm amplitude range, while avionics instruments are exposed to 30 μm to 100 μm amplitude vibration. A proper selection of a rubber blend or a wire mesh material for a given amplitude range may result in significant reduction of K/δ for a given installation and for a desirable higher static stiffness of the isolators, up to an order of magnitude.

Table A2.3. Values of K_{dyn}/δ for various isolator materials and vibration amplitudes.

Elastomeric materials		Double amplitude, mm			
Rubber type	Hardness	0.01–0.015	0.05	0.1	0.2
NR	41		4.6	4.3	3.3	2.6
	46		5.4	3.7	3.1	---
	56	5.4	4.2	3.6	---
	60	4.85	3.5	3.0	3.3
	61	5.9	4.4	4.4	3.0
	75	5.0	3.8	2.4	---
CR	42	4.6	3.85	3.75	3.0
	58	3.75	3.3	2.8	2.5
	74	6.2	3.1	**2.0**	**1.6**
	77	5.5	2.85	2.75	2.1
	78	5.8	**2.65**	2.2	1.85
NBR-26	42	4.0	3.6	3.5	3.3
	56	3.1	2.9	2.7	2.5
	69	**2.9**	**2.2**	**2.1**	1.95
NBR-40	45	3.45	3.7	2.8	2.05
	58	3.1	3.0	2.3	2.05
	62	3.5	2.8	2.6	2.25
	80	**2.6**	**2.3**	**1.9**	---
BR	50	**2.8**	**2.65**	2.2	---

Vibrachok wire-mesh isolating mounts					
Model	Load, N				
V439-0	400	30	9.5	3.4	**1.5**
	1,150	45		4.8	2.4	---
W246-0	870	13	4.5	3.4	1.8
	1,150	23	11.2	6.4	2.9
W246-5	2,300	27	4.1	2.9	**1.55**
Felt Unisorb		15	7	5.35	---

Table A2.3 compares ratios of K_{dyn}/δ_v for different vibration amplitude ranges for various materials used for making flexible elements of vibration isolators. The best (the smallest) ratios are indicated by bold print. It is clear that the best materials for isolators for high precision objects (low vibration amplitudes) are a Butyl rubber blend and the blend with hardness $H = 80$ of Nitrile-40% rubber. The best material for the intense vibrations of the floor is wire mesh. The differences between material criteria K_{dyn}/δ_v for different materials in a given amplitude range are quite significant.

A2.5 DISCUSSION

In many typical vibration control situations effects of changing stiffness (natural frequencies) and damping are not independent parameters and their interrelations can be expressed as criteria. Such criteria can be very useful for guidance during solving vibration control problems. They make the problem solving process more flexible, since the effects from changing stiffness and/or damping become more transparent.

These criteria often are applicable to solving general vibration control problems, such as preventing chatter vibration, reducing vibration level of an object caused by internal excitation, vibration isolation of vibration sensitive objects, etc. Accordingly, deriving the criteria may require generalization of the involved parameters. Examples of such parameters are equalization of amplitudes of internal excitation for the "vibration levels" problems, and of spectrum of floor vibrations for the problem of vibration isolation of vibration-sensitive objects. While these general approaches do not diminish usefulness of the derived criteria, a detailed analysis of specific critical cases may result in a somewhat better control of vibrations, e.g., for an ultra-vibration-sensitive equipment installed in a specially designed building having a unique vibratory environment.

REFERENCES

[1] Rivin, E.I., Kang, H.L., 1992, "Enhancement of Dynamic Stability of Cantilever Tooling Structures", *Int. J. of Machine Tools and Manufacture*, Vol. 32, No. 4, pp. 539–562.

[2] Kudinov, V.A., 1967, "Dynamics of Machine Tools", Mashinostroenie, Moscow, 360 pp [in Russian].

[3] Tlusty, J., 1985, "Machine Dynamics", in Handbook of High-Speed Machining Technology, ed. by R.I. King, Chapman and Hall. L/N.Y.

[4] Rivin, E.I., 2002, "Machine-Tool Vibration", in Harris' Shock and Vibration Handbook, 5th Edition, eds. C.M. Harris and A.G. Piersol, *McGraw-Hill, N.Y.*

[5] Rivin, E.I., Kang, H., 1989, "Improving Machining Conditions for Slender Parts by Tuned Dynamic Stiffness of Tool", *Intern. J. of Machine Tools and Manufacture*, vol. 29, No. 3, pp. 361–376.

[6] Rivin, E.I., Xu, L., 1994, "Damping of NiTi Shape Memory Alloys and Its application for Cutting Tools", in *Materials for Noise and Vibration Control*, ASME NCA-18/DE, pp. 35–41.

[7] Weck, M., Hennes, N., Krell, M., 1999, "Spindle and Tool Systems with High Damping", *Annals of the CIRP*, vol. 48/1, pp. 297–302.

[8] Hasem, S., Mori, J., Tsutsumi, M., Ito, Y., 1987, "A New Modular Tooling System of Curvic Coupling Type", *Proc. of 26th Intern. Machine Tool Design and Research Conf.*, MacMillan Publishing, N.Y., pp. 261–267.

[9] Rivin, E.I., 2003, "Passive Vibration Isolation", ASME Press, 420 pp.

APPENDIX

3

Influence of Axial Force on Beam Vibrations

The equation for deflection of beam in Fig. 7.4.7a subjected to distributed load, p, can be written as [1]:

$$\frac{d^2}{dx^2}\left(EI\frac{d^2y}{dx^2}\right) - T\frac{d^2y}{dx^2} = p. \tag{A3.1}$$

If the distributed load, p, is due to inertia forces of the vibrating beam itself, then:

$$p = -m\frac{\partial^2 y}{\partial t^2}, \tag{A3.2}$$

where m = mass of the beam per unit length. If the beam has a constant cross section and uniform mass distribution, m = const, and Eq. (A3.1) becomes the equation of free vibration:

$$\frac{EI}{m}\frac{\partial^4 y}{\partial x^4} = \frac{T}{m}\frac{\partial^2 y}{\partial x^2} + \frac{\partial^2 y}{\partial t^2} = 0. \tag{A3.3}$$

Substituting into Eq. (A3.3) an assumed solution with separate variables, $y = X(x)U(t)$, arrive at two equations with single variables:

$$\frac{\ddot{U}}{U} = -\omega^2 \tag{A3.4a}$$

$$\frac{T}{m}\frac{X''}{X} - \frac{EI}{m}\frac{X^{iv}}{X} = \omega^2. \tag{A3.4b}$$

The first equation is a standard equation describing a vibratory motion, the second one describes the modal pattern. It can be rewritten as:

$$X^{iv} - \alpha^2 X'' - k^4 X = 0, \tag{A3.4c}$$

where

$$\alpha^2 = T/EI, \qquad k = \frac{m\omega^2}{EI}.$$

The solution of Eq. (A3.4c) can be generally expressed as

$$X = C_1\sinh r_1 x + C_2 \cosh r_1 x + C_3 \sin r_2 x + C_4 \cos r_2 x, \tag{A3.5}$$

where

$$r_1 = \sqrt{\frac{\alpha^2}{2} + \sqrt{\frac{\alpha^4}{4} + k}}; \qquad r_2 = \sqrt{-\frac{\alpha^2}{2} + \sqrt{\frac{\alpha^4}{4} + k}} \tag{A3.6}$$

For the case of Fig. 7.4.7a of a double-supported beam, boundary conditions for $x = 0$ and $x = l$ are $X = 0$ and $X'' = 0$. Thus, $C_2 = C_4 = 0$ and:

$$C_1 \sinh r_1 l + C_3 \sin r_2 l = 0; \quad C_1 r_1^2 \sinh r_1 l - C_3 r_2^2 \sin r_2 l = 0 \qquad \text{(A3.7)}$$

Eqs. (A3.7) have non-trivial solutions if their determinant is zero or:

$$(r_1^2 + r_2^2) \sinh r_1 l \sin r_2 l = 0. \qquad \text{(A3.8a)}$$

Since for any magnitude of the axial force $T \neq 0$, $r_1 l > 0$, and $r_2 l > 0$, always $\sinh r_1 l > 0$ and Eq. (A3.8a) is equivalent to:

$$\sin r_2 l = 0. \qquad \text{(A3.8b)}$$

Since $r_2 l \neq 0$, Eq. (A3.8b) is satisfied when:

$$r_2 l = n\pi \ (n = 1,2,3,...). \qquad \text{(A3.9)}$$

Thus, natural frequencies, ω_n, can be determined from the following equation:

$$\sqrt{-\frac{\alpha^2}{2} + \sqrt{\frac{\alpha^4}{4} + \frac{m\omega^2}{EI}}} = n\pi . \qquad \text{(A3.10)}$$

The explicit expression for ω_n from Eq. (A3.10) is:

$$\omega_n = \frac{n^2 \pi^2}{l^2} \sqrt{\frac{EI}{m}} \sqrt{1 + \frac{Tl^2}{n^2 \pi^2 EI}} , \qquad \text{(A3.11a)}$$

or

$$\omega_n = \omega_{n0} \sqrt{1 + \frac{1}{n^2} \frac{T}{T_e}} , \qquad \text{(A3.11b)}$$

where ω_{n0} = the natural angular frequencies of the same beam without the axial load, T_e = the Euler force, and n = the number of the vibratory mode. If $T < 0$ (compressive force), then ω_1 becomes zero when the beam buckles. If $T > 0$, then any increase nin its magnitude leads to a corresponding increase of *all* the natural frequencies of the beam. However, this effect becomes less pronounced for higher frequencies (modes) because of a moderating influence of the factor $1/n^2$.

Although Eq. (A3.11a) was derived for a double-supported beam, Eq. (A3.11b) seems to be valid for any supporting conditions.

REFERENCE

[1] Craig, R.R., 1981, Structural Dynamics, John Wiley & Sons, N.Y. 528 pp.

APPENDIX

4

Characteristics of Elastomeric (Rubberlike) Materials

A4.1 BASIC NOTIONS

Rubber is a unique family of materials. Elastic modulus values of rubber are low, and rubber endures without failure static stretch up to 1000% (for soft rubber blends). After such deformation the rubber part quickly restores its initial dimensions. While rubber parts are easily deformable, rubber is a practically volumetric incompressible material (Poisson's ratio $\mu = 0.49$ to 0.4999, depending on type of rubber and on carbon black content. Fig. A4.1 [1] shows Poisson's ratio for some rubber types). As a result of the closeness of the Poisson's ratio to 0.5, deformation characteristics of rubber parts in different directions can be independently controlled by shaping the element. Rubber elements can be easily produced in various shapes and can be easily bonded to external and internal rigid (usually, metal or hard plastic) structural covers or inserts. Damping in rubber elements is of the hysteretic type, thus it can usually be characterized by magnitude of loss factor, η, or logarithmic decrement, $\delta = \pi \eta$.

A base elastomer or blend suitable for a specific application is selected depending upon the required dynamic and time stability performance and upon expected environmental conditions. Rubber compounds based on *natural* rubber (NR) or analogous synthetic *isoprene* rubber (IR) have high tensile strength, relatively low damping capacity ($\delta \approx 0.05$ to 0.7, depending on reinforcing fillers content), relatively low dynamic-to-static stiffness ratio K_{dyn}, good bonding properties with metals. However, this family of rubber is characterized by low resistance to mineral oils and gasoline. Always present in the air micro-quantities of ozone cause cracking in the stressed areas of parts made from NR and IR rubbers. The temperature range of use for natural rubber parts is low, usually below 90°C to 100°C (190°F to 212°F). Recently developed so-called *"epoxidized"* natural rubber family has much higher damping (two to four times higher than for NR) and higher temperature resistance. Highly polar elastomers, such as *acrylonitrile butadiene* (*nitrile* rubbers, NBR) are resistant to oils, have higher damping ($\delta \approx 0.4$ to 1.2), and have higher temperature resistance. *Chloroprene* (*chlorobutadiene, neoprene*) rubbers (CR) have good mechanical properties similar to natural rubber but also have a good resistance to ozone and mineral oil products. Their temperature resistance is rather poor (significant stiffening at temperatures below −10°C, which can be somewhat reduced by a proper blending). *Ethylene-propylene-diene* (EPDM) rubbers are resistant to oxidation and to ozone, have

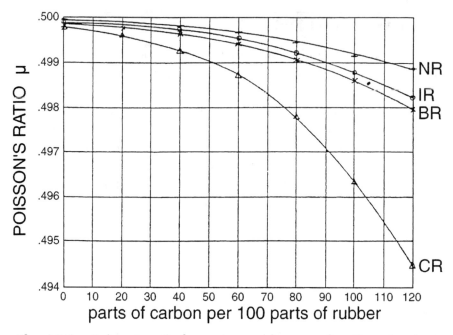

Fig. A4.1. Poisson's ratio for various rubbers as a function of carbon black content. NR — natural rubber; IR — cis-polyisoprene ("synthetic natural rubber"); BR — butyl rubber; CR — chloroprene rubber.

relatively low creep and compression set, may be blended for relatively high damping. *Butyl* elastomers (BR or IIR) have very high damping (δ ≈ 0.3 to 3.0), high aging resistance, are very temperature resistant, are practically impermeable for gases, but may have poor creep and compression set characteristics, are not oil-resistant, and have poor bonding properties. A BROMO modification of BR has even better temperature resistance and significantly lower compression set (creep), which is often an important characteristic for structural components. *Silicone* rubbers are extremely heat- and cold-resistant, but their mechanical characteristics and oil resistance are relatively low. *Fluoroelastomers* are very resistant to environmental conditions and can be blended for high damping. The above comments are very general since many parameters of rubber can be significantly upgraded by proper blending. For example, often high damping is associated with high creep rates. However, an excellent combination of both can be achieved by a judicious blending. Protection from the aggressive environments can be provided by embedding rubber elements in a close-cell foam matrix resistant to the environmental factors [2] or by other types of protective coatings.

Rubber properties can be changed in a very broad range by blending various additives with the base rubber compound. However, rubber properties may be sensitive to relatively minor deviations from quality and quantity of the ingredients and from the specified curing regime (temperature and/or time of curing). The main goal of the curing process is to bind separate molecules into a 3-D structure, thus transforming pliable raw rubber into an elastic material much more stable to aggressive fluids dissolving the raw rubber. The curing process is performed by adding *vulcanizing agents* to the raw rubber, usually 1% to 5% of sulfur or other compounds, and heating to 120°C to 150°C. Some *accelerators* are sometimes added to reduce the curing temperature or time. An increase of the vulcanizing agent content results in stronger intra-molecular connections and in higher stiffness (hardness). However, usually hardness is adjusted by adding *active* and *inert fillers* and *plasticisers*. The most common active fillers are *carbon blacks* with particle sizes 0.02 μm to 0.1 μm as well as special *clays*. The active fillers are enhancing tensile strength, wear resistance, and damping. Inert fillers do not have a strengthening effect but reduce cost of the rubber parts. Plasticisers (*softeners*) are usually liquids added to the rubber blend for improving processing characteristics and cold temperature properties. *Anti-aging agents* are added to the rubber blend to reduce oxidation and ozone cracking of the rubber parts, thus lengthening their useful life.

The volumetric coefficient of thermal expansion for vulcanized rubber is of the order of magnitude of 0.05% per 1°C, corresponding to reduction of volume by ~6% during cooling of the part from the curing temperature to the room temperature. Due to presence of some opposing effects resulting in some increase of the part volume, the overall volumetric shrinking is about 5%, corresponding to the linear shrinkage of 1.6% to 1.7%. If the rubber is bonded along a significant surface area to metal (or other rigid material) components having linear shrinkage 0.1% to 0.25% in the same temperature interval, significant tensile stresses may develop in the rubber part, sometimes detrimental for its durability and creep rate.

Some *highly damped* materials for vibration control devices use different polymers. EAR Corporation makes *Isodamp*™ and *Isoloss*™, which are polyvinyl chloride-based materials. Stiffness of these materials is highly dependent on temperature. One brand, EAR C-1002 has stiffness (moduli) increasing 8.6 times from room temperature (70°F) to 45°F, and decreasing to 0.15 of the room temperature value at 140°F. Sorbothane™ is a pure polyurethane characterized by extremely high damping and very low values of elastic modulus.

A4.2 STATIC DEFORMATION CHARACTERISTICS OF RUBBERLIKE MATERIALS

A rubberlike material may experience two basic types of deformation, described by two independent elastic moduli. The shear modulus G describes a shear deformation, not associated with a change in volume, Fig. A4.2a, and the bulk modulus K describes a volumetric deformation, not associated with a change in shape and describing an omnidirectional hydrostatic compression, Fig. A4.2b. The bulk modulus:

$$K = \frac{G}{1 - 2\mu}.$$

(A4.1)

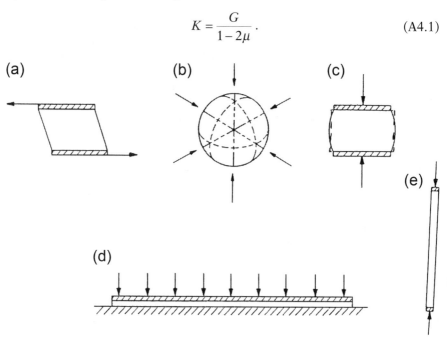

Fig. A4.2[3]. Typical modes of deformation of elastomeric elements. a-Pure shear; b-Omnidirectional (hydrostatic) compression; c-Compression of element bonded to rigid surfaces (dotted lines — compression of lubricated unbonded element); d-Highly restrained compression; e-Unrestrained compression.

Rubber compounds not having reinforcing fillers, such as carbon black, have G and K of the order of 0.5 MPa to 1 MPa and 1000 MPa to 5000 MPa (75 psi to 150 psi and 150,000 psi to 750,000 psi), respectively.

A rubber specimen sandwiched and bonded between parallel flat rigid surfaces like in Fig. A4.2c is not in a homogeneous compression governed by bulk modulus K, unless the lateral dimensions of the specimen are very large relative to its thickness as in Fig. A4.2d, so that the free surface for outward expansion (bulging) of the material during compression is relatively insignificant. In such a case, both volume and shape are changing and the stress/strain ratio is governed by a modulus [4]:

$$M = K + 4G/3 \approx K. \qquad (A4.2)$$

Properties of such thin-layered rubber-metal laminates are described in Chapter 3; they are very different from the deformation characteristics normally associated with rubber. In the other geometric extreme in Fig. A4.2e, the lateral dimensions are small relative to the specimen thickness or length so that the bulging is not restrained (provided that buckling stability of such bar or rod is assured). The similar effect of non-restrained bulging can be achieved in a more realistic case of the configuration as in Fig. A4.2c if there is no bonding and no friction between the rubber element and the rigid surfaces, so that the ideal sliding conditions exist, as shown by broken lines in Fig. A4.2c. In such cases the stress-to-strain ratio is governed by the Young's modulus, E_o, and the ratio of the lateral to axial strain is described by Poisson's ratio μ, ideally:

$$E_o = \frac{9KG}{3K + G} \approx 3G; \quad \mu = \frac{E_o}{2G} - 1 \approx 0.5 \qquad (A4.3)$$

The effective Young's modulus depends on shape and composition of the rubber element. Measuring of E_o or G on the finished rubber part without its destruction is often impossible, and mechanical characteristics of rubber elements strongly depend on their shape and size. Accordingly, the main practical characteristic of rubber stiffness or modulus is its *hardness* measured in Shore durometer units H, or in similar British Standard (BS 903) hardness degrees. The shear modulus, G, depends on H somewhat differently for different compositions of the rubber blend. Fig. A4.3 [4] shows correlation between BS hardness degrees and values of G and E_o as stipulated in the British Standard BS 903. Fabrication of

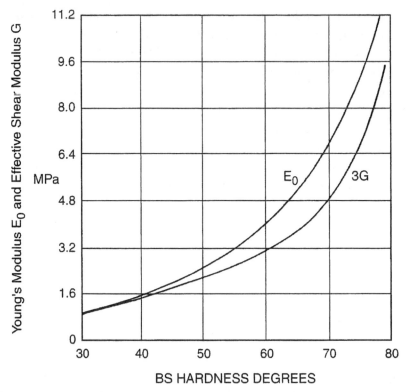

Fig. A4.3. Correlation between compression, E, and shear, G, moduli of rubber element and its hardness.

rubber parts involves many uncertainties, such as uniformity of properties and concentrations of various ingredients, uniformity of mixing these ingredients, deviations from nominal temperature and pressure in the mold, etc. Accordingly, some variation of rubber parameters, especially hardness, is inevitable. The "normal" variation is within ±5 durometer units (~±17% in stiffness/modulus variation). More stringent tolerances can be maintained, such as ±2 durometer units, but at a significantly higher price. The effective parameter variation can be reduced without tightening the tolerances in nonlinear rubber elements, such as constant natural frequency isolators as shown in Section 3.1.1.

The load-deflection characteristic of rubber in shear is softening nonlinear at relative shear $\gamma \geq 0.05$. For vibration control devices, it is important to know deformation characteristics at low relative shear $\gamma \leq 0.01 - 0,02$, which are significantly different from the characteristics at larger γ. Fig. A4.4 [5] shows values of G in two ranges of the relative shear:

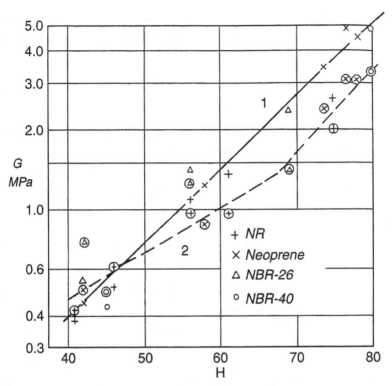

Fig. A4.4. Shear modulus, G, of rubber as function of relative shear, γ, plain marks — test data for $\gamma = 0$ to 0.014 fitted by line 1; encircled marks — $\gamma = 0.025$ to 0.035 fitted by line 2.

$\gamma = 0$ to 0.014 (plain marks) and $\gamma = 0.025$ to 0.35 (encircled marks). The measurements were performed on rubber cylinders (30 mm diameter, 14 mm high) made from various types (NR, Neoprene CR, NBR with 26% and 40% of acrylonitrile contents) and hardness of rubber, and precompressed to axial strain $\varepsilon = 0.07$. At these ranges of γ, there is no significant shear nonlinearity at $H = 45$ to 50; there is a weak hardening nonlinearity at $H \leq 45$; and there is a significant softening nonlinearity at $H > \sim 50$. For hard rubber blends the nonlinearity is very substantial, e.g., for $H = 80$, $G_{\gamma \approx 0.007} \approx 3G_{\gamma \approx 0.03}$. The shear modulus G depends on the relative compression. Shear stiffness is decreasing with increasing compression load/deformation due to developing of pre-buckling condition (see Section A4.3 below).

If a rubber element of a significant thickness h is bonded to rigid (metal) plates on both faces and is subjected to shear deformation, some bending deformation also occurs. It can be considered by using a bending correction term in calculating shear deformation x caused by shear force P_s [4],

$$x = \frac{P_s h}{GA}\left(1 + \frac{h^2}{36i^2}\right) \tag{A4.4}$$

where A = cross-sectional area of the loaded specimen and i = radius of inertia of the cross sectional area about its neutral axis.

A rubber element configured as in Fig. A4.2c is characterized by an effective modulus of elasticity E_{eff}, intermediate between magnitudes of K and M [3],

$$E_{eff} = \frac{E_o(1 + \beta S^2)}{1 + \frac{E_o}{K}(1 + \beta S^2)} \tag{A4.5}$$

Here S = the *shape factor*, the ratio of one loaded area of the element to the total of load-free surface areas, and β = empirical numerical constant dependent on rubber hardness, H, $\beta \approx 2.68$ to $0.025H$ at $H = 30$ to 55; $\beta \approx 1.49$ to $0.006H$ at $H = 60$ to 75.

The shape factor of a rubber cylinder (diameter D, height h) is $S_{cyl} = (\pi D^2/4)/\pi Dh = D/4h$; for a prismatic rubber block (sides, a, b, height, h), $S_{rec} = ab/h(a + b)$. For rubber elements with reasonably uniform dimensions (D/h, a/h, $b/h < \sim 10$), (A4.5) can be written as:

$$E_{eff} \approx E_o(1 + \beta S^2) = 3G(1 + \beta S^2). \tag{A4.6}$$

The elements with the uniform dimensions have linear load-deflection characteristics up to ~10% compression deformation, and quasi-linear characteristics up to ~15% deformation. If $b >> a$ for a rectangular shape element, expression (A4.6) should be replaced by [3]

$$E_{eff} = (2/3)E_o(2 + \beta S^2) = 2G(2 + \beta S^2), \tag{A4.7}$$

where $S = a/2h$.

While designers of metal parts are usually limited by allowable maximum stresses under specified loading conditions, it is customary to design rubber components using allowable strains, or relative deformations. For rubber components bonded to metal inserts or covers, a conservative strain limit for shear loading is relative shear, $\gamma = 0.5$ to 0.75 for the load applied and held for a long time. Some blends can tolerate higher values of relative shear, up to $\gamma = 1.0$ to 1.5. For shorter times of load application $\gamma = 1.5$ to 2.0 can be tolerated, up to $\gamma = 3.0$ for advanced rubber compositions and bonding techniques, especially for ultra-thin-layered rubber-metal laminates (see Section 3.3.3). A compression loading of a bonded rubber specimen is usually associated with the allowable relative compression $\varepsilon = 0.1$ to 0.15, with a somewhat larger limit for softer rubber blends.

A4.3 ELASTIC STABILITY OF RUBBER PARTS

When a rubber element is subjected to compression forces, there may be a possibility of its elastic instability or buckling. While buckling of a "squat", low profile isolator can not occur under the practically feasible load/deformation conditions, the danger of buckling is increasing with increasing ratio between the height h of the element and its smallest dimension in the lateral direction, a. Such elements are often used if low stiffness is required. A typical configuration of the buckled element is shown in Fig. A4.5 [6] for the typical case when lateral positions of the metal face plates bonded to both ends of the element can not shift. While a laminated rubber element is shown in Fig. A4.5, the similar buckling pattern is observed also for solid elements. It can be seen that the rubber part of the element is "shearing out" at the buckling event. Although there is a popular tendency to see the buckling process as a "bang-bang" process whereas the element has its initial characteristics at compression forces $P < P_{cr}$ and abruptly changes them at $P = P_{cr}$, where P_{cr} is *critical* or *Euler* force, the shear stiffness is actually decreasing even at $P << P_{cr}$. This effect is identical to the effect of compression force application to a metal column, e.g., see Section 1.3.1. The schematic in Fig. A4.5 illustrates a general case when in addition to compressive force, P_c, shear force $2P_s$ is also applied to the element (at its center). The critical force, P_{cr} is, of course, decreasing with increasing P_s.

The expression from which the critical compressive force, P_{cr}, can be determined is given in [4] as

$$\frac{P_{cr}}{T'}\left(1 + \frac{P_{cr}}{K'}\right) = \frac{4\pi^2}{L^2} \tag{A4.8}$$

where $K = AG/t$ is shear stiffness of a unit column of total thickness t_T (one rubber layer of thickness t, and cross-sectional area A, together

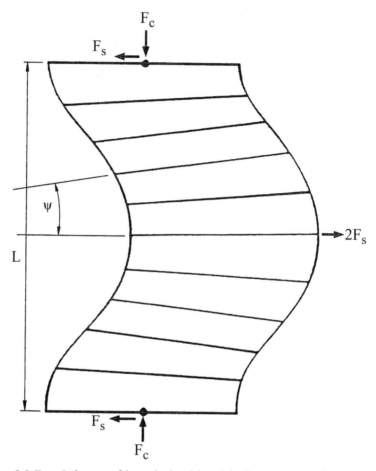

Fig. A4.5. Column of bonded rubber blocks between flat surfaces subjected to compression and shear.

with bonded to it metal end plate), and $K' = K t_T$ is "reduced" shear stiffness, T is experimentally determined bending stiffness of the unit column, and $T' = T t_T$ is "reduced" bending stiffness.

The overall shear stiffness, $K_o = K' \varepsilon$ (ε is the compression strain induced by force P_c) in the center of the column comprising n unit columns and compressed by force P_c is presented in Fig. A4.6 for columns with various number n of the unit layers (shown at the respective lines), each of which has the following parameters: cross-section $A = 76.5 \times 25.6$ mm^2; rubber thickness, $t = 4.4$ mm; total thickness of each unit, $t_T =$

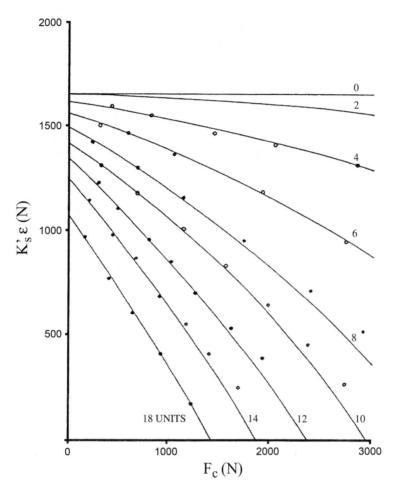

Fig. A4.6. Apparent shear stiffness, $K_s'e$, of column of bonded rubber blocks as function of compressive force.

7.5 mm; compression stiffness of one unit, $k_c = 7$ MN/m; shear stiffness of one unit, $K_s = 0.22$ MN/m; "reduced" shear stiffness $K_s' = K_s t_T = 1670$ N bending stiffness of one unit, $T = 11.4$ Nm/rad. The ordinate is expressed as "apparent" shear stiffness $K_s'e = K_s nt_T \varepsilon$. It is obvious that the shear stiffness can be controlled in a broad range by varying the compressive force, P_c.

For a column of a rectangular or elliptical cross section, the instability occurs in the direction of the shorter dimension of the cross-section. A specimen with a "tubular" cross section is generally more stable than a specimen with a solid cross section of a similar shape and similar dimensions. If the compressive load is below the critical load but close to it, the creep rate of the rubber is increasing, and after some time the specimen may buckle. The critical compression deformation at which the buckling event occurs does not depend significantly on hardness (elastic modulus) of the rubber. Instability is not observed if the dimension of the rubber body in the direction of compression is less than ~0.6 of diameter (for a cylindrical specimen) or of the smaller side of the rectangular cross-section. For the taller elements, the instability occurs at relative compression values decreasing with increasing height of the rubber body. For the column of given dimensions, the critical compression deformation (but not the compressive force!) is decreasing approximately proportionally to the number of rigid plates dividing the rubber column into the separate units/layers. For the rubber columns not bonded to the face plates, so that they may slide or otherwise expand in the direction transverse to the direction of the compressive force, the shape of the rubber body can change so much with the increasing compressive force, that the instability does not develop.

Stability of rubber components is often determined by the "critical strain", rather than by the critical load. For example, an axially loaded in compression hollow rubber cylinder (height $h = 75$ mm, OD and ID 75 and 50 mm, respectively), buckles at $\varepsilon = 0.25$ regardless of rubber blend.

While an unexpected collapse of a rubber component due to excessive compressive forces is highly undesirable, the effects accompanying compressive loading of rubber columns must be considered in the process of the component design; in some cases, a controlled use of these effects can be very beneficial. Two applications of these effects can be noted: adjustment/tuning of horizontal stiffness of rubber components and designing flexible elements to perform in the buckled condition to attain extremely low horizontal stiffness. The effect of decreasing shear stiffness in the compressed rubber columns should be considered in rubber elements which are loaded simultaneously in compression and in shear, e.g. a "chevron" rubber-metal element in Fig. 2.8.10. A judicious preloading in compression of elements containing rubber-metal laminates described in Section 3.3.3c may reduce their shear (horizontal) stiffness while further increasing their compression (vertical) stiffness due to nonlinearity.

A4.4 DYNAMIC CHARACTERISTICS OF RUBBERLIKE MATERIALS

Since energy dissipation in rubberlike materials has a hysteretic character, their dynamic characteristics are often expressed as *complex dynamic shear modulus* and *complex Young's modulus*,

$$G^*_{\omega,t,a} = G_{\omega,t,a}\left(1 + j\eta_{G_{\omega,t,a}}\right), \qquad E^*_{\omega,t,a} = E_{\omega,t,a}\left(1 + j\eta_{E_{\omega,t,a}}\right). \qquad (A4.9)$$

Here *dynamic moduli* $G_{\omega,t,a}$ and $E_{\omega,t,a}$ are the real parts of the complex moduli $G^*_{\omega,t,a}$ and $E^*_{\omega,t,a}$, and $\eta_{G_{\omega,t,a}}$ and $\eta_{E_{\omega,t,a}}$ are energy *"loss factors"* associated with shear deformations and with the Young's modulus of the element (log decrement $\delta = \pi\eta$). The subscripts ω, t, a indicate that both dynamic moduli and loss factors are functions of vibration frequency, temperature, and vibration amplitude. In most cases, the dynamic moduli in (A4.9) are experimentally found to increase with increasing frequency and decreasing temperature, and decrease with increasing vibration amplitude. Many rubberlike materials have their dynamic moduli and loss factors practically independent on frequency for frequencies below ~100 Hz (unless these materials are close to the glasslike state, see below). Dependencies of the loss factors are usually more complex. Figs. A4.7-9 [3] show typical dependencies of G and $\delta = \pi\eta$ on frequency ω, and temperature t. The "transition" frequency ω_t, and temperature t_t, refer to "glass" transition of rubberlike materials (at high frequencies or low temperatures) to a glasslike state, wherein they are loosing their elastomeric properties. The dynamic shear modulus and the corresponding log decrement are plotted in Fig. A4.7 for unfilled natural rubber ("gum"), in Fig. A4.8 for NR filled with 50% by weight of reinforcing carbon black, and in Fig. A4.9 for butyl rubber reinforced by 40% carbon black. One can see differences in behavior of damping δ plots for natural rubber and butyl, since for the latter the glass transition region is less abrupt. The data in Figs. A4.7 to A4.9 are shown from frequency 1 Hz. The transition from "static" to "dynamic" rate of loading is observed at about 0.05 Hz to 1.0 Hz, Fig. A4.10. The ratio K_{dyn} of dynamic k_{dyn} to static k_{st} stiffness is a very important parameter for vibration control applications. Generally, rubber blends with higher damping have larger values of K_{dyn}. However, a judicious use of ingredients can result in increasing damping without increasing K_{dyn}, e.g. see Fig. A4.11 where the correlation between K_{dyn}

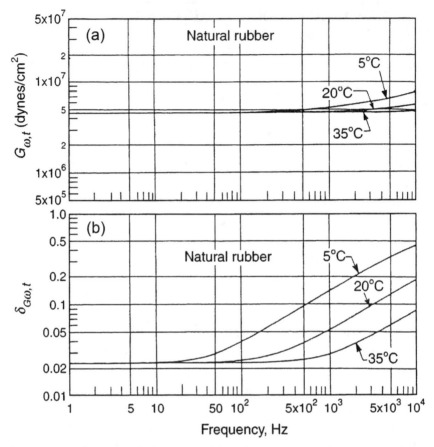

Fig. A4.7. Shear modulus, G, and log decrement, δ, for unfilled (gum) natural rubber.

and loss factor η is shown for carbon black filled NR specimens and specimens made from "epoxidized" natural rubber ENR [7]. Lower values of K_{dyn} result in smaller flexible elements having higher static stiffness for the given natural frequencies, thus reducing cost and weight and increasing stability of vibration control devices.

The data of Figs. A4.7 to A4.9 were obtained for very small vibration amplitudes, for which the rubberlike materials exhibit linear behavior. Unfilled and lightly filled rubber blends remain linear up to relatively large strains, but moderately and heavy filled (hard) rubber blends show strong amplitude dependencies for K_{dyn} and δ. Rubberlike materials comprise a polymeric base and both inert and active (usually, various brands

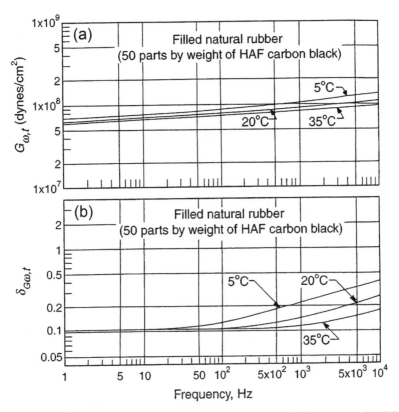

Fig. A4.8. Shear modulus G, and log decrement δ, for natural rubber with 50% carbon black.

of carbon black) fillers. The active fillers combine with the base polymer and develop reinforcing structures inside the rubber. These structures break down during deformation, and then quickly restore in somewhat different configurations, similar to the model analysed in Section 4.6.2. The destruction and reconstruction of these structures have a discrete character similar to dry (Coulomb) friction whereas the external force cannot move the body until it reaches the limit — the static friction force. When the body stops, the friction force quickly reaches the static friction force magnitude. Thus, the mechanism of deforming the carbon black structure is somewhat similar to the deformation mechanism of wire mesh structures. Dynamic characteristics of the filled rubbers are determined, like in fibrous felt-like materials, by dynamic characteristics of active carbon black structures (K_{dyn} and δ depend on the amplitude but not on the frequency of vibration) and dynamic characteristics of the

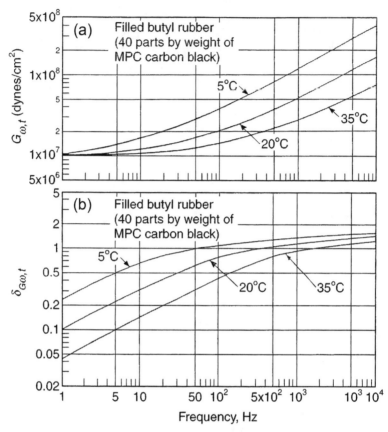

Fig. A4.9. Shear modulus, G, and log decrement, δ, for butyl rubber with 40% carbon black.

polymer base (K_{dyn} and δ do not depend on the vibration amplitude and do not strongly depend on frequency below ~100 Hz). Relative importance of these two factors is mostly determined by the concentration of the active fillers. Some rubber blends contain ingredients preventing development of the carbon black structures. In such blends amplitude dependencies of K_{dyn} and δ are weak even for large carbon black content values.

Fig. A4.12 [4] shows amplitude dependencies of the dynamic shear modulus G (a, b) and loss angle $\beta = \tan^{-1}\eta$ (c) for butyl rubber blends having the same composition and different active carbon black content. Fig. A4.13 shows amplitude dependencies of K_{dyn} and δ for various rubber blends developed for vibration isolators, obtained by testing in free vibration conditions on specimens 30 mm diameter, 13 mm thick loaded

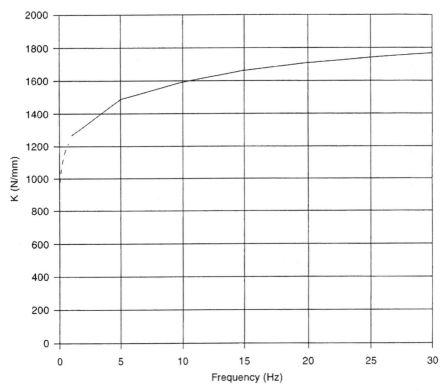

Fig. A4.10. Axial stiffness K for NBR rubber O-ring showing transition from "static" to "dynamic" stiffness.

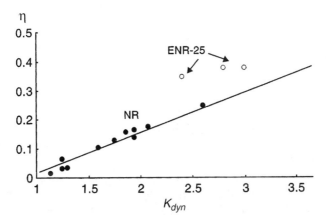

Fig. A4.11. Dynamic-to-static stiffness ratio versus damping (loss factor, η) for natural rubber (NR) and epoxidized natural rubber (ENR-25) blends.

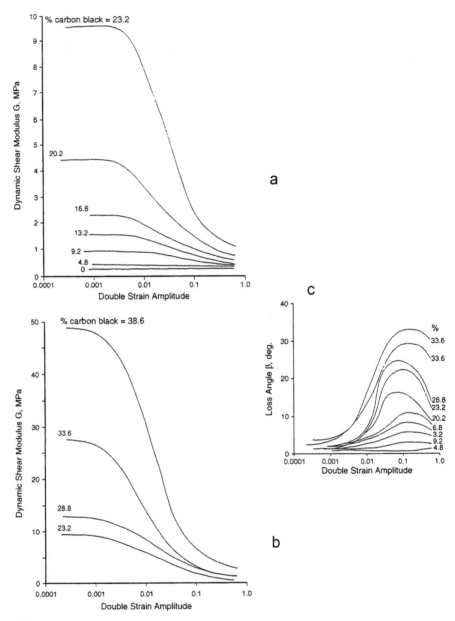

Fig. A4.12. Dependence of shear modulus, G, (a, b) and loss angle, β, (c) on vibration amplitude for butyl rubber blends with different carbon black content (shown in % by weight).

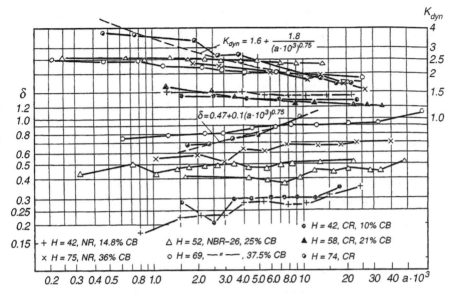

Fig. A4.13. Dependence of shear modulus, K_{dyn}, and log decrement, δ, on vibration amplitude for various rubber blends used in vibration isolators (CB = carbon black).

in compression. The concentration of the active carbon black corresponding to development of the uninterrupted carbon black structure (about 20% to 40%) can be detected by a sharp increase in electroconductivity; this condition is usually accompanied by sharp increases of both K_{dyn} and δ as well as by a steeper slope of their amplitude dependencies.

For some rubber blends their dynamic characteristics are not very sensitive to the static load P_{st} on the specimen. For other blends, such as ones based on NBR, K_{dyn} may significantly increase (and δ decrease) with increasing P_{st}, e.g., see [8]. In some cases, the "scale factor" may be of importance. Increase of the loaded surface A_l of the element was observed in some cases to be correlated with reduction of δ (in one case, increase of A_l by a factor of 5 resulted in 1.3 to 1.6 times reduction of δ).

A4.5 FATIGUE RESISTANCE OF ELASTOMERIC ELEMENTS

An important dynamic characteristic is *fatigue resistance* of rubber elements under vibratory loads. Usually, it is not as important as creep and/ or chemical degradation caused by environmental factors for vibration isolators used for protection of vibration-sensitive equipment due to small vibration amplitudes. However, fatigue failures should be considered for ielements exposed to large vibration amplitudes (elastic elements of vibration-generating production equipment, automotive engine mounts and rubber-metal bushings for steering/handling linkages, etc.). "Fatigue" of a rubber element characterizes gradual change in its performance characteristics caused by a prolonged action of vibratory stresses [6]. The most pronounced change is a gradual reduction of modulus (thus, increasing deflection) although sometimes the stiffness is increasing during the service life of the element. This may depend on the mode of loading (see *Example* below for compression loaded engine mounts). In some cases the fatigue criteria relate to a designated change in the dynamic stiffness, loss factor, etc. This is a form of relaxation similar to the gradual increase of strain/deflection under a constant stress (creep). A plot of modulus vs. logarithm of the number of stress cycles is a straight line, similar to the plot of creep vs. logarithm of time (or versus the number of stress cycles, e.g., see [9]). While a quantitative prediction of fatigue failure for rubber elements is a rather futile exercise due to major influences of the base rubber, additives, variations of production processes, etc., basic qualitative correlations related to fatigue of rubber elements are well represented in Fig. A4.14 based on an extensive experimental study [6]. Reduction of stiffness constant or modulus by 10% is considered as failure of the rubber specimen. Fig. A4.14 presents results of the study wherein rubber specimens were exposed to torsional shear, and the deformation cycle is described by the ratio of double amplitude of the shear deformation (strain) $\Delta\theta$ to the maximum shear strain, θ_m or $\gamma = \tan \theta_m$. Same maximum strain, θ_m (abscissa in Fig. A4.14) would result in very different number of cycles before the failure depending on $\Delta\theta/\theta_m$. Solid lines in Fig. A4.14 represent numbers of cycles before the failure (10% modulus reduction) occurs. The horizontal line $\Delta\theta/\theta_m$ represents the static situation (amplitude $\Delta\theta = 0$) and the point A represents the ultimate static strain ($\theta_m = 86°$ or shear strain $\gamma \approx 14$).

The worst condition is when $\Delta\theta/\theta_m = 1$, i.e. the minimum strain during the loading cycle is zero (complete unloading). Many studies confirmed the critical importance of avoiding such a condition. While it may occur only in a very severe loading environment, it emphasizes usefulness of preloading of rubber elements exposed to dynamic loading. Preloading is frequently used for rubber-metal bushings. A rubber bushing design

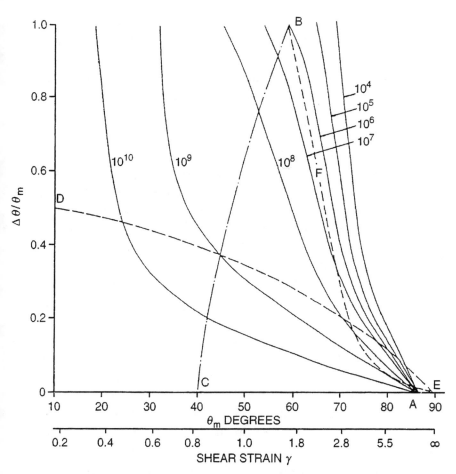

Fig. A4.14. Influence of maximum angular strain, θ_m, and double strain amplitude $\Delta\theta$ on fatigue life of mountings stressed in torsion.

is described in [10], in which the rubber element is molded between the inner and outer steel sleeves. After molding, the bushing is processed by swaging which reduces its outer diameter from 92 mm to 87 mm, thus creating compression preload of the rubber element. Some reduction of shear stiffness as a result of the compression preload of the rubber bushing should be considered.

Plots in Fig. A4.14 demonstrate also other effects of design changes and of various operating conditions on fatigue life (assuming linearity of the rubber element, $\tau = G\gamma$, where τ = shear stress). If the mean load is constant but the dynamic load amplitude varies, the fatigue life can be presented by line *BC*. As the amplitude increases from 0 at *C* to the maximum at *B* (momentary unloading), the fatigue life falls from 10^{10} to ~10^6 cycles.

If the ratio of the dynamic load amplitude to its mean value is constant, the fatigue life varies as shown by line *DE*, which represents the case when the ratio "dynamic load/mean load" is two-thirds (or the ratio "dynamic load/maximum load" is one-half). At the left side (*D*) the stresses are small and the fatigue life is > 10^{10} cycles. Point *E* represents an infinite stress ($\theta_m = 90°$, tan $\theta_m = \infty$), and the fatigue life is zero. Even if the stress distribution inside the component is non-uniform, $\Delta\theta/\theta_m$ is constant throughout the volume, thus the *DE*-like line represents the fatigue conditions for the whole part. Thus, the stress (θ_m) in some point determines fatigue endurance at this point. It can be seen that doubling of θ_m results in about ten-fold reduction of the fatigue life. This fact explains the experimentally observed superior fatigue life of streamlined rubber elements having relatively uniform stress distribution and low maximum stresses (see Fig. 3.3.8).

If the mean load varies while the vibratory load intensity does not change, line like *BFE* demonstrates the effects of such variation. It can be seen that the fatigue life increases with the increasing mean load (e.g., from 10^6 cycles for maximum deflection 59 deg. and minimum deflection 0 deg., point *B*, to 10^7 cycles for maximum deflection 72 deg., and to 10^8 cycles for maximum deflection 80 deg.). This also shows benefits of preloading.

Effect of vibration frequency on the fatigue life is not very significant although the higher frequency may result in somewhat faster reduction of the modulus. Of course, a higher frequency means a shorter fatigue life for a given number of cycles. Temperature variation in a not very broad range (like 40°C to 90°C) is also not a significant factor.

Example. While Fig. A4.14 gives a useful presentation of various factors influencing fatigue of elastomeric flexible elements, the deterioration process of dynamically loaded elastomeric components is not yet well understood. A practical study is described in [11]. First, dynamic stiffness and damping of a rather large batch of new truck engine mounts were measured as functions of frequency, Fig. A4.15, lines 1. The mounts are identified by their stiffness $k = 1,500$ N/mm at low frequency ~1 Hz. The standard deviation of stiffness from the mean value was found to be 4.8%, well within the 17% tolerance range (see Section A4.2 above).

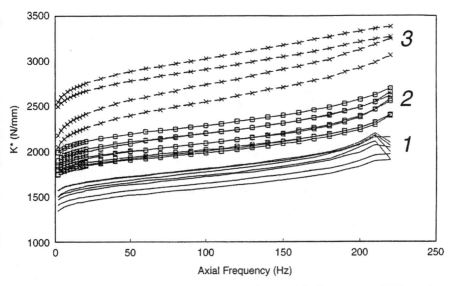

Fig. A4.15. Dynamic axial stiffness of new (1), fleet-tested (2) and durability-tested (3) engine mounts.

Then the mounts were installed on *fleet* trucks and on *"durability"* trucks. The fleet trucks are used on a daily basis for commercial business. The accumulated mileage may be from any part of the country, under a variety of load conditions and can be any combination of on- and off-highway driving. Durability vehicles are driven on a standard durability test track, normally located in one area of the country. The accumulated miles are more "severe" and are designed to compress the durability test duration. One durability test mile is equivalent to about four miles of customer usage. Both fleet and durability trucks accumulated between 30 miles and 40,000 miles.

It can be seen from Fig. A4.15, lines 2, that the mounts removed from the fleet-tested trucks have significantly higher stiffness than the new mounts (about 21% average increase). Even larger increase, about 31%, was exhibited by the mounts from the durability-tested trucks, lines 3 in Fig. A4.15. In addition, after such use the scatter of stiffness values within each tested group has also widened, up to 5.3% for the fleet trucks and up to 7.8% for the durability trucks, as compared with 4.8% for the new mounts. The 31% increase is equivalent to about 15% increase in natural frequencies, thus significantly degrading performance of the isolating mounts.

It can be expected that isolators employing streamlined elastomeric elements, Section 3.2 would exhibit slower aging due to lower stresses/ strains.

A4.6 CREEP OF RUBBERLIKE MATERIALS

A very important parameter of rubber blends for use in vibration isolators and other products is *creep*, i.e., gradual increase of deformation while loaded by a constant magnitude load (e.g., weight load). *Creep rate* is an important performance parameter both for rubber elements used for vibration isolators for precision machinery and equipment, as well as for tightly packaged installations (such as engine compartments in vehicles). The total deformation can be approximated as

$$\Delta' = \Delta + b \log t, \qquad (A4.10)$$

where t = time, min; Δ = relative deformation of the element at 1 min after load application; b = constant characterizing creep rate (intensity) and expressed as a per cent increase of deformation in one *decade of time* (i.e., from 1 minute to 10 minute, or from 8 days to 80 days, etc.). The creep rate depends on the rubber base type, composition of fillers (especially, type of carbon black), type of deformation, presence of rigid (metal) inserts, geometry of the element, and does not strongly depend on the absolute deformation provided that it does not exceed 10% to 15% for compression and 50% to 75% for shear for elements bonded

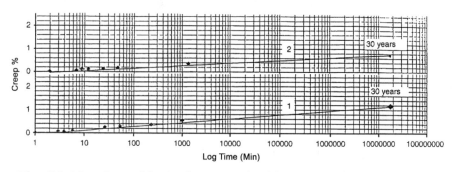

Fig. A4.16. Creep history for natural rubber O-ring (D = 63 mm; d = 13 mm; H = 45; P = 1400 N), (a), and for silicone rubber ball (D = 38 mm, H = 9, P = 75 N) (b); testing performed for 1000 minutes (three decades of loading).

to rigid load-carrying components. It was demonstrated experimentally in [9] and analytically in [12] that the creep rate of radially loaded unbonded cylindrical rubber specimens is 20% to 30% lower than the creep rate of the same elements axially loaded via rigid (metal) plates bonded to their faces. This effect was explained by reduced and more uniform stresses for the former mode of loading, see Fig. 3.3.8. Typical values of b are for NR-based rubbers ~2.5%; for CR (neoprene)-based rubbers ~11%; NBR ~7% [4]. However, these values can be radically different for properly formulated rubber blends. Thus, several NBR rubber blends used in vibration isolators were found to have $b = 0.5\%$ to 1.5%. While there is an opinion that the creep rate is correlated with damping of the rubber blend, in addition to the earlier quoted development of reasonably high damping/low creep NR-based rubber blends in [13], there are blends of very high-damping BR-based rubber having creep rate on the order of $b = 1\%$. The extensive creep tests of cylindrical specimens made from butyl rubber bonded to metal plates and loaded in compression with the specific load 129 psi (9 MPa) for more than 3.5 years are described in [14]. Five blends all having $H = 50$ demonstrated average creep rate ~0.76%. The lowest rate of creep was recorded for compression. If the creep rate for compression is 1.0, then it is usually ~1.2 to 1.25 for shear deformation, and ~1.5 to 1.6 for tension [4]. Fig. A4.16 shows creep test results for a low-filled NR O-ring specimen (63 mm diameter, 13 mm cross-sectional diameter, $H = 45$, compression load 1400 N), line 1, and for a very soft unfilled silicone rubber ball (38 mm diameter, $H = 9$, compression load 75 N), line 2. These curves indicate the creep rate to be only 0.1% to 0.2%, due to a combination of the elastomer properties and streamlined shapes of the test specimens.

REFERENCES

[1] Holownia, B.P., 1974, "Effect of Carbon Black on the Elastic Constants of Elastomers", *J. Instn Rubber Ind.*, vol. 4, No. 8, p. 157–161.

[2] Rivin, E., "Nonlinear Flexible Connectors with Streamlined Resilient Elements", *U.S. Patent 5,954,653.*

[3] Snowdon, J.C., 1965, "Rubberlike Materials, Their Internal Damping and Role in Vibration Isolation", *J. of Sound and Vibration*, vol. 2, pp. 175–193.

[4] Davey, A.B., Payne, A.R., 1964, "Rubber in Engineering Practice", McLaren & Sons, L.

[5] Rivin, E.I., 1965, "Horizontal Stiffness of Anti-Vibration Mountings," *Russian Engineering Journal*, No. 8, pp. 21–23.

[6] Freakley, P.K., Payne, A.R., 1978, "Theory and Practice of Engineering with Rubber", Applied Science Publishers, L.

[7] "NR/ENR High Damping Blends", Guthrie Latex, Inc.

[8] Lewis, T., 1991, "The Effects of Dynamic Strain Amplitude and Static Prestrain on the Properties of Viscoelastic Materials", *SAE Tech. Paper 911084.*

[9] Rivin, E.I., Lee, B.S, 1994, "Experimental Study of Load-Deflection and Creep Characteristics of Compressed Rubber Components for Vibration Control Devices", *ASME J. of Mechanical Design*, vol. 116, No. 2, pp. 539–549.

[10] Lee, S.-H., Lim, Y.-S., 1994, "The development and Performance Simulation of Polychloroprene High Temperature Bush Type Engine Mount", *SAE Techn. Paper 940888.*

[11] Gruenberg, S., Blough, J, Kowalski, D., Pistana, J., "The Effects of Natural Aging on Fleet and Durability Vehicle Engine Mounts from a Dynamic Characterization Prospective", *Proceed. of the 2001 SAE Noise and Vibration Conference (Noise 2001 CD), SAE Paper 2001-01-1449.*

[12] Lee, B.S., Rivin, E.I., 1996, "Finite Element Analysis of Load-Deflection and Creep Characteristics of Compressed Rubber Components for Vibration Control Devices", *ASME J. of Mechanical Design*, vol. 118, pp. 328–336.

[13] Lemieux, M.A., Kilgoar, P.C., 1984, "Low Modulus, High Damping, High Fatigue Life Elastomer Compounds for Vibration Isolation", *Rubber Chemistry and Technology*, vol. 57, pp. 792–803.

[14] Booth, D.A., 1969, "Butyl (IIR)", in *Rubber Handbook*, Morgan-Grampian (Publishers), pp. 135–144.

Power Transmission Couplings

A5.1 INTRODUCTION

Power transmission couplings are widely used for modification of stiffness and damping in power transmission systems both in torsion and in other directions (misalignment compensation). As stated in the Resolution of the First International Conference on Flexible Couplings [1], "... a flexible coupling, although it is relatively small and cheap compared to the machines it connects, is a critical aspect of any shaft system and a good deal of attention must be paid to its choice at the design stage." Technical literature on connecting couplings is scarce and is dominated by trade publications and commercial coupling catalogs. Many coupling designs use elastomers in complex loading modes; some couplings have joints with limited travel distances between the joint components accommodated by friction; often, couplings have severe limitations on size and rotational inertia, etc. These factors make a good coupling design a very difficult task which can be helped by a more clear understanding of the coupling's functions. Stiffness values of couplings in both torsional and misalignment directions as well as damping of couplings in the torsional direction have substantial, often determining, effect on the drive system dynamics. Torsionally flexible couplings are often used for tuning dynamic characteristics (natural frequencies and/or damping) of the drive/transmission by intentional change of their stiffness and damping.

The purpose of this Appendix is to distinctly formulate various couplings' roles in machine transmissions, as well as to formulate criteria for comparative assessment, optimization, and selection of coupling designs. To achieve these goals, a classification of connecting couplings is given and comparative analyses of commercially available couplings are proposed.

A5.2 GENERAL CLASSIFICATION OF COUPLINGS

According to their role in transmissions, couplings can be divided in four classes [2]:

1. *Rigid Couplings*. These couplings are used for rigid connection of precisely aligned shafts. Besides the torque, they also transmit bending moments and shear forces if any misalignment is present,

as well as axial force. The bending moments and shear forces may cause substantial extra loading of the shaft bearings. Principal application areas of rigid couplings are long shafting; space constraints preventing use of misalignment-compensating or torsionally flexible couplings; and inadequate durability and/or reliability of other types of couplings.

2. *Misalignment-Compensating Couplings*. These are required for connecting two members of a power transmission or a motion transmission system that are not perfectly aligned. "Misalignment" means that components coaxial by design are not actually coaxial, due either to assembly errors or to deformations of subunits and/or foundations. The latter factor can be of substantial importance for large turbine installations (thermal/creep deformations leading to drastic load redistribution between the bearings, e.g., [3]) and for power transmission systems on non-rigid foundations (such as ship, propulsion systems). Various types of misalignment as they are defined in AGMA Standard 510.02 [4] are shown in Fig. A5.1. If the misaligned shafts are rigidly connected, this leads to their elastic deformations, and thus to dynamic loads on bearings, to vibrations, to increased friction losses, and to unwanted friction forces in servo-controlled systems. Purely misalignment-compensating couplings have torsional deformations decoupled from misalignment-compensating deformations or movements associated with misalignments.

3. *Torsionally Flexible Couplings*. Such couplings are used to change dynamic characteristics (natural frequency, damping, and character/degree of nonlinearity) of a transmission system. The changes are desirable or necessary when severe torsional vibrations are likely to develop in the transmission system, leading to dynamic overloads. Designs of torsionally flexible couplings usually are not conducive to compensating misalignments.

4. *Combination Purpose Couplings* combine significant compensating ability with significant torsional flexibility. The majority of the commercially available connecting couplings belong to this group. Since the torsional deformation and deformations due to misalignments are not separated/decoupled by design, changes in torsional stiffness may result in changes in misalignment-compensating stiffness and vice versa.

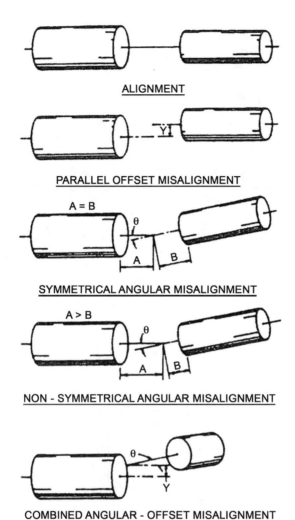

Fig. A5.1. Shaft misalignment modes.

A5.3 RIGID COUPLINGS

Typical designs of rigid couplings are shown in Fig. A5.2.

Sleeve couplings as in Figs. A5.2a,b are the simplest and the slimmest ones. Such a coupling transmits torque by pins, Fig. A5.2a, or by keys, Fig. A5.2b. These couplings are difficult to assembly/disassembly, they require significant axial shifting of the shafts to be connected/disconnected. Usually, external diameter $D = (1.5 - 1.8)\, d$, length $L = (2.5\ \text{to}\ 4.0)\, d$.

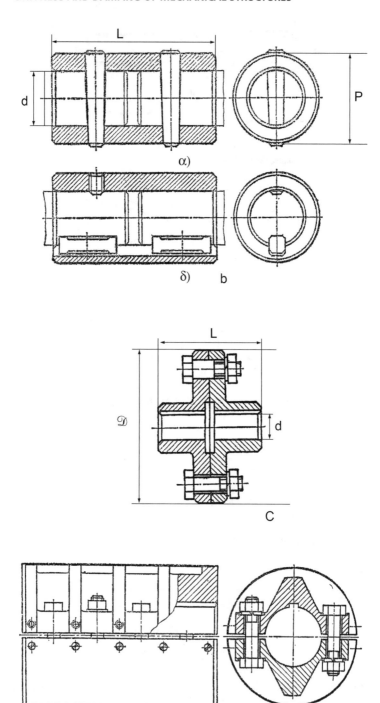

Fig. A5.2. Typical rigid couplings.

Flange couplings, Fig. A5.2c, are the most widely used rigid couplings. Two flanges have machined (reamed) holes for precisely machined bolts inserted into the holes without clearance (no backlash). The torque is transmitted by friction between the contact surfaces of the flanges and by shear resistance of the bolts. Usually, $D = (3$ to $5.5)$ d, $L = (2.5$ to $4.0)$ d.

Split-sleeve coupling in Fig. A5.2d transmits torque by friction between the half-sleeves and the shafts and, in some cases, also by a key. Their main advantage is easiness of assembly/disassembly.

A5.4 MISALIGNMENT-COMPENSATING COUPLINGS

Misalignment-compensating couplings are used to radically reduce the effects of imperfect alignment by allowing a non-restricted or a partially restricted motion between the connected shaft ends in the radial and/ or angular directions. Similar coupling designs are sometimes used to change bending natural frequencies/modes of long shafts. When only misalignment compensation is required, high torsional rigidity and, especially, absence of backlash in the torsional direction are usually positive factors, preventing distortion of dynamic characteristics of the transmission system. The torsional rigidity and absence of backlash are especially important in servo-controlled systems.

To achieve high torsional rigidity together with high compliance in misalignment directions (radial or parallel offset, axial, angular), torsional and misalignment-compensating displacements in the coupling have to be separated by using an intermediate compensating member. Typical torsionally-rigid misalignment-compensating couplings are Oldham coupling, Fig. A5.3a (compensates radial misalignments); gear coupling, Fig. A5.3b (compensates small angular misalignments); and universal (Cardan) joint, Fig. A5.3c (compensates large angular misalignments). Frequently, torsionally rigid "misalignment-compensating" couplings, such as gear couplings, are referred to in the trade literature as "flexible" couplings.

Usually, transmissions designed for greater payloads can tolerate higher misalignment-induced loads. Accordingly, the ratio between the load generated in the basic misalignment direction (radial or angular) to the payload (rated torque or tangential force) is a natural design criterion for purely misalignment-compensating (torsionally rigid) couplings.

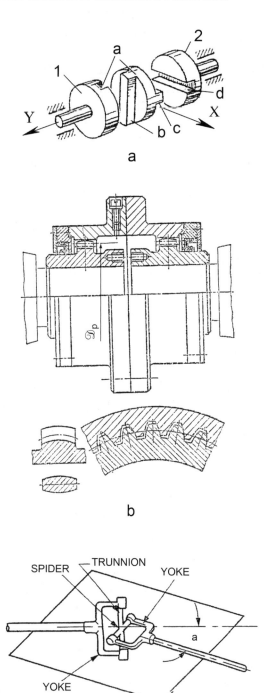

Fig. A5.3. Torsionally-rigid misalignment-compensating couplings; a-Oldham coupling; b-gear coupling; c-U-joint.

Selection Criteria for Misalignment-Compensating Couplings. The misalignment-compensating (torsionally rigid) couplings (Class 2 above) are characterized by presence of an intermediate compensating member located between the hubs attached to the shafts being connected and having mobility relative to both hubs. The compensating member can be solid or comprising several links. There are two basic design subclasses:

(A) Couplings in which the displacements between the hubs and the compensating member have a frictional character (examples: conventional Oldham couplings, gear couplings, and Cardan joints in Fig. A5.3); and

(B) Couplings in which the displacements are due to elastic deformations in special elastic connectors.

For Subclass (A), the radial force, F_{com}, or bending moment are acting from one hub to another and are caused by misalignment; only radial misalignments are addressed below. F_{com} is a friction force equal to the product of friction coefficient f, and tangential force F_t, at an effective radius R_{ef}, $F_t = T/R_{ef}$, where T is the transmitted torque,

$$F_{com} = f \frac{T}{R_{ef}} \tag{A5.1}$$

The force, F_{com}, does not depend on the misalignment magnitude. This is a negative feature since a relatively small real-life misalignment may generate high forces acting on bearings of the connected shafts. Since motions between the hubs and the compensating member are of a "stick-slip" character with very short displacements alternating with stoppages and reversals, f might be assumed to be the static friction ("stiction") coefficient.

When the rated torque, T_r, is transmitted, then the selection criterion is

$$\frac{F_{com}}{T_r} = \frac{f}{R_{ef}} \tag{A5.2}$$

or a lower friction and/or larger effective radius would lead to lower forces on the bearings.

For Subclass (B), assuming linearity of the elastic connectors,

$$F_{com} = k_{com}e,$$

where e = radial misalignment value, k_{com} = combined stiffness of the elastic connectors in the radial direction. In this case,

$$\frac{F_{com}}{T_r} = \frac{k_{com}}{T_r} e. \qquad (A5.4)$$

Of course, a lower stiffness of the elastic connectors would lead to lower radial forces.

Conventional Oldham and Gear Couplings and U-joints (Subclass A). Misalignment-compensating couplings are used in cases where a significant torsional compliance can be an undesirable factor and/or a precise alignment of the connected shafts cannot be achieved. Universal joints (U-joints or Cardan joints) are used in cases where the dominant type of shaft misalignment is angular misalignment. Use of a single joint results in a non-uniform rotation of the driven shaft, which can be avoided by using double joints or specially designed "constant velocity" joints. Compensation of a radial misalignment requires using two Cardan joints and relatively long intermediate shafts. If bearings of the U-joint are not preloaded, the joint has an undesirable backlash, but preloading of the bearings increases frictional losses and reduces efficiency. More sophisticated linkage couplings are not frequently used, due to the specific characteristics of general-purpose machinery, such as limited space, limited amount of misalignment to compensate for, and cost considerations.

While U-joints use sliding or rolling (needle) bearings, both Oldham and gear couplings compensate for misalignment of the connected shafts by means of limited sliding between the hub surfaces and their counterpart surfaces on the intermediate member. The sliding has a cyclical character, with double amplitude of displacement equal to radial misalignment e for an Oldham coupling and $D_p\theta$ for a gear coupling [5], where D_p = the pitch diameter of the gears, and θ = angular misalignment. If a radial misalignment e has to be compensated by gear couplings, then two gear couplings spaced by distance L are required, and $\theta = e/L$, where L = distance between the two gear couplings or the sleeve length. Such a motion pattern is not conducive to good lubrication since at the ends of the relative travel, where the sliding velocity is zero, a metal-to-metal contact is very probable. The stoppages are associated with increasing friction coefficients, close to the static friction values. This is the case for low-speed gear couplings and for Oldham couplings; for high-speed gear couplings the high lubricant pressure due to centrifugal forces alleviates the problem [5].

Fig. A5.3a shows a compensating (Oldham) coupling which, at least theoretically, allows to connect shafts with a parallel misalignment between their axes without inducing non-uniformity of rotation of the driven shaft and without exerting high loads on the shaft bearings. The coupling comprises two hubs 1 and 2 connected to the respective shafts and an intermediate disc 3. The torque is transmitted between driving member 1 and intermediate member 3, and between intermediate member 3 and driven member 2, by means of two orthogonal sliding connections a-b and c-d. By decomposition of the misalignment vector into two orthogonal components, this coupling theoretically assures ideal radial compensation while being torsionally rigid. The latter feature may also lead to high torque-to-weight ratios. However, this ingenious design finds only an infrequent use, usually for non-critical low speed applications. Some reasons for this are as follows.

For the Oldham coupling, radial force from one side of the coupling (one hub-to-intermediate member connection) is a rotating vector directed in the direction of the misalignment and with magnitude:

$$F_1 = f \frac{T}{R_{ef}}, \tag{A5.5}$$

whose direction reverses abruptly twice during a revolution. The other side of the coupling generates another radial force of the same amplitude, but shifted 90 deg. Accordingly, the amplitude of the resultant force is:

$$F_r = \sqrt{2} f \frac{T}{R_{ef}}; \tag{A5.6}$$

its direction changes abruptly four times per revolution. Similar effects occur in gear couplings. An experimental Oldham coupling with $T_r = 150$ N-m, external diameter $D_{ex} = 0.12$ m, $e = 10^{-3}$ m = 1 mm, $n = 1450$ rpm exerted radial force on the connected shafts $F_{com} = 720$ N.

The frequent stoppages and direction reversals of the forces lead to the high noise levels generated by Oldham and gear couplings. A gear coupling can be the noisiest component of a large power-generation system [6]. A sound pressure level $L_{eq} = 96$ dBA was measured at the experimental Oldham coupling described above.

Since a clearance is needed for normal functioning of the sliding connections, the contact stresses are non-uniform with high peak values (Fig. A5.4). Fig. A5.4a shows stress distribution between the contacting surfaces a, b of hub 1 and intermediate member 3 of Oldham coupling in Fig. A5.3a assembled without clearance. The contact pressure in each contact area is distributed in a triangular mode along the length $0.5(D - e) \approx 0.5D$. However, the clearance is necessary during the assembly and it is increasing due to inevitable wear of the contact surfaces. Presence of the clearance changes the contact area as shown in Fig. A5.4b, so that the contact length is $l \approx 0.3(D - e) \approx 0.3D$ [or the contact length is $0.5c \, (D - e)$, $c \approx 0.68$], thus significantly increasing the peak contact pressures and further increasing wear rate. This leads to a rapid increase of the backlash, unless the initial (design) contact pressures are greatly reduced. Such non-uniform contact loading also results in very poor lubrication conditions at the stick-slip motion. As a result, friction coefficients in gear and Oldham couplings are quite high, especially in the latter. Experimental data for gear couplings show $f = 0.3$ [7] or even $f = 0.4$ [8]. Similar friction coefficients are typical for Oldham couplings. The coupling components must be made from a wear-resistant material (usually, heat-treated steel) since the same material is used both for the hub and the intermediate disc structures and for the sliding connections. Friction can be reduced by making the intermediate member from a low friction plastic, such as ultra-high molecular weight (UHMW) polyethylene, but this may result in a reduced rating due to lowered structural strength.

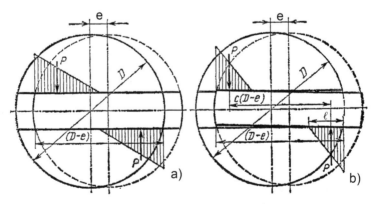

Fig. A5.4. Contact stress distributions in Oldham coupling; (a) Without backlash, (b) With backlash.

Because of high misalignment-compensation forces, deformations of the coupling assembly itself can be very substantial. If the deformations become equal to the shafts misalignment, then no sliding will occur and the coupling behaves as a solid structure, being cemented by static friction forces [8]. It can happen at misalignments below $e \approx 10^{-3}D_{ex}$. This effect seems to be one of the reasons for the trend toward replacing misalignment-compensating couplings by rigid couplings, such as the rigid flange coupling in Fig. A5.1c, often used in power generating systems.

Due to internal sliding with high friction, Oldham and gear couplings demonstrate noticeable energy losses. The efficiency of an Oldham coupling for $e/D_{ex} \leq 0.04$ is

$$\eta = 1 - 3.2 f \frac{e}{D_{ex}}. \tag{A5.7}$$

For $f = 0.4$ and $e = 0.01D_{ex}$, $\eta = 0.987$ and for $f = 0.3$, $\eta = 0.99$. Similar (slightly better due to better lubrication) efficiency is characteristic for single gear couplings.

The inevitable "backlash" in the gear and Oldham connections is highly undesirable for servo-controlled transmissions.

Oldham Couplings and U-joints with Elastic Connections (Subclass B). The basic disadvantages of conventional Oldham and gear couplings (high radial forces, jumps in the radial force direction, energy losses, backlash, nonperformance at small misalignments, noise) are all associated with reciprocal, short travel, poorly lubricated sliding motion between the connected components. There are several known techniques of changing friction conditions. Rolling friction bearings in U-joints greatly reduce friction forces. However, they do not perform well for small amplitude reciprocal motions. In many applications, shafts connected by U-joints are installed with an artificial 2 deg. to 3 deg. initial misalignment to prevent jamming of the rolling bodies.

Another possible option is using hydrostatic lubrication. This technique is widely used for rectilinear guideways, journal and thrust bearings, screw and worm mechanisms, etc. However, this technique seems to be impractical for rotating systems with high loading intensity (and thus high required oil pressures).

Replacement of sliding and rolling friction by elastically deformable connections allows to resolve the above problems. Fig. A5.5 shows a "K" or "Kudriavetz" coupling made from a high strength flexible material

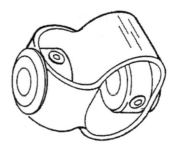

Fig. A5.5. Kudriavetz ("K") coupling.

(polyurethane) and kinematically similar to the Oldham coupling. Its middle membrane serves as an intermediate member and is connected with the hubs by two "tongues" each. If the connected shafts have a radial off-set, it is compensated by bending of the tongues thus behaving like an Oldham coupling with elastic connections between the intermediate member and the hubs. If the connected shafts have an angular misalignment, the middle membrane behaves as a "cross" while twisting deformations of the tongues create kinematics of an U-joint. As a result, this coupling can compensate large radial misalignments (~2.4 mm for a coupling with external diameter $D = 55$ mm), as well as angular misalignments ± 10 deg. The torque ratings of such couplings are obviously quite low, e.g. a coupling with outside diameter $D = 55$ mm has rated torque $T_r = 4.5$ N-m.

Since displacements in the sliding connections a-b and c-d in Fig. A5.3a are small (equal to the magnitude of the shaft misalignment), the Oldham coupling is a good candidate for application of the thin-layered rubber metal laminates, see Chapter 3. Some of the advantages of using laminates are their very high compressive strength, up to 100,000 psi (700 MPa), and insensitivity of their shear stiffness to the compressive force. This property allows to preload flexible elements without increasing their deformation losses.

Fig. A5.6 [9] shows such application, which was studied in [10]. Hubs 101 and 102 have slots 106 and 107, respectively; whose axes are orthogonal. The intermediate disc can be assembled from two identical halves 103a and 103b. Slots 108a and 108b in the respective halves are also orthogonally orientated. Holders 105 are fastened to slots 108 in the intermediate disc and are connected to slots 106 and 107 via thin-layered rubber-metal laminated elements 111 and 112 as detailed in Fig. A5.6b. These elements are preloaded by sides 125 of holders 105, which spread out by moving preloading roller 118 radially toward the center.

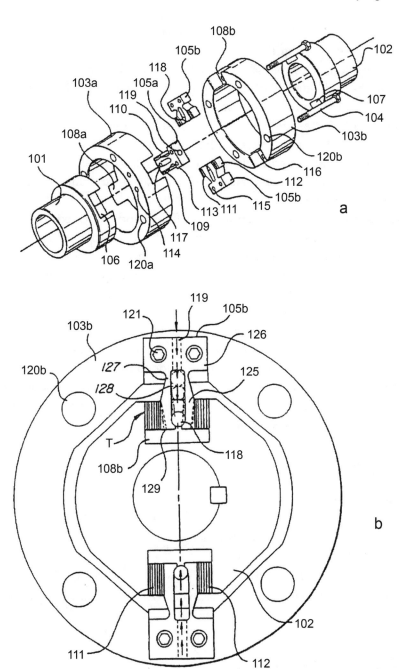

Fig. A5.6. Oldham coupling with thin-layered rubber-metal laminates.

This design provides for all kinematic advantages of the Oldham coupling without creating the listed above problems associated with conventional Oldham couplings. The backlash is totally eliminated since the coupling is preloaded. The coupling is much smaller for the same rated torque than the conventional one due to the high load-carrying capacity of the laminates and the absence of stress concentrations like ones shown in Fig. A5.4b. The intermediate disc (the heaviest part of the coupling) can be made from a light strong material, such as aluminum since it is not exposed to contact loading. This makes the coupling suitable for high-speed applications.

The misalignment compensation stiffness and the rated torque can be varied by proportioning the laminated elements (their overall dimensions, thickness and number of rubber layers, etc.). The loads on the connected shafts are greatly reduced and are not dependent on the transmitted torque since the shear stiffness of the laminates does not depend on the compression load.

To derive an expression for efficiency of the Oldham coupling with laminated connections, let the shear stiffness of the connection between one hub and the intermediate member be denoted by k_{sh} and the relative energy dissipation in the rubber for one cycle of shear deformation by ψ (see Appendix A1). Then, maximum potential energy in the connection (at maximum shear e) is equal to:

$$V_1 = k_{sh}\frac{e^2}{2}, \tag{A5.8}$$

and energy dissipated per cycle of deformation is:

$$\Delta V_1 = \psi k_{sh}\frac{e^2}{2}. \tag{A5.9}$$

Each of the two connections experiences two deformation cycles per revolution; thus the total energy dissipated per one revolution of the coupling is:

$$\Delta V = 2\Delta V_1 = 2\psi k_{sh}e^2. \tag{A5.10}$$

Total energy transmitted through the coupling per one revolution is equal to:

$$W = P_t \pi D_{ex} = 2\pi T, \tag{A5.11}$$

where $P_t = T/D_{ex}$ is tangential force reduced to the external diameter D_{ex} and T is the transmitted torque. Efficiency of a coupling is therefore equal to:

$$\eta = 1 - \frac{\Delta V}{W} = 1 - \frac{\psi k_{sh}}{\pi T} e^2 = 1 - \frac{k_{sh} \tan \beta}{T} e^2, \qquad (A5.12)$$

where β = loss factor of the rubber.

For the experimentally tested coupling ($D_{ex} = 0.12$ m), the parameters are: a laminate with rubber layers 2 mm in thickness; $\psi = 0.2$; $k_{sh} = 1.8$ x 10^5 N/m; $T = 150$ Nm; $e = 1$ mm $= 0.001$ m; thus $\eta = 1 - (0.2$ x 1.8 x 10^5 x $10^{-6})/150\pi = 1 - 0.75$ x $10^{-4} = 0.999925$, or the losses at full torque are reduced 200 times compared to the conventional coupling.

Test results for the conventional and modified Oldham couplings having $D_{ex} = 0.12$ m, have shown that the maximum transmitted torque was the same but there was a 3.5 times reduction in the radial force transmitted to the shaft bearings with the modified coupling. Actually, the coupling showed the lowest radial force for a given misalignment compared with any commercially available compensating coupling, including couplings with rubber elements. In addition to this, noise level at the coupling was reduced by 13 dBA to $L_{eq} = 83$ dBA. Using ultrathin-layered laminates for the same coupling would further increase its rating by at least one order of magnitude, and may even require a redesign of the shafts to accommodate such a high transmitted load in a very small coupling.

U-joints transmit rotation between two shafts whose axes are intersecting but not coaxial, Fig. A5.3c. It also has an intermediate member ("spider" or "cross") with four protruding pins ("trunnions") whose intersecting axes are located in one plane at $90°$ to each other. As in the Oldham coupling, two trunnions having the same axis are movably engaged with journals machined in the hub ("yoke") mounted on one shaft, and the other two trunnions - with the yoke attached to the other connected shaft. However, while the motions between the intermediate member and the hubs in the Oldham coupling are translational, in the U-joint these motions are revolute. This design is conducive to using rolling friction bearings, but the small reciprocating travel regime under heavy loads requires de-rating of the bearings.

A typical embodiment of the U-joint with thin-layered rubber-metal bearings [11] is illustrated in Fig. A5.7 for a large size joint. Fig. A5.7

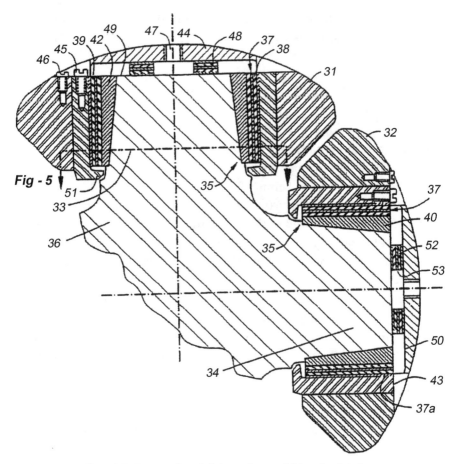

Fig. A5.7. Preloadable universal (Cardan) joint.

shows two basic units (out of four constituting the universal joint) "yoke 31–trunnion 33–elastomeric bearing sleeve 35" and "yoke 32-trunnion 34-elastomeric bearing sleeve 35". Sleeves 35 comprise rubber-metal laminates 37 having sleeve-like rubber layers 38 and separating them (and bonded to them) sleeve-like thin reinforcing intermediate metal layers 39 and inner (innermost) 40 and outer (outermost) 41 sleeve-like metal layers bonded to the extreme inner and outer sleeve-like rubber layers. The inner 40 and outer 41 metal layers of the laminated bearing sleeve are made thicker than the intermediate metal layers since they determine the overall shape of elastomeric bearing sleeves 35.

The inner surface of inner layer 40 is made tapered and conforming with the tapered outer surfaces of trunnions 33, 34. The outer surface of outer layer 41 is made cylindrical and conforming with the internal cylindrical surface of the bore in yoke 31. Bearing sleeve 35 is kept in place by cover 44 abutting end surface 43 of outer metal layer 41. Cover 44 is fastened to outer metal layer 41 by bolts 45, and to yoke 31 by bolts 46. Threaded hole 47 is provided in the center of cover 44.

Before assembly, wall thickness of elastomeric bearing sleeve 35 (a sum of total thickness of rubber layers 38, intermediate metal layers 39 and 39a, and inner and outer metal layers 40 and 41) is larger than the annular space between the inside surface of the bore in yoke 31 and the respective outside surface of trunnion 33. The difference between the wall thickness of the bearing sleeve and the available annular space is equal to the specified preloading compression deformation of the elastomeric bearing sleeve. To perform the assembly operation, the tapered bearing sleeve is inserted into the wider opening of the tapered annular space between the internal surface of the bore and the external surface of the trunnion and pressed into this space by a punch shaped to contact simultaneously both end surfaces: 42 of inner sleeve-like metal layer 40, and 43 of outer sleeve-like metal layer 41. Wedge action of the tapered connection between the conforming inner surface of metal layer 40 and outer surface of trunnion 33 results in expansion of metal layer 40, in compression (preloading) of rubber layers 38, and in gradual full insertion of bearing sleeve 35 into the annular space between the yoke and the trunnion. The simultaneous contact between the pressing punch and both end surfaces of inner metal layer 40 and outer metal layer 41 assures insertion of the bearing sleeve without inducing axial shear deformation inside bearing sleeve 35 which can cause distortion or even damage of the bearing sleeve.

To disassemble the connection, bolts 46 attaching cover 44 to yoke 31 are removed, and then a bolt is threaded into hole 47 until contacting end surface 49 of trunnion 33. The further threading of the bolt pushes outside cover 44 together with outer metal layer of the bearing sleeve, to which cover 44 is attached by bolts 45. The initial movement causes shear deformation in rubber layers 38, until disassembly protrusions 51 engage with inner metal layer 40, thus resulting in a uniform extraction of bearing sleeve 35.

It is highly beneficial that U-joints with the rubber-metal laminated bearings do not need sealing devices and are not sensitive to environmental contamination (dirt, etc.).

The efficiency analysis for such U-joint is very similar to the above analysis for the modified Oldham coupling. The efficiency of the joint is:

$$\eta = 1 - \frac{\Delta V}{W} = 1 - \frac{\psi k_{\text{tor}}}{\pi T} \alpha^2 = 1 - \frac{k_{\text{tor}} \tan \beta}{T} \alpha^2, \tag{A5.13}$$

where k_{tor} = torsional stiffness of the connection between the intermediate member and one yoke. It can be compared with efficiency of a conventional U-joint where d = the effective diameter of the trunnion bearing, $2R$ is the distance between the centers of the opposite trunnion bearings, and f is friction coefficient in the bearings

$$\eta = 1 - f \frac{d}{R} \frac{1}{\pi} \left(2 \tan \frac{\alpha}{2} + \tan \alpha \right), \tag{A5.14}$$

Comparison of Eqs. (A5.7) and (A5.14) with Eqs. (A5.12) and (A5.13), respectively, shows that while efficiencies of conventional Oldham coupling and U-joint for given e, α are constant, efficiency of the modified designs using rubber-metal laminated connections increases with increasing load (when the energy losses are of the greatest importance). The losses in an elastic Oldham coupling and U-joint at the rated torque can be one to two decimal orders of magnitude lower than the losses for conventional units. Due to high allowable compression loads on the laminate (in this case, high radial loads), the elastic Oldham couplings and U-joints can be made smaller than the conventional units with sliding or rolling friction bearings for a given rated torque. The laminates are preloaded to eliminate backlash, to enhance uniformity of stress distribution along the load-transmitting areas of the connections, and to increase torsional stiffness. Since there is no actual sliding between the contacting surfaces; the expensive surface preparation necessary in conventional Oldham couplings and U-joints (heat treatment, high-finish machining, etc.) is not required. The modified Oldham coupling in Fig. A5.6 and the U-joint in Fig. A5.7 can transmit very high torques while effectively compensating large radial and angular misalignments, respectively, and having no backlash since their laminated flexible elements are preloaded. However, for small rated torques there are very effective and inexpensive "alternatives" to these design whose kinematics are similar. These alternatives are also backlash-free.

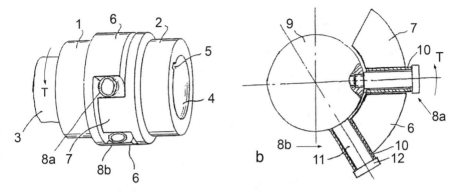

Fig. A5.8. Jaw coupling with tubular sleeve/coil spring elastic elements; a General view, b Flexible elements.

One alternative is Kudriavetz coupling shown in Fig. A5.5. Another alternative is a modified spider or jaw coupling whose cross section by the mid-plane of the six-legged spider is shown in Fig. A5.8 [12]. In this design the elastomeric spider of the conventional jaw coupling shown in Fig. A5.10a below is replaced with a rigid spider 9, 11 carrying tubular sleeves (or coil springs) 10 supported by spider pins 11 and serving as flexible elements radially compressed between cams 6 and 7 protruding from the respective hubs. Deformation and stress characteristics of the radially-loaded coil springs are given in Sec. 2.7.1. If the number of spider legs is four, at 90 deg. to each other, then the hubs have relative angular mobility, and the coupling becomes a U-joint with angular mobility greater than 10 deg., but with much higher rated torque than an equivalent size Kudriavetz coupling.

Purely misalignment compensating couplings described in this Section have their torsional and compensating properties decoupled by introduction of the intermediate member. Popular bellows couplings have high torsional stiffness and much lower compensating stiffness, but their torsional and compensating properties are not decoupled, so they are representatives of the "combination purpose couplings" group.

A5.5 TORSIONALLY FLEXIBLE COUPLINGS AND COMBINATION PURPOSE COUPLINGS

Torsionally flexible couplings usually have high torsional compliance (as compared with torsional compliance of shafts and other transmission components) in order to enhance their influence on transmission dynamics. An example of purely torsionally flexible coupling with elastomeric flexible element having low stiffness in torsional direction (shear of rubber ring 1) and high stiffness in misalignment-compensation directions (compression of rubber ring 1) is shown in Fig. A5.9a. Fig. A5.9b shows a torsionally flexible coupling with metal flexible element (Bibby coupling). The flexible element is a spring steel band wrapped around judiciously shaped teeth on each hub and deforming between the teeth. The deformations are becoming more restrained with the increasing transmitted torque, thus the coupling has a strongly nonlinear torsional stiffness characteristic of the hardening type. The lowest stiffness is at zero torque, Fig. A5.9c, increasing towards the rated torque, Fig. A5.9d, becoming very high at the allowed peak torque, Fig. A5.9e, and approaching a rigid condition at an overload/shock torque, Fig. A5.9f. Since some misalignment-compensating ability is desirable for many applications, use of purely torsionally flexible couplings rigid in the compensation directions is limited, with the combination purpose couplings being used as torsionally flexible couplings with more or less compensating ability.

For torsionally flexible and combination purpose couplings, torsional stiffness is usually an indicator of their payload capacity. In such cases, their basic design criterion can be formulated as a ratio between the stiffness in the basic misalignment direction and the torsional stiffness. In the following analysis, only radial misalignment is considered. Since couplings are often used as the cheapest connectors between shafts and the users do not have much understanding of what is important for their applications, it is of interest to analyze what design parameters are important for various applications.

A5.5.1 Roles of Torsionally Flexible Coupling in Transmission

Torsional flexibility is introduced into transmission systems when there is a danger of developing resonance conditions and/or transient dynamic overloads. Their influence on transmission dynamics can be due to one or more of the following factors: *torsional compliance; damping; nonlinearity of load-deflection characteristic.*

Reduction of torsional stiffness of the transmission and, consequently, shift of its natural frequencies. If a resonance condition occurs before installation (or change) of the coupling, then shifting of natural frequency

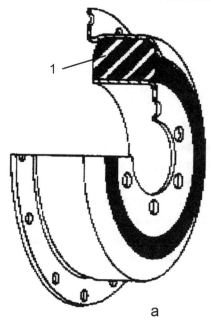

a

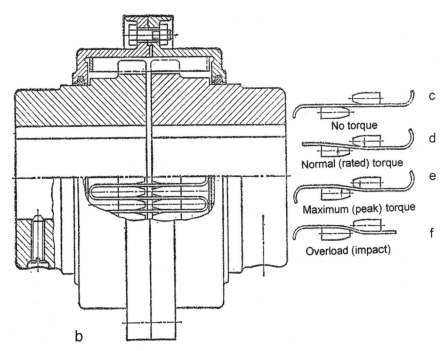

c

No torque

d

Normal (rated) torque

e

Maximum (peak) torque

f

Overload (impact)

b

Fig. A5.9. Torsionally-flexible radially-rigid couplings; (a) Dynaflex LCD, Lord Corp.; (b) All-metal Bibby coupling.

due to use of a high torsional compliance coupling can eliminate resonance, thus dynamic loads and torsional vibrations will be substantially reduced. However, in many transmissions (e.g., vehicle transmissions) frequencies of the disturbances acting on the system and, also, natural frequencies (especially in variable speed transmissions) may vary widely. In such instances, a simple shift of the natural frequencies of the drive may lead to a resonance occurring at other working conditions, but the probability of its occurrence is not sufficiently lessened. A reduction in the natural frequency of a drive, for example, is advisable for the drive of a milling machine only at the highest spindle speeds and may be harmful if introduced in the low-speed stages.

A shift of natural frequencies of the drive may be beneficial in transmissions with narrow variations in working conditions. If, however, a drive is operated in the pre-resonance region, an increase in torsional compliance would lead to increased amplitudes of torsional vibrations, and thus to a nonuniform rotation. In some cases excessive torsional compliance may lead to a dynamic instability of the transmission and create intensive self-excited torsional vibrations.

An important feature of multispeed (or variable-speed) transmissions is changing of effective torsional compliances of their components with changing output speeds due to changing reduction coefficients (although the physical condition of the components does not change), see Section 6.3. As a result, the role of the coupling as a compliant member can dramatically change depending on configuration of the drive. While compliance of a coupling of any reasonable size installed in the high-speed part of the system (close to the driving motor) would not have any noticeable effect at low output rpm, compliance of a coupling installed in the low-speed part of the system (close to the working organ, such as a wheel of the vehicle or a cutter of a mining combine) would be very effective, but the coupling size and cost might become excessive due to high torques usually transmitted at low rpm.

Increasing effective damping capacity of transmission by using high damping coupling or special dampers. When the damping of a system is increased without changing its torsional stiffness, the amplitude of torsional vibrations is reduced at the resonance and in the near-resonance zones. Increased damping is especially advisable when there is a wide frequency-spectrum of disturbances acting on a drive, e.g. for the drives of universal machines. The effect of increased damping in a torsionally flexible coupling of a milling machine transmission, whose mathematical models are shown in Fig. 6.3.6, is illustrated in Fig. A5.10 (natural frequencies f_{n1} = 10 Hz, f_{n2} = 20 Hz). Fig. A5.10a shows the resonance for an OEM coupling (flexible element made from neoprene rubber, log decrement $\delta \approx 0.4$). After this element was replaced by an identical element made from a high damping butyl rubber blend (same compliance, but $\delta \approx$ 1.5), the peak torque amplitude was reduced by ~ 1.8 times, the clearance opening (source of intensive noise) was eliminated, and oscillations with f_{n2} excited by the second harmonic of the cutting force, became visible (Fig. A5.10b). A common misconception about using high-damping elastomers for coupling elements is their alleged high heat generation at resonance. This misconception is disproved in Section 1.1.2, where it is shown that the resonance amplitude reduction caused by a high damping coupling translates in to a lesser heat generation. The influence of a flexible element on the total energy dissipation in a transmission increases with increasing of its damping capacity, of the torque amplitude in the element, and of its compliance. For maximum efficiency, the flexible element of a coupling must therefore have as high internal energy dissipation as possible; it must also possess maximum permissible compliance, and must be located in the part of the system where the intensity of vibrations is the greatest.

Introducing nonlinearity in the transmission system. A nonlinear dynamic system may automatically detune away from resonance at a fixed-frequency excitation, the more so the greater the relative change of the overall stiffness of the system in the course of the torsional deflection equal to the vibration amplitude. For example, when damping is low, a relative change of the stiffness by a factor of 1.3 reduces the resonance amplitude by ~1.7 times, but a relative change of stiffness by a factor of

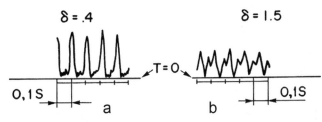

Fig. A5.10. Dynamic load in milling machine drive with:
(a) Manufacturer-supplied motor coupling (δ = 0.4); (b) High damping motor coupling (δ = 1.5).

2 reduces the resonance amplitude by ~1.85 times. Thus, nonlinear torsionally flexible couplings can be very effective in transmissions where high-intensity torsional vibrations may develop and where the coupling compliance constitutes a major portion of the overall compliance.

Vehicles usually have variable speed transmissions. The same is often true for production machines. In order to keep the coupling size small, it is usually installed close to the driving motor/engine, where it rotates with a relatively high speed and transmits a relatively small torque. At the lower speeds of an output member, the installed power is not fully utilized and the absolute values of torque (and of amplitudes of torsional vibrations) transmitted by the high-speed shaft are small. In vehicles, the

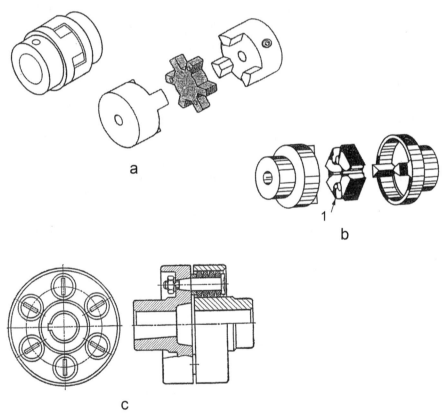

Fig. A5.11. Some combination purpose couplings; (a) Spider (jaw) coupling; (b) Modified spider coupling (1 — lip providing bulging space for the rubber element); (c) Finger-sleeve coupling; (d) Toroid shell coupling; (e) Rubber disc coupling; (f) Centaflex coupling.

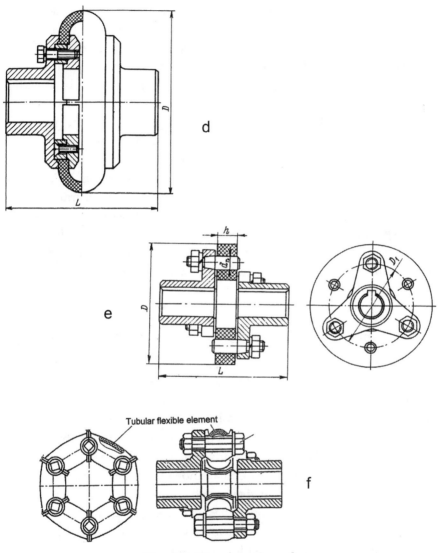

Fig. A5.11. (*continued*)

installed power is not fully utilized most of the time. Thus, an important advantage of couplings with nonlinear load-deflection characteristics is feasibility of making a resonably small coupling with low torsional stiffness and high rated torque. An overwhelming majority of power-transmission systems are loaded with less than 0.5 T_r for 80% to 90% of the total "up" time. A nonlinear coupling with a hardening load-deflection characteristic such as one in Fig. A5.14 below provides low torsional stiffness for most of the time, but since its stiffness at the rated torque is much greater its size can be relatively small.

Introducing additional rotational inertia in the transmission system. This is a secondary effect since couplings are not conventionally used as flywheels. However, when a large motor coupling is used, this effect has to be considered. As shown in Section 6.6.2, it is better to install a flywheel in a transmission on its output shaft (work organ of the machine). Since couplings are usually installed close to input shafts, their inertia can have a negative effect on transmission dynamics, by increasing non-uniformity of the work organ rotation, and also by reducing the influence of the coupling compliance, damping, and nonlinearity on transmission dynamics. The effects can be opposite (thus, positive) in dynamic systems with internal combustion engines. Accordingly, reduction in the coupling inertia is a beneficial factor in many designs with electric motors, but increase in coupling inertia is a benefitial factor in systems with IC engines.

A5.5.2 Compensation Ability of Combination Purpose Couplings

A huge variety of combination purpose couplings is commercially available. Unfortunately, selection of a coupling type for a specific application is often based not on an assessment of performance characteristics of various couplings, but on the coupling cost or on other, non-technical, considerations. As a result, bearings of the shafts connected by the coupling may need to be more frequently replaced than when an optimized coupling is used; the device might be noisier than it would be with a coupling type optimal for the given application; etc.

Fig A5.11 shows some popular designs of combination purpose couplings. The combination purpose couplings do not have a compensating member. As a result, compensation of misalignment is accomplished, at least partially, by the same mode(s) of deformation of the flexible element as used for transmitting the payload. To better understand the behavior of combination purpose couplings, an analysis of the compensating performance of a typical coupling with a spider-like flexible element is helpful. The coupling in Fig. A5.12 shows a schematic of the jaw coupling in Fig. A5.11a. It consists of hubs 1 and 2 connected with a rubber spider 3 having an even number $Z = 2n$ of legs, with "n" legs ("n" might be odd) loaded when hubs are rotating in the forward direction and the

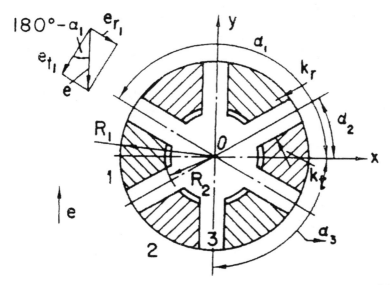

Fig. A5.12. Schematic of a spider coupling; 1, 2 — hubs; 3 — rubber spider.

other n legs loaded during the reverse rotation. Deformation of each leg is independent. Analysis of this coupling is presented in Appendix 6, Section A6.5.

The radial (compensation) stiffness of the coupling with $Z = 4$ is [2]:

$$k_{\mathrm{com}} = \frac{F}{e} = 2k_t \sqrt{\cos^2 \alpha + \frac{k_r^2}{k_t^2} \sin^2 \alpha}, \qquad \text{(A5.15)}$$

where F = radial force caused by the radial misalignment e and acting on the connected shafts, k_t is stiffness of one leg in compression (tangential direction), k_r is stiffness in shear (radial direction), and α is angle of rotation of the coupling. Expression (A5.15) shows that the total radial force, F, fluctuates both in magnitude and in direction during one revolution.

For a coupling with $Z \geq 6$,

$$k_{\mathrm{com}} = \frac{F}{e} = \frac{n}{2}(k_t + k_r), \qquad \text{(A5.16)}$$

or F is constant and is directed along the misalignment vector.

Since for the spider couplings usually maximum allowable $e \leq 0.01 D_{\mathrm{ex}}$, and the spider leg width $b = (0.2$ to $0.25)D_{\mathrm{ex}}$, the maximum relative shear of the leg does not exceed ~0.03. Accordingly, the value of the shear modulus

G has to be modified as compared with the nominal $G(H)$ relationship (H = rubber durometer) as shown in Fig. A4.4. Thus, ratio k_r/k_t varies with changing H, and for typical spider proportions, k_r/k_t = 0.26-0.3 for medium H = 40 to 50, and k_r/k_t = 0.4 for hard rubber spiders, H = 70 to 75.

A spring/tubular spider coupling modification per [12] is shown in Fig. A5.8. In this design each leg of the spider can be represented by a coil spring loaded radially by the transmitted torque. For the tightly coiled extension spring, Section 2.7.1:

$$k_t \approx \frac{\pi^2 E n d^4}{3.74 D^3}; \qquad k_r = \frac{G d^4}{8 D^3 n}, \qquad (A5.17)$$

thus, $k_r/k_t = 1/53n^2 = 0.02/n^2 \approx 0$.

The torsional stiffness of both spider coupling designs is [2]:

$$k_{tor} = n k_t R_{eff}^2, \qquad (A5.18)$$

where the effective radius, $R_{eff} = \sim 0.75 R_{ex} = 0.75(D_{ex}/2)$. The ratios between torsional and compensation stiffness values are:

for $Z = 4$, $\left(\dfrac{k_{com}}{k_{tor}} \right)_{max} = \dfrac{1}{R_{eff}^2} \approx \dfrac{1.8}{R_{ex}^2}$;

for $Z \geq 6$, $\dfrac{k_{com}}{k_{tor}} = \dfrac{1.15}{R_{ex}^2}, H = 40 - 50; \dfrac{k_{com}}{k_{tor}} = \dfrac{1.25}{R_{ex}^2}, H = 65 - 75;$

for the spring spider coupling per Fig. A5.8, $Z = 6$, $\dfrac{k_{com}}{k_{tor}} = \dfrac{0.9}{R_{ex}^2}$.

In general, the ratio of radial (compensating) stiffness and torsional stiffness of a combination purpose flexible coupling can be represented as

$$\frac{k_{com}}{k_{tor}} = \frac{A}{R_{ex}^2}, \qquad (A5.19)$$

where the "Coupling Design Index" A allows one to select a coupling design better suited to a specific application. If the main purpose is to reduce misalignment-caused loading of the connected shafts and their bearings for a given value of torsional stiffness, then the least value of A is the best, together with a large external radius. If the main purpose is to modify the dynamic characteristics of the transmission, then minimization of k_{tor} is important.

A5.5.3 Comparison of Combination Coupling Designs

The bulk of designs of torsionally flexible or combination purpose couplings employ elastomeric (rubber) flexible elements. Couplings with metal springs possess the advantages of being more durable and of having characteristics less dependent on frequency and amplitude of torsional vibrations. However, they may have a larger number of parts and higher cost, especially for smaller sizes. As a result, couplings with metal flexible elements, as of now, have found their main applications in large transmissions, usually for rated torques 1000 Nm and up. Use of the modified spider coupling in Fig. A5.8 may change this situation.

Couplings with elastomeric flexible elements can be classified in two subgroups:

(a) Couplings in which the flexible element contacts each hub along a continuous surface (shear couplings as in Fig. A5.9a, toroidal shell couplings, couplings with a solid rubber disc/cone, etc.). Usually, torque transmission in these couplings is accommodated by shear deformation of rubber;

(b) Couplings in which the flexible element consists of several independent or interconnected sections (rubber disk and finger sleeve couplings as in Fig. A5.11c, spider couplings as in Fig. A5.11a,b, couplings with rubber blocks, etc.). Usually, torque transmission in these couplings is accommodated largely by compression or "squeeze" of rubber, thus they are usually smaller for a given rated torque.

Comparative evaluation of the commercially available couplings based on available manufacturer-supplied data on flexible couplings is presented in Fig. A5.13 [2]. Plots in Fig. A5.13a to d give data on torsional stiffness k_{tor}, radial stiffness k_{rad}, external diameter D_{ex}, and design index A.

The "modified spider" coupling in Fig. A5.11b is different from the conventional spider coupling shown schematically in Fig. A5.11a by four features: its legs are tapered, instead of uniform width, and made thicker even in the smallest cross section, at the expense of reduced thickness of protrusions on the hubs; lips 1 on the edges provide additional space for bulging of the rubber when the legs are compressed; the spider is made of a very soft rubber. These features substantially reduce stiffness values while retaining small size characteristic of the spider couplings.

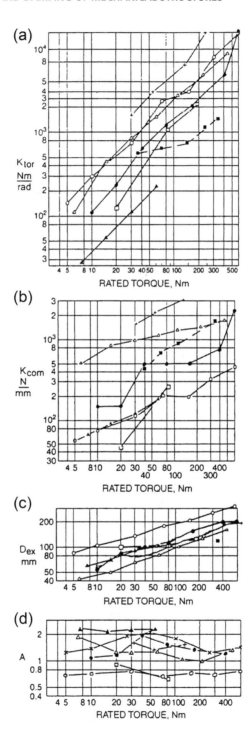

Data for "toroid shell" couplings in Fig. A5.13 are for the coupling as shown in Fig. A5.10d.

The "spider coupling" for $T_r = 7$ Nm has the number of legs $Z = 4$, while larger sizes have $Z = 6$ or 8. This explains differences in A ($A = 1.96$, close to theoretical 1.8, for $Z = 4$; $A = 0.98$ to 1.28, close to theoretical 1.15-1.25, for $Z = 6, 8$).

Values of A are quite consistent for a given type of coupling. Some variations can be explained by differences in design proportions and rubber blends between the sizes.

Plots in Fig. A5.13 help to select a coupling type best suited for a particular application, but do not address issues of damping and nonlinearity. Damping can be easily modified by proper selection of the elastomer. High damping is beneficial for transmission dynamics, and may even reduce thermal exposure of the coupling.

A coupling with hardening nonlinear characteristic may have high torsional compliance for the most frequently used sub-rated (fractional) loading in a relatively small coupling. Accordingly, the misalignment-compensating properties of a highly nonlinear coupling would be superior at fractional loads. The coupling in Fig. A5.14 [13] employs radially compressed rubber cylinders 117 for torque transmission in one direction and 118 – for the opposite direction, between hubs 111 and 113 attached to the connected shafts. This design combines the desirable non-linearity with a significantly smaller size for a given T_r. The coupling size is reduced due to the use of multiple cylindrical elements with the same relative compression in each space between protruding blades 112, 114, and due to high allowable compression of the rubber cylinders [14] and Section 3.3.2, thus allowing use of smaller diameters and, consequently, more sets of cylinders around the circumference). Test results for such coupling for $T_r = 350$ Nm are shown as ■ in Fig. A6.15; in this case the data does not refer to different T_r, but to the same coupling at different transmitted torques.

Fig. A5.13. Basic characteristics of some combination purpose couplings; (a) Torsional stiffness; (b) Radial stiffness; (c) External diameter; (d) Coupling design index. ▲— spider coupling in Fig. A6.10b; △ — spider coupling with straight rectangular legs; +— finger sleeve coupling; o — toroid shell (tire) coupling; □ — rubber disc coupling; ● — Centaflex coupling; ■ — test results for coupling in Fig. A5.14 (same coupling at various transmitted torque).

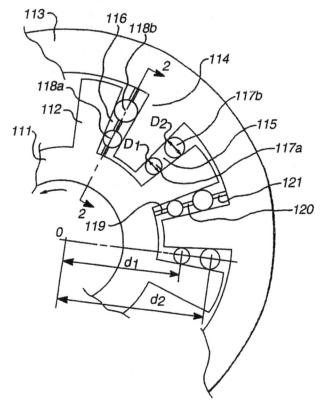

Fig. A5.14. Torsionally flexible coupling with flexible element composed of radially compressed rubber cylinders.

REFERENCES

[1] *Proceedings of International Conference on Flexible Couplings for High Powers and Speeds*, University of Sussex, England, 1977.

[2] Rivin, E.I., 1986, "Design and Application Criteria for Connecting Couplings," *ASME J. of Mechanisms, Transmissions and Automation in Design*, Vol. 108, No. 1, pp. 96–105.

[3] Ettles, C., Wells, D.E., Stokes, M., and Matthews, J. C., 1974. "Investigation of Bearing Misalignment Problems in a 500 MW Turbo-Generator Set," *Proc. of the Inst. of Mechanical Engineers*, Vol. 188, No. 35/74,

[4] "AGMA Standard Nomenclature for Flexible couplings," No. 510.02, AGMA, 1969.

[5] Calistrat, M.M., 1980, "Gear Couplings," *Wear Control Handbook*, Peterson, M.B. and Winer, W.O., eds., ASME, pp. 831–841.

[6] Pleeck, G., 1977, "Noise Control in the Turbine Room of a Power Station," *Noise Control Engineering*, Vol. 8, No. 3, pp.131–136.

[7] Crease, A.B., 1977. "Forces Generated by Gear Couplings," in *Proceed of International Conference on Flexible Couplings for High Powers and Speeds*, University of Sussex, England.

[8] Yampolskii, M.D., Palchenko, V.I., Gordon, E. Ya., 1976, "Dynamics of Rotors Connected with a Gear Coupling," *Mashinovedenie*, No. 5, pp. 29–34 [in Russian].

[9] Rivin, E.I., "Torsionally Rigid Misalignment Compensating Coupling", U.S. Patent 5,595,540.

[10] DeSouza, V., 1994, "Prototype Development and Performance Characteristics of a Novel Misalignment-Compensating Coupling", MS Thesis, Wayne State University.

[11] Rivin, E.I., "Universal Cardan Joint with Elastomeric Bearings", U.S. Patent 6,926,611.

[12] Rivin, E.I., "Spider Coupling", U.S. Patent 6,733,393.

[13] Rivin, E.I., "Torsional Connection with Radially Spaced Multiple Flexible Elements", U.S. Patent 5,630,758.

[14] Rivin, E.I., 1999, "Shaped Elastomeric Components for Vibration Control Devices", S)V *Sound and Vibration*, vol. 33, No. 7, pp. 18–23.

APPENDIX

6

Systems with Multiple Load-Carrying Components

A6.1 INTRODUCTION

There are numerous cases when the loads acting on a mechanical system are equally or unequally distributed between multiple load-carrying components. In these statically indeterminate systems evaluation of the overall stiffness becomes more involved. Several systems described below are typical examples of the systems with multiple load-carrying components. Understanding of these systems may help in analyzing other systems with multiple load-carrying components.

A6.2 LOAD DISTRIBUTION BETWEEN ROLLING BODIES AND STIFFNESS OF ANTIFRICTION BEARINGS

Radial load R applied to an antifriction bearing is distributed non-uniformly between the rolling bodies, Fig. A6.1 [1]. In the following, all rolling bodies are called "balls", although all the results would also apply to roller bearings. If the bearing is not preloaded, the load is applied only to the balls located within the arc not exceeding 180 deg. The most loaded ball is the central ball relative to the radial force, R. The problem of evaluating loads acting on each rolling body is a statically indeterminate one. The balls symmetrically located relative to the vector of action of R are equally loaded. The force acting on the most highly loaded central ball is P_o, the force acting on the ball located at one pitch angle γ to the load vector is P_1, the ball located at $2\gamma - P_2$, at angle $n\gamma - P_n$, etc. Here n is one half of the number of balls located in the loaded zone. It is assumed, for simplicity of the derivation, that the balls are symmetrical relative to the force R. The equilibrium condition requires that the force, R, is balanced by reactions of the loaded balls, or

$$R = P_o + 2P_1 \cos \gamma + 2P_2 \cos 2\gamma + \ldots + 2P_n \cos n\gamma. \qquad (A6.1)$$

In addition to Eq. (A6.1) describing static equilibrium, equations for deformations should be used in order to solve this statically indeterminate problem. Since the outer race is supported by the housing and the inner race by the shaft, their bending deformations can be neglected. It is assumed that there is no radial clearance in the bearing. In such case, it can be further assumed that the radial displacements between the races at each ball location, caused by contact deformations of the balls and the

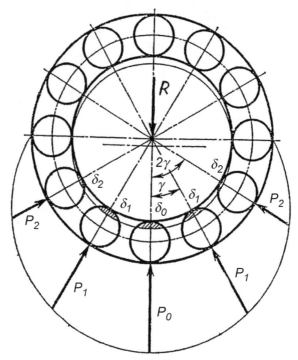

Fig. A6.1. Model of load distribution between rolling bodies in antifriction bearing.

races, are equal to projections of the total displacement, δ_o, of the inner race along the direction of R on the respective radii,

$$\delta_1 = \delta_o \cos \gamma; \qquad \delta_2 = \delta_o \cos 2\gamma; \qquad \delta_i = \delta_o \cos i\gamma, \qquad (A6.2)$$

where i = the number of a ball.

The total contact deformation, δ, of a ball and the races under force, P, is (see Chapter 4)

$$\delta = cP^{2/3}, \qquad (A6.3)$$

where c is a constant.

Substituting expressions (A6.3) into (A6.2), arrive to

$$P_1 = P_o \cos^{3/2} \gamma, \quad P_2 = P_o \cos^{3/2} 2\gamma, \quad \dots \quad P_i = P_o \cos^{3/2} i\gamma \qquad (A6.4)$$

Substitution of Eq. (A6.4) into equilibrium Eq. (A6.1) gives:

$$R = P_o \left(1 + 2 \sum_{i=1}^{i=n} \cos^{5/2} i\gamma \right) \qquad (A6.5)$$

From this equation, P_o can be determined. If a coefficient k is introduced,

$$k = \frac{z}{1 + 2\sum_{i=1}^{i=n} \cos^{5/2} i\gamma} \tag{A6.6}$$

then

$$P_o = kR/z, \tag{A6.7}$$

where z = the total number of balls. If P_o is known, radial deformation and stiffness of the bearing can be determined by using Eq. (A6.2).

For bearings with the number of balls $z = 10$ to 20, $k = 4.37 \pm 0.01$. If the bearing has a clearance, then the radial load is accommodated by balls located along an arc lesser than 180 deg., which results in the load, P_o, on the most loaded ball being about 10% higher than given by Eq. (A6.7). Because of this, for single row radial ball bearings it is assumed that $k = 5$ and $P_o = 5R/z$.

In spherical double-row ball bearings there is always some nonuniformity of radial load distribution between the rows of balls. To take it into consideration, it is usually assumed that $P_o = 6R/z\cos\alpha$, where α = angle of tilt of contact normals between the balls and the races.

Radial force on each ball for the radial loading in angular contact ball bearings is greater than in radial ball bearings by a factor $1/\cos\beta$, where β = the contact angle between the balls and the races.

For roller bearings the solution is similar, but contact deformations of the rollers and the races are approximately linear,

$$\delta = c_1 P, \tag{A6.8}$$

where c_1 is a constant (see Chapter 4). Similarly to ball bearings, for roller bearings

$$P_o = kR/z \tag{A6.9}$$

But, obviously,

$$k = \frac{z}{1 + 2\sum_{i=1}^{i=n} \cos^2 i\gamma} \tag{A6.10}$$

For roller bearings having $z = 10 - 20$, average value of $k = 4$; considering radial clearance, it should be increased to $k = 4.6$. For double-row bearings $k = 5.2$ to take into consideration nonuniformity of load distribution between the rows of rollers.

Load distribution between the rolling bodies can be made more uniform (thus reducing P_o and δ_o, i.e. *enhancing stiffness*) by modification of the bore in the housing into which the bearing is fit. The bore should be shaped as an elliptical cylinder elongated in the prevailing direction of the radial load.

The above derivations were performed with an assumption that all rolling bodies have the same dimensions. Inevitable dimensional scatter between the rolling bodies in one bearing may significantly alter the load distribution. Fig. A6.2 [2] compares the load distribution for a taper roller bearing for the cases when dimensions of all rollers are identical ("Optimal stress distribution", plot 1) and the case when two rollers are oversized ("+") and three rollers are undersized ("--"), plot 2.

If the bearing is preloaded, each rolling body is loaded by a radial force P_{pr} caused by the preload, even before the radial load, R, is applied. As a result, *all rolling bodies* (along the 360 deg. arc) are participating in the loading process, with the bodies in the lower half of Fig. A6.1 experiencing increase of their radial loading, and the bodies in the upper half

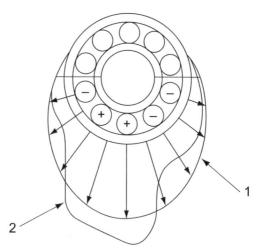

Fig. A6.2. Influence of dimensional variations between rollers on load distribution on the outer race; 1 — high precision rollers; 2 — less precision rollers (oversized rollers "+", undersized rollers "–").

experiencing reduction of their radial loading. The process is very similar to the preloading process of a belt drive, see **Section 3.4** above. Such pattern of the load sharing will continue until at least one of the rolling bodies in the upper 180 deg. arc becomes unloaded. The deformation of a preloaded bearing under a radial force R can be analyzed using the same approach as for not preloaded bearings above, but derivations become more complex since the expression for deformation of the ball caused by forces acting on each bearing becomes, instead of Eq. (A6.3),

$$\delta_i = c \, (P_{\mathrm{pr}} \pm P_i)^{2/3} \qquad \text{(A6.11)}$$

A6.3 LOADING OF SPOKED WHEELS

The spoked bicycle wheel is subjected to high static (weight of the rider) and dynamic (inertia forces on road bumps; torques for acceleration, braking and traction; etc.) loads [3]. Its predecessor — a wagon wheel, Fig. A6.3 — had relatively strong wooden spokes and rim. The spokes in the lower part of the wheel accommodate (by compression) the loads transmitted to them by the rim and the hub, just like bearings discussed

Fig. A6.3. Spoked wagon wheel.

above. The bike's wheel must be much lighter, thus wood was replaced by high strength metals which allowed to dramatically reduce cross sections of both the spokes and the rim. In fact, the spokes possess necessary strength while having very small cross sections equivalent to thin wire (Fig. A6.4). Consequently, such thin spokes cannot withstand compression due to buckling. The solution was found in prestressing the spokes in tension, so that the tensile preload force on each spoke is greater than the highest compressive force to be applied to the spoke during ride conditions. With such a design, the spoke would never lose its bending stiffness if the specified loads are not exceeded. However, the lateral stiffness of the spoke is decreasing when the tension is reduced by high radial compressive forces. In such condition, an excessive lateral force, e.g. caused by turning of the bike, may lead to collapse of the wheel.

Contrary to the bearing races which were discussed above, the rim is a relatively compliant member. Since the total force applied to the rim by the prestressed spokes can be as great as 5000 N (1100 lb), the rim is noticeably compressing, thus reducing the effective preload forces on

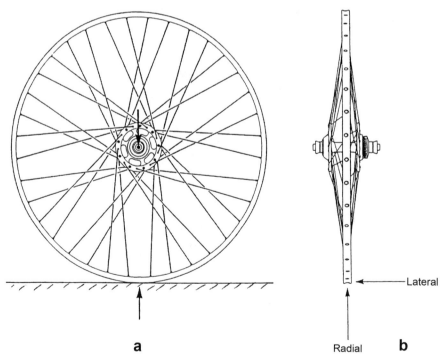

Fig. A6.4. Spoked bicycle wheel; a Front view; b Side view.

the spokes. Pressurized tires also apply compressive pressures to the rim, equivalent to as much as 7% to 15% of the spoke load. The tension of the spokes is changing due to driving and braking torques which cause significant pulling and pushing forces on the oppositely located spokes. Combination of the vertical load with torque-induced pushing and pulling loads results in local changes in spoke tension which appear as waves on the rim circumference. These effects are amplified by the spoke design as shown in Fig. A6.4: to enhance the torsional stiffness, the spokes are installed not radially but somewhat tangentially to the hub, Fig. A6.4a; to enhance lateral stiffness and stability of the wheel, the spokes are installed in a frusto-conical manner, not in a single plane, Fig. A6.4b.

When a vertical load is applied to the wheel hub, the spokes in the lower part of the wheel are compressed (i.e., their tension is reduced). The spokes in the upper half of the wheel are additionally stretched, but tension is increasing even in the spokes in the mid-section of the wheel,

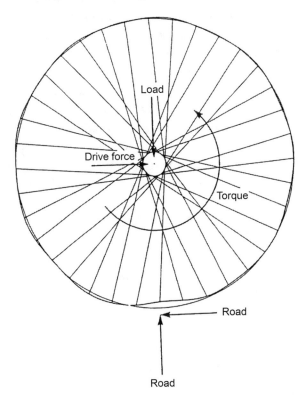

Fig. A6.5. Distortion of bicycle wheel rim under radial and torque loads.

since the rim becomes somewhat oval. This ovality is not very significant, the *increase of the horizontal diameter* is about 4% of the deformation at the contact with the road, but it has to be considered. Since the rim is not rigid, its flattening at the contact with the road leads to reduction of effective stiffness of the wheel.

It is important to note that performance loads (radial, torque, braking and turning loads) cause significant distortion of the rim (e.g., see Fig. A6.5) which in its turn results in very substantial deviation from load distribution for an idealized model, e.g., like the bearing model in Fig. A6.1.

A6.4 ANALYTICAL SOLUTION FOR BICYCLE WHEEL

While the problem of load distribution in and deformations of the real life spoked bicycle wheel is very complex, there is a close form solution for load and bending moment distribution along the wheel rim and for rim deformations with some simplifying assumptions [4]. The assumptions: the road surface is flat and rigid; the spokes are radial and co-planar with the rim; the number of spokes n is so large that they can be considered as a continuous uniform disc of the equivalent radial stiffness. Then the problem becomes one of a radially loaded ring with elastic internal disc. Every point of the rim would experience a radial reaction force from the spokes proportional to deflection w at this point. There are $n/(2pR)$ spokes per unit length of the rim circumference (R = radius of the wheel). Deformation w and bending moment M along the rim circumference due to vertical force, P, acting on the wheel from the road are:

$$w = \frac{PR^3}{4\alpha\beta E_1 I}\left(\frac{2\alpha\beta}{\pi a^2} - A\cosh\alpha\phi\cos\beta\phi + B\sinh\alpha\phi\sin\beta\phi\right) \quad \text{(A6.12)}$$

$$M = -\frac{PR}{2}\left(\frac{1}{\pi a^2} + A\sinh\alpha\phi\sin\beta\phi + B\cosh\alpha\phi\cos\beta\phi\right) \quad \text{(A6.13)}$$

The force acting on a spoke is, obviously,

$$P_s = \frac{E_2 F}{R}w. \quad \text{(A6.14)}$$

Here E_1 and I are Young's modulus and cross-sectional moment of inertia of the rim, E_2, F, and $l \approx R$ = Young's modulus, cross-sectional area,

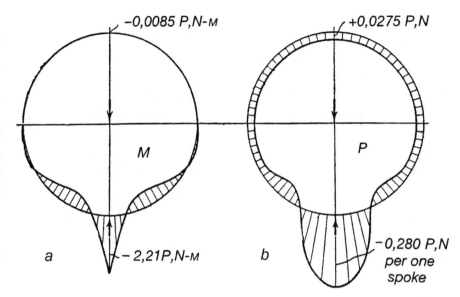

Fig. A6.6. Calculated distribution of (a) bending moment and (b) radial forces along the circumference of a bicycle wheel.

and length of the spoke; $a = \dfrac{R^2 n}{2\pi} \dfrac{E_2 F}{E_1 I}$; $\alpha = \sqrt{\dfrac{a-1}{2}}$; $\beta = \sqrt{\dfrac{a+1}{2}}$; angular coordinate f of a point is counted from the top point of the wheel. Formulas (A6.12) to (A6.14) allow one to evaluate , at least in the first approximation, influence of various geometric and material parameters on force and bending moment distributions. Fig. A6.6 shows these distributions for $R = 310$ mm; $I = 3000$ mm^4; $n = 36$; spoke diameter $d = 2$ mm ($F = 3.14$ mm^2); $E_1 = E_2 = 2 * 10^5$ N/mm^2 (steel). After the maximum static and dynamic forces on the wheel are estimated/measured, the data from Fig. A6.6b can be used to specify the necessary initial tension of each spoke P_t. The tension should be safely greater than the maximum possible compressive force acting on the spoke, which for listed parameters is $0.280 P_{max}$.

A6.5 TORSIONAL SYSTEMS WITH MULTIPLE LOAD-CARRYING CONNECTIONS

Such connections are typical for various types of power transmission couplings. Many coupling designs have hubs attached to the shafts being connected by the coupling, with the driving hub transmitting torque to the driven hub by means of multiple identical radially placed flexible elements. Frequently these elements are designed as one integral flexible member. Each flexible element which is radially placed along the misalignment vector, transmits its share of the torque by compression, while compensating inevitable radial misalignments between the connected shafts by shear. A typical representative of such couplings is so-called "jaw" or "spider" coupling whose cross section is shown in Fig. A6.7 [5]. The coupling in Fig. A6.7 consists of hubs 1 and 2 connected with a rubber spider 3 having an even number $Z = 2n$ of legs, with "n" legs ("n" might be odd) loaded when hubs are rotating in forward direction and the other n legs loaded during reverse rotation. Deformation of each leg is independent. Assuming that the radial misalignment, e, of the coupled shafts is in Y direction, then the following relationships exist for the i-th leg:

$$e_{ti} = e \cos a_i; \qquad e_{ri} = e \sin a_i;$$
$$F_{ti} = k_t e_{ti} = k_t e \cos a_i; \; F_{ri} = k_r e_{ri} = k_r e \sin a_i;$$

$$F_{xi} = -F_{ti} \sin a_i + F_{ri} \cos a_i = k_t e \sin a_i \cos a_i$$
$$+ k_r e \sin a_i \cos a_i = e(-k_t + k_r)\sin a_i \cos a_i \qquad (A6.15)$$

$$F_{yi} = F_{ti} \cos a_i + F_{ri} \sin a_i = k_t e \cos^2 a_i + k_r e \sin^2 a_i$$
$$= e(k_t \cos^2 a_i + k_r \sin^2 a_i),$$

where subscripts t, r denote tangential and radial components, respectively; k_t, k_r is stiffness of a leg in compression (tangential direction) and shear

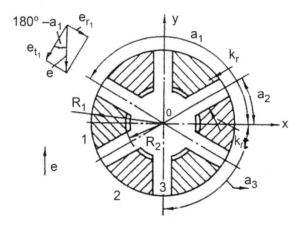

Fig. A6.7. Schematic of a spider coupling; 1 and 2 are hubs, 3 rubber spider.

(radial direction); e_t, e_t are components of deformation of the i-th leg produced by misalignment e; F with subscripts are corresponding components of compensating force from the i-th leg. Overall components of the compensating force in x, y directions are sums of F_{xi} and F_{yi} for all loaded legs. For a four-leg spider (two loaded legs, $n = 2$) $a_1 = a$, $a_2 = a + 180$ deg., and

$$\Sigma\, F_x = 1/2e(k_r - k_t)[\sin 2a + \sin (2a + 360 \text{ deg.})] = e\,(k_r - k_t)\,\sin 2a$$
$$= - k_t\,(1 - k_r/k_t)\,e\,\sin 2a;$$

(A2.16)

$$\Sigma\, F_x = e\{k_t[\cos^2 + \cos^2(a + 180 \text{ deg.})] + k_r\,[\sin^2 a + \sin^2 (a + 180 \text{ deg.})]\}$$
$$= 2k_t\,e[\cos^2 a + (k_r/k_t)\,\sin^2 a];$$

(A2.17)

$$F = \sqrt{(\Sigma F_x)^2 + (\Sigma F_y)^2} = 2k_t e\sqrt{\cos^2 \alpha + \frac{k_r}{k_t}\sin^2 \alpha}, \qquad (A6.18)$$

thus the total radial force fluctuates both in magnitude and in direction. The magnitude of fluctuation decreases with decreasing ratio, k_r/k_t. The compensation (radial) stiffness of the coupling also varies, and is

$$k_{com} = \frac{F}{e} = 2k_t\sqrt{\cos^2 \alpha + \frac{k_r}{k_t}\sin^2 \alpha}. \qquad (A6.19)$$

When $n \geq 3$,

$$\Sigma F_x = e\sum_{k=0}^{n-1}(-k_r + k_t)\sin\left(\alpha_1 + k\frac{360\,\text{deg.}}{n}\right)\cos\left(\alpha_1 + k\frac{360\,\text{deg.}}{n}\right) =$$
$$= \frac{e}{2}(k_r - k_t)\sum_{k=0}^{n-1}\ \sin\left(2\alpha_1 + 2k\frac{360\,\text{deg.}}{n}\right) = 0;$$

(A6.20)

$$\Sigma F_y = e\sum_{k=0}^{n-1}\left[k_r\cos^2\left(\alpha_1 + k\frac{360\,\text{deg.}}{n}\right) + k_t\sin^2\left(\alpha_1 + k\frac{360\,\text{deg.}}{n}\right)\right] =$$
$$= e\sum_{k=0}^{n-1}\left[k_r\frac{1+\cos\left(2\alpha_1 + 2k\frac{360\,\text{deg.}}{n}\right)}{2} + k_t\frac{1-\cos\left(2\alpha_1 + 2k\frac{360\,\text{deg.}}{n}\right)}{2}\right]$$
$$= \frac{n}{2}e(k_r + k_t)$$

(A6.21)

thus for $Z \geq 6$ the total radial force, F, is constant and directed along the misalignment vector,

$$k_{com} = F/e = n/2(k_t + k_r). \tag{A6.22}$$

Since maximum allowable radial misalignments e for conventional spider couplings with rubber flexible elements (and other types of couplings with rubber flexible elements having radial protrusions loaded in compression by the tangential forces) do not exceed 0.007 to 0.01 of the outer coupling diameter, D_o, maximum shear of the radial protrusions (spider legs) does not exceed ~0.03 [5]. Accordingly, the value of shear modulus should be taken from Fig. 2 and ratio, k_r/k_t varies with changing rubber durometer H. For typical spider proportions $k_r/k_t = 0.26$ to 0.3 for medium durometer $H = 40$ to 50, and $k_r/k_t = $ ~0.4 for hard rubber spiders, $H \, k_r/k_t = 70$–75.

Torsional stiffness of the couplings with radial protrusions loaded in compression by the tangential force is [5]

$$k_{tor} = \frac{T}{\varphi} = \frac{F_t R_{eff}}{\Delta / R_{eff}} = \frac{F_t}{\Delta} R_{eff}^2 = n k_t R_{eff}^2 \tag{A6.23}$$

where $T = $ the transmitted torque, $\varphi = $ angular deformation of the flexible element, R_{eff} is the effective radius of the coupling, $R_{eff} = 0.75(D_o/2)$ for the spider coupling with a rubber spider [5].

In many applications the ratio, k_{com} / k_{tor}, can be important. For $Z = 4$, $n = 2$,

$$\frac{k_{com}}{k_{tor}} = \frac{\sqrt{\cos^2 \alpha + \left(\dfrac{k_r}{k_t}\right) \sin^2 \alpha}}{R_{eff}^2} \; ; \quad \left(\frac{k_{com}}{k_{tor}}\right)_{max} = \frac{1}{R_{eff}^2} \approx \frac{1.8}{R_{eff}^2}. \tag{A6.24}$$

For $Z \geq 6, n \geq 3$,

$$\frac{k_{com}}{k_{tor}} = \frac{1 + \dfrac{k_r}{k_t}}{2 R_{eff}^2}. \tag{A6.25}$$

The stiffness ratio of a "combination purpose" flexible coupling, combining both torsional and radial compensation flexibilities, can be expressed as:

$$\frac{k_{com}}{k_{tor}} = \frac{A}{R_o^2}, \tag{A6.26}$$

where R_o = the outer radius of the coupling, and A = "Coupling Design Index," which allows one to select a coupling design better suited for a specific application. If the main purpose is to reduce misalignment-caused loading of the connected shafts and their bearings for a given torsional stiffness, the smallest value of A is the best, together with a large external radius. If the main purpose is to modify dynamic characteristics of the transmission, minimization of k_{tor} for given R_{os} and rated torque is important.

REFERENCES

[1] Reshetov, D.N., 1974, "Machine Elements", Mashinostroenie, Moscow, 656 pp. [in Russian].

[2] SKF, *Evolution Magazine*, 2003, No. 4, p. 30.

[3] Brandt, J., 1995, "The Bicycle Wheel", *Avocet Inc.*, Menlo Park.

[4] Feodosiev, V.I., 2005, *"Advanced Stress and Stability Analysis: Worked Examples"*, Springer.

[5] Rivin, E.I., 1986, "Design and Application Criteria for Connecting Couplings", *ASME J. of Mechanisms, Transmissions, and Automation in Design*, pp. 96–105.

APPENDIX

7

Compliance Breakdown for a Cylindrical (OD) Grinder

Although modal analysis of complex structures is very useful for determining weak links in complex mechanical structures, it can be complemented by measuring static deformations of various components and their connections under loads, simulating the working conditions of the system. Evaluation of the static compliance breakdowns allows us to better understand the role of small but important components, to more readily simulate diverse working conditions/regimes, to detect and study nonlinear deformation characteristics of some components that can confuse the modal analysis procedure, and more. Since measuring the compliance breakdown under static loading is much more time-consuming than the dynamic evaluation, the former should be undertaken only in cases of critical importance or when the nonlinear behavior is strongly suspected. This Appendix is a summary of a detailed analysis of static deformations for a precision cylindrical OD grinder for grinding parts up to 140 mm in diameter and up to 500 mm long [1]. However, this study is rather generic, and its techniques and conclusions can be very useful for other mechanical systems.

The machine is sketched in Fig. A.7.1 and the most stiffness-critical units are identified as follows. Wheelhead 3 houses spindle 1 supported in hydrodynamic bearings 2 and is mounted on a carriage comprising upper 4 and lower 8 housings. Hydraulic cylinder 6 and ball screw 15 with nut 16 are mounted in housing 8. The machined part is supported by headstock 21 and by tailstock (not shown in Fig. A.7.1) installed on angular (upper) table 19, which is attached to longitudinal (lower) table 18 moving in guideways along bed 11.

One end of ball screw 15 is supported by bracket 17 fastened to housing 4; the other end is driving pusher 13. The force between pusher 13 and ball screw 15 is adjusted by compression spring 12 and by threaded plug 10, which are housed in bracket 9 fastened to housing 4. Thus, spring 12 generates preload (up to 1,000 N) in connection: bracket 17 - ballscrew 15.

Wheelhead 3 may perform setup motions along the guideways on housing 4, with its final position secured by set screw 5; fast motion (before and after machining) is executed by piston 7 moving in hydraulic cylinder 6; and feed motion — by ballscrew 15 driven by worm gear 14. During the feed motion, piston 7 is touching the left face of cylinder 6.

The principal contributors to stiffness/compliance breakdown are: spindle 1 in bearings 2; joint between wheelhead 3 and housing 6; joint

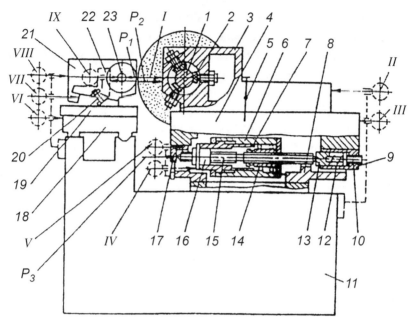

Fig. A7.1. Test forces P_1, P_2, and P_3 and measurement positions/ deformation sensors I-IX for cylindrical (OD) grinder.

between bracket 17 and screw 15; connection between ball screw 15 and ball nut 16; joint between table 18 and guideways; headstock 21 (or tailstock); and joint between supporting center 23 and housing of headstock or tailstock.

In some similar grinders the wheelhead can be installed directly on rolling friction guideways of the lower housing of the carriage, not on the upper housing as shown in Fig. A.7.1. It does 'not change the compliance breakdown; it is shown below that displacement of the wheelhead relative to the upper housing is only about 0.5 µm under 600 N load.

Test forces were applied (through load cells) in several locations (Fig. A.7.1): between non-rotating wheel and part (P_1); between bracket 22 attached to table 19 and wheel head (P_2); and to end face of screw 15 (P_3). The forces were varied in two ranges 0 N to 300 N and 0 N to 600 N. Displacement transducers were located in positions I to IX: I and II measured displacement of wheel head 3 relative to table 19 and bed 11; III measured displacement of housing 4 relative to bed 11; IV measured the joint between end face of nut 16 and housing 8; V measured the contact between end of screw 15 and housing 8; VI and VII measured displacements of tables 18 and 19 relative to bed 11; VIII measured displacement of tailstock (headstock) relative to bed 11; and IX measured displacement of supporting center relative to table 19.

Figure A7.2, shows measuring setups for displacements of the part as well as of the headstock and tailstock and their components. Transducers

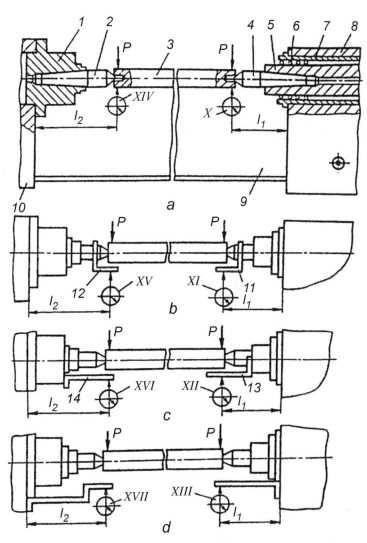

Fig. A7.2. Part support schematics used for measuring deformations of various components under radial load, P; X to XVII are position/deformation sensors.

X to XIII are on the tailstock side and measure displacements of part 3, supporting center 4, holder 5 which is mounted in sleeve 7 on balls 6, and tailstock 8 relative to upper table 9. Transducers XIV to XVII are on the headstock and measure displacement of part 3, supporting center 2, central bushing I, and headstock 10 relative to table 9, respectively.

Stiffness of the part as well as of components of headstock and tailstock were determined under force, P, applied to an end of the part from the headstock or tailstock (12 and 11 are, respectively, arm lengths of load application). To determine influence of stiffness of various components of headstock and tailstock on the total part stiffness relative to table 9, measurements were performed not only in setup of Fig. A.7.2a, but also as shown in Figs. A.7.2b, c, and d. For example, to identify influence on the part displacements of stiffness of the supporting centers relative to table 9, bracket 11 or 12 was attached to the front or rear center, respectively, and the load was applied to the part. Displacements of ends of these brackets were measured at distances l_1 and l_2 from the head-/tailstock. Influence on effective stiffness of the part of stiffness of center holder 5 relative to the table was measured in a similar way. To perform this measurement, bracket 13 was attached to the holder (Fig. A.7.2c) and displacements of the bracket end were recorded.

Supporting centers 2 and 4 have Morse taper #4. Depending on grinding conditions, the centers could be short or long (the cylindrical part is 20 mm longer). Overhang of the center holder of the tailstock was varied during grinding within 20 mm. Distance\ l_1 = 85 mm for the short center and minimum holder overhang; 105 mm for the short center and the maximum overhang as well as for the long center and minimum overhang; and 125 mm for the long center and maximum overhang. Bushing 1 is fixed stationary in the headstock; thus 12 depends only on length of center l_2 = 160 mm for the short center and 180 mm for the long center.

Plots in Fig. A7.3 show displacements δ versus radial load for wheel head, tailstock and headstock, and other components. Spindle stiffness (245 N/μm) was determined from computed stiffness of its hydrodynamic bearings since the measurements were performed without spindle rotation. The broken line in Fig. A7.3a is plotted by adding computed spindle displacements to the measured (transducer II) displacement of the wheelhead. This line represents the total effective compliance of the spindle due to compliance of all components sensing the forces from the

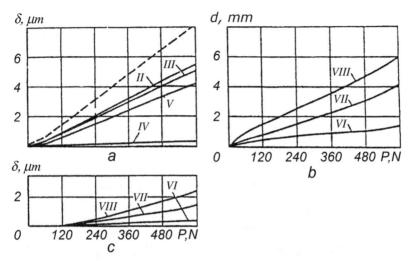

Fig. A7.3. Deformations, δ, μm, of components of (a) – wheelhead, (b) – tailstock, and (c) – headstock relative to bed under radial force, P, at different measuring points. The dashed line is the calculated displacement of spindle.

wheelhead. These plots show that the largest contributor to compliance of the wheelhead is the ball screw/nut transmission (the distance between lines IV and V). The next contributor is spindle (the distance between the broken line and line II), the third is joint screw 15-bracket 17 (the distance between lines V and III). Other displacements/compliances are very small and can be neglected.

Stiffness of the upper and lower tables is different when measured at the tailstock and at the headstock (Fig. A7.3b and c). This is mostly due to manufacturing imperfections of the guideways resulting in their non-uniform fit along the length and thus in stiffness variations. Since the radial force is applied at the centerline level, it also generates angular displacements of the lower table in the guideways in the transverse vertical and horizontal planes. As a result, displacements of tables as well as of the tailstock and headstock become uneven.

Stiffness of the headstock and tailstock depends on the attachment method of each unit to the table. The headstock is fastened by two short bolts 20 (Fig. A7.1). The tailstock is fastened by one long bolt in the middle of its housing. This arrangement simplifies resetting of the tailstock but reduces its stiffness. It can be concluded from lines VIII in Fig. A7.3b and II in Fig. A7.3a that compliance of the tail stock (taking into account also deformations of lower and upper tables) is close to compliance magnitude of the wheelhead. Figure A7.4 gives deformations of components of both headstock and tailstock for various combinations of lengths of the centers and overhang of the sleeve. Lines X to XIII in Figs. A7.4a-d

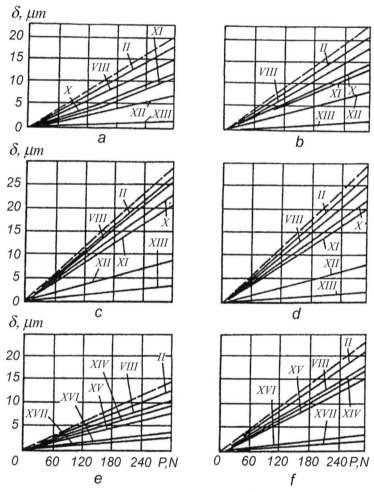

Fig. A7.4. Displacements, δ, μm of components of tailstock (a–d), and headstock (e–f) at: (a) – short support center with minimal sleeve overhang; (b) – short center with maximum overhang; (c) – long center with minimal overhang; (d) – long center with maximum overhang; (e) – short center; and (f) – long center. The dashed line is spindle displacement.

represent deformations of the machined part, supporting center, sleeve, and tailstock, relative to the upper table measured in positions indicated in Figs. A7.2a-d, respectively. Lines VIII in Fig. A7.4a-d show the total displacement of the part in relation to the bed; lines II show displacement of the part in relation to the wheelhead; and the broken lines show displacement of the part in relation to the wheel spindle. Lines XIV to XVII in Figs. A7.4d-e show displacement of the part, the supporting center sleeve, and headstock, respectively, in relation to the upper table.

These plots give an understanding of influence of deformations of each main structural component of the machine tool on the spindle-part deformations under forces up to 300 N. The data in Table A7.1 shows that the most compliant elements of the tailstock are the supporting center and the sleeve. Depending on the length of the center, it is responsible for 21.7% to 39.7% of the total displacement of the spindle, and the sleeve is responsible for 18.8% to 29%.

The tailstock displacement in relation to the upper table (measuring position XIII) and in relation to the bed (position VIII), as well as displacement of the wheelhead in relation to the bed have close magnitudes. The most significant component of the headstock deformation is the supporting center (the short center accounts for 32.8%, the long center for 54.3% of the total deformation of the tailstock spindle). The spindle displacement accounts for 29.8% of the total displacement with the short center and 18.8% with the long center.

Table A7.1. Compliance Breakdown of OD Grinder.

Set up Fig. A7.4	Spindle displacement, %	Contribution of unit deformation, %									
		II	VIII	X	XI	XII	XIII	XIV	XV	XVI	XVII
a	7.7	14.2 26.0	17.4	3.2	21.7		9.8				
b	6.8	12.5 29.0	15.5	4.8	21.7		9.7				
c	4.6	9.4 11.1	11.2	5.2	39.7	18.8					
d	4.4	9.1 12.1	10.7	6.0	38.9	18.8					
e	9.6	20.1	5.2	-	-		-	11.1	32.8	2.5	18.7
f	6.1	12.7	3.3	-	-		-	8.5	54.3	1.4	13.7

The plots in Fig. A7.4 indicate that stiffness of the same machine tool (not considering deformations of the machined part) varies, for loading forces to 300 N, from 16.3 N/μm to 10.1 N/μm at the headstock.

Thus, the most significant component in the compliance breakdown is compliance of the supporting centers. Displacement of the loaded center is:

$$\delta = \delta_b + \delta_c + \delta_a \tag{A7.1}$$

where δ_b = bending deformation; δ_c = radial displacement due to contact deformations; and δ_a. = displacement at the load application point due to angular contact deformations. From [1]:

$$\delta = P(Ak^{0.5} + Bk^{0.71}); \qquad \delta = P(Ck^{0.25} + Dk^{0.5}) \tag{A7.2}$$

where P = radial force; A, B, C, and D = constants depending on the Young's modulus of the center and on the sleeve taper diameter, and distance from the load application point to the sleeve face; k = parameter depending on quality of the tapered connection sleeve = 0.1 μm mm^2/N to 1.0 μm mm^2/N.

For cases illustrated in Fig. A7.4 at the force 300 N, δ_b = 1.6 μm for the short center and δ_b = 4.2 μm for the long center. Separate tests performed on the centers measured $\delta_c + \delta_a$ = 3 μm to 4. μm for the short center and 5.8 μm to 6.8 μm for the long center, and k = 0.4 μm mm^2/N to 0.8 μm mm^2/N. Additional tests on other machine tools and on other supporting centers resulted in values k = 0.4 μm mm^2/N to 1.2 μm mm^2/N. Displacements of the wheelhead relative to the upper table (measuring position I) under the force from bracket 22 to wheel head 3 (Fig. A7.1) are close to displacement of the wheelhead relative to the bed (line II in Fig. A7.3a) and to displacement of the tailstock relative to the bed (line VIII in Fig. A7.3b).

For enhancement of the total stiffness of the grinder, the most important are stiffness values of the supporting centers, the sleeve, and the tailstock, as can be seen from Fig. A7.4. Reducing values of parameter k can enhance stiffness of the centers. This can be achieved by improving the fit in the Morse taper connection between the center and the sleeve or by preloading this connection. The contact area in the tapered connections of the grinder on which the tests were performed was about 70% to 80% of the nominal contact area; it was not possible to obtain $k < 0.4$

μm mm^2/N. Mutual lapping of the tapered connections was out of consideration since the centers have to be frequently replaced depending on the grinding conditions. Axial preloading of the tapered connection resulted in bulging of the sleeve (as in Fig. 4.32) and in over preloading of the balls guiding the sleeve motion. To maintain accessibility of the grinding wheel to the part, the cylindrical part and the 60 deg. "center part" of the supporting center can be made like in the standard Morse taper #4, while the tapered seat should be dimensioned as the standard Morse taper #5. Such arrangement would reduce $\delta_c + \delta_a$ up to 2.5 times.

An analysis using data on contact deformations (see Chapter 4) has shown that 20% to 30% (depending on the overhang) of the sleeve deformation is due to compliance of the balls guiding the sleeve in the holder. The other 80% to 70% is due to bending of the sleeve inside the ball bushing. Reduction of the bending deformation can be achieved by increasing diameter of the sleeve or the number of guiding balls, or by reduction of the overhang. The first approach is unacceptable since the tailstock would interfere with the wheelhead in case of grinding of a tapered part between the centers with angular displacement of the upper table in relation to the lower table. The third approach cannot be realized since the overhang in the present design is the minimum acceptable one. Thus, some reduction of deformations and the resulting stiffness enhancement can be achieved by increasing the number of balls and by their optimal packaging. Another approach, using rollers instead of balls, would enhance stiffness but would result in a significantly more complex and costly system.

Stiffness of the tailstock could be enhanced by using stiffer attachment to the table, e.g., by using two bolts instead of one. However, it would lengthen the setup time, which might not be desirable.

REFERENCE

[1] Marcinkyavichus, A.-G. Yu, 1991, "Study of stiffness of cylindrical OD grinders," Stanki i instrument, No. 2, pp. 2–4 [in Russian].

ABOUT THE AUTHOR

Eugene Rivin (Sc.D. 1972, PhD 1962, PE) spent three years in a manufacturing plant, then joined the Experimental Research Institute for Machine Tools in Moscow (ENIMS). In 1968, he formed the Vibration Control and Advanced Machine Elements Lab in the Research Institute for Standardization.

Dr. Rivin emigrated from the USSR in 1975; in 1976, he joined the Research Staff of Ford Motor Co. as Principal Staff Engineer responsible for Plant Noise Abatement Technology and NVH Reduction in Light Weight Vehicles Projects. Since 1981, Dr. Rivin has been Professor at Wayne State University. In 1987 he spent four months with the Institute for Sound and Vibration Research of the University of Southhampton (UK), and in 1988 Dr. Rivin worked for six months with the Advanced Engineering Laboratory of GM Corp. In 1994-95, he led a successful NVH - related project at Ford Motor Co.

His major professional achievements include: algorithms for constructing computational models for multi-DOF driveline systems; reduction of DOF without distorting natural frequencies and modes in the specified frequency range, analysis of transient overloads, etc.; vibration isolation (theory and widely used means for isolating precision and vibration-producing objects); vibration/noise control; machine tools/tooling; robotics (mechanical design, trajectory measuring, correcting, and simulation); advanced machine elements; and creative problem solving.

Dr. Rivin has published 15 monographs and book chapters, as well as 150+ articles. His latest books are: *Mechanical Design of Robots* (1988); *Stiffness and Damping in Mechanical Design* (1999); *Passive Vibration Isolation* (2003); *Innovation on Demand* (2005, with Victor Fey). Dr. Rivin has 60+ patents (26 US Patents); some widely implemented in the US, former USSR, and worldwide. He is a co-founder of The TRIZ Group, actively involved in TRIZ training and problem solving for US and European companies. TRIZ is a Russian acronym for the Theory of Inventive Problem Solving.

Dr. Rivin is Fellow of The International Academy of Production Engineering (CIRP), of ASME, and of SME; he is a past Member of ASME Publications Committee.

INDEX

Note: Italicized page numbers refer to figures and tables

7/24 taper connections, 498–512
  definition of problem in, 498–500
  standard connection, 500–502
  toolholder/spindle interfaces, 502–518
    alternative designs, 503–510
    giant superelasticity effect, 510–512
    requirements, 502–503

A

accuracy enhancement factor, 39
active systems, 389–399. *See also* stiffness enhancement techniques
  antivibration devices, 390, 391*f*
  axis-alignment for boring bar, 390, 390*f*
  chatter resistance, 391*f*, 392
  comprehensive active installation system, 397–399
  deflection correction for spindle head, 392–393, 393*f*, 394*f*
  hydraulic actuators in, 389–390
  hydraulic vs. piezoelectric actuators, 393–394
  laser-based active systems, 390–392

  vibration isolators, 396–397
  vibration suppression in surface grinder, 394–395
alumina, 45*t*
aluminum, 49*t*
angular deflection, 189–190
angular stiffness, 104, 105, 108–109, 190–191
anisotropic elastic elements, 541–560
  elastic kinematic connections, 543–547
  for rotational motion, 543–544
  elastic transformers, 546–547
  for translational motion, 545–546
  overview, 541–543
  thin-layered rubber-metal laminates, 548–560
anisotropic stiffness, 527
antagonist actuators, 154–155
arches, 58*f*
assembled structures, 161–174
  dynamics of, 172–174
  frequency response function, 173
  integrity of, 161–166
    beam, 161–174
    preloaded structure, 163–166
  stiffness of, 166–172
axial clamping, 494
axial compression forces, 17

axial displacement, 232*f*
axial force, 609–610
axial shear stiffness, 108, 109
axial tension, 412–416

B

balanced forces, 282–283
ball bearings, computational
  models of shafts, 262*f*
ball screws, 307–310
  antifriction design, 307*f*
  axial compliance of, 308
  cross-sectional profiles, 307
  gothic arch profile, 307–308
  non-preloaded screw, 308–309
  preloaded screw, 310
  semicircular profile, 308
  stiffness of, 307–310
  thread profiles, 307–308, 308*f*
ball-spline connection, 299–302,
  301*f*
beam, assembled, 161–174
beam design, 64–72
  composite/honeycomb beams
    and plates, 70–72
  helically patterned tubular
    beams, 67–69
  round vs. rectangular cross
    section, 65–67
beam vibrations, axial force and,
  609–610
bearings, antifriction, 353–365
  axial compliance factor, 355
  axial stiffness, 359
  ball bearings, 152
  centrifugal forces, 357–358
  contact angle, 356–357

damping in, 360–364
elastic displacement of shafts
  in, 151
Hertzian deformations in,
  152
mutual elastic approaching of
  races, 355
preloading of, 151–154
radial compliance coefficient,
  355, 356*f*
radial stiffness of, 353*f*,
  354–355
roller bearings, 152
stiffness of, 681–685
bearings, compliant, 531–534
"bell mouthing", 209
Bellevile springs, 124–126
belt drives
  compliance of, 305
  preloading of, 146–149
belt-block stick-slip, 20*f*
bending deflections, 55
bending moment, 16–17
bending stiffness, 16–17, 285
beryllium, 45*t*, 46–47
beryllium copper, 46*t*
Bessel's points, 263
Bibby coupling, 665*f*
bicycle wheels, load distribution
  in, 688–689
bi-material effect, 291
bolted joints
  components of, 167*f*
  test setup for, 166*f*
  tested position relative to base,
    168*f*
boring bar, with high tangential
  compliance, 466*f*

boron, 45*t*, 46*t*
boron carbide, 45*t*
brass, 46*t*
buckling, 16, 402–405, 539, 540*f*
bulging, 131–135
bulk modulus, 618–619
bushings, 106–107
butyl elastomers, 615–616

C

cantilever
    inverse, 259*f*
    reducing reaction forces in,
        259*f*
cantilever beam
    stiffness change of, 16*f*
    stresses and deflections in, 58*f*
cantilever components, 427–440
    combination link, 435–440
    rotating around transverse
        axis, 433–439
    solid component, 433–435
    stationary/rotating around
        longitudinal axis, 430–431
cantilever loading, 379*f*
cantilever shafts, alternative
    designs of, 378*f*
cantilever tools, clamping devices.
    See also "clamping devices" for,
    475–481
carbide cutting inserts, 460
carbon-fiber reinforced plastic
    (CFRP), 47
Cartesian frame manipulator, 21*f*
    dynamic path deviations,
        24–25*f*
cast iron, 46*t*, 291

CAT connection, 222–223
Centaflex coupling, 669*f*
C-frames
    deformations of, 375*f*
    schematics of, 374*f*
    structural optimization of,
        373–374
chain drives
    compliance of, 306
    preloading of, 149–151
chain sprocket with compliant
    teeth, 531*f*
chain transmission, stiffness
    reduction of, 530–531
chatter resistance, 14, 596–598
chatter stability, 5
circumferential displacement,
    231*f*, 234*f*
clamping devices, 481–492
    axial clamping in, 494
    collet chuck, 483–487
    general purpose, 482–487
    overview, 481–482
    performance characteristics of,
        482
    solid state, 487–492
    power shrinking toolholders,
        488–491
    thermal shrink-fit toolbars,
        488
    Tribos chuck, 489–490, 491*f*
    stiffness of, 481–492
    Weldon clamp, 481–482, 481*f*,
        482–483
clearance-fit cylindrical
    connections, 194–202
CNC controller, 413
CNC machine tools, 458

CNF vibration isolator, 133–134
  deformation patterns, 134*f*
  load-natural frequency of, 134*f*
coil springs, 86–92. *See also*
  springs
  compression, 87
  correction factor, 87
  deflection, 91
  equivalent normal stress, 91–92
  extension, 87
  natural frequencies, 90
  nonlinear stiffness in, 122–124
  shear stress, 86–87
  spring index, 86
  torsional stiffness of, 90
collet chucks, 483–487
combination purpose couplings,
  646, 668–669*f*, 670–675
  comparison of, 673–675
  spider coupling, 670–672
compensation, passive, 380–389
  forward compensation,
    382–383
  in front spindle bearing,
    388–389
  hydraulic counterbalancing in,
    386–388
  opposing bending moments in,
    384*f*
  in radial drill press, 380–381
  in thread-rolling machine,
    380–381
  in vertical boring mill, 382
  weight counterbalancing in,
    385–386, 385*f*
compliance, 6
  calculation of, 302–305
  generalized, 320

compliance breakdowns, 329–348
  electromechanical jointed
    robot, 335–339
  electromechanical
    parallelogram robot,
    339–342
  electromechanical spherical
    frame robot, 342–347
  hydraulically driven robot,
    330–335
compliance reduction coefficient,
  303*f*
compliant bearings, 531–534
composite beams and plates,
  70–72
composite materials, 46*t*
compressed liquid, compliance of,
  313
compressed rubber sphere,
  deformation characteristics of,
  141
compression springs, 87
compression stiffness, 105, 107,
  109–110
conformal gears, 528
conforming surfaces, contact
  deformations between,
  183–186
conical connections, 203–223.
  See also contact deformations
  axial preload, 205–207
  CAT connection, 222–223
  computational evaluation,
    210–213
  damping of, 249–250
  elastic displacements in,
    203–204
  finite element model, 217–218

load-angular deflection, 205
load-deflection characteristics,
    208
manufacturing deviations,
    207–208
manufacturing errors, 210–217
Morse taper connections,
    221–222
preloaded taper connection,
    209–210
radial position accuracy,
    208–209
radial stiffness, 208
shallow-type, 221–222
short, 219–220
test data, 204–210
types of, 221–223
constant force vibration isolation
    systems, 534–541
automatic height adjustment,
    538f
buckling phenomenon, 539,
    540f
cams for compensating
    devices, 539, 539f
hand-held impact machines,
    535, 537f
linkage-based compensating
    devices, 537–538, 539f
load-deflection characteristics,
    540–541, 540f
self-adjusting for changing
    static loads, 535–536
static deflection, 534
stiffness compensator in, 535,
    536f
constant natural frequency (CNF)
    characteristic, 116–119, 140f

contact deformations, 120
between conforming/quasi-
    conforming surfaces, 183–
    186
cast iron and/or steel parts,
    184–185
cast iron/steel and plastic
    materials, 185
contact stiffness, 185–186
Hertzian deformations, 184
large contacting surfaces, 185
positive effects, 184
conical (tapered) connections,
    203–210
axial preload, 205–207
CAT connection, 222–223
computational evaluation,
    210–213
elastic displacements in,
    203–204
finite element model, 217–218
load-angular deflection, 205
load-deflection characteristics,
    208
manufacturing deviations,
    207–208
manufacturing errors,
    210–217
Morse taper connections,
    221–222
preloaded taper connection,
    209–210
radial position accuracy,
    208–209
radial stiffness, 208
shallow-type, 221–222
short, 219–220
types of, 221–223

contact deformations (*continued*)
  cylindrical connections,
    194–203
  clearance fits, 194–202
  interference fits, 202–203
  between non-conforming
    surfaces, 180–183
  displacements in point and
    line contacts, 182*t*
  parallel axes cylinders, 181
  spheres, 180
  overview, 179–180
contact stiffness, 186–193
  angular deflection, 189–190
  angular stiffness, 190–191
  recommendations from, 193
coordinate coupling, 14
copper, 46*t*
CoroGrip system, 489–491, 492*f*
Coulomb friction dampers,
  442–443
Coulomb's law, 235
couplings, 645–676
  combination purpose, 646,
    668–669*f*, 670–675
  comparison of, 673–675
  spider coupling, 670–672
  misalignment-compensating,
    646, 649–663, 650*f*
  Oldham couplings/U-joints
    with elastic connections,
    655–662
  Oldham/gear couplings/
    U-joints, 652–655
  selection criteria, 651–652
  rigid, 647–649, 648*f*
  torsionally flexible, 646,
    663–670

flange couplings, 649
  sleeve couplings, 647
  shaft misalignment modes,
    647*f*
  Bibby coupling, 665*f*
  damping with, 666–667
  Dynaflex LCD, 665*f*
  non-linearity of load-
    deflection, 667–670
  roles in transmission, 664–670
  rotational inertia in
    transmission, 670
  with rubber flexible elements,
    676*f*
  torsional compliance with,
    664–666
crankshaft, effective stiffness of,
  376–377
creep, 638–639
critical compressive force,
  623–624
cubic boron nitride (CBN), 457,
  461
Curvic coupling connection,
  513–514
cutting forces, reduction of,
  459–461
cutting process, 35–37
  damping of, 35–37
  stiffness of, 35–37
cutting stiffness coefficient, 35
cutting tools, 457–481
  cutting forces, 461–463
  damping characteristics, 460,
    461–463
  high stiffness in, 457–458
  increasing damping in,
    471–481

modification of clamping
systems for cantilever tools,
475–481
for turning of low stiffness
parts, 472–475
materials used in, 458
mathematical model of, 473*f*
overview, 457–458
reduced normal stiffness in,
463–465
reduced tangential stiffness in,
465–471
reducing cutting forces in,
459–461
stiffness characteristics,
461–463
tool life, 461–463
cylindrical (OD) grinder,
compliance breakdown for,
697–705
cylindrical connections, 194–203
clearance fits, 194–202
damping of, 249–250
interference fits, 202–203
cylindrical springs, 122–124

D

dampers, 441–448
Coulomb friction, 442–443
elasto-damping materials, 443
felt, 445
high-modulus elasto-damping
materials, 447
rubber, 445–446
viscous, 441–442
volumetric wire mesh
materials, 443–445

damping, 8–10. *See also* stiffness
in antifriction bearings,
360–364
cutting forces and, 461–463
of cutting process, 35–37
definition of, 8
heat generation and, 9
of helically patterned tubular
beams, 67–69
hysteresis-induced, 578–583
impact, 583–584
influence on vibration/
dynamics, 13–14
of mechanical contacts,
246–250
causes of, 246
cylindrical connections,
249–250
energy dissipation in power
transmission components,
250
flat joints, 247–249
tapered connections, 249–250
negative, 14–15
in power transmission systems,
364–368
of structural materials, 42–49
tangential contact compliance
and, 234–242
tool life and, 461–463
viscous, 575–578
damping coefficients, 329
damping effectiveness product,
49*t*
damping enhancement techniques,
440–449
approaches, 440
dampers, 441–448

damping enhancement techniques
(*continued*)
  Coulomb friction, 442–443
  felt, 445
  high-modulus elasto-damping
    materials, 447
  rubber, 445–446
  viscous, 441–442
  volumetric wire mesh
    materials, 443–445
damping impact, 583–584
dead stop, 229*f*
deflection, 91
deflection, generalized, 320
deflection ratio, 57–58
deformation phases, 19*f*
deformations, of long machine
  bases, 288–291
diamond, 45*t*, 47
disc springs, 124–126
double-supported beam, 58*f*, 59*f*
drag-link extrusion press, linkage
  schematic of, 28*f*
dual nonlinearity, 121
Dynaflex LCD, 665*f*
dynamic models, 560–566
  inertia components of, 561–564
  stiffness of, 561–564
  vibration response control,
    564–566
dynamic shear module, complex
  627
dynamic stiffness, 6, 120–121, 240
dynamic stiffness coefficient, 6,
  121
dynamic vibration absorbers
  (DVAs), 448–449, 469–471, 590
dynamic-to-static stiffness ratio, 241

E

E glass/1002 epoxy, 46*t*
EAR C-1002, 617
effective cutting damping
  coefficient, 36
effective cutting stiffness
  coefficient, 36
effective modulus of elasticity, 98,
  622
effective rigidity, 285
elastic deformations, 116–119
elastic instability, 16–18
elastic kinematic connections,
  543–547
  for rotational motion, 543–544
  thin-layered rubber-metal
    laminates, 142–146, 548–560
    as anisotropic elastic
      elements, 548–550
    as compensators, 560
    as limited travel bearings,
      550–555
    wedge mechanisms, 558–560
  motion transformers, 546–547
  for translational motion,
    545–546
elastic stability, 623–626
elasto-damping materials, 443,
  447
elastomeric materials, 615–639
  additives, 617
  accelerators, 617
  active fillers, 627
  anti-aging agents, 617
  inert fillers, 617
  plasticisers, 617
  softeners, 617

bulk modulus, 618–619
compression loading, 100–101
creep of, 638–639
critical compressive force,
    623–624
dynamic characteristics of,
    627–633
effective modulus of elasticity,
    622
elastic stability of, 623–626
fatigue resistance of, 633–638
hardness of, 604*t*
high damping, 100
nonlinear, 135–146
  constant natural frequency,
      139–140
  deformation characteristics,
      141
  packaging, 139*f*
  streamlined elements,
      135–141
  stress distribution, 137–138*f*
  thin-layered rubber-metal
      laminates, 142–146, 548–560
overview, 97–110
shape factor, 622
shear modulus, 618
shear stiffness, 624–625
small deformations, 98
static deformation of, 618–623
stiffness of bonded rubber
    blocks, 102–111
angular stiffness, 104, 108–109
compression stiffness, 107,
    109–110
torsional stiffness, 103–104,
    105
twist stiffness, 109

ethylene-propylene-diene (EPDM)
    rubber, 615–616
  nitrile rubbers, 615
  silicone rubbers, 616
epoxidized natural rubber, 615
  vibration amplitudes, 604*t*
  vulcanizing agents, 617
  Young's modulus, 619
electric motors, 316–318
  damping coefficient of, 317
  direct drive, 318
  dynamic parameters of, 316–318
  dynamic stiffness coefficient
      of, 317
  negative damping, 317
  non-direct drive, 318
energy dissipation, 9, 250
Euler force, 16–18, 402
extension springs, 87

F

fatigue resistance, 633–638
felt, 445
felt Unisorb, 604*t*
fibrous mesh, 120
finger-sleeve coupling, 668*f*
finite element method (FEM), 305
flat joints
  damping of, 247–249
  preloading of, 150–151
  tangential contact compliance
      of, 224–234
flexible hoses, volumetric
    expansion of, 315*f*
flexures, 543–547
  for rotational motion, 543–544
  motion transformers, 546–547

flexures (*continued*)
  for translational motion, 545–546
fluid compressibility, 314
forced vibrations, 13–14, 598–603
  vibration isolation criteria for vibration-sensitive objects, 601–603
  vibration level criteria, 599–601
forward compensation, 382–383
foundations, structural deformations and, 280–288
frame/bed components, 76–83
  contour distortion, 82*f*
  design influence on stiffness, 77*f*
  holes, influence on stiffness, 80–81
  local deformations of, 81–83
  reinforcing ribs, 78, 79*f*
  stiffness of, 76–83
free body diagrams, 561
frequency response function (FRF), 173, 564–566
friction springs, 92–94
friction-induced position uncertainties, 497–498
front spindle bearing, 388–389
fusible alloys, 426

G

gantry frames, 373–374
  deformations of, 375*f*
  schematics of, 374*f*
gear coupling, 652–655
gear reducer, vibration of, 160–161*f*
gear train, configurations of, 302*f*

generalized deflection, 320
giant superelasticity effect (GSE), 94–97, 510–512
glass, 45*t*
goose neck tool, 463
gothic arch profile, 307–308
granite, 46*t*
graphite, 45*t*, 47
grinding wheel, elastic attachment, 465*f*
guideways, 376–377

H

hardening load-deflection characteristics, 115, 116*f*
hardening nonlinear stiffness, 6
heat generation, damping, 9
helically patterned tubes, stiffness and damping of, 67–69
Hertzian deformations, 184
high carbon gray iron, 49*t*
high speed steels (HSS), 457
high-modulus elasto-damping materials, 447
high-speed robots, compliant bearings of, 531–534
hollow bearings, stiffness reduction in, 520–523
honeycomb beams and plates, 70–72
Hooke's Law, 120, 235
horizontal milling machine, 270*f*
HSK system, 460, 516–518
HTS graphite, 46*t*
hydraulic counterbalancing, 386–388
hydraulic cylinders, 315

hydraulically driven robot, 331–335
hydrostatic bearings, compliance of, 305
hydrostatic compressibility, 146
hysteresis loop, 578–579, 580*f*
hysteresis-induced damping, 578–583. *See also* damping
    hysteresis loop, 578–579, 580*f*
    kinetic energy, 580
    logarithmic decrement, 581
    loss factor, 583
    potential energy, 580–581
    response amplitude, 582
    transmissibility, 582–583

I

inert fillers, 617
interference-fit cylindrical connections, 202–203
invar, 46*t*
inverse cantilever, 259*f*
Isodamp™, 617
Isoloss™, 617

J

jaw coupling, 663*f*
joint stiffness
    effect of beam width on, 170*f*
    effect of contact surface on, 171*f*

K

Kevlar, 45*t*
Kevlar 49/resin, 46*t*

key connections, 299
    equivalent torsional compliance of, 296
    helical spring, 301*f*
    no-play, 299
kinematic coupling, 497–498
kinematically induced damping, 27
kinetic energy of mass, 13
KM system, 514–516
Kudriavetz coupling, 656*f*, 663

L

lanxide, 46*t*
laser carbon dioxide, 460
Laws of Evolution of Technological Systems, 524
lead, 46*t*
leaf springs, 122–124
leveling isolators, 396–397
linear cylinder-piston system, 311
linear stiffness, 6
load distribution, 376–379
    component deformations and, 16*f*
load-deflection characteristics, 115–116
lockalloy, 45*t*
logarithmic decrement, 576–577
long beam, 282
long machine bases, deformations of, 288–291
loss angle, 576
loss factor, 576
loudspeaker diaphragms, material properties for, 47–49
low carbon steel, 49*t*
lubricoolants, 460

M

machine tools, 457–481
  cutting forces, 461–463
  cutting tools with reduced
    normal stiffness, 463–465
  cutting tools with reduced
    tangential stiffness, 465–471
  damping in cutting zone,
    461–463
  high stiffness in, 457–458
  increasing damping in,
    471–481
    modification of clamping
      systems for cantilever tools,
      475–481
    by turning of low stiffness
      parts, 472–475
  materials used in, 458
  mathematical model of cutting
    system, 473f
  overview, 457–458
  reducing cutting forces in,
    459–461
  stiffness in cutting zone,
    461–463
  tool life, 461–463
  tooling structure in, 458
machining system
  accuracy enhancement factor,
    39
  compliances, 40f
  stiffness of, 37–42
magnesium, 45t, 49t
magnesium alloy, 49t, 447
marble, 45t
material-related nonlinearity,
  120–122

maximum deformation, 255
maximum stress ratio, 57–58
mechatronic systems. See servo-
  controlled systems
Mclram, 46t
mesh-like materials
  damping of, 121f
  dynamic stiffness of, 121f
micro electro-mechanical systems
  (MEMS), 94–97
misalignment-compensating
  couplings, 646, 649–663, 650f
  Oldham and gear couplings
    and U-joints, 652–655
  Oldham couplings and U-joints
    with elastic connections and
    U-joints, 655–662
  selection criteria, 651–652
modal synthesis, 564–566
modes of loading, 55–64
  parallel kinetic machines, 61–63
  of rod-like structure, 56f
  types of, 55
modified spider coupling, 668f
modular tooling system, 242–246
  angular deformation, 244
  axial force, 243
  axial tightening, 243f
  inclination, 244
  joint stiffness, 493–494
  number of modules/joints in,
    494–496
  radial clamping in, 494
  runout of, 245f
  stiffness of, 492–496, 495f
  vibration amplitudes, 495f, 496
molybdenum, 45t
Morse taper connections, 221–222

multi-degrees-of-freedom systems, 561–564
multi-pad bearings, compliance of, 305
multiple-load carrying components, 681–693
    analytical solution for bicycle wheels, 688–689
    loading of spoked wheels, 685–688
    stiffness of antifriction bearings, load distribution between rolling bodies, 681–685
    torsional systems with multiple load-carrying connections, 689–693

N

natural frequencies, 90, 116–119
negative damping, 14–15. *See also* damping
    in induction motors, 317
    mechanical linkages, 20–23
    mechanisms with nonlinear position functions, 23–31
    stick-slip, 19–20
negative stiffness, 14–15. *See also* stiffness
    elastic instability, 16–18
    electromechanical systems, 31–35
    mechanical linkages, 20–23
    mechanisms with nonlinear position functions, 23–31
    stick-slip, 19–20

nickel, 49t
nickel aluminide, 45t
nickel-based superalloys, 49t
nickel-titanium (NiTi) alloys, 49t, 94–97, 447
niobium, 46t
Nitinol, 49t
nitralloy, 45t
Nivco, 49t
non-conforming surfaces, contact deformations between, 180–183
nonlinear rubber elements, 130–146
    compressed elements with controlled bulging, 131–135
    constant natural frequency, 139–140
    deformation characteristics, 141
    packaging, 139f
    streamlined elements, 135–141
    stress distribution, 137–138f
    thin-layered rubber-metal laminates, 142–146
nonlinear spring elements, softening nonlinear characteristics in, 126–130
nonlinear stiffness, 6
    changing part/system geometry and, 120
    Constant Natural Frequency characteristic, 116–119
    contact deformations and, 120
    elastic deformations and, 120
    geometry-related nonlinearity, 122–130

nonlinear stiffness (*continued*)
    Bellevile springs, 124–126
    coil springs, 122–124
    leaf springs, 122–124
    nonlinear spring elements,
        126–130
    hardening load-deflection
        characteristics, 115
    material-related nonlinearity,
        120–122
    softening load-deflection
        characteristics, 115
non-symmetrical structures,
    optimization of, 373–374
normal stiffness, reducing,
    463–465
nylon, 46*t*

O

Oldham couplings, 652–663, 660*f*.
    See also couplings
    preloadable universal joint,
        660*f*
    with thin-layered rubber-metal
        laminates, 657*f*
one-cylinder engine, components
    of, 172*f*
overconstrained systems,
    273–280
    design mistakes and, 278–279
    guideways, 274–275
    intentional reduction of
        stiffness, 279*f*
    symmetrical driving of guided
        components, 275
    thermal deformations,
        277–278

P

paper, 46*t*
parallel cylinders, 182*t*
parallel kinematic machines,
    61–63
parametric vibration, 14
passive compensation, 380–389
    forward compensation,
        382–383
    in front spindle bearing,
        388–389
    hydraulic counterbalancing in,
        386–388
    opposing bending moments in,
        384*f*
    in radial drill press, 380–381
    in thread-rolling machine,
        380–381
    in vertical boring mill, 382
    weight counterbalancing in,
        385–386, 385*f*
PCD cutting tools, 460
penetration rate coefficient, 35
perpendicular cylinders, 182*t*
piezoelectric actuators, 393–394
piston, displacement of, 313–314,
    314*f*
plano-milling machine tool, base
    deformations of, 289*f*, 290–291
pneumatic actuators, 311–313
Poisson's ratio, 43, 143, 282
polycrystalline diamond (PCD),
    457, 461
polymer concrete, 46*t*
polypropylene, 46*t*
power shrinking toolholders,
    488–491

power transmission components, energy dissipation in, 250

power transmission couplings, 645–676
  combination purpose couplings, 646, 668–669f, 670–675
  comparison of, 673–675
  spider coupling, 670–672
  misalignment-compensating couplings, 646, 649–663, 650f
  Oldham couplings and U-joints with elastic connections and U-joints, 655–662
  selection criteria, 651–652
  torsionally rigid couplings, 645–649, 648f
  torsionally flexible couplings, 646, 663–670
  flange couplings, 649
  sleeve couplings, 647
  shaft misalignment modes, 647f
  Bibby coupling, 665f
  damping with, 666–667
  Dynaflex LCD, 665f
  non-linearity of load-deflection, 667–670
  roles in transmission, 664–670
  rotational inertia in transmission, 670
  with rubber flexible elements, 676f
  torsional compliance with, 664–666
stiffness reduction in, 523–530

power transmission systems and drives, 294–368
  compliance breakdowns, 329–348
  electromechanical jointed robot, 335–339
  electromechanical parallelogram robot, 339–342
  electromechanical spherical frame robot, 342–347
  hydraulically driven robot, 330–335
  compliance of
  basic components, 296–307
  dynamic properties of electric motors, 316–318
  hydraulic actuators, 313–316
  pneumatic actuators, 311–313
  stiffness of ball screws, 307–310
  damping in, 364–368
  effective stiffness/compliance, 294–295
  parameter reduction in models, 317–329
  breakdown of inertia, 325
  damping coefficients, 329
  gear box, 326–329
  potential energy of elastic system, 320
  reduction procedure, 319, 324f
  transmission ratio, 321
  stiffness factors, 294
  stiffness of spindles, 348–353
  stiffness/damping of antifriction bearings, 353–364
  structural optimization of, 378–379

pre-deformed frame parts, 383
preload force, 116
preloaded structure, integrity of,
163–166
preloading, 146–161
  of antagonist actuators, 154–155
  of antifriction bearings, 151–154
  ball bearings, 152
  roller bearings, 152
  positive/negative effects of
  of belt drives, 146–149
  of chain drives, 149–151
  of flat joints, 150–151
  of flexible elements with
    variable stiffness, 155–159
  of linear systems, 146–149
  of nonlinear systems, 149–151
  overview, 146–147
prestressing, 12
prismatic rubber blocks,
  compression stiffness of, 106t

Q

quasi-conforming surfaces,
  contact deformations between,
  183–186

R

radial drill press, 380–381
radial stiffness, 107
radially loaded tube, deformation
  phases, 19f
reinforcing ribs, stiffening effect
  of, 78f
relative angular displacement, 304
relative energy dissipation, 9

response amplitude, 577
reverse buckling, 392, 405–412
rigid beam, 282
rigid couplings, 645–649, 648f. See
  also couplings
  flange couplings, 649
  sleeve couplings, 647
Rivin's compression method, 561
robot electromechanical jointed,
  335–338, 336f
  compliance, 335
  elbow joint, 337
  forearm, 336–337
  shoulder joint, 337
  steel spur gear mesh, 337–338
  upper arm, 337
  waist, 337
robot electromechanical
  parallelogram, 340–342
  compliance breakdown,
  340–342
robot electromechanical spherical
  frame, 342–347
  compliance breakdown,
  343–347
  "super ball" bushing, 344f
Robot NEOS
  design schematic and work
    zone of, 64f
  specifications of, 64t
roller bearings
  approach of contacting bodies,
  182t
  auxiliary coefficients, 182t
  Hertzian deformations, 152
  with hollow rollers, 520–523
roller between plates, 182t
roller chains, 149f

rolling bodies, load distribution between, 681–685
Routh-Hurvitz criteria, 466
Rubber
  types of, 615–616
    acrylonitrile butadiene, 615
    butyl elastomers, 615–616
    chlorobutadiene, 615
    chloroprene rubber, 615
    epoxidized natural rubber, 615
    ethylene-propylene-diene, 615–616
    fluoroelastomers, 616
    isoprene rubber, 615
    neoprene, 615
    natural rubber (NR), 604t, 615
    NBR-26 rubber, 604t
    NBR-40 rubber, 604t
rubber disc coupling, 669f
rubber flexible elements, 97–110
  axially loaded bars, 101
  compression deformation, 99–100
  compression loading, 100–101
  Constant Natural Frequency characteristic, 116–119, 139–140
  effective modulus of elasticity, 98
  high damping, 100
  low block, 101
  nonlinear, 130–146
    compressed elements with controlled bulging, 131–135
    deformation characteristics, 141
    packaging, 139f
    streamlined elements, 135–141

stress distribution, 137–138f
  thin-layered rubber-metal laminates, 142–146
  shape factor, 98
  small deformations, 98
  stiffness of bonded rubber blocks, 102–111
    angular stiffness, 104, 105, 108–109
    compression stiffness, 105, 107, 109–110
    shear stiffness, 102–103, 105, 108, 109
    torsional stiffness, 103–104, 105, 108, 109
  variable stiffness in, 155–159
rubber sphere, deformation characteristics of, 141
rubber torus
  deformation characteristics of, 141
  natural frequency vs. axial compressive load, 140f
rubber-like materials, 615–639
  creep of, 638–639
  critical compressive force, 623–624
  dynamic characteristics of, 627–633
  effective modulus of elasticity, 622
  elastic stability of, 623–626
  fatigue resistance of, 633–638
  shape factor, 622
  shear modulus, 618
  shear stiffness, 624–625
  static deformation of, 618–623
  types of, 615–616

rubber-like materials (*continued*)
    acrylonitrile butadiene, 615
    butyl elastomers, 615–616
    chlorobutadiene, 615
    chloroprene rubber, 615
    epoxidized natural rubber,
      615
    ethylene-propylene-diene,
      615–616
    fluoroelastomers, 616
    isoprene rubber, 615
    neoprene, 615
    nitrile rubbers, 615
    silicone rubbers, 616
    vulcanizing agents, 617
    Young's modulus, 619
rubber-metal laminates, 142–146
    as anisotropic elastic elements,
      548–550
    anisotropy of stiffness, 146
    bulk modulus for rubber
      blends, 143
    as compensators, 560
    effective compression modulus,
      143–144
    effective shear modulus, 145
    hydrostatic compressibility, 146
    as limited travel bearings,
      550–555
    specific compression load,
      143–144
    types of, 142*f*
    wedge mechanisms, 558–560

S

sapphire, 45*t*
scraping, 383

self-excited vibrations-dynamic
    stability, 590–596
    cutting force coefficient, 591
    for machine tools, 593
    for spindle bearings, 592–593
    toolholder/spindle interfaces,
      593–595
self-leveling isolators, 396–397
servo-controlled systems, 389–399
    antivibration devices, 390, 391*f*
    axis-alignment for boring bar,
      390, 390*f*
    chatter resistance, 391*f*, 392
    comprehensive active
      installation system, 397–399
    deflection correction for
      spindle head, 392–393, 393*f*,
      394*f*
    hydraulic actuators in, 389–390
    hydraulic vs. piezoelectric
      actuators, 393–394
    laser-based, 390–392
    vibration isolators, 396–397
    vibration suppression in
      surface grinder, 394–395
7/24 taper connections, 498–512
    definition of problem in,
      498–500
    standard connection, 500–502
    toolholder/spindle interfaces,
      502–518
      alternative designs, 503–510
      giant superelasticity effect,
        510–512
      requirements, 502–503
shaft misalignment modes, 647*f*
shafts, torsional compliance of,
    296, 297–298*f*

shallow tapered connections, tapered connections, 221–222
shape factor, 98, 622
shape memory alloys, 447, 476
shear stiffness, 43, 102–103, 105, 109, 618, 624–625
shear stress, 86–87
short beam, 282
short tapered connections, tapered connections, 219–220
silent chains, 149*f*
silentalloy, 49*t*
silicon carbide, 45*t*
silicon nitride, 45*t*
silicone rubbers, 616
single-degree-of-freedom (SDOF) systems, 575–584, 575*f*
  hysteresis-induced damping, 578–583
  impact damping, 583–584
  transmissibility of, 579*f*
  viscous damping, 575–578
slender parts, stiffening of, 412–416
slotted springs, 92
softening load-deflection characteristics, 6, 115, 116*f*, 126–130
solid state tool clamping devices, 487–492
  CoroGrip system, 489–491
  power shrinking toolholders, 488–491
  thermal shrink-fit toolbars, 488
  Tribos chuck, 489–490, 491*f*
Sonoston, 49*t*
Sorbothane™, 617
sphere and cylinder contact, 182*t*

sphere and cylindrical groove contact, 182*t*
sphere and plane contact, 182*t*
spider coupling, 668*f*, 671*f*
spindle interfaces, 497–518
  7/24 taper connections, 498–512
  alternative designs, 503–510
  Curvic coupling connection, 513–514
  giant superelasticity effect, 510–512
  HSK system, 516–518
spindles, 348–353
  angular deflection, 351
  effective overhang, 352
  optimal dimensioning of, 348–353
  radial deflection of, 350–351
  radial stiffness of, 349, 352*f*, 353*f*
  stiffness of, 348–353
  typical loading schematics, 350
spline connections, 299
  equivalent torsional compliance of, 296
  load-deflection characteristics, 300*f*
split beam, 162–163
spoked wheels, loading of, 685–688
spring index, 86
springs, 85–94
  Bellevile, 124–126
  bending-induced stresses in, 60
  coil, 86–92, 122–124
  compression, 87
  correction factor, 87

springs (*continued*)
  deflection, 91
  equivalent normal stress, 91–92
  extension, 87
  friction, 92–94
  leaf, 122–124
  natural frequencies, 90
  nonlinear, 126–130
  overview, 85–86
  shear stress, 86–87
  slotted, 92
  spring index, 86
  torsional stiffness of, 90
spring-shaped inserts, face milling
  with, 464*f*
stainless steel, 45*t*
stainless steel martensitic, 49*t*
static deformation, 534, 618–623
statistically indeterminate
  systems, 273–280
steel, 45*t*
steel band drive, compliance of,
  305
steel cylinders, 315
Stewart platform, 61, 62*f*
stick-slip, 19–20
stiffness, 3–8, 255. *See also*
  damping
  of assembled structures,
    172–174
  compliance, 6
  cutting forces and, 461–463
  of cutting process, 35–37
  dynamic, 6–8
  excessive, 10
  of frame/bed components,
    76–83
  generalized, 320

of helically patterned tubular
  beams, 67–69
importance of, 4
inadequate, 5, 10
influence of accuracy and
  productivity, 37–42
influence on uniformity of
  stress distribution, 10–13
influence on vibration/
  dynamics, 13–14
linear vs. nonlinear, 6
in mechanical design, 50
mode of loading and, 55–64
modification of dynamic model
  parameters, 560–566
negative, 14–15
overview, 3–8
principal axes, 4–5
of spindles, 348–353
static, 6
structural contact, 5
of structural materials, 42–49
structural proper, 5
of tool clamping devices,
  481–492
tool life and, 461–463
torsional, 72–76
stiffness enhancement techniques,
  282, 373–440
  for cantilever components,
    427–440
  combination link, 435–439
  rotating around transverse
    axis, 433–439
  stationary/rotating around
    longitudinal axis, 430–431
  compensation of structural
    deformations, 379–399

passive compensation,
380–389
servo-controlled systems,
389–399
strength-to-stiffness
transformation, 401–425
buckling, 402–405
reverse buckling, 405–412
self-contained devices,
416–425
stiffening of slender parts by
axial tension, 412–416
stress concentration reduction,
400–401
structural optimization,
373–379
load distribution, 376–379
replacing C-frames with
gantry frames, 373–375
temporary stiffness
enhancement, 425–427
stiffness reduction, 457–566
7/24 taper connections,
498–512
anisotropic elastic elements,
541–560
chain transmission, 530–531
tool clamping devices, 481–492
axial clamping in, 494
collet chuck, 483–487
general purpose, 482–487
overview, 481–482
performance characteristics
of, 482
solid state, 487–492
stiffness of, 481–492
Weldon clamp, 481–482,
481f

compliant bearings of high-
speed rotors, 531–534
constant force vibration
isolation systems
automatic height adjustment,
538f
buckling phenomenon, 539,
540f
cams for compensating
devices, 539, 539f
hand-held impact machines,
535, 537f
linkage-based compensating
devices, 537–538, 539f
self-adjusting for changing
static loads, 535–536
specially shaped static
elements, 537, 538f
stiffness compensator in, 535,
536f
cutting edge/machine tool
structure interface, 457–481
cutting forces, 461–463
damping characteristics, 460,
461–463
high stiffness in, 457–458
increasing damping in,
471–481
materials used in, 458
mathematical model of,
473f
overview, 457–458
reduced normal stiffness in,
463–465
reduced tangential stiffness
in, 465–471
reducing cutting forces in,
459–461

stiffness reduction (*continued*)
  stiffness characteristics,
    461–463
  hollow bearings, 520–523
  modular tooling, 492–496
  power transmission gears,
    523–530
springs, 85–94
  coil, 86–92
  compression, 87
  correction factor, 87
  deflection, 91
  equivalent normal stress, 91–92
  extension, 87
  friction, 92–94
  natural frequencies, 90
  overview, 85–86
  shear stress, 86–87
  slotted, 92
  spring index, 86
  torsional stiffness of, 90
stiffness-damping criteria, 603–604
strength-to-stiffness
  transformation, 146–161, 401–425
  antagonist actuators, 154–155
  antifriction bearings, 151–154
  ball bearings, 152
  elastic displacement of shafts
    in, 151
  Hertzian deformations in, 152
  positive/negative effects, 153
  positive/negative effects of
    preloading in, 153
  preloading of, 151–154
  roller bearings, 152
  belt drives, 146–149
  buckling, 402–405
  chain drives, 149–151

flat joints, 150–151
flexible elements with variable
  stiffness, 155–159
linear systems, 146–149
nonlinear systems, 149–151
overview, 146–147
reverse buckling, 405–412
self-contained devices, 416–425
stiffening of slender parts by
  axial tension, 412–416
stress concentrations, 76
  reduction of, 400–401
structural compliance
  breakdowns, 329–348
  electromechanical jointed
    robot, 335–339
  electromechanical
    parallelogram robot, 339–342
  electromechanical spherical
    frame robot, 342–347
  hydraulically driven robot,
    330–335
  on grinder, 703–714
structural components, stiffness
  enhancement of, 83–85
structural contact stiffness, 5
structural deformations, 280–288
  balanced forces, 282–283
  foundation beams, 282
  location of supporting mounts
    and, 280–281
  machines installed on
    individual foundations,
    284–288
    bending stiffness, 285
    horizontal boring mill, 287–288
    monolithic floor plate, 286
    torsional stiffness, 285

overview, 280
stiffness enhancement
coefficients, 282
torsional stiffness
enhancement coefficient, 284
structural materials, 42–49
composite, 46t
damping of, 42–49
homogenous, 45–46t
stiffness of, 42–49
structural proper stiffness, 5
superelastic alloys, 447
superelastic materials (SEM),
510–512
superelasticity effect, 94–97
supporting systems, 255–291
influence of support
characteristics, 255–263
components' design, 257f
contour design, 260f
fit, 258f
load pattern, 256f
reaction forces, 259f
selection and configuration of
supporting elements, 258f
overconstrained systems, 273–280
overview, 255
rational location of supporting/
mounting elements, 263–273
balancing of deformations,
263–264
bi-directional central support,
265–266
cylindrical (OD) grinder,
266–268
kinematic support, 268
reaction-cause deformations,
268–269

round table system, 264–265
unidirectional central
support, 265
swan neck tool, 463
synchronous belts, compliance of,
306

T

T50 graphite/2011 Al, 46t
tangential contact compliance,
223–242
tangential stiffness, reducing,
465–471
tap-bolting assembly, 173f
tapered connections, 203–223,
496–519
7/24 taper connections,
498–512
axial preload, 205–207
CAT connection, 222–223
computational evaluation,
210–213
damping of, 249–250
elastic displacements in,
203–204
finite element model, 217–218
friction-induced position
uncertainties, 497–498
load-angular deflection, 205
load-deflection characteristics,
208
manufacturing deviations,
207–208
manufacturing errors, 210–217
Morse taper connections,
221–222
preloaded taper connection,
209–210

tapered connections (*continued*)
  radial position accuracy,
    208–209
  radial stiffness, 208
  shallow-type, 221–222
  short, 219–220
  test data, 204–210
  toolholder/spindle interfaces,
    498–518
    7/24 taper connections,
      498–512
    Curvic coupling connection,
      513–514
    HSK system, 516–518
    KM system, 514–516
    types of, 221–223
tension-compression
  vs. bending for structural
    components, 59f
  machine tool structures, 61–63
  robot manipulator, 63–64, 64f
thermal deformations, 277–278
thermal expansion coefficients,
  291
thermal shrink-fit toolbars, 488
thin film gaskets, 186
thin-layered rubber-metal
  laminates, 142–146
    as anisotropic elastic elements,
      548–550
    anisotropy of stiffness, 146
    bulk modulus for rubber
      blends, 143
    as compensators, 560
    effective compression modulus,
      143–144
    effective shear modulus, 145
    hydrostatic compressibility, 146

as limited travel bearings,
  550–555
Oldham couplings with, 657f
specific compression load,
  143–144
types of, 142f
wedge mechanisms, 558–560
thread-rolling machine, 380–381
three-jaw chucks, 482
through-bolting assembly, 173f
titanium carbide, 45t
tool clamping devices, 481–492
  collet chuck, 483–487
  general purpose, 482–487
  overview, 481–482
  performance characteristics of,
    482
  solid state, 487–492
    power shrinking toolholders,
      488–491
    thermal shrink-fit toolbars, 488
    Tribos chuck, 489–490, 491f
  stiffness of, 481–492
  Weldon clamp, 481–482, 481f,
    482–483
toolholder/spindle interfaces,
  497–4518
  7/24 taper connections, 498–512
  alternative designs, 503–510
  Curvic coupling connection,
    513–514
  giant superelasticity effect,
    510–512
  HSK system, 516–518
  KM system, 514–516
  requirements, 502–503
  self-excited vibrations-dynamic
    stability, 593–595

tooling system, modular, 242–246
  angular deformation, 244
  angular displacement, 244
  axial force, 243
  axial tightening, 243*f*
  inclination, 244
  joint stiffness, 493–494
  number of modules/joints in,
    494–496
  radial clamping in, 494
  runout of, 245*f*
  static stiffness, 494–496, 495*f*
  stiffness of, 492–496
  vibration amplitudes, 495*f*, 496
toroid shell coupling, 669*f*
torsional compliance, 296
torsional stiffness, 72–76, 90,
  103–104, 105, 108, 109, 285
torsional stiffness enhancement
  coefficient, 284
torsionally compliant head (TCH),
  468–471
torsionally flexible couplings, 646,
  663–670. *See also* couplings
  Bibby coupling, 665*f*
  damping with, 666–667
  Dynaflex LCD, 665*f*
  non-linearity of load-deflection,
    667–670
  roles in transmission, 664–670
  rotational inertia in
    transmission, 670
  with rubber flexible elements, 676*f*
  torsional compliance with,
    664–666
total linear displacement, 303
total vector deflection, 302
transfer functions, 26

Tribos chuck, 489–490, 491*f*
truss bracket, 58*f*
truss bridge, 58*f*, 59*f*
truss structures, 57
tungsten, 45*t*
tuning parameters, 448
turning of slender parts, 414
turning tool, with high tangential
  compliance, 466*f*

U

U-joints, 652–662
uniformity of stress distribution,
  influence of stress on, 10–13
urea, 426

V

variable stiffness, 14
  dynamic effects of, 159–161
  in preloaded flexible elements,
    155–159
Varilock modular tooling system,
  242–246
V-belts, tension modulus of, 306
vector displacement, 302
vertical boring mill, 382
vertical milling machine, 270*f*
Vibrachok wire-mesh isolating
  mounts, 604*t*
vibration control, 589–590
  dynamic vibration absorbers, 590
  self-excited vibrations-dynamic
    stability, 590–596
  stiffness-damping criteria,
    603–604
  vibration isolation, 596–603
    chatter resistance and, 596–598
    forced vibrations, 598–603

vibration isolation, 596–603
    chatter resistance and, 596–598
    forced vibrations, 598–603
vibration isolation systems, zero-
    stiffness, 534–541
    automatic height adjustment, 538*f*
    buckling phenomenon, 539, 540*f*
    cams for compensating
        devices, 539, 539*f*
    hand-held impact machines,
        535, 537*f*
    linkage-based compensating
        devices, 537–538, 539*f*
    load-deflection characteristics,
        540–541, 540*f*
    self-adjusting for changing
        static loads, 535–536
    static deflection, 534
    stiffness compensator in, 535, 536*f*
vibration isolator, 18*f*
vibration level criteria, 599–601
vibratory behavior, 564–566
viscous dampers, 441–442
viscous damping, 575–578. *See
    also* damping
    equation of motion, 577
    logarithmic decrement, 576–577
    loss angle, 576
    loss factor, 576
    response amplitude, 577
volumetric compressibility
    modulus, 314
volumetric wire mesh materials,
    443–445

W

water/ice, temporary stiffness
    enhancement with, 425–426

weight counterbalancing, 385*f*
Weldon clamp, 481–482, 481*f*,
    482–483
Wildhaber/Novikov gears,
    528–529
wood, 45*t*
Wood alloy, 426

Y

Yaskawa Electric Corp., 389
Young's modulus, 42–43, 47, 282,
    401–402, 619
Young's modulus, complex 627

Z

zerodur, 46*t*
zero-stiffness vibration isolation
    systems, 534–541
    automatic height adjustment,
        538*f*
    buckling phenomenon, 539,
        540*f*
    cams for compensating
        devices, 539, 539*f*
    hand-held impact machines,
        535, 537*f*
    linkage-based compensating
        devices, 537–538, 539*f*
    load-deflection characteristics,
        540–541, 540*f*
    self-adjusting for changing
        static loads, 535–536
    specially shaped static
        elements, 537, 538*f*
    static deflection, 534
    stiffness compensator in, 535,
        536*f*
zirconia, 45*t*